THE GREAT BARRIER REEF
HISTORY, SCIENCE, HERITAGE

One of the world's natural wonders, the Great Barrier Reef stretches more than 2000 kilometres in a maze of coral reefs and islands along Australia's north-eastern coastline. This book unfolds the fascinating story behind its mystique, providing for the first time a comprehensive cultural and ecological history of European impact, from early voyages of discovery to the most recent developments in Reef science and management. Incisive and a delight to read in its thorough account of the scientific, social and environmental consequences of European impact on the world's greatest coral reef system, this extraordinary book is sure to become a classic.

After graduating from the University of Sydney and completing a PhD at the University of Illinois, **James Bowen** pursued an academic career in the United States, Canada and Australia, publishing extensively in the history of ideas and environmental thought. As visiting Professorial Fellow at the Australian National University from 1984 to 1989, he became absorbed in the complex history of the Reef, exploring this over the next decade through intensive archival, field and underwater research in collaboration with **Margarita Bowen**, ecologist and distinguished historian of science. The outcome of those stimulating years is this absorbing saga.

*To our grandchildren,
with hope for the future in the hands of their generation*

THE GREAT BARRIER REEF
HISTORY, SCIENCE, HERITAGE

JAMES BOWEN

AND

MARGARITA BOWEN

PUBLISHED BY THE PRESS SYNDICATE OF THE UNIVERSITY OF CAMBRIDGE
The Pitt Building, Trumpington Street, Cambridge, United Kingdom

CAMBRIDGE UNIVERSITY PRESS
The Edinburgh Building, Cambridge CB2 2RU, UK
40 West 20th Street, New York, NY 10011–4211, USA
477 Williamstown Road, Port Melbourne, Vic 3207, Australia
Ruiz de Alarcón 13, 28014 Madrid, Spain
Dock House, The Waterfront, Cape Town 8001, South Africa

http://www.cambridge.org

© James Bowen & Margarita Bowen 2002

This book is in copyright. Subject to statutory exception
and to the provisions of relevant collective licensing agreements,
no reproduction of any part may take place without
the written permission of Cambridge University Press.

First published 2002

Printed in Australia by Ligare

Typeface Times (Adobe) 10.5/14 pt.   System QuarkXPress® [PH]

A catalogue record for this book is available from the British Library

National Library of Australia Cataloguing in Publication data
Bowen, James (James Ernest).
The Great Barrier Reef: history, science, heritage.
Bibliography.
Includes index.
ISBN 0 521 82430 3 hardback.
1. Human ecology – Queensland – Great Barrier Reef –
History. 2. Nature – Effect of human beings on –
Queensland – Great Barrier Reef – History. 3. Great
Barrier Reef (Qld.) – History. I. Bowen, Margarita Jean.
304.209943

ISBN 0 521 82430 3 hardback

Contents

Abbreviations and Acronyms	x
Preface	xii
Acknowledgments	xv
Introduction: An Overview	1

PART ONE
NAVIGATORS AND NATURALISTS IN THE AGE OF SAIL

1	QUEST FOR THE GREAT SOUTH LAND	11
	Navigators in the New World	12
	Portugal and Spain: first navigators of Terra Australis?	13
	New explorers: Dutch, English, French	20
	South Pacific navigation and the Transit of Venus, 1769	23
2	VOYAGE OF THE *ENDEAVOUR*: COOK AND THE 'LABYRINTH'	27
	The expedition to Tahiti	27
	Encounter with the Reef: through the 'Labyrinth', 1770	29
	An error of judgment? *Endeavour* holed on a reef	31
	The *Endeavour* voyage published: the Reef mystique develops	35
	A bizarre controversy: did Cook discover the Reef?	37
3	*ENDEAVOUR* NATURALISTS: 'A SEPARATE CREATION'	41
	Eighteenth century natural science: a ferment of ideas	42
	Endeavour voyage: first scientific study of the Reef	44
	Banks' *Florilegium*: the great compendium	51
	Fate of the *Endeavour* zoological specimens	54
4	MATTHEW FLINDERS: VOYAGE OF THE *INVESTIGATOR*	56
	British settlement of New South Wales	57
	Bligh and Flinders: early Reef surveys	58
	Flinders' surveys 1795–1799	59
	The urgent issue: survey of the colony	61
	New Holland: missing links in nature's chain?	65
	Robert Brown: plant taxonomy in a new world	68

Science assessed: Brown and Bauer in England 1805–1814 — 71
A Voyage to Terra Australis: the journal published, 1814 — 74

5 THE REEF EXPLORED: EARLY SURVEYS, 1821–1844 — 77

Expansion of the colony: first decades — 77
Australian coastline charts completed: Jeffreys and King 1815–1822 — 79
Geological observations: rise of controversy — 83
Accurate reef charts: *Beagle* survey 1837–1844 — 84
Jukes and MacGillivray: naturalists on the *Fly* 1843–1845 — 88

6 EARLY REEF CHARTS COMPLETED: 1846–1862 — 94

Reef surveys of the *Rattlesnake*: 1847–1850 — 94
Rattlesnake naturalists: MacGillivray and Huxley — 98
Australia in a new era: changing international relations — 101
Coral Sea surveys of the *Herald*: 1853–1860 — 102

7 THE REEF AS A MARITIME HIGHWAY: COLONY OF QUEENSLAND, 1859–1900 — 107

Queensland exploration and expansion: 1859–1870 — 110
Coastal settlement of the Reef: towns, ports, railways — 112
Regulating the Reef: coastal shipping — 118

8 FROM NATURAL HISTORY TO SCIENCE, 1850–1900: VOYAGES OF THE *CHALLENGER* AND THE *CHEVERT* — 124

Cabinet collectors and museums: Darwinism resisted — 124
Oceanography and politics: *Challenger* and *Chevert* voyages, 1872–1876 — 129
Queensland Museum: marine science lags, 1859–1880 — 135
Scientific change 1888–1900: Darwinism established — 137

9 EXPLOITATION AND RESOURCE RAIDING: 1860–1890 — 141

Pearls and pearlshell: resource depletion — 144
A climate 'unfit for Europeans': slave trading — 147
Advent of the Japanese — 152

10 FOR MAXIMUM YIELD: REEF BIOLOGY — 155

Reef science: achievements of Saville-Kent — 156
Collapse of the pearling industry: 1906–1916 — 161
Charles Hedley's 'Marine Biological Economics' — 165

PART TWO
A New Era in Reef Awareness: From Early Scientific Investigation to Conservation and Heritage

11	ORIGIN AND STRUCTURE OF CORAL REEFS: FROM FORSTER TO DARWIN	173
	Mystery of coral: the ancient quest	173
	Another mystery: formation of coral reefs	176
	Lyell's solution of 1832	179
	Charles Darwin and the voyage of the *Beagle*: 1831–1836	182
	Formation of barrier reefs and atolls: 'so deductive a theory'	185
	Darwin's subsidence theory: 1839–1842	187
	The problem solved?	191
12	DARWIN'S LEGACY: CORAL REEF CONTROVERSY 1863–1923	193
	Darwin's opponents: Semper and Murray, 1863–1880	193
	Alexander Agassiz: pursuing a solution, 1881–1910	196
	Funafuti subsidence theory tests, 1896–1898	200
	Voyages of Agassiz: Great Barrier Reef to Fiji, 1896	203
	Geology and biology: coral research expanded, 1902–1920	207
	Mayor's ecological surveys: Torres Strait and Samoa, 1913–1920	208
13	EXPLOITATION CHALLENGED: RISE OF ECOLOGY	214
	Nature and the Inner Sea	215
	'Beachcomber' Banfield: a different drum	218
	Rise of ecology: the subversive science	224
	Respecting nature: conservation and sanctuary	225
14	REEF RESEARCH AND CONTROVERSY: 1920–1930	231
	The Pan-Pacific Union 1920	231
	A research body founded: the Great Barrier Reef Committee 1922	233
	Second Pan-Pacific Congress, Melbourne 1923	240
	Boring Michaelmas Reef: GBRC research and controversy, 1925–1927	242
15	THE LOW ISLES EXPEDITION, 1928–1929: PLANNING AND PREPARATION	249
	Issue of Reef biological studies, 1923–1927	249

First major biological study planned: Potts Expedition of 1927	252
Third Pan-Pacific Science Congress, Tokyo 1926	254
Expedition reorganised: Yonge as leader	257
Expedition begins: London to Cairns	260

16 BIOLOGICAL RESEARCH OF THE LOW ISLES EXPEDITION   264

Research organisation and procedure	264
Experiments with corals by Yonge and Nicholls	270
Reef geology: Geographical Society survey attempt	273
Low Isles Expedition: results and significance	277

17 FROM DEPRESSION TO WAR: TOURISM, CONSERVATION AND SCIENCE, 1929–1939   283

Naturalists, turtles and early tourism	283
Growth of the conservation movement	289
Geology revived: the Heron Island bore, 1937	295

18 THE PACIFIC WAR AND ITS AFTERMATH   300

Japanese invasion fears: naval survey of the Reef	301
Aerial photography: an innovation in Reef survey	303
War comes to the Reef	305
Coral reef problem solved: Bikini and Enewetak	308
Quest for a marine research station: the background	310
Reef research stations: 1948–1973	312
Sydney's reef stations: One Tree and Lizard Islands, 1965–1973	314

19 A NEW PROBLEM: THE CONSERVATION CONTROVERSY, 1958–1972   317

Conservative ascendancy: unrestrained exploitation, 1953–1968	317
Reef resources: devising a legal regime	320
Environmental anxiety for the Reef	322
Commonwealth concern for reef conservation	328

20 CRISIS RESOLUTION: FORMATION OF AN ENVIRONMENTAL MANAGEMENT AUTHORITY   337

From resolution to implementation: first drafts of an Act	337
The Whitlam government: a radical development	343
Planning an authority	346
Report of the Royal Commission 1971–1974	348
A formula for Reef management: the Emerald Agreement, 1979	355

21	A NEW ERA: RESEARCH BASED MANAGEMENT	357
Reef research: James Cook University and the Australian Institute of Marine Science		357
Towards a national marine science and technology policy		360
Planning the Great Barrier Reef Marine Park		363
Management through research: the essential function		369
A conservation climax: World Heritage listing of the Reef		370
Reef management and public relations		372
Crown of Thorns: conflict and controversy		374

22	THE REEF UNDER PRESSURE: RESEARCH AND MANAGEMENT	379
Environment and economic growth in the greenhouse decade		379
Oceans of Wealth? marine science under review		381
Coral reef research: into the sustainability era		383
Coral reef science at the turn of the century		386
Warning signals: a 'particularly sensitive area'		388
The Reef as a maritime superhighway		391
World Heritage protection: a twenty-five year strategy		393
Cultural heritage: recognition of indigenous rights		395
Great Barrier Reef Marine Park Authority management reform		398
A wave of concern: maintaining the heritage value of the Reef?		402

23	THE REEF AS HERITAGE: A CHALLENGE FOR THE FUTURE	404
Management reform and the 'dugong wars'		405
The Oyster Point controversy		408
Issues for resolution: cooperative management of the Reef		412
Heritage management in a warming world		418
Conserving biodiversity: the innovative Representative Areas Program		421
Heritage: a sustainable ideal?		425

References	429
Index	446

Colour plates can be found after pages 110 and 302.

ABBREVIATIONS AND ACRONYMS

AAO	Australian Associated Oilfields
ABARE	Australian Bureau of Agricultural and Resource Economics
ACC	Aboriginal Co-ordinating Council
ACF	Australian Conservation Foundation
ACSI	Australian Council for Science and Industry
AFZ	Australian Fishing Zone
AIMS	Australian Institute of Marine Science
AMPOL	Australian Motorists Petroleum Organisation Limited
AMS	Australian Museum Series archives
AMSA	Australian Maritime Safety Authority
AMSTAC	Australian Marine Scientific and Technology Advisory Committee
ANZAAS	Australia and New Zealand Association for the Advancement of Science
APEA	Australian Petroleum Exploration Association
ASN	Australasian Steam Navigation Company
ASTEC	Australian Science and Technology Council
Au.Arch.	Author's Archives: documents held in the author's possession
AUC	Australian Universities Commission
BA	British Association (for the Advancement of Science)
BISN	British India Steam Navigation Company
BRIAN	Barrier Reef Imaging Analysis
CCD	Correspondence of Charles Darwin
CFC	chlorofluorocarbons
CISI	Commonwealth Institute of Science and Industry
COAG	Council of Australian Governments
COTSAC	Crown of Thorns Starfish Advisory Committee
CRC	Cooperative Research Centre
CSIRO	(Archives of) Commonwealth Scientific and Industrial Research Organisation
DITAC	(Commonwealth) Department of Industry, Technology and Commerce
dwt	deadweight (total tonnage mass of a ship and all contents)
ENSO	El Niño Southern Oscillation
EEZ	Exclusive Economic Zone
EMBL	Enewetak Marine Biological Laboratory
FAO	(United Nations) Food and Agriculture Organisation

Abbreviations and Acronyms

GBRC	Great Barrier Reef Committee
GBRMPA	Great Barrier Reef Marine Park Authority
GEEP	Group of Experts on the Effects of Pollution
GPS	Global Positioning System
HRA	Historical Records of Australia
HR.NSW	Historical Records of New South Wales
IDC	(Commonwealth) Interdepartmental Committee
IMO	International Maritime Organisation
IPCC	Intergovernmental Panel on Climate Change
IUCN	International Union for the Conservation of Nature and Natural Resources, now the World Conservation Union
LIE	Low Isles Expedition. The six Reports to the GBRC by Yonge in the author's archives are catalogued as follows: Yonge:LIE, followed by the sequence in Roman numerals from I to VI, followed by month and year, and finally page number
MPBL	Mid-Pacific Marine Laboratory
mya	million years ago
NOAA	National Oceanic and Atmospheric Administration
NPA	National Parks Association (of Queensland)
OPEC	Organisation of Petroleum Exporting Countries
PPC	Pan-Pacific Congresses, 1923, 1926
PPSC	Pan-Pacific Scientific Conference, 1920
PSA	Prices Surveillance Authority
QDoE	Queensland Department of the Environment
QFBP	Queensland Fisheries and Boating Patrol
QFMA	Queensland Fisheries Management Authority
QNPWS	Queensland National Parks and Wildlife Service
QPD	Queensland Parliamentary Debates
QPWS	Queensland Parks and Wildlife Service
RAP	Representative Areas Program
RGSAQ	Royal Geographical Society of Australasia, Queensland
TrawlMAC	Trawl Management Advisory Committee
UQA	University of Queensland archives
UNESCO	United Nations, Educational, Scientific and Cultural Organisation
VLCC	Very Large Crude Carriers (ships over 60,000 tonnes dwt)
VMS	Vessel Monitoring system
VOC	United East India Company (Vereenigde Oostindische Companie)
WCU	World Conservation Union

Preface

The Great Barrier Reef, Australia's most outstanding natural feature, has captured the interest of scientists and tourists from around the world. Yet surprisingly, despite its immense attraction, scientific importance and heritage value, no single, comprehensive account of its fascinating history has ever been published.

My own interest in the Reef, arising from a lifetime of involvement with coastal and marine environments, was initially aroused by the Great Barrier Reef conservation conflict of the 1960s. During an academic career that included extensive publishing in the history of ideas and environmental thought, the present study was commenced as a visiting Professorial Fellow in the Centre for Resource and Environmental Studies of the Australian National University from 1984 to 1989. In that stimulating context the task was conceived as a project to bring into the public record the history of the Great Barrier Reef since its discovery by Europeans.

This became a challenging collaborative research project with Dr Margarita Bowen, scientist and historian. Following the original conception we worked closely together, guided by her wide experience in ecological studies and competence in the study of the development of scientific thought, originally presented in her impressive study of scientific ideas in the eighteenth and nineteenth centuries in *Empiricism and Geographical Thought* (1981). Published in the prestigious Cambridge Geographical Studies series, that work still challenges much ecological theory today. The conceptual structure for this book has also depended on her concern to ensure indigenous justice, and understanding of heritage issues gained while working in the Commonwealth Department of the Environment in Canberra.

Throughout the following decade this led us to undertake extensive and memorable field studies over the entire Reef area: on research vessels and patrol boats; in aerial surveys, and scuba diving with scientific colleagues and Marine Park rangers on reefs from Lady Musgrave and the outer Swains complex in the south to Lizard Island in the north; with further travel on ships to Torres Strait, the Great North East Channel and reefs in Papua-New Guinea, New Britain and across the Coral Sea to New Caledonia and Vanuatu. Margarita, however, drew the line at joining me on several expeditions to the outer Swain Reefs to catch sea snakes for ecological surveys, perhaps a wise precaution since on one occasion I was struck on the leg by an aggressive olive sea snake, whose fangs fortunately did not penetrate my wetsuit. A significant part of our time was spent with research teams at the field stations of Heron, One Tree and Orpheus Islands, and in residence for numerous periods at the Australian Institute of Marine Science and at the Sir George Fisher Centre for Tropical Marine Studies of James Cook University in Townsville. In

Preface

every case our inquiries met with an encouraging enthusiasm for this project and a generous sharing of information.

As research on this study progressed, it was dismaying to discover how little of the historical record was freely available: the numerous scientists with whom we had sustained contact knew almost nothing of the background to their own specialties. Modern science is characterised by the dominant ethos of what I call 'presentism': ten years ago for many is ancient history and considered of no use in current research. It is perhaps a sad irony that the very nature of the sciences as they developed in the nineteenth century – reductionist, specialised, materialist and with a pronounced rejection of the past in favour of progress – has frequently resulted in a loss of knowledge, with an exclusion of history and the search for a broader vision. That is a deprivation for science, and for the human spirit: without knowledge of the past we lack any sense of continuity with it and with the world of ideas where our minds can travel, survey and enjoy the incredible richness of human experience.

The quest then, became one of attempting to return that historical heritage, in a coherent, integrated account, to public view. Along with extensive field experience this led me to the equally fascinating search for early primary sources, including travel journals, archive records and in many cases the personal recollections of key figures in reef research and management since 1975. On many occasions I was evidently the first person to open boxes long stored on repository shelves, or to ask questions of significant participants in more recent events. Those searches have enabled the later chapters to be documented from previously unpublished archival material covering the period from the late 1890s to the Low Isles Expedition of 1928–29. The chapters recounting the final decades of the twentieth century are based, in part, on previously obscure Commonwealth records, and valuable interviews with persons directly involved in the formation of the Great Barrier Reef Marine Park Authority and subsequent events leading to its World Heritage listing by UNESCO. In the case of several of those distinguished men and women who entrusted me with their records and documents, their voices, regrettably, are now silent.

As the book developed Margarita Bowen undertook the demanding task of scientific revision and editing to ensure the highest standard of accuracy. Throughout the text she wrote new sections and substantially revised others to correct errors of fact or interpretation and provide greater clarity of expression, particularly in regard to the conceptual framework of current theory of science, ecological debate and heritage issues. Her expertise has brought the narrative to a standard that otherwise could not have been attained.

During the process of editing the very lengthy draft, due to limitations of space, had to be extensively abridged for publication. As a result a large amount of contextual detail had to be removed, including a number of topics closely related to Reef history, notably the century long quest for an Australian marine research centre as revealed in the fascinating story of the Miklouho–Maclay station of the 1880s and Sydney's aborted Watsons Bay station of the 1930s. Likewise excluded were fuller accounts of MacGillivray's untimely dismissal from the survey by the Herald, and the furious continuation of the controversy on coral reef formation with the defence of Darwin by the great American geologist James Dana in the late nineteenth century. Moreover, given the explosion in scientific investigation of coral reefs worldwide, and associated conservation activity, it has been possible to report only selectively from the huge quantity of publications.

From the extensive literature of relevant studies – in addition to numerous colourful travel books on the wonders of the Reef – many informative works have been consulted. These, however, invariably concentrated on a single issue such as exploration, settlement, resource extraction, or on scientific aspects, chiefly topics in biology and geology. Despite the considerable volume of publications, the Great Barrier Reef so far has not had the fascinating saga behind its complex tapestry of history brought together in a composite narrative. This book, then, is our pioneer project to provide a continuous – but by no means complete – contextual history of the Reef in a single volume.

James Bowen
Research Fellow
Environmental Science & Management
Southern Cross University

ACKNOWLEDGMENTS

This project has depended on many persons and institutions: it is both a pleasure and responsibility to acknowledge the valuable help received. Initial support was generously provided by the Centre for Resource and Environment Studies (CRES) at the Australian National University, with an appointment as Visiting Professorial Research Fellow 1984-89. The encouragement given by former Directors, Professor Stuart Harris and Professor Henry Nix, and by distinguished colleagues such as Professor Stephen Boyden and the late Dr H. C. ('Nugget') Coombs, is deeply appreciated.

Much subsequent field work was undertaken during periods of residence at the Australian Institute for Marine Science (AIMS) near Townsville, and thanks are extended to former Director, Dr Joe Baker OBE and to many staff members, including Dr John (Charlie) Veron, Dr Paul Sammarco and Dr Barry Clough who gave freely of their expertise and provided opportunities to join them in reef research at sea and in mangrove restoration projects on Hinchinbrook Island and in the Daintree River. Professor David Hopley, Director of the Sir George Fisher Centre for Tropical Marine Studies at James Cook University, kindly arranged a year as Visiting Research Fellow in 1990, with invaluable access to their Orpheus Island Research Station and surveys aboard the RV *Kirby*. Professor Hal Heatwole of the University of New England provided opportunities to join several voyages to the outer Swain Reefs to assist in projects such as ecological surveys of sea snakes, and Dr Ove Hoegh-Guldberg of Sydney University facilitated a period of research residence on One Tree Island and Heron Island Research Stations.

Acknowledgment must be made of the valuable assistance received throughout from the Great Barrier Reef Marine Park Authority: in particular from former Chairman of the Authority, Professor Graeme Kelleher AO; former Chief Executive Officers Dr Don Kinsey AM and Dr Wendy Craik, and Education Officer, Mr Ray Neale. Mr Jon Day, as Director of its Conservation, Biodiversity and World Heritage Group kindly helped revise sections on current management, while many staff were helpful in providing ready access to recent data.

The Queensland Parks and Wildlife Service also generously provided field experience in aerial surveying of the Marine Park northern Section and with QNPWS Rangers on visits to Lizard Island. In the Cairns area Rangers Robert Zigterman and Gary Barnes arranged both informative discussions with staff on joint management of the Reef, and valued opportunities to observe monitoring of reefs and islands in the patrol boat *Caretta*. Dr Col Limpus welcomed participation in the Turtle Research and Conservation Program at the Mon Repos research station; and Peter Ogilvie, Manager of the World Heritage Unit of QPWS contributed valuable advice on recent developments in Reef management.

Important insights into problems of navigation in Reef waters were gained from seamen on traverses across the Coral Sea past Willis Island to Townsville, and through the North East Channel to Torres Strait. On MV *Norwegian Star*, Captain Konstantinos Fafalios and his officers were most helpful in assisting studies from the bridge, and the Torres Strait pilot on board, Captain John Foley, generously shared his considerable navigational expertise and knowledge of Reef maritime history. Expert assistance on historical aspects and current procedures in Reef hydrography was willingly provided at the Australian Hydrographic Office of the Royal Australian Navy by Hydrographer Captain Bruce Kafer, assisted by Nautical Information Officer Kevin Slade and archivist Arnie Larden. Former RAN Deputy Hydrographer, Commander Reg Hardstaff, author of *Leadline to Laser*, was particularly generous in sharing his extensive surveying experience.

This project has also benefited from the assistance of many persons who were centrally involved in the development of Reef science and conservation in Australia and who courteously gave time for lengthy interviews, throwing new light on recent decades. It is a pleasure to express appreciation for the kind cooperation of: Dr Max Day AO, FAA, chairman of the Interim Council of AIMS; Professor Ken Back AO, Foundation Vice Chancellor of James Cook University; the late Mr Henry Higgs, and Dr Donald McMichael CBE, former Chairs of the Great Barrier Reef Marine Park Authority; Professor Malvern Gilmartin, Foundation Director of AIMS; the late Emeritus Professor Dorothy Hill AC and the late Dr Robert Endean of the University of Queensland; Dr Patricia Mather AO at the Queensland Museum, and Emeritus Professor Burdon-Jones, formerly Foundation Professor of Biology at James Cook University; Professor John Coll, formerly of James Cook and now Vice Chancellor of the Australian Catholic University. Valuable advice was also provided by Emeritus Professor Frank Talbot of Macquarie University, formerly Director of the Australian Museum and then of the Smithsonian National Museum of Natural History, and by Dr Ian McPhail, a former Chair of GBRMPA and now Deputy Director-General of the Queensland Environmental Protection Agency.

As in all historical research this study, which often required access to rare or obscure materials, has relied constantly on libraries and the dedicated service so willingly given by their staff – in every case with skill and enthusiasm. In Canberra valuable information and archival assistance came from the National Library of Australia and the Australian National University libraries, including the library of the Australian Institute of Aboriginal and Torres Strait Islander Studies and its bibliographer Mr Barry Cundy; while interesting and highly relevant archives from 1900 to 1920 were located with the kind help of CSIRO archivist Mr Rodney Teakle.

Courteous and expert assistance was also provided in Sydney by the Mitchell Library; by Mr Julian Holland of the Macleay Museum in the University of Sydney; and by Ms Jan Brazier, archivist of the Australian Museum, with Ms Carole Cantrell and Ms Ann Pinson, current Records Officer. In Lismore the librarians of Southern Cross University kindly provided access to early Australian materials in the Manning Clark collection.

Likewise, in Brisbane thanks are extended to: Ms Megan Lyneham, archivist in the University of Queensland; Mss Kath Buckley and Megan Lloyd, librarians of the Queensland Museum; and the staff of both the Oxley Library and State Library of Queensland. In Townsville excellent assistance was provided by the superb library at AIMS with its Yonge historical collection, under the skilled attention of former librarian Inara Bush; and by the library of James Cook University, particularly its rare book collection. The most sustained support in Reef research documents over a number of years came unfailingly from Ms Suzie Davies, librarian of the Great Barrier Reef Marine Park Authority.

In addition, grateful acknowledgment is made to the Great Barrier Reef Marine Park Authority for their generous provision of the colour illustrations, to the photographers who are acknowledged within the captions, and to Julie Jones and Holly Savage, GBRMPA Image Collection officers, for their competent assistance. Sincere appreciation is also extended to Cambridge University Press for their enthusiasm and guidance through the process of publication.

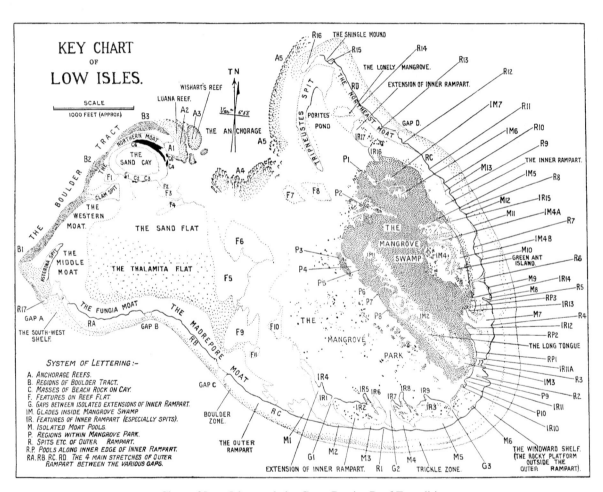

Chart of Low Isles made by Great Barrier Reef Expedition

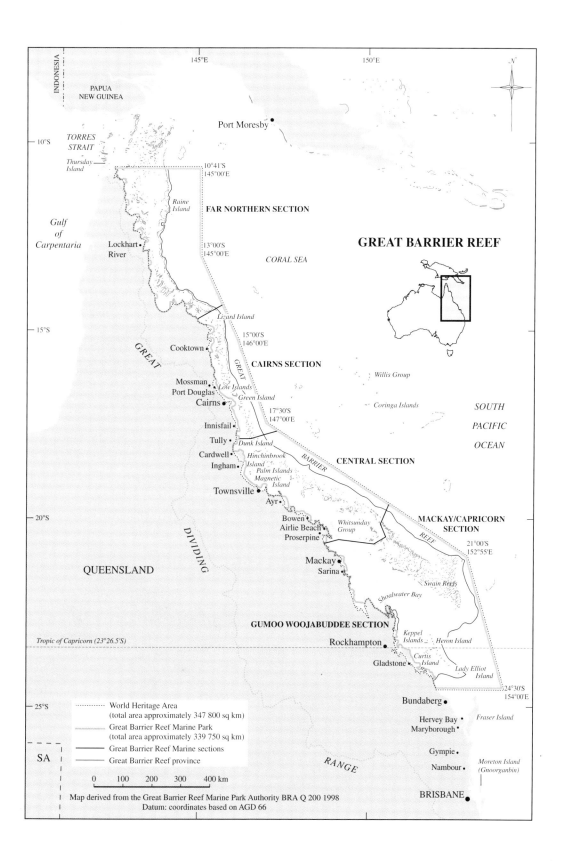

INTRODUCTION: AN OVERVIEW

Ever since Captain James Cook charted the Great Barrier Reef in 1770 it has exerted a fascination that shows no sign of diminishing. Over those centuries its chequered history has moved through a sequence of phases: from a navigation hazard to be feared and then conquered, to a geological challenge and a realm of extraordinary plants and animals offering a seemingly inexhaustible range of natural resources for scientific study and exploitation.

In recent decades the Reef has come into the international spotlight as the world's greatest marine park with its listing by the United Nations Educational, Scientific and Cultural Organisation (UNESCO) in 1981 as a special World Heritage Area of 'superlative natural phenomena' containing 'formations of exceptional natural beauty [with] superlative examples of the most important ecosystems'. It also was recognised as an 'outstanding example of the major stages of the earth's evolutionary history', and 'of significant ongoing geological processes, biological evolution and man's interaction with the natural environment'. Of profound relevance today is its further listing as a site of 'the foremost natural habitats where threatened species of animals or plants of outstanding universal value from the point of view of science or conservation still survive' (UNESCO 1980:22–23).

This cultural and environmental history, then, has a special objective. It takes the reader through the endlessly absorbing story of the impact of Western discovery and settlement on the Great Barrier Reef, and equally, the response of Western science to that encounter with the world's greatest living natural feature. Within that epic narrative, it also brings into focus the story of coral and coral reefs, the fear they held for early navigators, and the fascination for naturalists, scientists and tourists of recent centuries, leading, in the case of the Great Barrier Reef, to intense efforts since the 1960s to ensure its protection for the future.

An appreciation of the complex history of the Great Barrier Reef must begin with its awesome dimensions. Stretching over 2000 kilometres along the tropical northern coast of eastern Australia it is by far the largest of the world's barrier reefs and the only living structure visible from space. Its southern limit, where Cook first entered, begins at 24°6'S with Lady Elliot Island, a sand cay in the Bunker Group of islands, while the great coral masses of the barrier itself begin further north some 200 kilometres offshore in the Swain Reefs at 22°S. From there the main coral formations curve north-westerly to approach the coastline near Cape York, then swing north-easterly to reach the Gulf of Papua around 8°S latitude.

Encompassing a vast area of almost 350 000 square kilometres – an area comparable with Japan and larger than the combined landmass of Great Britain and Ireland – it contains over 2900 separate coral reefs, some fully submerged, others only visible at low tide. In structure it is not a continuous barrier but a vast and almost impenetrable assemblage of reefs: long strips of ribbon reefs, large areal spreads of patch reefs, and circular formations, known as cays, with a central lagoon and an elevated area of pulverised coral 'sand'. Many cays, like Heron Island, are vegetated and a haven for flocks of sea birds. In addition, fringe reefs surround the tips of some 200 submerged mountains that rise above water to form rocky outcrops and the larger continental islands such as the Whitsunday group and Hinchinbrook, Dunk and Magnetic islands.

When Cook encountered the 'great wall of Coral Rock' and came to grief with the *Endeavour* in what he called its 'Labyrinth', the mystery of its origins was still unresolved. If such reefs had been 'formed in the Sea by animals', as he pondered over elevated relict reefs observed during his second Pacific voyage, then, he asked, 'how came they [to be] thrown up to such a height?' (Cook 1774:438). It is still astonishing to realise that such massive structures, which provide habitat for the brilliantly coloured and teeming life of coral reefs, have been created by colonies of tiny animals – usually less than 5 millimetres in diameter – the polyps of a large range of hermatypic ('reef building') corals of the order Scleractinia.

Moreover, as confirmed by research in the mid twentieth century, polyps of these 'hard' corals are themselves dependent on a symbiotic interaction with a myriad of single-celled algae, called zooxanthellae, in their tissues. That interaction, in warm sunlit waters, can supply most of the polyps' nutrition. It also provides carbon enabling them to build their characteristic hard calcified cups, known as corallites, which over millennia can fuse together into extensive colonies and form immense buttresses rising hundreds of metres to the ocean surface. The processes of earth subsidence and sea level change which can cause this to occur, as discussed later in Part II, became a focus of geological debate throughout the nineteenth century and well into the twentieth, until the 1950s when Cook's query was finally answered.

The geophysical origins of the Reef began with tectonic plate activity more than 50 million years ago that broke up the great southern landmass of Gondwanaland and moved the Australian plate northwards. Around 17 million years ago it reached the tropic zone, where reef-building corals had become established, as revealed in the world fossil record, more than 200 million years earlier, during the Triassic geological period. Over the ensuing periods, the continental shelf of eastern Australia became consolidated as deposits from terrestrial degrading and carbonate sediments were laid down, forming the substrate for extensive coral formations in tropical waters, predecessors of the present Great Barrier Reef. In recent geological times, at least as early as 50 000 years ago, a broad land platform appeared north of the continent, supporting open forest growth and connected to New Guinea. Then came sea level fluctuations during the last of the Ice Ages, around 11 000 years ago, which began to close the land bridge, perhaps finally by 8000 years ago. As the sea progressively inundated the coastal plain, forming the present Torres Strait, it created at the same time the submarine platform on which the coral communities of today's Reef system, in their thousands of species, were able to grow.

The relatively recent formation of the present Reef within the last 10 000 years is dwarfed, however, by the formidable time span of Aboriginal settlement on the continent, estimated to date back more than 50 000 years, with recent archaeological research in the Cape York region reporting dates ranging from 29 000 to 13 000 years ago (Morwood 1993:175–76). Given that the Torres Strait land bridge region, like the now submerged shelf of the Barrier Reef, largely disappeared around 8000 years ago, evidence of earlier settlement in those areas remains severely constricted. It must be with such awareness that the Australian continent and the entire Great Barrier Reef province is to be understood as having had a long period of settlement by Aborigines and Torres Strait Islanders prior to, and continuing in sadly diminished form alongside, European involvement. For them, many regions of the Reef remain significant heritage cultural landscapes, rich in associative values. The recording of that experience, however, is beyond the scope of this work and requires a major study in itself. Only with that caveat in mind, then, can we speak in the following pages of a European 'discovery' of Australia and the Great Barrier Reef.

Forming a barrier to coastal shipping and to the pounding waves of the Pacific Ocean, the great Reef became a source of dread to early sailors as the millennia of indigenous occupation were followed by Western exploration and settlement. With the arrival of Europeans the Great Barrier Reef also became an intriguing natural realm that demanded investigation, as did the entire continent of Australia. That process is examined here within the wider context of the ferment of ideas throughout Western science. With Cook's charting of the Reef, his claiming of 'New South Wales' for Britain, and the explorations of those who came after him, it seemed that here was found yet another New World. Unlike the Americas, Australia was so curiously different that at first it was considered a separate Creation. What sense could be made of the strange biota, especially the unique egg-laying monotremes, the platypus and echidna? The 'opossum' and the kangaroo gave London scientists considerable taxonomic difficulty, while the platypus when it first arrived there was considered a cunningly contrived hoax. As more new plants and animals were found, the traditional acceptance of the earth as a Divine Design, already under severe strain as the Biblical account of an unchanged Creation was increasingly questioned, threw natural science into even greater turmoil.

Indeed, so perplexing was the natural landscape that Barron Field, Supreme Court judge and Australia's first poet, wrote in the poem 'Kangaroo' that the country, although a 'fifth part of the Earth' was so bizarre that it 'would seem an after-birth, not conceived in the beginning' but later as a result of Original Sin, 'When the ground was therefore crust, And hence this barren wood'. And the kangaroo itself was so peculiar, Field jested, that it must have been created while God was resting on the Seventh Day:

> Join'd by some divine mistake
> None but Nature's hand can make
> Nature in her wisdom's play
> On Creation's holiday. (Field 1819:3)

Even a century after Cook, when publishing *The Naturalist in Australia* (1897) for a British readership, Reef scientist William Saville-Kent could write in the Preface that 'Australia is, *par excellence* . . . the land of topsy-turveydom. Christmastide is a midsummer festival; the swans are black; cherry stones grow outside the fruit; flies eat the spider, and oysters grow on [mangrove] trees, along with many other things' (vi).

Throughout the nineteenth century, as the continent was mapped, explored and opened to European settlement, the Reef became a phenomenon of major world interest. As its biota were described and its geology explored, major contributions to world science accumulated. In geology, coral reefs became central to new theories of terrestrial formation as the challenging hypotheses of Charles Lyell and Charles Darwin helped undermine the foundations of the received tradition. Equally important was the impact of Darwin's theory of evolution by natural selection that eventually triumphed over the Creationist beliefs of early gentlemen naturalists and cabinet collectors. By the twentieth century the so-called heresies of Lyell and Darwin became the orthodoxies of our present era. In discussing these developments, the narrative ranges widely to an informative account of the broader context within which the Reef, its values, exploration, scientific research and heritage values were considered through time.

The history of the Reef, however, encompasses much more than the ferment in natural history as it evolved into more rigorous science. It provides yet another international example of the heedless exploitation of natural resources and indifference to nature that were leading to its despoliation, even though voices of protest were raised at the time. That too is an important theme traced throughout, finding its most eloquent advocates in 'Beachcomber' Banfield, Charles Hedley and the later activists of the 'Save the Reef' campaign of the 1960s and 1970s, Australia's most politically vigorous and successful early environmental protest movement. From those efforts the Reef came to be designated by the Australian government as a marine park in 1975, and internationally protected in 1981 as a World Heritage Area.

Today the concept of 'heritage', moreover, as it relates to the Reef, has become much wider than the earlier conventional definition that identified only natural features and cultural artifacts. Heritage now recognises the significance that localities can hold for communities through association with their daily lives, past and present, and Reef management is sensitive to the value of such continuing and associative landscapes. That expanded concept also enables a greater understanding of the perceptions of indigenous peoples in their close identification with their lands (Cotter, Boyd & Gardiner 2001).

At the same time, it is important to understand that the Great Barrier Reef Marine Park, which covers most of the Reef province, to the tip of Cape York, is necessarily managed according to the principle of multiple use. It contains an international shipping route used by 25 per cent of Australia's overseas shipping commerce, while its waters support a large fishing industry. In addition, management has to contend with problems from increasing agricultural and urban residue pollution, while tourism, now exceeding 2 million visitor-days a year, imposes tremendous pressure. Indeed, this book traces in dramatic form the pattern of Western impact on the Reef from the enchanting 'other Eden' of Cook's day to the markedly changed environment by the beginning of the twenty first century.

Even so, the Reef maintains its powerful, enduring mystique, and that, too, is of central interest in this study. In response to the pressures of modern living we have come to appreciate the importance of places that inspire a sense of awe and mystery and offer the simple joy of the contemplation of nature. In a world of increasing technologically driven development, alienation from nature and the dehumanisation of so much in everyday life, the preservation of such places is vital. Too often, however, they are seriously endangered, and can easily be destroyed. Public awareness of this was aroused only in the late twentieth century with the emerging recognition that humans and nature are all part of one great ecosystem, which must be considered and managed as an inseparable whole. It was, therefore, one of the great advances of recent decades, as worldwide environmental destruction gathered momentum and began to leave a devastating swathe across the natural landscape, that communities and governments came to see the need for more effective management and protection of natural environments. Reinforcing that view, the emergence of ecological perspectives and complexity theory in science has led to a more comprehensive understanding of coral reef ecosystems, the manifold interrelations involved in their development over time, the problems of management, and the consequences of human interaction: all essential knowledge if reefs are to be secured for the future, not only for human enjoyment but for their value as natural systems.

The historical trail for this research – through rare books, archives, interviews with significant individuals and active field investigation – has identified three main themes: the history of European engagement and colonisation, the scientific study of coral reefs, and rising concern over the last century for their heritage value and protection. The approach adopted here aims to identify and present those strands as an interacting complex of contemporaneous events, producing as it were, a composite portrait. Moreover, since the Reef, from its first discovery by Europeans, has never been isolated from world affairs, this account has been set within an international context of exploration, scientific debate, commercial pressures and political action.

Part I, Navigators and Naturalists in the Age of Sail, begins with chapters covering the early voyages of discovery and charting of the dangerous Reef waters in the task of locating safe passages to develop secure lines of communication between Britain and the new colonies. Proceeding in parallel with those early surveys are accounts of the naturalists who sailed aboard the naval vessels from the late eighteenth century, initially under British Admiralty instructions to find new raw materials, both botanical and mineral, in the service of economic imperialism.

Those chapters are followed by an account of the subsequent foundation and expansion of the new colony of Queensland and the rapid, and in many cases disastrous, exploitation of natural resources in the later nineteenth century and the early decades of the twentieth. As the focus of Reef activities moved from discovery to European settlement and exploitation, a changed emphasis is evident in natural science as it moved from cabinet collecting into applied research for commercial benefit.

Part II, A New Era in Reef Awareness, in a departure from strict chronology, returns to examine Darwin's epochal study of coral reef formation presented in the years from 1839 to 1842, and to trace the ensuing international controversy over the origin and structure of coral reefs that continued until the mid twentieth century. Concurrently, the rise of ecological theory, research and conservation ideals is explored in the compelling works of 'Beachcomber' Banfield and his predecessors, and in Mayor's paradigm ecological surveys of coral reefs in Torres Strait and Samoa (1913–20). Following those surveys a considerable advance in Reef biological research was achieved with the British expedition to the Low Isles in 1928–29.

Despite those developments in biological investigation, however, geological issues continued to dominate the scientific agenda throughout the 1920s and 1930s under the influence of Henry Richards, Professor of Geology at the University of Queensland. Although such early reef geology, which involved deep drilling of several reef sites, was ostensibly aimed at resolving fundamental issues of coral reef formation, concern grew that economic interests were involved. Indeed, public alarm that the Great Barrier Reef had become a prime target for petroleum exploration precipitated the conservation controversy of the 1960s and 1970s.

That significant movement is examined in the context of the simultaneous development of interest in the Reef as a tourist destination and its promotion, stimulated by Banfield

Introduction: An Overview

earlier in the century, as a magic wonderland far removed from the banalities of everyday life. When from that heightened awareness the vigorous public campaign for the protection of the Reef as a national marine park was successful in 1975, it created a management problem of the greatest magnitude. After some difficult negotiation that task was undertaken jointly by the Commonwealth and Queensland governments. Responsibility for formulating management policy under the Commonwealth *Great Barrier Reef Marine Park Act 1975* was to be the prime function of the new Great Barrier Reef Marine Park Authority (GBRMPA), with day-to-day field operations to be discharged by the Queensland Government through various agencies, mainly the Queensland Parks and Wildlife Service, the Queensland Fisheries and Boating Patrol and the Queensland Fisheries Service (as they are now designated). All these responsibilities make increasing demands on scientific support and effective policy development and implementation. The present account, then, looks at the conservation challenge that lies ahead in providing effective heritage management, maintenance and remediation, consistent with the revolutionary social awareness in our own time that coral reefs and their ecosystems must be protected as structures of incredible complexity, beauty and fragility.

In that task the value of an understanding of Reef history becomes clear when current needs for base data are considered. An outstanding case occurred in 1998 when, after twenty years of stewardship, the GBRMPA conducted a thorough assessment of the condition of the Reef to confirm that it was continuing to meet UNESCO's exacting World Heritage classification criteria. The disturbing conclusion reached was that 'for most environmental categories, it is not possible to say with certainty if they are in a satisfactory or unsatisfactory condition' (Wachenfeld et al. 1998:1). Any meaningful understanding of the Reef's current state and projected trends cannot be based simply on short term data over one or two decades. To assess long term changes, we will benefit enormously from evidence in the journals of Cook, Banks and those who continued after them, such as Jukes, MacGillivray, Saville-Kent, Hedley, Banfield, Mayor and Yonge.

While the primary focus of this account is on the Great Barrier Reef, it also provides a case study of the generation of those relentless pressures on the world environment that are leading to international conferences and conventions for remediation. Coral reefs, in effect, are early warning indicators of global imbalance, now evident in serious biodiversity loss, ecological and water disturbance and, on reefs worldwide, persistent coral bleaching, an ominous signal of world climate change. Still relatively well preserved, the Great Barrier Reef forms a stark contrast to the damaged Caribbean and seriously degraded Southeast Asian coral reefs where unremitting human impact has devastated them, at least within a human time frame, perhaps irreversibly. With the further threat, then, that global warming could accelerate damage to coral reefs worldwide, it will be vital in this century for management to consider a greatly extended range of available data, beginning with the fossil record, and traced through succeeding periods as revealed in palaeoecological studies, indigenous oral history and historical records of colonial and recent periods, to the present global phase shift. Use of such resources is now being urged by numerous forward-looking scientists and managers (Jackson et al. 2001).

No such extended range of data was available to the early GBRMPA planners, nor, possibly, was it thought necessary. Perhaps, in the tradition of current scientific positivism, such historical knowledge was considered irrelevant to understanding the 'real condition' of the Reef. Yet, within the journals and narratives of early investigators we have accounts of the Reef environment at given occasions before and after European colonisers began processes of unrelenting change. A fundamental requirement for Reef managers and scientists, especially at the higher decision-making levels, should be at least a clear understanding of the progress of scientific knowledge and the historical processes of Reef disturbance at human hands over the past two centuries. And for us all, access to these rich sources of information promotes an enduring awareness of the Great Barrier Reef as an essential feature of the Australian National Estate and the World Heritage, requiring the greatest care in its management as a uniquely valuable part of our heritage to transmit to future generations.

PART ONE

Navigators and Naturalists in the Age of Sail

CHAPTER 1

QUEST FOR THE GREAT SOUTH LAND

The Great Barrier Reef burst suddenly into European consciousness in 1773. In that year the sensational account of James Cook's amazing voyage and discovery was released to the public as part of a huge three volume edition entitled *An account of the voyages undertaken by order of his present Majesty for making discoveries in the southern hemisphere*. Two years earlier, when he returned to England on 13 August 1771 after a three year voyage around the world, Cook reported that he had discovered and traversed the eastern and northern shores of the mysterious Great South Land which for centuries had been a quest for navigators. What became a central feature of the voyage was his description of a reef that beggared belief at the time: 'a wall of Coral Rock rising all most perpendicular out of the unfathomable Ocean ... the large waves of the vast Ocean meeting with so sudden a resistance make a most terrible surf breaking mountains high'.

By Cook's time coral reefs had already become well known and had acquired an extensive folklore, but nothing in the literature equalled the account of his nightmare travel through dangerous waters unmatched anywhere else in the world. For two years his journals were embargoed by the Admiralty to preserve their sensitive commercial information, especially from the French who were anxious to beat the British in the race to create an overseas empire. For both nations, the aim of creating a colonial network was one of economic imperialism: the search for new lands and new resources for manufacturing and trade. Yet, when Cook's discoveries were popularised in the impressive 1773 publication edited by John Hawkesworth, it was not the economic possibilities that intrigued the reading public, but accounts of exotic peoples with strange customs, and the unbelievable phenomenon of the great coral reef, to which Cook on his charts gave the sinister name of the 'Labyrinth'.

NAVIGATORS IN THE NEW WORLD

Probably the first person to create for European readers a sense of the mystique and dangers of coral reefs was the French navigator François Pyrard de Laval, whose ship, crossing the Indian Ocean in the first decade of the seventeenth century, was wrecked on a coral reef in the Maldives. Laval lived there for five years and on his return wrote a narrative of his adventures that was translated into English and included in Samuel Purchas' famous 1626 edition of dramatic travel tales. Part II is an accurate description of that archipelago, comprised entirely of atolls (a word of Malay origin). With a touch of hyperbole Laval described the Maldives as 'twelve thousand Iles', noting that 'c'est une merveille de voir chacun de ces atollons, environné d'un grand banc de pierre tout autour, n'y ayant point d'artifice humain'. The Purchas translation is more brief: 'It is admirable to behold, how that each of these Atollons are invironed around with a huge ledge of rocks', omitting Laval's interesting further comment, 'with no human construction whatever' (Purchas 1626:IX.508).

Throughout, Purchas keeps the prose heightened and the region threatening: the outer edges of atolls are 'a very fearfull thing even to the most couragious to approach this ledge, and so the waves come from afarre off and breake furiously on every side', around them swim 'sharkes [which] devoure men and breake their legges and armes' while in the 'depths of the Sea are generally very keene and sharpe Rockes' which include 'a certain thing not unlike Corall, ... branched and piercing, ... all hollow and pierced with little holes and passages, yet abides hard and ponderous as a stone' (509).

Ever since Purchas, whose account did not reach the dramatic intensity of Cook's record of the impending disaster in the Labyrinth, a formidable record of shipwrecks from hidden obstacles and coral reefs had become commonplace in navigating circles. In increasing numbers these were becoming published in sensational stories of voyages to exotic, strange lands in the era of imperial expansion from the sixteenth through to the eighteenth century. The historical origins of European interest in coral reefs, then, begins in that period of exploration when they were hazards to be feared, dangerous obstacles to voyaging at a time when one of the enduring puzzles was the existence of a Great South Land, depicted on early maps as Terra Australis. Many geographers had long maintained the belief that it certainly existed, since it was argued that the southern half of the globe must have comparable landmasses to balance the continents in the northern hemisphere. Supporting evidence came from the increasing number of world maps where Dutch and Spanish discoveries, along with that of the Englishman William Dampier, in the Indian Ocean region showed evidences of a landmass somewhere in the South Pacific. The question was: where exactly?

The first known map to assert the existence of a Great South Land is the famous world map of 1533 by the Paris mathematician Orontius Finaeus, teacher of Mercator. His innovative projection, known as a double cordiform (literally two heart-shaped adjoining hemispheres), has the northern hemisphere, centred on the North Pole, on the left of the chart, the southern hemisphere on the right, joined along the coast of Africa. The South Pole is shown within a very large landmass extending out beyond the Antarctic Circle, the coastline generally lying between 50°S and 60°S. The depiction, however, has two striking features: firstly, the land is labelled *Terra Australis, recenter inventa sed nondum plene cognita* ('Terra Australis, recently discovered but not yet fully known'). Secondly, the coastline, although lying some 20 degrees too far north, is astonishingly close to the outer limits of the Antarctic pack ice that, in many places, reaches nearly 55°S. In addition, there is remarkable similarity to the great indentation of the Ross Ice Shelf at longitude 180°. One supposition is that it was based, in part, on the voyage of Magellan in 1519–22 since the South Orkney and South Shetland Islands, just 5 degrees south of Tierra del Fuego, are shown within the edge of the pack ice. Unfortunately, there is no firm evidence for this. What is worth noting, however, is that the phrase *Terra Australis, incognita*, was reproduced on every world map thereafter for nearly two hundred years. Finaeus depicted a continent that was to become the object of continuing speculation and many voyages of exploration.

PORTUGAL AND SPAIN: FIRST NAVIGATORS OF TERRA AUSTRALIS?

While the sensational account of Cook's voyage through the 'labyrinth' brought the continent of Australia and its great reef to the forefront of European consciousness in 1773, there are serious doubts that he was the first person to chart its coastline. Today, an increasing conviction among some Western historians is that the first European explorers to visit and chart the location of the Reef were Portuguese, probably in the later fifteenth or the early sixteenth century. In the fifteenth century Portugal and Spain had emerged as the dominant European powers and both accelerated shipbuilding technology and, in the case of Prince Henry's maritime college at Sagres, navigation. Under their stimulation the carrack evolved into the exceptionally seaworthy caravel, which, along with Portuguese improvements in the science of navigation, enabled mariners to move into the open sea. In the Europe of that century's early decades only Portuguese and Spanish ships were equipped to venture far away from the safety of the coast, in the process traversing the same regions in the Atlantic where they came into sustained and bitter conflict.

As their exploration of the North Atlantic – the European 'known world' – increased each continued to dispute the other's territorial claims. In order to mediate between the two Catholic nations, Pope Nicholas V, after two decades of dispute, finally provided a compromise in 1479 with the Treaty of Alcaçovas, whereby a maritime latitudinal demarcation line of 28°N was drawn between Madeira and the Canaries in which the Spanish acknowledged the Portuguese monopoly to the south. However, once Columbus had returned from his explorations in the Caribbean in 1493, Spain claimed the whole region for itself. After many Byzantine machinations between the successor pope, Alexander VI, and the joint Spanish monarchs, Ferdinand of Aragon and Isabella of Castile, in the Treaty of Tordesillas (1494) a new longitudinal demarcation line was drawn 370 leagues west of the Cape Verde Islands (46°W, Greenwich) dividing the world into two spheres of influence, with Spain claiming the western half, Portugal the eastern.

By 1488 the Portuguese had already moved down the entire western coast of Africa, trading with powerful native chiefs in red coral beads, among other goods, enslaving weaker tribes as labourers for agricultural industries in their colonies, and establishing a series of fortified trading posts. In 1497 Vasco da Gama, travelling down this sea route, rounded the Cape and entered the Indian Ocean, reaching India the following year where he founded a base at Goa.

Spain, meantime, was not idle. Unable to cross the Isthmus of Panama from the Caribbean, and restricted by the Treaty of Tordesillas to the Western Atlantic, it ignored the demarcation meridian and sought to reach India by sailing down the east coast of South America in search of a passage to the west. That was achieved in the years 1519–22 when Magellan, in his ill-fated voyage, rounded Cape Horn and entered the Pacific, claiming that ocean for Spain.

By this time, in the 1520s, Spain was in financial and naval difficulties trying to enforce its authority not only in the Caribbean and the Americas against other European interlopers, but also in the Pacific and Indian Oceans against the Portuguese who were disputing the exact location of the Tordesillas antemeridian in the eastern hemisphere. In 1529, for the payment by Portugal of 350 000 ducats, Spain agreed in the Treaty of Saragossa to an antemeridian demarcation line (at modern Greenwich 144°E) which put all of southeast Asia, the East Indies, the Philippines and the western half of the Australian continent – from Arnhem Land to the Eyre Peninsula – in the Portuguese sphere.

Since the Philippines, New Guinea and Australia were apparently still unknown to Europeans, a most interesting historical puzzle emerges which has direct bearing upon the European discovery of eastern Australia and the Great Barrier Reef. Why did Portugal willingly pay an enormous sum for rights to a part of the globe that was still unexplored, and which, conceivably, could be as empty as the rest of the Pacific? Throughout the sixteenth century in their period of great power, the Portuguese continued voyages of exploration from bases in the East Indies, and there is the strong possibility they had prior knowledge of the great landmasses of New Guinea and Australia which influenced them to pay for the demarcation line to be moved further east.

Did the Portuguese know, secretly, of the existence of a continent of Terra Australis? We will never know. On 1 November 1755 in the greatest European earthquake of recent times, Lisbon was severely damaged. Many public buildings, including the Casa da India where Portuguese records and charts of the region were kept, were totally destroyed by a combination of tremor, tsunami and fire.

Despite their precautions, however, the Portuguese were never able to keep their discoveries and charts secret. Competition for commercial information was intense, industrial espionage was commonplace. Meanwhile, in growing competition with the Portuguese and Spanish, the Norman French had established a navigation and cartographic centre at Arcques near the port of Dieppe, modelled on that at Sagres. By the mid sixteenth century it was producing maps based on Portuguese discoveries, of which one chart – and a few derived from it – bear directly, and most intriguingly, on the charting of eastern Australia. One such map, some 2.5 by 1.2 metres, on vellum, richly decorated and inscribed in French, and titled, in block letters, *Jave la Grande*, is embellished in the lower left corner with the legend: *'Faicte a Arcques par Pierres Desceliers L'an 1550'* ('Made in Arques by Pierre Desceliers in 1550'). Two cartographical features of the map are of profound significance: firstly, Europe, Africa, Arabia, India and Indo-China, as well as North and South America are reasonably accurately delineated. Secondly, Australia is located roughly in its correct geographical position, but distorted to the point of being unrecognisable. The eastern coast of the presumed continent, however, has intriguing designations for geographical features of which three are arresting. In the north along a row of dotted shapes in the sea, is the term *'Coste dangereuse'*; to the south near a river entering a bay with several islands is *'R. beaucoup d'isles'* ('River of many isles'); further south on a bend in the coastline is *'Coste des herbaiges'* ('Vegetated coastline'). If the row of dots represents a coral reef, some have conjectured that the *'Coste dangereuse'* could well be the Great Barrier Reef, the 'many isles' the Whitsundays, and the *'Coste des herbaiges'* the coastline southward to Botany Bay.

Throughout the second half of the sixteenth century, maps of the south-east Asian region had become increasingly accurate, and Antwerp in the Netherlands displaced Dieppe as the European centre of advanced cartography, as evidenced by the publications of Abraham Ortelius, chiefly his atlas of the world, *Theatrum orbis terrarum* of 1570. One of his maps, *Indiae orientalis, insularumque adiacientium typus* ('Islands and other territories of the East Indies') is particularly intriguing for two reasons. Firstly, on the bottom margin, immediately south of Java and just visible above the lower border, is a tiny area labelled 'Beach, pars continentis Australis' ('Beach, part of the continent of [Terra] Australis'), 'Beach' being a misprint for Boeach, variant spelling of Lohac, confusedly thought by cartographers to be the term used by Marco Polo for the unknown south land in his famous account of travels to India. More interesting is the labelling of the western part of New Guinea in the lower south eastern corner, the only part showing, as *'Nova Guinea'*, followed by the intriguing statement 'An insula sit, an pars continentis Australis incertum est' ('Whether it is an island, or part of the continent of [Terra] Australis is uncertain'). Now this kind of Latin construction carries the force that the second clause is to be preferred, suggesting that

New Guinea had not been circumnavigated at the time. By 1589, however, Ortelius had published *Maris pacifici* in which New Guinea is shown as an island lying west of the Solomons archipelago. Despite the rather ovoid shape of New Guinea it depicts a river where the Fly is located, and nearby islands, conjecturally Daru and Saibai.

Around 1855, Richard Major, Keeper of Maps in the British Museum, began to investigate and catalogue its Dieppe collection. Having become particularly interested in Portuguese history, and drawing upon documents in his care, in 1859 he published *Early Voyages to Terra Australis*. In 1861, writing in the journal *Archaeologia* (V.38,459), he argued that 'the land described as Jave-la-Grande on the French maps to which I have referred can be no other than Australia; and that it was discovered before 1542 may be almost accepted as demonstrable certainty' (cited, McIntyre 1982:200, 224). Major's general line of inquiry was pursued by his friend George Collingridge in Sydney, a skilled draftsman and amateur geographer, competent in Portuguese language and history. For more than a decade he worked on his exhaustively researched and argued geographical history, *Discovery of Australia*, published in 1895, which presented Major's thesis in detail. Collingridge, however, went further: he attempted a reconstruction of one of the Dieppe maps, known as the 'Dauphin' of 1536 – originally charted with poor instruments without benefit of sextant or chronometer – according to the improved techniques of the nineteenth century. His reconstructed map placed the 'Coste dangereuse' between Cape Tribulation and Hinchinbrook Island, approximately at latitude 17°S, not far from where Cook foundered on Endeavour Reef (Collingridge 1895:307).

At that time, the first collection of Australia's foundation documents, *Historical Records of Australia*, was published in 1893 under the editorship of Frank Bladen who wrote in his Introduction that 'Since Dalrymple's time [1786] several old charts [the Dieppe maps] of the Sixteenth Century have been brought to light, which indicate a knowledge of our eastern coast more than two hundred years before Cook visited it'. These charts, when allowance is made for the current imperfect method of ascertaining longitude, 'give so correct a representation of the eastern coast that it is impossible to regard them as the creations of fancy' (Bladen 1893:xxiii).

An opposing view was taken by George Wood, professor of history at the University of Sydney, in an influential book of 1922, *The Discovery of Australia*, where he totally rejected Portuguese discovery of the east coast of Australia. In the sixth chapter, 'Was Australia known in the sixteenth century?', he dismissed the entire issue of Portuguese priority with the sweeping assertion that 'while I find it very difficult to believe that these maps represent the results of voyages of ships, I find it very easy to believe that they represent the results of voyages of imagination' (Wood 1922:120).

Following Wood, John Beaglehole, one of the most erudite historians of the period, in his volume *The Voyage of the Endeavour 1768–1771* (1968), in a discussion of that debate, drew attention to the legend on the '*mappa mundi*' (World Map) of Cornelius Wytfliet of 1597:

> Terra Australis is the most southern of all lands. It is separated from New Guinea by a narrow strait. Its shores are hitherto but little known, since, after one voyage and another, that route has been deserted, and seldom is the country visited unless when sailors are driven there by storm.

Beaglehole adds, while not taking sides on the issue of the Portuguese discovery of eastern Australia and the Great Barrier Reef, 'we cannot ignore' that inscription of Wytfliet. He does, however, take a cautious approach by suggesting 'there is a strong *a priori* case against the Portuguese being able to carry out the voyages necessary, with the ships and manpower available to them' (Beaglehole 1968:xxxi–xxxii).

That, however, is a curious argument: the Portuguese had already demonstrated superb navigating and exploring skill using exceptionally seaworthy caravels for their voyages of exploration. Although no caravel remains exist, reconstruction from drawings and descriptions such as we find in the log kept by Columbus who sailed in a caravel on his voyage to the New World in 1492, indicate they were between 20 and 30 metres in length, with a draft of 2 or 3 metres. Lateen rigged with either two or three masts, their low topsides reduced windage and allowed them to sail close to the wind. Between November 1492 and the final weeks of January 1493, Columbus recorded that in negotiating the maze of islands and coral reefs in the Caribbean (possibly between Hispaniola and Puerto Rico) on the return voyage he was able to make headway against currents, swirling tides and the strong prevailing easterly headwinds (Columbus, *Log* for Nov. 22, 27, 1492, et seqq., trans. Fuson 1987:113f.). The Portuguese were equally capable of a similar feat.

Oskar Spate, a leading authority on the Portuguese and Spanish era in the Pacific, was far more supportive. In a short but incisive introduction to a posthumous reprinting of *The Discovery of Australia*, Spate criticised Wood's dismissal, asserting that 'the arguments for possible Portuguese priority are stronger than Wood allows'; and further that 'the Dieppe maps are much more serious evidence . . . which can only [have been] derived from Portuguese originals'. Indeed, Spate's final, telling thrust was to criticise 'Wood's somewhat cavalier treatment of his sources' (Spate, in Wood 1969:xi). A decade later, in a further review of the evidence, Spate concluded that the Dieppe maps, based on lost Portuguese originals, definitely show that 'the northern projections of Australia, Arnhem Land, and the Cape York peninsula, and the long eastern and western coastlines of Jave la Grande are in truth those of Australia itself' (Spate 1979:221). Wood, he added, 'on Portuguese matters . . . is at his weakest . . . [and] on the whole faulty geographically, linguistically, and in logic' (227).

More recently, Kenneth McIntyre has examined the Major-Collingridge thesis in fascinating detail in *The Secret Discovery of Australia* (1982). His complex argument, from evidence contained in French copies of early Portuguese maps kept highly secret from Spain and the newer European maritime powers, confirms the Collingridge thesis, with some significant corrections. Most importantly, for the first time he identified the explorer whose charts provided the basis for the Dieppe maps as Christovao de Mendonça. Knowing of Magellan's voyage to the South Pacific the Portuguese administration centred in Goa had dispatched three caravels under the command of Mendonça to sail east to intercept and destroy Magellan. It was too late; Magellan had travelled north to the Phillipines, where he was later killed in a local conflict. Mendonça, McIntyre argued, sailed east through what later became designated as Torres Strait and then, following the coastline south, charted Australia's eastern coastline which provided the data for the Portuguese maps, of which the Dieppe maps are pirated versions (McIntyre 1982:145).

McIntyre's revised book of 1982 attracted the attention of Lawrence Fitzgerald, formerly a brigadier commanding the Australian Army Survey Corps. Turning his considerable cartographic experience to the Dieppe maps, and armed with a wealth of knowledge concerning earlier methods of charting and navigation before the introduction of the sextant in 1731 and Harrison's marine chronometer of 1773, Fitzgerald dissected the Dauphin map. Using a reconstruction technique to correct the original bearings, he produced a very convincing chart of Australia from Darwin to Cape York and then to Melbourne. Departing from McIntyre, Fitzgerald identified the Dauphin features of the '*Coste dangereuse*' as indeed the Great Barrier Reef around Cape Tribulation, the '*Rivière de beaucoup d'Isles*' as Broad Sound, and '*Coste des herbiages*' not as Botany Bay, but Fraser Island. Claiming 'a definitive solution of the gross distortions of the Dieppe maps', Fitzgerald argued that his reconstruction established beyond any doubt that eastern Australia and the Great Barrier Reef were charted by the Portuguese early in the sixteenth century. The navigator, perhaps, but not certainly, was Mendonça, some 250 years before Cook (Fitzgerald 1984:134).

In the same period, the Spanish, with no respect for the Treaty of Saragossa, had continued to send scouting expeditions across the Pacific demarcation line, seeking all the evidence they could of Portuguese discoveries. From 1542 on several attempts were made to establish a base and they eventually succeeded in doing so on the island of Cebu in the middle of an archipelago, which they named the Philippines, after Philip II. By 1571 Miguel Lopes Legaspi had founded Manila on the largest, and most northerly island of Luzon. Portugal was powerless to resist. Her power was waning and in 1578, after a period of conflict, Spain occupied and absorbed Portugal as a vassal state and sought to exercise dominion over the entire Pacific Ocean – one third of the Earth's surface. Claiming it, in the evocative metaphor of Oskar Spate as a 'Spanish Lake', their ships traversed freely in search of Terra Australis to add to their other New World conquests.

One of these expeditions, led by Pedro Fernando de Quiros, sailed from Callao, Peru, on 21 December 1605 with two ships, one captained by himself as General, the other by Luis Váez de Torres, his deputy or Almirante, along with a smaller vessel, *Los Tres-Reyes*. Six months later, anchored in a bay in Espiritu Santo (modern Vanuatu), Torres reported that on the night of 11 June 1606, the flag vessel, or Capitana, of de Quiros, *San Pedro y Paulo*, 'departed at one hour past midnight, without any notice given to us, and without making any signal' (Torres 1607; document in Collingridge 1895:233). Earlier, the crew of de Quiros was restless to the point of mutiny, Torres recorded, and it later transpired, that such was the reason.

Torres, now alone in the Pacific with the crew of the *San Pedrico*, had to reach Manila alone, and departed two weeks later on 26 June, eventually reaching the eastern islands of the New Guinea archipelago, correctly given as '11½°S'. His description of the New Guinea coast 'which runs W by N and E by south' is followed by his unambiguous statement: 'I could not weather the east point, so I coasted along to the westward on the south side' (234). The supposition is that Torres knew of the strait between Australia and New Guinea since de Quiros had specifically referred to it – presumably from Portuguese maps and documents that have not survived – in his statement that mid-year on the south side of New Guinea 'would be a bad time of the year to make the voyage being the season of the south-west winds' (Markham 1904:356; cited, McIntyre 1982:181). Even so, Spanish knowledge of the strait was kept secret and in fact does not appear on European maps until the earlier decades of the eighteenth century.

While it is now generally accepted that Torres traversed the Strait, the puzzle remains over exactly where Torres made his passage. Central to the argument against a prior Portuguese passage from the west are the weather and topography factors. Firstly, Torres Strait has strong prevailing easterly winds, the north-east trades, which would make sailing from the west more difficult in the much larger, less manoeuvrable galleons than in the earlier trim caravels. Secondly, the strait is very shallow: in the words of one of its modern shipping pilots that 'area of over 80 000 square miles . . . [is] studded with literally thousands of reefs, coral cays, islands, rocks, and shoals – between which swirl tidal currents of varying intensity and direction [measured up to 13 knots] . . . attempting entry near the Torres Strait proved fatal for hundreds of sailing ships in the 1800s' (Foley 1982:xviii).

Although the route taken by Torres through the strait is unknown, several attempts have been made to chart it, the most plausible being that of Brett Hilder, a ship's captain and maritime historian, with nearly fifty years experience in Australia's northern waters. While accepting that Torres sailed through from the east, he argued strongly against any prior traverse from the west, citing evidence that Dutch explorers from Batavia [Carstenz in 1622, Tasman in 1644] 'were deterred each time by the prevailing winds, strong currents and the maze of reefs. Later explorers, like Dampier in 1699 and Bougainville in 1768, kept clear of the area and left the problem to be finally solved by Cook' (Hilder 1980:136). Debate and conjecture will continue and perhaps will never be settled; regardless, we do know that the discovery by Torres remained secret, the Spanish and other navigators kept away, possibly as Spate argued because Spain was now so overextended that further discoveries were unappealing.

One mystery, however, will always intrigue historians. The Dieppe map of 1550 by Desceliers, despite its distortions, is incontrovertibly a depiction of Australia, so the question will always remain: who made the voyages that led to its charting? If the Portuguese did so, how did they navigate Torres Strait? Did they, as one possibility, circumnavigate New Guinea from the north and make an easterly approach? These questions will always excite curiosity and, perhaps, lead one day to more conclusive answers. One aspect, however, is certain: whoever first sighted and charted the coastline of Jave la Grande, those discoveries had no practical bearing whatsoever on the British acceptance that the discovery of eastern Australia and the Great Barrier Reef by Cook had led to the subsequent occupation and settlement of the continent. That remains solely Cook's achievement.

NEW EXPLORERS: DUTCH, ENGLISH, FRENCH

Already, a century before Cook, developments in Europe were under way that would destroy both Portuguese and Spanish domination of the newly explored world and have a profound effect on the Indo-Pacific regions. In an age when economic power depended upon command of the high seas, England, France and the Dutch United Provinces (a former Spanish territory) were determined to increase their share of world trade and break the Portuguese and Spanish dominance of three hundred years. The turning point came at the beginning of the seventeenth century when the Dutch ignored the Portuguese claim to most of the Atlantic and all of the Indian Ocean as their '*mare clausam*' (closed sea) and licensed a single cartel, the United East India Company (the VOC: Vereenigde Oostindische Companie) with a twenty-one year charter to trade in Indo-Pacific waters. The days of international corporations had commenced.

The Portuguese attempted to prevent Dutch incursions by force; in response, the superior Dutch vessels attacked and confiscated the cargoes of Portuguese ships whenever they met them on the high seas. A crisis was precipitated in 1604 when Jacob van Heemskirk captured a valuable Portuguese cargo ship and sailed it back to the Netherlands. In the ensuing complex debate modern international maritime law has its beginnings. The Portuguese cited the papal decrees as their legitimising authority; the Protestant Dutch countered with the concept of 'freedom of the seas', argued with persuasive force by their eminent jurist Hugo Grotius in 1609 in '*Mare liberum*' ('Freedom of the Seas'). Perhaps in exasperation as well as defiance, he dismissed their claim in the words: '*Vindicant subi Lusitani quicquid duos Orbes interjacet*' ('the Portuguese claim everything for themselves [lying] between the two worlds') (text in Fulton 1911:342). Arguing for public rights under natural law, which he asserted began in the earliest of historical times, Grotius declared the seas to be '*res nullius*': nobody's possession. An alternative term of the period, with the same meaning, was '*territorium nullius*': nobody's territory. His argument, which came to be generally accepted by the European colonising powers, was that a possession, to be regarded as private property, had to be enclosable and occupied ('*occupatio*'), in the European sense of worked, tilled land. The sea, by its nature, meets neither of those criteria. That development gave great encouragement to the British in their efforts to discover the mysterious Great South Land.

Early in the seventeenth century the Dutch became dominant in the Indo-Pacific region centred on the East Indies where the VOC had a monopoly of all trade. They established ports and factories in Ambon, the Molluccas and Java, and continued the search for additional profitable trading locations. Despite their skilful espionage of Portuguese and French maps, they had no clear idea of the limits of the Australian landmass. The route from the Netherlands via the Cape took them into the trade winds of the Roaring Forties, from where, estimating their longitude, they were able to take a northerly heading to reach their main base in Batavia (modern Jakarta). If their calculations were out they reached the arid western coast of Australia. Encountering the anomalous south-flowing Leeuwin current (the Coriolis effect normally makes oceans circulate anticlockwise in the southern hemisphere) they were, therefore, probably forced to make a regressive loop in the Indian Ocean to get north again to Java.

As a result of these events, such as the landing of Dirck Hartog in the *Eendracht* in 1616, the western coast of the continent gradually became known, but most of the charting has not survived. These coastal waters today are the graveyards of many VOC ships. In 1606 in an attempt to chart the northern coastline, Willem Janz in the *Duyfken* sailed into the Gulf of Carpentaria and reached Prince of Wales Island in western Torres Strait, and in 1623 Jan Carstenz repeated the voyage. Both mariners retreated from the confusing currents, reefs and prevailing easterlies of Torres Strait. Even Abel Tasman in his two voyages of 1642–43 and 1644, when he circumnavigated Australia and charted the southern tip of Tasmania, was forced to abandon the westerly approach to Torres Strait and return to Batavia. Clearly, if the Portuguese had mapped Australia, and had discovered Torres Strait, that achievement was unknown to the Dutch. For their part, nothing the Dutch saw impressed them: neither western deserts nor tropical mangrove coastlines. Only small groups of Aborigines were seen, some quite hostile. They lost interest in the Great South Land and concentrated their energies on the very profitable trade in the East Indies.

Throughout the seventeenth century, while the Dutch were establishing supremacy in the East Indies, the enigma of the Great South Land remained an absorbing concern in England. Public interest was intensified by a growing body of widely published travel literature – much of it published under titles beginning with 'Voyages . . .', or 'Travels . . .' – which, regardless of provenance, were usually translated promptly into other European languages. Travel literature multiplied, especially from the late seventeenth century on, becoming second only to theology in popularity: between 1660 and 1800 more than a hundred collections of voyages appeared in print, many of them in multiple editions and translations. By 1694 Tasman's *Journal* had been translated into English. Three years later in 1697 appeared the most popular travel book yet, *A New Voyage around the World* by the English buccaneer William Dampier, recording his experiences in the Pacific. In a second voyage of 1699–1701, from England via the Atlantic to the Indian Ocean, Dampier sighted and charted the west coast of New Holland (later, Western Australia) from Shark Bay (26°S) to North West Cape (22°S), naming 'Sharks Bay', and evidently for the first time in any exploration voyage, making a collection of biota for scientific purposes.

Following the conflict of the Seven Years War between Britain and France that began in 1756, and despite a temporary cessation of hostilities, both nations remained intractable commercial and military rivals. French activity in the Pacific, with the dispatch of Louis-Antoine de Bougainville in command of the *Boudeuse* and *Etoile* in 1766, created alarm in England. The French, looking ahead for new colonies with economic possibilities, appointed the botanist Philibert Commerson to the complement of the *Boudeuse* – the first scientist ever assigned to accompany an exploring expedition – who collected and described one thousand new species. These were used by the great French taxonomist Antoine-Laurent de Jussieu, at the famous Jardin des Plantes in Paris, for his innovatory classification in *Genera plantarum* in which he went beyond morphology – which provides only a key to identification – and used both ovules and pollen as indicators. Although a distinct advance, this, rather curiously, was not followed through. One reason may well have been that Commerson had neither artist nor draftsman to make accurate visual records of plants in their living condition, and although he was sufficiently competent to make drawings himself, he returned with mostly verbal descriptions and dried pressings.

To advance its position in the Pacific it had become imperative for Britain to locate, chart and claim Terra Australis; accordingly, preparations were made for the dispatch of Captain Samuel Wallis in 1766 in command of the *Dolphin*, accompanied by Lieutenant Philip Carteret in the *Swallow*. An Admiralty document prepared for the voyage states its purpose:

> Whereas there is reason to believe that Land or Islands of Great Extent, hitherto unvisited by any European power may be found in the Southern Hemisphere between Cape Horn and New Zeeland, in Latitudes convenient for navigation, and in Climates adapted to the produce of Commodities useful in Commerce, [the king requires] that an attempt should forthwith be made to discover and obtain a complete knowledge of the Land or Islands supposed to be situated in the Southern Hemisphere (Carrington 1948:xxii f.).

Wallis was instructed specifically not to set out merely in search of the yet uncharted east coast depicted on Tasman's map of New Holland but to look for the mysterious landmass which had appeared in early maps occupying most of the central South Pacific.

The latitudes where the Great South Land was believed to lie, between 45°S and 60°S, are exactly the counterpart location to Europe in the northern hemisphere. For all of these explorers, however, voyaging in the southern oceans was far more hazardous. Due to the vast ice masses of Antarctica and the absence of any continents except the southern tip of Patagonia to block winds and retain warmer air, temperatures below 45°S are lower and the freezing winds and cold currents made conditions very difficult for square-rigged ships, facts unknown to the early explorers. Such difficulties hindered Wallis and Carteret. Scarcely had they cleared the Straits of Magellan than the *Dolphin* and *Swallow* became separated; Wallis continued alone and eventually reached the Tuamotu Archipelago where he landed at Tahiti, at 17°S, situated in a far more comfortable climate. The first English explorer to do so, Wallis described it as a 'luxuriant paradise'.

While Wallis, as required, kept his captain's log for the Admiralty it has never been published. The English discovery and reaction were also carefully recorded, in voluminous detail, by George Robertson, sailing master of the *Dolphin*, in a separate journal, such meticulous record keeping being mandatory on naval ships of all seafaring nations, and becoming particularly important as expansion continued, both for commercial and scientific reasons. Robertson's *Journal*, like that of Wallis, was embargoed to withhold valuable intelligence from rival nations and not published until 1948 when it was edited by Hugh Carrington for the antiquarian Hakluyt Society under the title *The Discovery of Tahiti: a journal of the second voyage of HMS* Dolphin *round the world by George Robertson*. Like all such records, however, material was leaked at the time – often by bribery of the crew – to publishers, anxious to satisfy the demands of the 'travel crazy' public. Versions quickly found their way into newspapers and popular journals, especially the salacious descriptions of the women of Tahiti among whom, in a shorter, edited account published a few years after the voyage by John Hawkesworth, 'chastity does not seem to be considered a virtue among them, for they not only readily and openly trafficked with our people [*Dolphin*'s crew] for personal favours, but were brought down by their fathers and brothers for that purpose' (Robertson, 27 July 1767; ed. Hawkesworth 1773:I.481). The travel literature was rapidly building a highly imaginative, florid and confusing scenario for European readers, especially of the exotic tropic islands of Tahiti and Mururoa, and of the still tantalising puzzle of the mysterious '*Terra Australis incognita*'.

SOUTH PACIFIC NAVIGATION AND THE TRANSIT OF VENUS, 1769

Despite the disappointments coming from those voyages of exploration, the British government realised that impetus had to be maintained. The industrial revolution with its capitalist economy was already exerting an influence; equally, France was still a powerful rival and political advantage had to be sought. Whereas France, with a population of some

24 million in 1750, possessed extensive fertile lands and a more temperate climate on the continent of Europe and was capable of comfortable self-sufficiency, England (and Scotland after the union of 1707 to form Great Britain) faced serious problems. Before enclosure practices from the 1760s onwards the less predictable, and more limited, agricultural yield of the nation was low, while England's population, around 5 million in 1650, nearly doubled to almost 10 million by the late 1700s, causing serious problems. Occupation of new colonies, both for emigration and raw materials, and further trade therefore, was to become England's chief hope for a secure future. Fundamental to that was the need to improve maritime technology, especially the instruments and tables necessary for accurate navigation throughout the rapidly expanding empire.

While latitude problems had been solved with the development of the sextant, longitude determination remained intractable, and for centuries had led to voyaging disasters. Throughout the seventeenth century Spain, the Dutch United Provinces, France and Venice all offered huge cash rewards for an accurate means of determining longitude. In 1714 the British government established a Board of Longitude to offer the immense prize of £10 000 for a successful method, which when eventually won by William Harrison for his marine chronometer, was not awarded until 1773 after years of controversy and negotiation. After several prototypes, the earliest having a mass of 33 kilograms and measuring 90 centimetres in all three dimensions, the final version of his chronometer, some five inches (125 mm) in diameter, and known as the H4, revolutionised ocean voyaging.

The principle involved is simplicity itself. Since the earth rotates through 360 degrees every 24 hours, each degree of longitude represents four minutes of elapsed time; one hour corresponds to 15 degrees. So, with nothing more than two reliable timepieces, one set at a datum point, known as the prime meridian, wherever designated, and left unchanged, and the other reset each day from the sun at noon (when the nautical day began) ascertained by a sextant, the time difference between the two instruments allows the calculation of the number of degrees longitude east or west. Harrison's genius lay in recognising that simple principle and then surmounting the challenge of devising an absolutely accurate timepiece. Despite the ingenuity of the design, and the meticulous accuracy of the construction, slight errors could creep in, and so it became necessary for several to be carried, some set to the prime meridian and others re-calibrated daily, from which mean times could be calculated. It was the K1 version of Harrison's chronometer, made by craftsman Larcum Kendall, that Cook carried aboard the *Resolution* on his second voyage of 1772–75.

At the time of the *Endeavour* voyage however, only the less satisfactory astronomical tables method was available by which longitude was determined from observations of the positions of all heavenly bodies throughout the year – Sun, Moon, planets and stars – along with tables of tides and currents, used in conjunction with existing instruments: compass, log, quadrant, and later, the sextant. The Astronomer Royal, Nevil Maskelyne, was obsessed with 'lunar distances' (Moon tables) which depend upon the observable apparent distances between the Sun, Moon, and fixed stars. Maskelyne was also responsible for establishing

Greenwich in his *Nautical Almanac* of 1767 as the prime meridian that became increasingly accepted by ships of many nations.

The lunar distances method while useful – but never accurate – was fiendishly difficult to apply. All maritime powers recognised the inadequacy of current tables, and knew they could be improved if the distance of the Earth from the Sun were able to be determined accurately. One way of effecting this had been proposed as early as 1639 by an amateur astronomer, the Reverend Jeremiah Horrocks (1617–41) who suggested that the infrequent passage of the planet Venus across the face of the sun could be used to measure the distance by means of solar parallax. Unfortunately, he died at a tragically young age and was never able to take his ideas further. That possibility was revived by the astronomer, Edmond Halley (1656–1742), who had completed a catalogue of the southern skies from St Helena. In 1691 he proposed a similar solution to the Royal Society of London, which recorded in its *Journal Book* for 23 September 1691 that 'Mr. Halley read his Paper about the means of finding the distance of the Sunn from the Earth by help of two observations of Venus in the sun observed at two convenient places, which he supposed might determine the Sun's parallax to the 500th part'. Pursuing the quest for a method of determining longitude accurately, Halley further proposed, in a paper read to the Royal Society six years later, 20 May 1696, that by observation of 'the Occultations and Transits of the fixt stars by the Moon, there [would be] no surer method to determine the Longitude of Places than by such Observations, especially by the Occulations of Stars on the dark limb of the Moon, which are instantaneous' (Halley 1696:224, 238).

Halley's approach was based on knowledge that the planet Venus, which lies between the Earth and the Sun, passes across the disk of the Sun, as seen from the Earth, on very rare occasions – four times, in irregular periods between 8 and 120 years every 243 years – this passage being known as the Transit of Venus. By timing the transit, observing the angles subtended at entry and exit, and knowing the diameter of the Earth, Halley asserted that the distance from the Earth to the Sun could be calculated and the measurements used to improve the astronomical tables. Although Halley's approach seemed a promising way forward, difficulties abounded due to infrequency of the transits and the fact that it relied upon physical observations that were easily subverted by unpredictable weather events. Virtually no practical improvements were made, however, and throughout the following decades navigation remained a difficult task for mariners.

The French astronomer, Alexander-Gui Pingré, when attempting an observation of the transit in 1761 at Rodriguez (near Mauritius, then known as Isle de France) had been frustrated by cloud, which led him in 1767 to write a *Mémoire sur les lieux où le passage de Venus, le 3 Juin 1769, pourra être observé avec le plus d'avantage* ('Memorandum on the most suitable locations to observe the [next] Transit of Venus due on 3 June 1769'). There was sense of urgency at the time: after 1769 the next transits would not be until 1874 and 1882.

Many other nations had begun sending astronomers to 77 separate locations in preparation for observations in 1769. The Admiralty, therefore, was easily persuaded by the astronomers of the Royal Society of London to send expeditions to suitable, and as far as possible, cloudless locations, finally settling on Port Royal Harbour in King George's Island (Tahiti). Negotiations were commenced with the Royal Society for planning the expedition.

CHAPTER 2

VOYAGE OF THE *ENDEAVOUR*: COOK AND THE 'LABYRINTH'

THE EXPEDITION TO TAHITI

The Royal Society, in fact, had begun discussing the scientific advantages of the Venus observations early in 1767 and that year invited to their meetings Alexander Dalrymple, an experienced navigator and, briefly, ship's captain of the *Cuddalore* while in the employ of the British East India Company in Madras from 1752 to 1767. Dalrymple was active in promoting the theory of the counter-balancing continent; he was also a well-informed historian, an avid collector of Portuguese and Spanish exploration and colonising documents, maps and charts. Moreover, he was keen to implement the Royal Society's recommendation to lead the expedition to Tahiti. The Admiralty, however, insisted upon a naval officer to command the vessel that had already been selected by Dalrymple, the *Earl of Pembroke*, re-commissioned as the *Endeavour*, and extensively rebuilt. Instead, they chose James Cook, a warrant officer known for his navigational ability and meticulous hydrographic skills in Newfoundland and the St Lawrence region during the conflict with France in Canada. Dalrymple declined to travel as anything other than overall commander, although Joseph Banks, a gentleman naturalist, readily accepted the opportunity to take charge of the scientific aspects of the voyage.

Promoted to lieutenant, Cook supervised the provisioning of the ship at Deptford, and then set out from Plymouth on Friday, 26 August 1768, recording in his *Journal* that he 'got under sail and put to sea having on board 94 persons including Officers Seamen Gentlemen and their servants' (Cook 1770:4). The 'Gentlemen and their servants' were the scientific party of six led by the 25 year old fellow of the Royal Society, Joseph Banks, already experienced in ocean voyaging after sailing in 1766 aboard the naval ship *Niger* in eastern Canadian waters. With him were his Swedish colleague Carl Daniel Solander, and another Swede, Herman Spöring. Also aboard, in those days before photography, were the artists Sydney Parkinson and Alexander Buchan.

Cook's instructions from the Admiralty came in two sections, both marked 'Secret'. The first required him to proceed to Tahiti and build an observatory for the astronomer, Charles Green, in order to make the essential observations of the transit of Venus. Once completed, the second required him to engage, as others before him, in a search for Terra Australis by sailing this time 'Westward between the Latitude before mentioned [40°S] and the Latitude of 35° until you discover it, or fall in with the Eastern side of the Land discover'd by Tasman and now called New Zealand' (Beaglehole 1968:cclxxxii). To provide maximum assistance, Cook carried aboard the *Endeavour* all available intelligence for the voyage, much of it updated material from the writings of Hakluyt, Purchas and their followers. Cook was well aware of Tasman's map, and possibly the widely distributed 'Complete Map of the Southern Continent' taken from Tasman by Emanuel Bowen, an American, and published in 1744. As an enthusiastic and highly competent hydrographer, Cook would have been equally aware of the proliferating number of Pacific maps of the period. Dalrymple, despite his manifest disappointment, very generously lent his books and materials to Banks for the voyage – bringing the total aboard to some forty volumes – along with an advance copy of his yet unpublished book on Pacific exploration which came out in 1769 after the *Endeavour* sailed. That collection supplemented the unpublished, and classified, journal of Wallis, along with John Campbell's 1744–48 redaction of Harris' *Navigantium atque Itinerarium Biblioteca*, and John Callender's translation of Charles de Brosses' *Histoire des Navigations aux Terres Australes* of 1756. This last book contained a map of the Pacific by Robert Vaugondy, based on the Tasman map, showing, quite clearly, New Guinea separated from New Holland by a wide strait.

Although Cook's mission in Tahiti was satisfactory, he encountered the same problems with the natives as did Wallis. These, when recounted in England in numerous publications, although true, created a wholly distorted view of the South Pacific and served to heighten curiosity and intensify the mystique of these unusual exotic regions. Especially compelling for English – and their European – readers were the titillating accounts of amatory encounters between locals and the ships' crews. Wallis, for example, had apparently not realised for some time that Tahitian women had quickly established the price of one iron nail – highly valued by their menfolk – for an evening's delights ashore. Indeed, Wallis recorded in his *Journal* that not only was chastity unvalued by them, but also that each woman had established a trading relationship whereby, 'conscious of the value of [her] beauty . . . the size of the nail was always in proportion to her charms'. Only when nails became conspicuously absent from the ship's topsides – those from which hammocks were slung were the first to disappear – did Wallis realise the situation and recorded in his Journal that 'when I was acquainted with [the practice] I no longer wondered that the ship was in danger of being pulled to pieces for the nails and iron that held her together' (ed. Hawkesworth 1773:226).

So, when the *Endeavour* sailed, its crew, well acquainted by hearsay from previous mariners, were only too eager to experience the lush sensuality offered by Tahitian women. In a popular account written up later from Cook's records by John Hawkesworth, the public would have been agog to learn that in Tahiti the houses 'have no walls. Privacy, indeed, is little wanted among people who have not even the idea of indecency, and who gratify every appetite and passion before witnesses, with no more sense of impropriety than we feel when we satisfy our hunger at a social board with our family and friends' (Hawkesworth 1773:III.20). At the same time, Cook was highly conscious of the serious health problems created by inadequate ship's diet, chiefly scurvy – then unknown, resulting from Vitamin C deficiency – which begins as swollen gums that prevent chewing and adequate digestion, leading in turn if not corrected to muscle cramps, bone decalcification, swelling of the joints, anaemia, and eventually death. Consequently, the fresh tropical fruit, meat and vegetables of Tahiti were very restorative. His Admiralty instructions required him to show 'every kind of civility and regard' to the natives, and evening shore leave, therefore, was considered necessary for good diplomacy as well as rest and recreation. While Cook was unable to prevent fraternisation, he had to maintain discipline and recorded in his *Journal* for Sunday, 4 June 1769, that he 'Punished Arch'd Wolf with two Doz'n Lashes for theft, having broken into one of the Store rooms and stolen from thence a large quantity of Spike nails, some few of them were found upon him'. It is interesting that this was twice the usual punishment and that Cook recorded his concern as economic: Wolf's free use of the nails as currency would undermine the value of the trading 'trinkets' carried for that purpose by leading to inflation (Beaglehole 1968:98).

At the same time, acutely conscious of health problems, Cook became aware by July of the spread of what was probably gonorrhoea, recording on Thursday, 13 July 1769, that the crew from 'the too free use of women were in a worse state of hilth than they were on our first arrival, for by this time full half of them had got the Venereal disease' (Cook 1770:138). Since Wallis' crew on an earlier visit had not become infected, the supposition was that a Spanish ship, following Wallis, and preceding Cook, was the vector. The news back in England only served to intensify hostility to Spain.

ENCOUNTER WITH THE REEF: THROUGH THE 'LABYRINTH', 1770

The transit observations completed, the *Endeavour* weighed anchor from the Society Islands, as Cook named them, on 9 August 1769, and set out to fulfil the second of his secret *Instructions*. Over the ensuing two months the *Endeavour* sailed south, looped back to the north-west, and on Saturday, 7 October 1769, Cook sighted the east coast of New Zealand which he circumnavigated by the end of March 1770, establishing it as two islands by sailing through what later became known as Cook Strait. No great south land, however, was discovered. The mission was now considered complete and Cook therefore sailed for home on Sunday, 1 April 1770, 'with an intention to steer to the westward . . . in the Latitude 40°30's and longitude 185°58'w from Greenwich'. Cook believed, from Tasman's chart, and subsequent redactions, that the east coast of New Holland could be reached from such a bearing, although unknown to him it was a risky venture. Tasman charted the coastline of Van Diemen's Land only from its southern extremity at 43°37'S to Flinder's Island, the northern tip of which lies in the middle of Bass Strait, with the then unknown continental mainland beginning further north at 39°07'S. Cook's bearing, obviously taken from Tasman's map, had it been followed, would have given a landfall near the east mainland coast of Tasmania. He would then have been obliged to head north as Tasman had already charted the south, and most likely would have passed through Bass Strait, then unknown, missing the east coast of the continent altogether. In this respect, unlike those before him, he was indeed a fortunate explorer since, bearing a little more to the north on Thursday, 19 April 1770, the long unconfirmed eastern coast of New Holland was finally sighted and Cook recorded that 'I have Named it *Point Hicks*, because Lieut't Hicks was the first who discover'd this land' (Cook 1770:299).

Clearly anxious to return to England, the 'Indians' of New Zealand having been generally hostile – unlike the Tahitians – and the crew now seriously depleted, Cook's transit north along the eastern coast of New Holland was relatively rapid. The first landing was made at Botany Bay on 29 April 1770, where Banks and Solander, having been collecting, describing and illustrating their finds industriously all the way from South America, found Botany Bay, as they named it, an exciting, if strange treasure trove, containing what they considered to be an entirely novel flora – indeed, it led Banks to consider it, following the dominant Divine Design philosophy of the period, almost totally anomalous.

Leaving on 6 May 1770, Cook made careful topographic descriptions along the way, with illustrations by the artists, and named prominent features – Pigeon House Mountain, Smoky Cape, Solitary Isles, Cape Byron, Glasshouse Mountains, Noosa Head. Cook then reached the largest sand island in the world (now Fraser Island) although, standing well out to sea he was unaware that it was an island, and where, seeing Aborigines on a promontory, named it Indian Head. Further north, reaching the sandbar at the tip of the island, he named it Break Sea Spit. Bearing north-west into the open sea, on Wednesday, 23 May 1770, a landfall was made where a shore party shot and cooked for dinner a large bird, designated *Ardeotis kori* (today, *A. australis* and endangered), weighing 17½ pounds (8 kilograms); Cook saw a resemblance to the English bustard, so the bird received its common name and the location 'occasioned my giving this place the name of *Bustard Bay*', which it bears to this day. In further remembrance of this landing the later settlement was named the Town of Seventeen Seventy.

In this region the Great Barrier Reef has its southern limits, at 24°S, slightly south of the Tropic of Capricorn. The voyage north proceeded rapidly, following the coastline closely, where Cook continued to name topographic features. In this southernmost area only isolated, highly visible coral cays are found, first in the Bunker and Capricorn group, while the dangerous submerged masses of ribbon and patch reefs are 200 kilometres further out in the Swain Reefs, beginning around 22° S. The coastal waters through which Cook sailed have mostly continental islands and when he reached the first major group by Monday, 4 June 1770 – the day of the Christian religious festival of Pentecost, known as Whitsunday – he recorded that 'this Passage I have named Whitsunday's Passage'. Still unaware of the great barrier lying offshore, although it would come progressively closer to the mainland in the north, Cook continued northward to England, passing and naming Cape Upstart, Cape Bowling Green, Cape Cleveland and nearby '*Magnetical head* or *Isle*'. Thereafter he named Halifax Bay, Dunk Island, Cape Sandwich, among many others – obviously honouring every member of the English nobility and political establishment he could – names the coastline is obliged to carry as a legacy to this day. Reaching and naming Cape Grafton (modern Cairns) he anchored off a true coral sand cay. In honour of his now seriously ill astronomer Charles Green – who also carried a considerable burden of navigational duties – he named it Green Island. On the same day, Sunday, 10 June 1770, since it was one week after Whitsunday, known as Trinity, he named Trinity Bay where the *Endeavour* was anchored.

AN ERROR OF JUDGMENT? *ENDEAVOUR* HOLED ON A REEF

In this latitude, between 17°S and 16°S, the coral cays of the Reef begin to come closer to the coastline and form a more continuous barrier and it was there that Cook made what some subsequent studies consider a serious error of judgment. The numerous cays and reefs, many submerged at high tide, become risky to navigation, a hazard to any sailing ship, especially the slow, less manoeuvrable cat-built (that is, blunt bowed) ex-collier *Endeavour* with a best top speed of barely 6 or 7 knots. Cook had seen these reefs with increasing frequency as he made his way north, where they rise almost vertically out of the deep blue depths to form translucent turquoise patches, with a fringe of broken white surf on their eastern windward sides. At night, especially with a high tide, they are invisible. On Monday, 11 June 1770, at 6 p.m., Cook noted 'two low woody Islands' off the port bow, yet inexplicably he decided to continue sailing 'all night as well to avoid the dangers we saw ahead as to see if any Islands lay in the offing'. Under shortened sail, the *Endeavour* kept a steady course 'having the advantage of a fine breeze of wind and a clear moonlight night . . . from 6 untill near 9 oClock'. He had, of course, a seaman (known as a leadsman) 'in the chains' standing on one of the small platforms fitted to the sides of the ship which hold the mast stays, sounding with a lead line. At one point when the bottom rose quickly 'into 12, 10 and 8 fathom' the ship was quickly put about and then continued. Banks and companions, meantime, had enjoyed their dinner and gone 'to bed in perfect security, but scarce were we warm in our beds when we were called up with the alarming news of the ship fast ashore on the rock' (Banks 1770:II.77). Cook's entry is more explicit: 'a few Minutes before 11 when we had 17 [fathoms; that is, 31 metres] and before the man at the lead could heave another cast the Ship Struck and stuck fast' (Cook 1770:344).

In the frenzy that ensued everything dispensable was jettisoned to lighten the ship, including the cannon: three of the four serviceable pumps were manned continuously in which even the officers and Banks' team took their turn, the midshipman Monkhouse led a team to fother, or patch, the area beneath the damage with a canvas sail and various fillers of wool and oakum (unravelled rope) dragged underneath the hull from the bow which the water forced up into the hole. Once stabilised and with the pumps gaining on the inflow, a week later Cook managed to get the ship to a nearby inlet for careening and repairs at what is now Endeavour River, with its modern settlement of Cooktown. The hull was not seriously damaged, although inspection revealed that the impact had broken through a strake under the starboard bow, Cook's journal recording that when he 'examin'd the leak . . . 4 feet of the plank [was] cut thro' by the rocks. Found a piece of rock sticking in her bottom and several other streaks damag'd'; in a later emendation Cook added 'as large as a Man's fist' (Beaglehole 1968:350, n.6). Altogether four planks had been cut through, three more were damaged, part of the copper hull sheathing had been ripped from the port bow, and some of the false keel (6 inches deep and designed to protect the main keel above) had been lost along with some of the main keel and the forefoot. What really surprised Cook was the precision with which the damage occurred: the manner in which the 'planks were . . . cut out as I might say is hardly creditable, scarce a splinter was to be seen but the whole was cut away as if it had been done by the hands of Man with a blunt edge tool' he recorded in his Log for Friday, 22 June.

While relatively minor, the extent of damage was still serious and repairs were to take nearly two months since the ship had to be run up on to the bank, bow first, at high tide and secured with anchors, to allow the carpenters to work at repairs during the low tides each day. During that unexpected delay, the ship's pinnace was sent to seek an opening through the Reef into the Pacific, while Banks, Solander and Parkinson, who had their own small boat, along with a shore party, continued scientific collecting and recording. Cook himself apparently saw his first kangaroo while there, describing it as 'of a light Mouse colour and the full size of a grey hound . . . with a long tail; but for its walking or runing in which it jumped like a Hare or a dear' (Cook 1770:352). He made quite a few listings of terrestrial biota, and in one entry for Thursday, 23 August 1770 – his only account of marine animals – he listed 'Fish of various sorts, such as Sharks, Dog-fish, Rock-fish, Mullets, Breames, Cavallies, Mackarel, old wives, Leather-Jackets, Five-fingers, Sting-Rays, &c – all excellent in their kind'. He also recorded 'oysters of several kinds, cockles and clams, crabs, mussels, crayfish, green turtles and alligators' (394–95).

The ship repaired, on Monday, 6 August 1770, the *Endeavour* sailed north again, after naming Cape Tribulation, proceeding very carefully indeed. As his *Journal* records, giving an explicit account of the dangers of the Reef, he claimed some kind of a record since

> having been intangled among them more or less ever sence the 26th of May, in which time we have sailed 360 Leagues [1800 kilometres] without ever having a man out of the cheans heaving the lead when the ship was under way, a circumstance that I dare say never happen'd to any ship before and yet here was absolutely necessary (375).

Cook's bewilderment is highlighted in the name he gave for the region from Cape Grafton (off Cairns) to Cape York, shown on his charts in large capitals as the 'Labyrinth', an allusion to the classical Greek legend of the deliberately contrived complexity of passages guarding the underground lair of the dreaded Minotaur in Crete. In the Greek legend the hero Perseus, having slain the monster, was able to return because as he entered his lover Ariadne had given him a ball of twine to unwind to find his way back through the maze. Cook had no such assistance and was forced to search blindly for a passage. Clear water appeared again and five days later they anchored in a bay off a high continental island (14°40'S) where Banks and Solander again went collecting. The only land animals noted were lizards (since identified as the sand monitor *Varanus gouldii*) and this, Cook wrote, 'occasioned my nameing the Island *Lizard Island*' (373). Anxious to be free of those endless hazards, and with supplies running low, the pinnace found a channel a little to the north (14°32'S) and with great relief Cook – a modern Perseus – sailed out into the deep Pacific, heading north to search for a direct passage west in order to reach Batavia for repairs and supplies.

Four days later, through a lucky discovery, Cook found a passage back into Reef waters at 12°42'S. Understandably, he named it Providential Channel. Again, his awareness of Reef dangers was always heightened as his graphic account of the dangerous 'wall of coral rock', quoted earlier, with its rapid tides and steep reef walls on the ocean side which he encountered while searching for a re-entry point, indicates: 'A Reef such as is here spoke of is scarcely known in Europe' (378). Fortunately, that transit through the final northern waters of the 'Labyrinth' was incident free, although the ship was still leaking and the pumps faulty. On Tuesday, 21 August 1770, he named York Cape, and then, the following day, on Possession Island – as he had done several times previously along the coastline – he claimed all of the east coast territory of New Holland for King George III and named it South Wales, and then, for reasons never really clear to this day, changed it to New South Wales (Beaglehole 1968:388,n.1).

Having then charted and named Prince of Wales Island and Booby Island, Cook set the *Endeavour* on course for Batavia, now with a leaking ship and a depleted, unhealthy crew, extremely anxious for the safety of both. Reflecting in his *Journal* on the many difficulties of his long journey, especially the ever-present possibility of losing his crew, his ship, and his own life – which had come close the night of his grounding – he speculated on the inevitable conflict every explorer into unknown territories must experience, between determination and timidity, or as he described it, between temerity and timorousness. Too much of either, he wrote, 'can lead to the charge of want of conduct' (Cook 1770:380).

Sailing into open waters, now clear of coral reefs, in recollecting events of the previous weeks, Cook recorded on Thursday, 23 August 1770, a very penetrating and moving account of the Aborigines at Endeavour River, who, although appearing 'to be the most wretched people on earth' were,

> in reality . . . far more happier than we Europeans; being wholly unacquainted not only with the superfluous but necessary conveniences sought after in Europe, they are happy in not knowing the use of them. They live in a Tranquillity which is not disturb'd by the Inequality of Condition: The Earth and sea of their own accord furnishes them with all things necessary for life (399).

Cook noted that 'they covet not Magnificent Houses . . . they live in a warm and fine Climate and enjoy a very wholsome Air . . . in my opinion [this] argues that they think themselves provided with all the necessarys of Life and that they have no superfluities' (399). Three days later Banks penned very similar views and described the Aborigines as 'almost happy' (Banks 1770:II.108).

By the time the *Endeavour* left Possession Island and set sail for repairs at Batavia, all the travellers were in poor physical health. Despite Cook's careful precautions to keep a clean, well-scrubbed ship, and the best available food to minimise scurvy and other diseases, illness continued to take its toll. Already Green, the astronomer, was worsening, Parkinson and Spöring were little better, Banks and Solander were also ailing. Subsequently Banks and Solander became seriously ill and one after another sailors began to succumb to malaria, exacerbated by malnutrition and exhaustion. Cook, himself ill with fever, was obliged to remain for nearly three months. Banks and Solander moved to a rented house upland and eventually recovered sufficiently to do some further collecting while Parkinson, ailing more, rallied to make further illustrations. By the time of departure, however, Tupeia the Tahitian, and his servant Taieto – possibly his son – who had accompanied them, were dead, along with six others. With the 'number of sick on board . . . 40 or upwards and the rest of the Ships company . . . in a Weakly condition', and having signed on nineteen additional crew, the *Endeavour* left for the Cape of Good Hope, where a further ten were signed on (Cook 1770:441).

Proceeding to St Helena for further rest and supplies, Cook then departed on the final leg for England. Illness was raging: 'Many of our people at this time [24 January 1771] lay dangerously ill of Fevers and Fluxes' Cook recorded (447), and that was no understatement. His *Journal* entries for January and February are a melancholy record of those who 'Departed this Life': Spöring on 25 January, Parkinson, Green and four of the crew in the following days. On 6 February the surgeon Monkhouse died and around the same period another ten of the crew, 'of the flux'. When the *Endeavour* docked at Deal, a port on the Downs, on 13 July 1771, barely half of those who set out two years earlier remained. Even so, those losses were accepted stoically as the will of Divine Providence. The last entry in Cook's *Journal*, for Saturday, 13 August 1771, reads 'At 3 oClock in the PM Anchor'd in the Downs, and soon after I landed in order to repair to London'.

THE *ENDEAVOUR* VOYAGE PUBLISHED: THE REEF MYSTIQUE DEVELOPS

Like those of Byron, Wallis and Carteret before him, all records kept by Cook and his crew were surrendered to the Admiralty, for security and investigation. No single journal records the voyage, and of many documents, some are extant, others, in whole or part, lost. The Admiralty allowed none of the journals, diaries, logs and other documents from any of its voyages to be published immediately and Cook's records were no exception. Not until the mid twentieth century were all surviving documents – by Cook and his officers, and by Banks and his assistants – brought together into a critical edition when the massive project was undertaken by John Cawte Beaglehole of Victoria University in Wellington, New Zealand. Like Carrington's editing of Robertson's voyage on the *Dolphin*, it was produced for the Hakluyt Society in London. It is from Beaglehole's superb collation and editing of thirty-five separate sources, mostly holograph or original transcript sources, others secondary but near contemporary accounts, on which all historians depend, and from which the foregoing account has drawn freely. The 'travel crazy' public at the time, however, was eager for news, and that followed quickly in the form of titillating, sensationalised accounts in the popular press.

The Admiralty, mindful of continuing French activity in the region, needed to promote significant British voyages, both as a record of great achievement and to make clear to all European nations British claims in the Pacific. Accordingly, they commissioned the professional writer John Hawkesworth to prepare popular editions – shorn of any naval intelligence – of the journals and records of Byron, Wallis, Carteret, Cook and Banks, and other records kept by their subordinates. Recognising the sales potential, the London publishers Strahan and Cadell advanced Hawkesworth the enormous sum of £6000 for the rights. In 1773 they released Hawkesworth's prodigious effort: three huge volumes of 700 to 800 pages each, with numerous illustrations, in large format (29 by 22 cm) under the omnibus title *An account of the voyages undertaken by order of his present Majesty for making discoveries in the southern hemisphere and successfully performed by Commodore Byron, Captain Carteret, Captain Wallis and Captain Cook . . . drawn up from the journals kept by the several commanders, and from the papers of Joseph Banks by John Hawkesworth*. These volumes were an instant success and were reprinted the same year;

the following year, 1774, they were published in the United States, and in translation in France and Germany, and later in other European languages. In later years they came out in pocket-sized versions (18 by 10 cm), such as the Morison/Mudie edition of Edinburgh in 1789, and were serialised in sixty weekly sections. They became widely reviewed and, inevitably, heavily criticised by many. Hawkesworth, however, died the year of the first edition, his death possibly having been hastened by the considerable controversies that followed, as described in the next chapter.

There is no question that Hawkesworth did anything less than a thoroughly professional job in abridging and converting official mariners' records into lively, readable prose; obviously, the continuing publication success testified to that. Aware of the public interest, he gave full accounts of all the Tahiti episodes of Wallis and Cook, including the problems of the missing nails from the *Dolphin* and the *Endeavour*. With heightened effect he also reported on the sensuality of Tahitian women who, when Wallis and the *Dolphin* arrived 'came down to the beach stripping themselves naked [and] endeavoured to allure [the crew] by many wanton gestures, the meaning of which could not possibly be mistaken' (Hawkesworth 1773:I.433).

It is from Hawkesworth's version of Cook's records, and not directly from Cook himself, that the public perception of the exotic allure of the South Pacific, and the mystique of coral reefs – long a mariners' nightmare – was generated. The great Reef, in fact, is brought vividly into focus in Book III of the final, third volume of the original 1773 edition, in five consecutive chapters. Illustrated with engravings of a Maori war canoe and war clubs, and of the careened *Endeavour* after the mishap, it included three larger maps (75 × 50 cm) of 'NSW or the East Coast of New Holland' (pp.481, 589) with the Reef labelled, in capitals, 'THE LABYRINTH'. These serve to guide the reader through the narrative and bring into sharper relief the ever-present hazards Cook faced in navigating totally unknown and treacherous waters. Whereas Cook wrote in a deceptively disarming, even detached style – a master, really, of understatement and self-abnegation – Hawkesworth gives Cook's achievements a more deserved prominence.

Although the texts correspond closely in factual detail, Hawkesworth's reads with far more verve: in that near-disastrous passage from Cape Tribulation to the holing on Endeavour Reef, the hazard was portrayed as 'a rock of coral, . . . more fateful than any other'; then, 'the leak increased to a most alarming degree' and 'gained upon us so considerably, that it was imagined she must go to the bottom as soon as she ceased to be supported by the rock . . . a dreadful circumstance'. The reader is carried along to anticipate 'when the dreadful crisis should arrive . . . [the fate of those] left on board to perish in the waves . . . [who] would probably suffer less upon the whole than those who should get ashore' since they could not defend themselves against the natives 'who perhaps were some of the most rude and uncivilised on earth'. The land offered no subsistence, and 'they must be condemned to languish out the rest of their lives in a desolate wilderness' (I.545–49).

Cook's dilemma after the successful repairs was brought out forcefully by Hawkesworth for eighteenth-century readers: 'Being now convinced that there was no passage to the sea, and through the labyrinth formed by the shoals, I was altogether at a loss which way to steer . . . we plainly heard the roaring of the surf [at Endeavour Strait] and saw it foaming to a great height . . . Our distress now returned upon us with double force' (I.593–604). Accurate, if dramatised, the real dangers are emphasised with far more effect than Cook's dry, erratically spelled, though readable prose. Hawkesworth's account, however, was less overwritten than those that appeared in the popular press, much earlier, from the crew's stories. The *Public Advertiser* for 6 August 1771, in contrast, informed its readers that after hitting the Reef the *Endeavour* sailed 'many hundred leagues with a large Piece of Rock sticking in her bottom, which had it fallen out must have occasioned inevitable destruction'; in the 19 August edition that was altered to 'a large trunk of coral', something more than Cook's record 'about the size of a man's fist'.

A BIZARRE CONTROVERSY: DID COOK DISCOVER THE REEF?

Hawkesworth's version, however, did not completely enhance Cook's reputation. Instead, it created a series of controversies – and a continuing historical problem – that arose in the first instance from strong criticism of Hawkesworth by Dalrymple for minimising his own contribution to the success of the voyage, and one that generated a most bizarre chain of events. Cook, too, was dissatisfied with the Hawkesworth version, and it is possible that the initial reactions of Cook and Dalrymple helped to hasten Hawkesworth's death. The controversy, however, continued after that and took another direction centring on one main issue, asking: was Cook really as great a navigator as public acclaim made him out to be? Did he really discover and explore the Reef without help from former sources? Or, did Cook know, secretly, of the Portuguese charting of the Reef from one of the Dieppe series, the Dauphin Map, so that, his actual achievement was far less than the public believed?

That controversy did not envelop Cook personally: he was killed in Hawaii on 14 February 1779, in command of *Resolution* and *Discovery* on his tragic third voyage in the North Pacific after yet another unsuccessful British quest to find an Arctic passage from the Pacific to the North Sea. It did not erupt, in fact, until 1786 when, in *A Memoire concerning the Chagos and adjacent islands* Dalrymple wrote that 'I have a manuscript in my possession, belonging to Sir Joseph Banks . . . painted on parchment with the Dauphin's Arms; it contains much lost knowledge; . . . the east coast of New Holland, as we name it, is expressed' (Dalrymple 1786:4). At that time Dalrymple was well respected in maritime circles; in 1779 he had been appointed to the London office of the British East India Company as hydrographer, and in 1795 when the Admiralty established a hydrography office Dalrymple was appointed the first official hydrographer to the British navy.

The Dauphin Map he referred to in 1786 already had a mysterious background since it did not emerge from obscurity until early in the eighteenth century when it came into the possession of Edward Harley, Earl of Oxford. Stolen in 1724 it apparently disappeared a second time for several decades and then, in some way, came to the attention of Daniel Solander who informed Banks who then bought it in 1790 and presented it to the British Museum where it remains, known as the 'Harleian Map'.

The French, now in active competition in the South Pacific, probably piqued by the British success and envious of Cook's commanding reputation gained from Hawkesworth's eulogising, were less adulatory. In a letter to *La Revue ou Décade philosophique, litteraire et politique* (11 November 1805:Vol.47,261–66), Frédéric Metz insinuated that Cook was familiar with the Dauphin Map, even though Metz himself had evidently not actually read Dalrymple's *Memoire* (Fry 1979:55,239,n.45). This was followed by Barbié de Bocage who read a paper in July 1807 to the Institut de France in which he asserted that the Dauphin Map was held in well-known libraries in England. Those aspersions were followed, uncritically, by a succession of writers who perpetuated the belief that Dalrymple was jealous of Cook's achievements and was hostile to him personally because he believed that Cook had encouraged Hawkesworth, in editing the account of the voyage of the *Endeavour*, to minimise Dalrymple's considerable contribution to making the expedition possible. Hawkesworth defended himself vigorously in the preface to the second printing of his account issued just three months later. He stated most emphatically that he had never read Dalrymple's work and, therefore, 'his charge that I wilfully supressed whatever I thought could do him credit, is wholly without foundation' (Hawkesworth 1773:9). So the legend grew that Dalrymple produced the Dauphin Map in 1786 – seven years after Cook's death – in order to belittle Cook's achievements.

This theme, Fry points out, was followed by Henry Major in his study of the Dieppe maps in which he sought to assign credit for discovering the Reef to the Portuguese. When the first volume of the series of *Historical Records of Australia* was published in 1893, the editor, Frank Bladen, wrote in his Introduction that, although the Dieppe maps seem to indicate a prior Portuguese discovery of the Reef, and despite the claims of Barbié de Bocage, there is no evidence that either Cook or Dalrymple knew of the existence of any Portuguese maps or of the Harleian Map. Some historians today hold the view that neither Cook, nor the Admiralty generally, knew of Torres' documents describing his traverse in 1606. However, it is known that many maps were in existence at the time, including those of Cornelius Wytfliet of 1597, Abel Tasman of 1644, Melchisidec Thévenot of 1663 and Emanuel Bowen of 1744, which showed an indeterminate charting of the area. That of Robert de Vaugondy of 1756, which was published in Charles de Brosses' *Histoire des voyages aux Terres Australes* by Durand of Paris in 1756 – which was on board the *Endeavour* – however, is in a very different category. Two versions of this map were bound into de Brosses' book: a world map, '*Carte Générale qui represente Les Mers des Indes, Pacifique et Atlantique, et principalement Le Monde Australe*' ('General Map of the

Indian, Pacific and Atlantic Oceans, and principally of the South Land') and a sectional reduction, '*Carte Réduite de l'Australasie, pour servir à la lecture de l'Histoire des Terres Australes*' ('Sectional Map of Australasia to assist study of the history [of the quest for] the South Lands').

While it is a virtual copy of the Tasman Map (as are many other subsequent publications), this map has one significant exception. Whereas the east coasts of earlier Spanish maps of the New Hebrides (their '*Austrialia del Espiritu Santo*') are firmly drawn and are linked to New Holland by dotted lines from Van Diemen's Land in the south to Cape York in the north, in the Vaugondy map New Guinea is shown as a separate island. It has been argued that the map in this regard is pure fancy; at the same time New Guinea and New Britain are sufficiently well depicted to indicate that they had been explored and charted. What is even more intriguing is the accurate charting of the Gulf of Carpentaria. Groote Eylandt and its satellites are in position and the Wellesley group in the centre is drawn as an archipelago (a reasonable assumption), the west coast of Cape York is well presented with its numerous rivers. Most interesting are the islands in the region. One is drawn in the Gulf of Papua close to modern Saibai, a second group of three marked '*Hoge landt*' (Dutch, 'high land') corresponds very closely to the Prince of Wales group. Even the western coastline from Cape York to Van Spoult Head, marking the north-eastern limit of the Gulf, is depicted accurately in direction, all of this region having been visited, and, presumably charted, by Janz in 1606 and Carstenz in 1623.

Although the strait is clearly marked, at correct latitude, 10°S, what is missing is the eastern approach: the Reef region is only dotted (indicating uncertainty) right to the tip of the Cape itself. Cook certainly knew this map: he states so quite explicitly (Cook 1770:411). In addition, Banks had Dalrymple's chart of the region showing the passage through the strait taken by Torres and Prado, although Beaglehole has questioned Dalrymple's sincerity since his map has all of the Solomons joined to New Guinea and the map could have been merely an attempt at an 'undemonstrated theory' (Beaglehole 1968:clxii).

In a vigorous defence of both Dalrymple and Cook, Howard Fry has argued very persuasively that there was no hostility between Dalrymple and Cook: on the contrary, they were good, if not close, friends (Fry 1979:41–57). In opposition to the question of Cook's prior knowledge of the Reef and of Torres Strait, Fry cites Cook's *Journal* entry for Monday, 3 September, 1770, when, having successfully passed through the strait into the Gulf of Carpentaria, he made a lengthy entry on the topic. Referring to the map of Vaugondy and the two others bound into the copy on board of de Brosses, Cook recorded that 'I allways understood before I had a sight of these Maps that it was unknown whether or no New-Holland and New-Guinea was not one continued land, and so it is said in the very History of Voyages these maps are bound in' (Cook 1770:411).

When leaving the Reef on Tuesday, 14 August 1770, through Cook's Passage, he wrote quite explicitly that he had to exit the Reef waters 'with great regret' since 'I firmly believe that [New Holland] doth not join New Guinea', adding, however, that he hoped to settle the issue when the chance arose (375). Continuing north, still outside the Reef, his anxiety begins to show in an entry for Friday, 17 August 1770, when he reveals that he always had the intention of settling the existence of a strait 'from my first coming upon the Coast'. The necessity of finding safer waters outside, however, he mused, may preclude that opportunity since he could 'be carried so far from the coast as not to be able to determine whether or no New Guinea joins to or makes a part of this land' (380). Fortunately, the following day he was able to re-enter Reef waters at Providential Channel and six days later, on Thursday, 23 August, having cleared Cape York and entered the Gulf, he wrote of gaining 'no small satisfaction' in establishing, finally, that a strait did exist. His last comments on the issue, contained in an entry for Monday, 3 September, though, and cited above by Fry to defend Cook, are somewhat confusing. If Cook regarded the earlier travel books and maps as conjectural and unreliable, and if as an explorer he was fired with the ambition to settle the confusion – having explicit instructions from the Admiralty to do so – then his comments are consistent, except for his 'firm belief' that a strait did exist and that New Holland did not join New Guinea.

Why did Cook make this assertion? The inference can be drawn that he based his belief on an overall assessment of all the evidence from previous voyages and records and that, on balance of probabilities, there was a separation between New Guinea and New Holland. A further inference is that, while personally unaware of the Portuguese discoveries, their charts, having been incorporated into Spanish and Dutch maps and used for exploration, along with the explosion of travel literature, provided Cook with evidence to persuade him that New Holland did not join New Guinea.

Whatever the situation, and present indications are that final clarification will never be reached unless some earlier hidden store of documents comes to light – some Pacific equivalent to the Dead Sea Scrolls – James Cook must remain the world's greatest navigator and without doubt, the explorer who clearly first settled and extinguished the enduring problem of the existence of the counterbalancing Terra Australis, and went on to complete the eastern section of Tasman's map of New Holland. All of his first voyage in the *Endeavour*, in which he circumnavigated the globe and made incredibly accurate charts, was performed entirely through direct observation and navigation tables without benefit of the recently developed chronometer. As if that were not achievement enough, Cook also retains the kudos of having navigated, described and charted the 'labyrinth' of the great Reef. With his chart which now provided a thread for future navigators to follow, Cook had brought it to the forefront of exploration literature as a mysterious and hazardous region, quite unlike any other natural phenomenon yet encountered on Earth.

CHAPTER 3

ENDEAVOUR NATURALISTS: 'A SEPARATE CREATION'

The voyage of the *Endeavour* was to have profound political and scientific consequences in Europe. The immediate result was an almost complete cartography of the world's oceans: the myth of a single great southern landmass was dispelled and Cook's careful mapping of the Pacific, amplified in his second and third voyages, 1772–75 and 1776–79, with the aid of William Harrison's new marine chronometer, enabled the first accurate world maps to be published. Such accuracy had a stunning impact on Europeans: it revealed, finally, the vastness of the Pacific, reaching some 16 000 kilometres from the Arctic to the Antarctic – from Bering Strait to the Ross Sea – and from Panama to the Philippines, some 18 000 kilometres. Even so, in the days of square-rigged ships when overseas voyages were reckoned in many months and often years, the geographical extent of the Pacific remained virtually incomprehensible. Along with scientific discoveries of strange biota, the reports from French and British voyagers of previously unknown exotic societies in the Pacific gave a stimulus to the newly emerging science of ethnology. Once relayed to Europe and heightened by the popular and sensationalising press, each new report dazzled the imagination.

Cook's voyages occurred in the same period as the apogee of the French Enlightenment, promoted by Rousseau, Voltaire, Diderot and the Encyclopedists. Against a background of the evident corruption of a decaying *ancien régime* as depicted by the *Philosophes*, the returning voyagers, so vividly reported by Hawkesworth and by others, opened a window to a new world, imagined as an ante-Diluvian paradise, some other Eden, sadly lost from Europe. Rousseau's promotion of the ideal of the 'noble savage', uncorrupted by the decadence of civilisation, first observed in the New World of the Americas, was given support by Cook's observations on the admirable life of Australian Aborigines on the Reef coast. It found expression also in Bank's short paper, 'Thoughts on the Manners of Otaheite'. Here, departing from the favoured classical tradition (in which many shipboard artists were trained) that looked back to the classical perfection of Greek sculptors such as Phidias, Banks considered the Tahitian women to be as beautiful as 'the Grecians were from whose model the Venus [de Medici] was copied'; their bodies 'might even defy the imitation of the chizzel of a Phidias or the Pencil of an Appelles', noting also that they lived in an unsullied Arcadia (cited in Smith 1985:43). At the same time, Europeans had received an alternative ethnographic vision as well, focused on the perceived wretchedness and brutishness of other indigenous peoples.

EIGHTEENTH CENTURY NATURAL SCIENCE: A FERMENT OF IDEAS

Equally as influential in Europe as the geographical discoveries of the *Endeavour* voyage were the natural history collections made by Banks and Solander along the Australian coast, mostly from within the region of Cook's 'Labyrinth'. These, when they reached Europe, were responsible for some major revisions of scientific knowledge. In order to appreciate the impact of their discoveries it is necessary first to place them in the context of eighteenth century scientific activity which itself was a ferment of ideas, having undergone one rapid revolution after another following the immense stimulation coming from the Italian renaissance of the fifteenth and sixteenth centuries.

A strong stimulus came when Francis Bacon (1561–1626) published his stringent criticisms of existing science in his seminal work of 1620, *Novum Organum*, literally, a 'new instrument' for achieving knowledge and control of nature. Rejecting the outmoded medieval tradition that knowledge is to be gained by logical deduction, with deference to past authorities, Bacon declared his reforming intention in the subtitle as 'True Directions concerning the Interpretation of Nature'. Calling for a renovation of all knowledge he advocated new methods of precise observation and experiment, stating further that all generalisations must be derived by induction from experience, that is, as he saw it, purely from the evidence of the senses. From Greek *empeiria*, 'experience', the term 'empirical' came into use for his method.

Meanwhile across the Channel the French rationalist Descartes (1596–1650) with his dualism between mind and matter, asserted the dominance of human reason over the material world, governed by mechanistic laws. In the following years efforts were made to reconcile conflicting elements and link these two views, at a time when the term 'science' (from Latin *scire*, 'to know'), indicating the rational investigation of nature using empirical methods, began to replace 'natural history', a term now associated with natural theology and its defence of creation teachings (Bowen 1981).

Stimulated in large part by astronomy, this quest for accurate knowledge, despite the risk of church condemnation, continued to spread, particularly in the increasing number of learned societies. Of these, two became pre-eminent: the first was the Royal Society of London, founded in 1660, explicitly on Baconian principles, and chartered by Charles II in 1662. Its motto set the theme: *Nullius in verba* ([trust] 'nothing in words'), and its first historian described its purpose as being, by means of empirical observation and experiment on nature, 'to follow all links of this chain, till all their secrets are open to our minds' (Sprat 1667:110). The second, a French equivalent, came soon after, founded by Chief Minister Colbert in 1666 as the Académie des Sciences, and built upon the work of a number of earlier French societies. Through these societies the new spirit of inductive empiricism now became dominant and it was the consequent burst of activity on so many fronts from the mid seventeenth century that gathered momentum, creating the ferment of ideas in the emerging natural sciences.

An additional stimulus to science during this period came from developments in mathematics and new material technology, including telescopes and microscopes, along with improved instruments of navigation. Equally influential in the burgeoning study of life forms – named 'biology' by Jean Baptiste de Lamarck (1744–1829) – was a new system of classification. Termed 'taxonomy' (Gk *taxis* 'arrangement') by the Swiss naturalist Augustin de Candolle in 1813, it had been developed by Carl von Linné (1707–78) of Sweden – known by his latinised name of Linnaeus – who first presented it in his *Systema naturae* published in Lyons in 1736. Linnaeus followed in the tradition of the Englishman John Ray (1627–1705) in his *Methodus plantarum* where, aiming to put natural history into some kind of 'methodical' order, Ray was especially concerned to throw out the fanciful and anthropocentric categories of previous eras, basing his classification on morphology (structure) rather than utility.

Like Ray, Linnaeus worked within the unquestioned 2000 year old system originally set out by Aristotle who saw all nature as a continuum, arranged in a hierarchy of forms, as outlined in his *Historia animalium* (*Inquiry into Nature*). By the time of Linnaeus that system, known as the '*scala naturae*' (ladder of nature), had been comfortably accommodated within the Christian doctrine of a divinely designed universe, with all species in place and unchanged since the moment of Creation. Linnaeus, consequently, tried to classify all 'links of the chain' – the Great Chain of Being as it became known – by identifying essential characteristics and genuine affinities. The approach of Linnaeus, as developed in his *Philosophia botanica*, the first modern exposition of a philosophy of nature, was in many ways close to the controversial pantheism of Spinoza who viewed nature as a divine design, indeed as the immanent Deity, in complete and perfect order. The task, Linnaeus believed, was to discover and reveal to humans the order in which species had been created, and were fixed for all time. Yet he never seemed to grasp the inherent contradiction that within an Aristotelian *scala naturae* of infinite gradations there could not, at the same time, be a strict separation of species.

The philosophical problem with which both Ray and Linnaeus grappled remains one of the most intractable confronting naturalists: can we identify an implicit order in nature? In effect, is taxonomy a classification of phenomena (appearances) or noumena (mental constructions)? The issue proved unsolvable, and the question of whether species have a separate identity or are actually mental constructions remains an area of considerable debate. Linnaeus was unable to devise a natural 'method' based upon functional relationships and was finally forced to settle for an effective, but admittedly artificial, 'system' of information retrieval. For plants this reached its definitive form in the fifth edition of the *Genera plantarum* in 1754, and for animals in the tenth edition of *Systema naturae* in 1758, a work that provided the basis of a new taxonomy in which he created the binomial system of species nomenclature, in operation to this day.

Other aspects of his system, however, became highly controversial, arousing violent moral condemnation from conservative bourgeois readers since the relationships within nature that he identified would never have been allowed by the Creator. Although the visible generative organs of plants, stamens and pistils seemed obvious choices, in the first edition of *Systema naturae*, in his 'Key to the Sexual System' he devised a most unfortunate metaphor of the 'marriages' of plants as the organising criterion for sexual union. Flowers can have one to twenty or more stamens, which he identified as male organs, and usually have one pistil to hold the seeds, which he called a *gynoecium* (Greek for 'women's quarters'). So his schema listed such scandalous relationships as females in bed with one or more males, even 'many males' (Gk *polyandria*: 'many men') and 'clandestine marriages' while plants with scarcely visible flowers were called 'hidden husbands' (Gk *cryptogamia*) (Linnaeus 1735). Barely published, it was condemned as licentious and 'loathsome harlotry' (Stearn 1969: 96–100). Even today, his system has been criticised by radical feminists for his choice of breasts (Lat. *mammae*) as the identifying criterion for 'mammals', instead of some equally distinctive criterion such as 'hair covered' (Lat. *pilosus*), for the class of what could be termed 'pilosids' (Schiebinger 1999).

ENDEAVOUR VOYAGE: FIRST SCIENTIFIC STUDY OF THE REEF

In this controversial framework, then, Linnaeus' pupil, Daniel Carl Solander (1733–82), also of Sweden, but living in England, was promoting the theories of Linnaeus. In 1764 he met Joseph Banks at Oxford to discuss the proposed voyage to Tahiti to observe the transit of the planet Venus. As an additional venture, they planned to use the opportunities presented to extend the study of natural history in the New World, particularly the previously unexplored regions. For his part, Joseph Banks (1743–1820) epitomised the gentleman amateur with an overwhelming interest in natural history. From a wealthy London family, and inheriting a huge fortune at the age of eighteen, Banks had a privileged education at two of the famous grammar schools, Harrow and Eton, and then at Christ Church, one of the constituent colleges of Oxford. At that time England had only the two ancient universities, still unreformed: both were dominated by classical studies although Cambridge also taught mathematics. The curriculum at the grammar schools was mainly classical, with an emphasis on Latin and Greek language and authors, the young Banks thereby becoming inducted into the grand culture of antiquity, and the thought of Aristotle.

An enthusiastic student of the new science of botany (Gk *botáne*: 'plant' or 'pasture') at the age of twenty-three Banks became a fellow of the Royal Society and through the influence of Lord Sandwich, First Lord of the Admiralty, was enabled to travel with a suite of six, at his own expense, aboard the *Endeavour*. In addition to Solander, the suite consisted of Herman Dieteric Spöring (1735–71), professor of medicine from Abo (then in Sweden, now in Finland) and a skilled draftsman; along with two trained artists: for botanical specimens, the exceptionally talented Sydney Parkinson (1745–71), a Quaker from Edinburgh (but probably not a Scot), and for topography and landscape the very young Scot, Alexander Buchan (d.1769); as well as two assistants to Banks. In addition to the travel books of previous voyages, Banks took with him a number of scientific works in natural history for reference and a more than adequate supply of scientific instruments and other equipment for collecting, dissecting, storing and preserving specimens including seine trawls, jars with ground stoppers, and transport boxes (since glass-panelled Wardian cases, which retained moisture and warmth, were not yet in use).

For pressing flower specimens he took a large supply of paper, having obtained printers' remaindered sheets, which – ironically, as later European settlement of Australia demonstrated – were from Milton's *Paradise Lost*. One of Banks' assistants, in fact, had the sole responsibility of dealing with the pressings for the entire voyage, as Banks recorded in his journal. 'Our plants dry better in paper books than in sand' he wrote on 2 July 1770,

> with this precaution that one person is entirely employed in attending them who shifts them all once a day, exposes the Quires in which they are, in the greatest heat of the day and at night covers them most carefully up from any damps, always careful not to bring them out too soon in the morning or leave them out too late in the evening' (Banks 1770:II.87).

From the outset Banks and Solander were active in their collecting, especially since the Admiralty instructions specifically enjoined Cook, and therefore Banks as the supernumerary scientist,

> to observe the Nature of the Soil, and the Products thereof; the Beasts and Fowls that inhabit or frequent it, the fishes that are to be found in the Rivers or upon the Coast and in what Plenty; and in case you find any Mines, Minerals or valuable

stones you are to bring home specimens of each, as also such Specimens of the Seeds of Trees, Fruits and Grains as you may be able to collect, and Transmit them to our Secretary that We may cause proper Examination and experiments to be made up them (Beaglehole 1955:cclxxxii).

Indeed, so zealous were Banks and Solander in meeting that requirement they were criticised by envious opponents as mere 'species hunters' or 'Kew collectors' (Hoare 1976:135; Brockway 1979:189). There was a ring of truth in the charge: by that time the Royal Gardens at Kew, in London, developed since 1728, had become a centre for British botanical imperialism in a process today termed genetic piracy. As Lucile Brockway has reported in *Science and Colonial Expansion: the role of the British Royal Botanic Gardens* (1979), at Kew, seeds and plants from around the world were being collected, propagated and then transferred to commercial plantations in new colonies, of which the most dubious was the theft of rubber tree seeds from Brazil and their smuggled export to Kew, and then, after propagation, to British plantations being established in Malaya.

The specimens, mostly botanical, collected in the first stage of the voyage of the *Endeavour* down the eastern Atlantic coast of South America, were exciting in themselves. Here were lush tropical plants still undescribed in large part since neither the Portuguese nor the Spanish had been interested in science. The scientific revolution since the seventeenth century, in fact, due in large part to the repressive activities of the Catholic Inquisition, had not reached Iberia. Banks and Solander experienced difficulties all the way with various authorities although in Madeira they were able to do some collecting, listing 230 species and describing 25 new ones, of which Parkinson illustrated 19 (Stearn 1969:67). In Brazil, however, the Portuguese governor – determined to prevent further theft of economic plants for Kew – denied them landing rights and ignored Banks' written protests. Solander recorded in a letter of 1 December 1768 to Lord Morton, President of the Royal Society of London, that at Rio de Janeiro they were continually subjected to 'a Guard-boat of soldiers rowing around our Ship, and were afterwards told that none of our Ship had leave to go on shore' (Solander 1768:278).

Determined to collect, Banks and Solander bribed some locals to bring them specimens: 'Our few botanical Collections have been made by clandestinely hiring people; and we have got them on board under the name of Greens for our Table' (279). On one occasion, in addition, they organised a clandestine foray past the sentries at night in which they managed to collect and describe some 300 species (Stearn 1969:69). In Tierra del Fuego, not garrisoned, they were able to continue collecting energetically although in the final excursion, into the hinterland, 16 January 1770, the party became so exhausted in the harsh cold climate that two black servants died from exposure and a consequent excess of rum in an effort to keep warm. Later on, in Tahiti, despite the warmer climate, the young artist Alexander Buchan, subject to bouts of epilepsy, died. From then on, Parkinson, with help from Spöring, had to assume most of the illustrating responsibility, a task he discharged with brilliance. Tahiti was not just sensuous evenings with Venus-like paramours: the scientists spent their days in relentless pursuit of botanical knowledge, collecting, pressing, illustrating and preserving, with Solander doing most of the describing. One notable discovery was the breadfruit (*Artocarpus altilis*), later to figure in the disastrous episode of William Bligh and the mutiny on the *Bounty* of 1787.

Having left Tahiti, the scientists took the opportunity for exploration in New Zealand where a wealth of discoveries were made, even though the 'Indians' were less than welcoming in many places, especially in the north island. One of the most promising commercial finds was a species of flax (*Phormium tenax*), previously unknown, observed as the fibre used for the cloak of a Maori chief. Flax in that era was mainly *Linum usitatissimum*, grown and processed in Baltic regions and vital for the manufacture of ships' sailcloth. Due to the frequent wars in that region, and interruptions to supply, the alternative source in New Zealand seemed promising.

After making the initial sighting of the eastern Australian coast at Point Hicks, the *Endeavour* party made their first landing at 'Stingrays Bay' (34°S) on Sunday, 29 April 1770, named after the numerous barbed-tail rays (family Dasyatidae) seen there. The intriguing wealth of new plants collected, however, led Banks to change its name to Botany Bay; later its northern headland was named Cape Banks, the southern, Cape Solander. For the next four months the entire length of the coast of New South Wales, as Cook was to rename the eastern region of New Holland, became an overwhelming joy for the scientists: scarcely anything fitted into the *scala naturae* of the Great Chain of Being, and Linnaeus' taxonomies were of little help. There were, of course, familiar ocean birds, pelagic fish and coastal vegetation – the Hibiscus and the palm-like Pandanus, for example – which had been seen elsewhere in the Pacific; at the same time, Banks and Solander were excited by the seemingly inexhaustible range of new botanical species.

Most of their Australian collections came from the Great Barrier Reef shoreline. After Botany Bay the *Endeavour* made all of its shore excursions within the Reef, in sequence, at Bustard Bay, Thirsty Sound, Great Palm Island, Cape Grafton, Endeavour River, Lizard Island, and finally Possession Island off Cape York. It was the Endeavour River region, where they were forced to wait during repairs to the *Endeavour* after the holing on the reef, that provided the greatest yield of new and undescribed specimens, both botanical and zoological, much more than Botany Bay which had first stimulated their scientific interest with its novelties. At the same time, all collecting was entirely littoral, intertidal or marine from seine nets and lines. Forays inland were few and limited, since they had to travel on foot, and plagued with uncertainty regarding the Aborigines. Consequently, although they saw the kangaroo and the possum, much of Australia's unique Gondwanan heritage from a 50 million year separate evolution, such as the koala, the platypus, the wombat, and most marsupials, were not discovered. Solander, however, did observe that most distinctive genus, the *Eucalyptus*, but failed to recognise it (80). The industry of Banks and Solander, with the superb support of Parkinson and Spöring, was prodigious. Even while the ship was being repaired, the scientists were fully engaged ashore where they collected and described to the point of repletion. 'The Plants were now intirely compleated and nothing new to be found, so sailing is all we wish for if the wind would but allow us', Banks wrote for Saturday 28 July 1770 (Banks 1770:II.84–85). Cook at that moment was standing by to leave the river but gales and contrary currents kept the ship anchored for another week until fairer weather allowed a departure.

By 17 August 1770 the *Endeavour* was able to leave Reef waters at Cook's Passage and sail north through the open Pacific until it reached latitude 12°28'S where Cook chose to turn west and re-enter the Reef through a treacherous passage in the turbulence of the reef fronts and fast tides, Providential Channel. Once inside, the ship was anchored in calm shoal water, and the pinnace was sent to explore the vicinity. Banks and Solander, consequently, seized the chance for some marine collecting of fish, molluscs, ascidians and coelenterates which were mentioned in his *Journal* with the comment that 'the sea, however, made amends for the Barreness of the Land, Fish tho not so plentyfull as they generaly are in higher latitudes were far from scarce' (Banks 1770:II.120).

Equally intriguing is the ensuing comment on coral; in fact, the only such biological comment in Banks' *Journal* since all other coral references are to their physical hazards for navigation. Without describing how the coral was collected – probably at times caught on a dragline or anchor chain – he wrote that in exploring they found 'many curious fish and molluscs besides Corals of many species, all alive, among which was the *Tubipora musica*'. Perhaps Banks, who was in correspondence with the two leading authorities of the day, Linnaeus and John Ellis, would have liked to study coral formations, but, as his findings reveal, he was more than overloaded with the numerous tasks in hand. His regret, however, is expressed in his statement that 'I have often lamented that we had not time to make proper observations upon this curious tribe of animals but were so intirely taken up with the more conspicuous links of the chain of creation as fish, Plants, Birds &c &c. that it was impossible' (108).

A wealth of information can be unpacked from that one sentence. Immediately striking, of course, is the reference to 'the chain of creation' which reveals Banks' thoroughly Aristotelian heritage and the scientific framework – 'paradigm' is the current term – that constrained him to work within the Linnaean taxonomy. Banks and Solander, in fact, were the first to try and use Linnaean binomial nomenclature for Australian descriptions, even though they had already experienced its limitations in the Atlantic voyage when Solander wrote to Linnaeus himself that when barrels of mollusca were dredged they had 'been obliged to divide them into several more genera than are mentioned in [Linnaeus'] *Syst. nat.*' (Solander 1768:282).

Even more interestingly, we learn how completely Banks was absorbed in botany. While he did at least know that coral formations are created by animals, he made no reference at all to the intense scientific debates since the eighteenth century regarding the biological nature of coral reef communities, and the equally profound geological disputes concerning the processes by which coral reefs are formed. Banks certainly would have been aware of these as was Cook, who recorded his understanding, that 'Coral rockes' had been 'first formed in the Sea by animals', in his *Journal* for June 1774, during his second Pacific voyage to the islands of Polynesia and the New Hebrides, all of which have extensive coral reefs, and in many instances visible elevated relict reefs. Cook most likely would have discussed such current geological issues, and the possible role of polyps in reef formation, with Banks during the *Endeavour* voyage where both men regularly exchanged their journals and copied passages from each other, as well as discussing the science of the voyage at the captain's table.

This particular practice of exchanging observations often appears in their journals as, for example, when the *Endeavour* was about to re-enter the Reef at Providential Channel. In describing the awesome phenomenon of the steeply rising reef fronts Cook wrote in his *Journal* for 16 August 1770 that the wall of coral was rising almost perpendicularly out of the 'unfathomable ocean' (Cook 1770:378). This passage is repeated in Banks' *Journal*, almost word for word, of the same day,[*] 15 August 1770, describing 'a wall of Coral rock rising almost perpendicularly out of the unfathomable ocean . . . making a most terrible surf Breaking mountain high' (Banks 1770:II.105). It was in such an exchange of ideas that Cook is likely to have learned much of his natural history from Banks and Solander. Given the great interest in corals, and their rare opportunity to study them in one of the world's richest coral regions, it is both puzzling and disappointing that Banks missed such an opportunity to contribute to the rapidly accelerating interest in marine zoology.

Despite the absence of coral studies, the scientific impact of their biological Reef discoveries was considerable. Writing to the Comte de Lauraguais in December 1771, five months after his return, Banks informed him they had collected on the entire voyage 'about 1000 Species of Plants that have not been described by any Botanical author; 500 fish, as many birds, and insects Sea and Land innumerable' (Banks 1771:II.328). From his enormous botanical collection along the Reef of the eastern seaboard of Australia, the endemic *Banksia* genus, in particular, fascinated Banks, after whom the genus is named. He collected four species of banksias in the first foray at Botany Bay, and Parkinson produced some of the finest flower paintings ever done, not only of different species of *Banksia*, but also of many flora of the Reef islands and shoreline.

When the specimens of the voyage, accompanied by Parkinson's superb artworks, arrived in London, it was the botanical series that excited most interest. England was enjoying a period of intense involvement in botany and 'natural' landscaping, exemplified in the demand by the wealthy for the services of Lancelot 'Capability' Brown. For his part, Banks revelled in the admiration of the privileged set that gathered around him in London, where the immense project of an illustrated edition of all the flora collected was conceived and commenced. The amount of material gathered was vast; on the voyage back Solander had made substantial progress on the description of some 1300 new species and about 110 new genera, while Parkinson had made 955 drawings: 675 as sketches, the other 280 as superb finished drawings (Ebes 1988:9). The specimens themselves were Banks' personal property and he kept them in his home. By early 1777 he was forced to move to a more commodious residence at 32 Soho Square. At that fashionable address the greatest collection of floral specimens in all Europe was assembled, with close links to the Royal Gardens at Kew, along with Banks' unparallelled library, where they were freely accessible to all scientists.

Banks' residence was now the epicentre of all progressive botanical activity, guided by the talented Swede, Jonas Dryander (1748–1810), who arrived at Soho Square in 1777 and remained as librarian and cataloguer until his death in 1810. After Banks' death in 1820 the botanical specimens passed to Robert Brown, his curator and librarian as successor to Dryander – who Banks had sent earlier to accompany Flinders on his voyage to New South Wales in 1801–03 to augment the original collection – and Brown continued to pursue intense research on them, before passing them, as stipulated in Banks' will, to the British Museum, where they remain to this day.

Word of Banks' successful voyage on the *Endeavour* spread rapidly throughout European scientific circles; Linnaeus was enthralled, and anxious. He wrote back to Solander in October 1771 asking him to prod Banks into immediate and vigorous action: 'nothing should be trusted to futurity' he wrote, 'I earnestly beseech you to urge [Banks to proceed with] the publication of these discoveries' (Whitley 1970:46). As full a list as can be determined of the botany of the Australian and Reef section of the voyage has been collated by William Stearn in the *Notes and Records of the Royal Society* (24:1,1969; 71–73,79–80) while the zoological findings are recorded in Gilbert Whitley's *Early History of Australian Zoology*. Whitley, at the Australian Museum in Sydney, revealed the profound significance of the zoology of the *Endeavour* voyage in a statistical analysis. When Banks and Solander left England in 1769 the *Systema naturae* of Linnaeus for the entire world totalled but 4380 animal species: 184 mammals, 554 birds, 218 amphibia, including some reptiles and fish, 2110 insects and 936 'vermes', that is, worms, molluscs, jellyfish and corals. When the *Endeavour* collection was analysed, it contained an additional 616 species, a 14 per cent increase, comprising: 22 mammals (12 per cent increase), 93 birds (16 per cent), 17 amphibia (8 per cent), 65 fish (17 per cent), 244 insects (11.5 per cent) and 161 'vermes' (17 per cent) (Whitley 1970:64–72).

The impact on natural science was profound. Belief in an anthropocentric Divine Design that presumably accommodated all of creation was now brought into question: had there been a separate creation? How could these strange findings from New South Wales be incorporated into the comfortable taxonomies so far constructed? How could the economy of nature be explained when nature in the southern hemisphere was found, as so many believed, to be perverse?

Already the Linnaean taxonomy was being challenged by the talented de Jussieu family (Antoine, Bernard and Antoine-Laurent) at the Jardin des Plantes in Paris, who were attempting to develop a 'natural' system. Simultaneously Jean-Baptiste de Lamarck (1744–1829) was questioning the concept of a single, static Great Chain of Being, directed by an immanent purposeful force known as '*natura naturans*'. His thought was tending towards the newer concept of '*natura naturata*', nature conceived solely as the evolution of interactive processes governed by physical laws as developed in the Newtonian philosophy. Evolution – by a still unknown mechanism – along with processes of development and change in nature were being actively discussed, and the *Endeavour* voyage contributed significantly to the emerging debate. Even though *scala naturae*, the Great Chain of Being, or Divine Design – the concept had several names – was still the accepted basis for natural history, it was beginning to come under severe strain. Banks and Solander made, unwittingly perhaps, a major contribution to that process which enabled natural history to become a very different scientific activity.

BANKS' *FLORILEGIUM*: THE GREAT COMPENDIUM

The illustrated natural history, based upon Parkinson's exquisite craftsmanship, however, was to have a much more chequered career. Full of enthusiasm, in 1771 Banks commissioned five skilled artists and sixteen assistant engravers to complete Parkinson's sketches and make copperplates of the entire collection from Parkinson's watercolours, as well as the specimens and pressings brought back, for a monumental *Florilegium* (literally, 'a binding of flowers') in large coloured format, at an estimated personal cost to him of the then astronomical sum of £7000. By 1784, some thirteen years later, 743 superb engravings had been prepared for printing. Parkinson's actual drawings survive, comprising 260 signed and 16 unsigned watercolours, and 676 sketches, along with proof pulls from finished plates, and are bound into eighteen volumes in the Botanical Library of the British Museum (Beaglehole 1968:cclxviii).

Tragically for Banks' superb vision, a most unpleasant and eventually fatal controversy developed over the ownership of the illustrations and their translation into the grand *Florilegium*. One consequence of Sydney Parkinson's death during the return journey was that his brother Stanfield claimed ownership of the artworks, and, therefore, if Banks were to use them for the project, he would be entitled to some financial consideration. This led to a bitter confrontation with Banks who had, after all, paid Sydney Parkinson's expenses throughout the voyage. Banks settled the issue with a gratuity of £500 to Stanfield Parkinson who, none the less, went ahead and planned an account of the voyage of the *Endeavour* using his brother's papers and illustrations.

Stanfield Parkinson's production was a mediocre effort. The preface was a bitter diatribe against both Banks and Hawkesworth 'from whose superior talents and situation

---

* The journals were written up each evening in the Great Cabin. While it was still 15 August for Banks, for Cook, since the ship's day began at noon, when the overhead sun gave an exact time, it was the 16th.

in life better things might be expected' instead of being 'misemployed in striving to baffle a plain, unlettered man' (Parkinson 1773:v). The preface was followed by 212 pages of loosely constructed text, with lengthy descriptions of the 'Indians' along the Reef littoral, much of it pirated from the journals of Cook and Banks, as transcribed by Hawkesworth, and freely illustrated with engravings based on the work of his brother Sydney. The dramatic moment of the encounter with the Reef near Cape Tribulation was given even greater exaggeration: when 'we examined the ship's bottom' he had Cook comment, we 'found a large hole; through the planks into the hold, which had a piece of coral rock, half a yard square, sticking in it' (143).

The controversy became even more vicious because Hawkesworth, in preparing his eagerly awaited *Voyages* of 1773, and having used some of Parkinson's drawings for black and white engraved illustrations, heard of Stanfield Parkinson's competing production, and consequently obtained a court injunction to prevent Stanfield's proposed publication of *A Journal of a Voyage to the South Seas* until after his own *Voyages* came out. Perhaps to spite Stanfield, Hawkesworth did not acknowledge Sydney Parkinson as the artist responsible for the illustrations in his production. The controversy became even more heated since it involved Banks whose response was to increasingly distance himself. At the same time, disappointed, even piqued, at the refusal of the Admiralty to allow him the extra accommodation he required on the *Resolution* for Cook's second voyage on which he had planned to sail in another botanising expedition, Banks withdrew, lost interest in the *Florilegium*, and turned his interests in other directions. By the time he abandoned the project, expenses had escalated to more than £10 000 and so the final cost of the planned ten folio volumes could have been beyond even Banks' finances.

Solander, however, continued to work on the proposed *Florilegium*, at the same time editing Ellis' *Natural History of some curious and Uncommon Zoophytes*, until his own premature death in 1782. Two years later all work on the *Florilegium* ceased and publication was postponed indefinitely, creating what Banksian scholar Hank Ebes has described as a 'tragedy of science and a puzzle of history' (Ebes 1988:110). The superb plates were not even cleaned of ink from the proof pulls: they were wrapped in paper and put aside, slowly deteriorating in the British Museum throughout the entire nineteenth century. The visual and aesthetic botanical results of Banks' and Solander's voyage along the Reef were lost at large to science, and to aficionados, even though the specimens remained available for reference.

In 1895, George Murray, Keeper of Botany in the Museum, urged the Trustees to consider rescuing the plates and to finally issue the *Florilegium*. Under the title *Illustrations of Australian Plants Collected in 1770*, 317 black and white photolithographic illustrations were issued in three separate printings: the first 100 in 1900, another 143 the next year, and a final 74 in 1905 (Ebes 1988:28 and *passim*). The results, however, were disappointing, the engravings having been described as 'disagreeably mechanical, though of high technical skill' with, however, the saving grace that 'many of the drawings, both in their rough and finished states, are of considerable beauty and interest' (Blunt & Stearn 1994:168).

A second effort to bring the beauty of the *Florilegium* to the public was made in 1973 when two connoisseurs and botanical authorities, Wilfred Blunt and William Stearn, joined with the Lion and Unicorn Press (teaching press of the London Royal College of Art) to bring out a small selection of the finest of the illustrations under the title *Captain Cook's Florilegium*. Only 29 plates covering the entire voyage were issued, 12 being of Australian flora from the Reef coast in Bustard Bay, Endeavour River and Lizard Island. The engravings, however, were delicately executed, and had been selected for both their aesthetic and scientific value. In a very limited edition of 100 copies, it is a very large format book, in oversize typeface, reproducing exactly the original page size of 24 by 18 inches (61 × 45.7 cm).

Yet a third attempt to bring the *Florilegium* to life occurred in 1978 when Alecto Editions began negotiations with the British Museum for a limited edition of 110 copies, 100 for sale. By that time, however, the 743 metal plates had deteriorated from the uncleaned ink and poor storage conditions and were unsuitable for direct printing. Instead, lithographic plates, measuring 46 by 31 cm, were prepared from the 738 in sufficiently good condition, using a process developed in the late seventeenth century by the Dutch printer Johannes Tayler, known as the 'à la poupée' multicolour (up to twenty colours simultaneously) in a single-pull process. Using a ball of cloth – hence the 'rag doll' epithet – the plate was inked with the required number of colours, a process taking several hours, and passed through the hand-turned press in one pull. Each plate then had to be recoloured for the next print. The aesthetic aim was to match the beauty of the original brushwork and at the same time reveal the calligraphic strength of the engraved lines (see Ebes 1988:47; Adams 1986:*passim*).

The printing was a triumph of graphic art and vindicated the skill of Parkinson, the vision of Banks and the beauty of the distinctive Australian flora. Today, each plate is extremely valuable, and when occasionally they reach auction rooms, each individual illustration is sold for many thousands of dollars, given even greater rarity value by the decision of the Trustees of the British Museum not to allow another printing for at least fifty years. Altogether, 738 plates were printed, and these illustrate most of the Reef region flora. Their publication was a superb achievement, but, unfortunately, came far too late for the illustrations to make their fullest contribution to science.

Perhaps the ultimate irony is the fact that today Banks' catalogues provide an index of the immense continuing destruction and extinction of Australian biota since European settlement in 1788 which in the early twenty first century is creating serious alarm. Indeed, the massive loss since Banks' day has been described as a movement from 'frontier to fragments' (Humphries & Fisher 1994:344, 3–9).

FATE OF THE *ENDEAVOUR* ZOOLOGICAL SPECIMENS

The zoological specimens fared badly. In the 1770s there was no suitable place for a repository since the Natural History Department of the British Museum, founded in 1753, was inadequately housed and was entering a period of lethargy. Indeed, its foundation collection bequeathed by Hans Sloane (1660–1753) was deteriorating and even disappearing: by 1833 nothing remained. Many of the Museum's specimens had been so badly preserved they had to be destroyed, and were periodically burned as a means of disposal. Banks kept his collection safely at Soho Square where they were still to be fully catalogued: botany and the *Florilegium* had been the priority. Solander, however, did considerable work on the collection, beginning the task with binomial descriptions on small pieces of paper – known as 'Solander slips' – which came together in twenty-seven volumes, three relevant to Australia and the Reef, along with 512 foolscap (20 × 33 cm) sheets describing the specimens in the customary Latin.

For still obscure reasons, Banks suddenly wanted nothing more to do with the zoological collection and it was dispersed by 1806, much of it lost forever, the fish specimens entirely. Banks gave some specimens away and auctioned the remainder in 1806, the staggering total of 7879 lots going to purchasers at 123 separate locations, including the Museum of the Royal College of Surgeons along with the Bullock, Ashmolean and Leverian Museums. The shells and insects went to the Museum of the Linnean Society of which Banks had been a founding member in 1788; other lots went to various continental museums and to the collection of an enthusiastic amateur natural historian, the Duchess of Portland, who had provided one of the microscopes carried aboard the *Endeavour*. Some even found their way to Russia and provided in their day the most extensive available collection for the study of the Pacific region (Whitehead 1968:xix, 1978:52–55; Whitley 1970:48–49). No publications appeared, then or subsequently, and scientists still experience almost insuperable difficulties in trying to determine the exact status of the zoology of the *Endeavour* voyage. By 1846 no actual specimen of any of the fish was known to exist.

Fortunately, illustrations do survive in addition to Solander's papers, now housed in the Zoological Library of the British Museum. Most are by Parkinson, some by Spöring, a few by Buchan; even so, they are mainly of terrestrial fauna. Bound into three volumes, also shelved in the Library, they contain 83 finished watercolours signed by Parkinson, and one by Buchan; 87 finished but unsigned by Parkinson, and 125 of his both unfinished and unsigned (Beaglehole 1968:cclxviii). Like the botanical illustrations, Parkinson displays a graphic virtuosity that defies words and has provided such an accurate record of Reef fauna that the species can be readily identified.

In one sense the scientific achievement did not produce the dramatic effects Banks imagined, especially in the visual realm. Certainly it contributed decisively to scientific thought at the time, and the careful cataloguing by Solander made the discoveries widely available. However scientists depend heavily on type specimen descriptions for continuing revision and these are now no longer readily available; unfortunately, for many species, both plant and animal, the preserved type specimens often give little or no indication of the living form. Careful and accurate delineation by skilled artists plays a significant role in identification and taxonomy, and that, regrettably, was denied to the world after the voyage of the *Endeavour*.

Such a scientific loss was made even more tragic given the terrible privations of the crew, and the death of so many, including Parkinson, Buchan and Spöring, during the voyage. For the general public, however, the discoveries served to intensify the mystique of Australia and the Reef: the strange, pouched marsupials were 'curiosities'; the vegetation, although considered drab by comparison with Europe, had most intriguing forms which were to advance scientific knowledge considerably; the inhabitants were a race apart, incomprehensible. Meanwhile, coral reefs themselves continued to remain one of nature's greatest mysteries.

CHAPTER 4

MATTHEW FLINDERS: VOYAGE OF THE *INVESTIGATOR*

On his return from the *Endeavour* voyage, Banks was the dominant figure in British botany. Well connected to the governing establishment, his London home commanded resources unrivalled in Britain, and equal to many in Europe. In 1778 Banks was elected president of the Royal Society and until his death in 1820 was its autocratic and unquestioned leader, which enabled him to promote natural history as a counterbalance to the previous emphasis on mathematics. Three years after election, in 1781, he was created a baronet with the title Sir Joseph, gaining further stature in society.

The decision to abandon the *Florilegium* in 1784 must have been a great disappointment to all concerned, although by then Banks was involved in a multitude of activities that took all of his energies. His major preoccupation was directing the Royal Kew Gardens which were playing a major role in the rapidly changing economy of Britain, as the nation became increasingly dependent on its colonies for economic survival in the face of French and Dutch competition. Following the occupation of the New World by Portugal, Spain, and then Holland in the seventeenth century, tropical crops had become essential to the European economy, and intense rivalry ensued. The range of introduced products was considerable: new food staples – rice, sorghum, buckwheat and millet – along with legumes, coconuts, yams; spices including cloves, nutmeg and pepper; sugar, tea, coffee, opium and cinchona; the fibres of flax and hemp, and later wool; along with indigo dye, became well-established trading commodities – virtual necessities – in European markets (Brockway 1979:37).

To secure its share of tropical and warm temperate lands to produce these crops, Britain, along with other European powers, invaded suitable territories already occupied by indigenous peoples, forcing them into political and economic servitude. By the late eighteenth century a large number of colonies had been occupied, all for either economic or strategic benefits, and every British expedition was given, in the Sailing Instructions to the commander, specific orders to investigate commercial possibilities. Waiting in London to collect the botanical specimens at Kew Gardens was Joseph Banks. By that time, botanical gardens had become well established throughout Europe, having begun in the sixteenth century in the grounds of dispensaries for pharmaceutical purposes. In 1728, Kew House, on the banks of the Thames above London, was leased by Frederick, Prince of Wales, and so became Royal Kew. The collection grew steadily; by 1764 superintendent William Aiton had published the first catalogue, *Hortus kewensis* ('Kew Gardens'), which listed 5000 plants in Linnaean taxonomy. Followed by his son, William Townshend Aiton, in 1793, with whom Banks worked closely, the collection had increased considerably as a result of continuing additions. When revised by Banks and Aiton in 1810–13 the new catalogue ran to five volumes and listed 11 000 species. With the rapid development of botanical knowledge, such gardens moved from being pharmaceutical adjuncts into major facilities for research and, significantly in the case of Kew, for propagation and distribution in the service of economic imperialism.

The leading botanical research institution in Europe was then in Paris, founded in 1635 by Louis XIII as the Jardin Royal des Plantes Médecinales; in 1718 it was reconstituted independently of the Faculty of Medicine at the University of Paris as the Jardin Royal des Plantes. After the 1789 Revolution its name was changed in July 1793 to the Jardin des Plantes, and at the same time the Muséum Nationale de L'histoire Naturelle was founded in Paris. Under the brilliant direction of Antoine-Laurent de Jussieu, using specimens brought in by French expeditions, beginning with the superb collection made by Philibert Commerson, naturalist aboard the *Boudeuse* on Bougainville's voyage of 1766–69, botany – as the new science was beginning to be termed – developed rapidly. The French, indeed, eschewed the practical, horticultural emphasis followed at Kew. At that time the French, unlike the British, had not yet acquired any major colonies. Kew, by contrast, was dominated by the processes of reception, propagation and transfer of plants to and from its overseas domains. Despite being derided in some quarters as a mere 'Kew collector', Banks' efforts led to the vast extension of British agricultural and horticultural activities and the parallel development of gardens in the colonies – Jamaica, St Vincent, St Helena, Calcutta and Sydney – to supply and receive plants from Kew. From this were created the great tea plantations of India, those for rubber and cinchona (for anti-malarial quinine) in Malaya, for sugar-cane in the West Indies, and cocoa in Africa, along with many others.

BRITISH SETTLEMENT OF NEW SOUTH WALES

As part of the British program of global expansion New South Wales also came under scrutiny for settlement and by the 1780s discussions were under way concerning the future of that region, claimed by Cook in 1770 but not yet occupied. For their part, the French were actively exploring the Pacific, but so far had not sighted the continent. Banks, among others, recommended New South Wales for settlement and in 1786 a decision was made to do so. The reasons will always remain obscure, and debatable, since few records exist. A range of possible reasons – but not the convict 'dumping ground' theory – were first set out in a Memorandum of 23 August 1783 to the British government by James Matra, an American of Corsican origin who had sailed as a midshipman on Cook's *Endeavour*. In that document he urged the settlement of New South Wales for both economic and strategic reasons. The east coast of the region, he wrote, had much fertile soil and but 'a few black inhabitants . . . in the rudest state of society' (*HR.NSW*:I.2.1). Similar in climate to the Spice Islands, this region, Matra suggested, would be suitable for sugar-cane, tea, coffee, silk, cotton, indigo and tobacco, while in nearby New Zealand, as the *Endeavour* botanists had discovered, there was a 'hemp or flax plant' (*Phormium tenax*) of such strength that a cable of 10 inches (25 cm) circumference was equal in strength to one of 18 inches (46 cm) made from European hemp (*Cannabis sativa*). In addition, New Zealand and the nearby Norfolk Island promised a plentiful supply of pines for masts and spars. At all times, it must be remembered, the economic and political power of nations depended on their sailing ships, both naval and merchant, and ships in turn needed hemp for cables and rigging, flax for woven sailcloth, and tall trees for masts. As a final argument, Matra pointed out the strategic location of New South Wales as a base for the growing trade with China, Japan and other Asian centres, and for the interdiction and harassment of competitive Dutch and Spanish ships if necessary (I.2.2–5).

Whatever the motives, active plans for settlement were begun at least as early as 1786 and in 1788 the First Fleet arrived in Botany Bay to begin the European occupation of New South Wales. Certainly the settlement of New South Wales was anything but the careless expedition often alleged. Along with a carefully chosen range of clothing, tools, building materials, domestic animals and food supplies were brought fruit trees, berry vines, herbs, nut trees, kitchen vegetables, and commercial plants: hemp, flax, tobacco, wheat, barley and corn (Frost 1993:2). With Kew ever the centre of his concerns, Banks gave explicit instructions to the gardeners and collectors on the succession of supply voyages to the new colony regarding plant care and watering at sea, seed collecting 'wherever possible' and their recording in the Correspondence Book (25).

BLIGH AND FLINDERS: EARLY REEF SURVEYS

Despite initial difficulties, the colony under Governor Arthur Phillip became firmly established and almost immediately began to expand on to the adjoining Sydney plain. Phillip's tour finished in 1795, and in February of that year his replacement John Hunter left England in the *Reliance*, accompanied by the brig *Supply*, arriving on 7 September 1795.

The master's mate of the *Reliance*, already impressing Captain Hunter with his competencies, was 21-year-old Matthew Flinders, already a veteran of the 1794 great naval defeat of the French on 'The Glorious First of June'.

Stimulated as a youth by the story of Robinson Crusoe, Flinders (1774–1814) had an early vocation for the sea and sought the patronage of Captain Pasley who, in 1791, recommended the 17-year-old as a midshipman to William Bligh on the voyage of *Providence* in the second expedition to collect breadfruit (*Artocarpus altilis*) – the starchy fruit of a tropical plant which grows in Tahiti – for West Indian cotton planters as a cheap food supply for African slaves. Flinders would have been well aware of the widely discussed mutiny on the first expedition in 1787 when Captain Bligh had left England in the *Bounty* via Cape Horn for Tahiti to collect plants for propagation at Kew and then for transplanting to the West Indies.

In Tahiti, on 28 April 1789, on learning that the return voyage via Torres Strait and the East Indies was to pass through the dangerous coral reef, and knowing of Cook's earlier disaster, the crew had mutinied, setting Bligh and eighteen loyal crew adrift in the ship's longboat. In that incredible feat of seamanship, despite almost insuperable difficulties, Bligh navigated the 23 foot (7 m) longboat westbound in search of the Reef and an opening. A month later, Bligh recorded in the log he kept meticulously, that he had been 'determined to look for a passage & take the first opening. About 1am fell in with a reef wh. broke dreadfully' (Bligh 1789:90; ed. Bach 1987:111). Standing off for the night in its lee, the next morning he passed through the main reef without incident in what is now Bligh's Passage (12°50'S). The novice Flinders would have been more than impressed with the navigating skill of the captain whose command he was about to join.

It was, then, on Bligh's second breadfruit voyage in 1791 in command of the *Providence* that Flinders, sailing as a midshipman, became familiar not only with the superb – if tyrannical – seamanship of Bligh, of whom he held a less than flattering opinion, but also had his first experience of the Reef. Bligh, as instructed, having secured the cargo of breadfruit in Tahiti, carried it westbound to the Caribbean colonies through Torres Strait again, and on this voyage, made extensive charts of those extremely hazardous waters, giving Flinders considerable experience of coral reefs. On that occasion, Bligh's skill added yet another valuable increment to the hydrographic knowledge of one of the most dangerous navigation waters in the world.

FLINDERS' SURVEYS 1795–1799

Upon arrival in New South Wales, with his ambitious master's mate aboard, Governor Hunter immediately made good use of Flinders' talents, authorising him to accompany his close friend, ship's surgeon George Bass, on two short surveys of the coastline in the immediate vicinity of Port Jackson. Three years later in the sloop *Norfolk*, from 7 October

1798 to 12 January 1799, Flinders sailed down the coastline of the mainland and, heading west, entered what came to be designated Bass Strait. Proceeding south in an anticlockwise traverse he then circumnavigated Van Diemen's Land (modern Tasmania), proving it an island, thereby charting the incomplete section of Tasman's map of 1642.

Flinders' considerable navigational and leadership skills continued to be employed by Governor Hunter in exploring the coastline north of Sydney. In 1799 in command of the *Norfolk*, aided by Cook's charts of nearly thirty years earlier, he reached the location of modern Brisbane by rounding the north cape of what he took to be an island at Cape Moreton, named by Cook after James, Earl of Morton and president of the Royal Society. Entering Glass House Bay, now Moreton Bay, he believed Stradbroke Island to be a peninsula, noting that the passage between it and Moreton Island was 'blocked by many shoals of sand' which had required them to continue to the cape. Forced to retreat from further exploration by the mangrove flats in the southern reaches of the bay, Flinders retraced his passage northwards, and named it '*Moreton Island*, as supposing it would have received that name from Cook, had he known of its insularity' (Flinders 1814:I.cxcvi,cxcix). Passing Breaksea Spit at the northern tip of what he charted as the Great Sandy Peninsula (today, Fraser Island) he continued searching for large rivers or waterways leading to the inland regions of the continent until he reached latitude 24°S where the Great Barrier Reef begins near Lady Elliot Island. Not sighting that island, he broke off with considerable regret, recording that he was 'disappointed in not being able to penetrate into the interior of New South Wales . . . but, however mortifying the conviction might be' he declared that no river of importance existed. In the same passage, however, he adumbrated his future plans with the comment that

> on the east coast from Bass Strait to Bustard Bay in latitude 24° the shore might said to be well explored; but from thence northward to Cape York, there were several portions which had either been passed by captain Cook in the night or at such distance in the day time, as to render their formation doubtful: The coast from 15°30' to 14°30' [Cooktown to Cape Melville] is totally unknown (I.ccii).

On 3 March 1800, Flinders, now promoted to lieutenant, returned to England to await another posting. Already a determined – if frustrated – explorer, fired with a zeal to chart more of that still unknown continent, he wrote of his ambitious plans to Banks who was already aware of the exploits of the young lieutenant, having read his reports to the Admiralty. Flinders had, in fact, dedicated the most important of these, *Observations on the coast of Van Diemen's Land* to Banks. For his part, Banks was keen to patronise Flinders as useful to his own wide interests in science since, after the *Endeavour* voyage, and similar French initiatives, all naval ships now included naturalists in their complement. His conversations with Flinders gave every indication the young lieutenant would be of great value.

THE URGENT ISSUE: SURVEY OF THE COLONY

Writing to Under-Secretary King of the Colonial Office on 15 May 1798, Banks lamented that after ten years of settlement still 'not one article has hitherto been discovered by the importation of which the mother country can receive any degree of return for the cost of founding and hitherto maintaining the colony. It is impossible to conceive that such a body of land, as large as all Europe does not produce vast rivers, capable of being navigated into the heart of the interior; or if properly investigated, that such a country, situate in a most fruitful climate, should not produce some native raw material of importance to a manufacturing country as England is' (*HR.NSW*:III.382–83). The immensity of the continent, in fact, had become a paramount concern in the Colonial Office, a feature magnified when the only means of travel was by sea, and at the slow speed of 5 to 10 knots. Ground travel, given the lack of roads, and the forbidding nature of the eastern Great Divide, was virtually impossible: only great navigable rivers would provide a ready means of access. On 10 July 1799 Governor Hunter wrote to the Home Secretary, William Bentinck, Duke of Portland, insisting that the only practicable means would be to begin with coastal exploration: 'I am of the opinion . . . that a knowledge of its sea coast' should take priority: by using Cook's charts, the 'innumerable appearances of harbours which were observed and carefully marked' should first be examined more closely to see whether 'there may be extensive rivers or arms of the sea' (*HR.NSW*:III.693).

For his part, Flinders also kept arguing the case for a voyage under his command, writing to Banks on 6 September 1800 urging him to initiate further action. By the end of the year Banks had persuaded Robert Spencer, First Lord of the Admiralty, to send an exploring expedition to New Holland to chart not only the eastern coast of New South Wales, and the dangerous Reef – now recognised as a potentially useful sea lane to Asia – but also the entire continent. It was especially important, apart from the possible economic advantages for imperial trade, to discover whether, as early charts seemed to indicate, New South Wales was a separate landmass from the western region of New Holland, and whether the poorly charted Gulf of Carpentaria in the north extended right through the continent to the Bight in the south. On 28 September 1800 Hunter's successor, Governor Philip Gidley King, had written to Banks about the high level of competence of Flinders and Bass in circumnavigating Van Diemen's Land, and of his support for the new plan to explore most of the continent, especially since it would 'solve the doubt whether the mountains are separated from the other part of New Holland by a sea or strait running from the Gulf of Carpentaria into the Southern Ocean, which is a very favourite idea in this country' (*HR.NSW*:IV.1896:207). Since there were large, easily navigable rivers in Europe, the Americas, Africa and Asia which enabled interior settlement, it seemed logical to assume a similar situation in New Holland. So far, however, the charts of neither Cook nor Flinders showed any such possibilities.

It was his failure to find any such rivers, in voyages both south to Van Diemen's Land, and north past Breaksea Spit to the beginning of the Reef, that had so troubled Flinders.

The new expedition, he suggested to Banks, could well offer greater knowledge. Banks was taken by the idea, which he pushed to the limit for political purposes to get the expedition going, although his real interest was in the natural science of the continent, for which he began to plan in detail. What would be needed was a first class botanist, supported by a competent gardener, along with recording artists: in effect, another Solander and Parkinson team. By the end of the year Banks had selected a suitable group. Impressed with the capabilities of Mungo Park, who had become a minor celebrity in London when he returned from exploring Gambia in the years 1795–99, Banks offered him the position of scientific leader. Park declined, however, probably because of his approaching marriage.

Seeking an alternative botanist of exceptional competence, Banks wrote on 12 December 1800 to Robert Brown, an ensign and assistant surgeon in a Scottish regiment, offering him the position. Brown, already a keen and very promising young botanist, had met Banks previously when on leave in London, seeking access to the facilities at Soho Square. Banks' letter explained that a ship was 'to be fitted out for the purpose of exploring the natural history (among other things) of New Holland, and it is resolved that a naturalist and botanic painter shall be sent in her'. The pay was to be £400 per annum: would Brown like Sir Joseph to recommend him? (*HR.NSW*:IV.265). Little could Banks anticipate that not only was Robert Brown to fulfil the task beyond all expectations, but, beginning with that voyage, would become one of the world's greatest botanists; acclaimed also as a front rank microscopist and celebrated throughout Europe by its leading scientists. In addition to Robert Brown he selected the Austrian natural history artist Ferdinand Bauer as 'botanic painter', William Westall as landscape and figure draftsman, Peter Good as gardener, and John Allen as geologist – then termed 'miner' – to report on mineral possibilities.

Plans for the expedition continued to mature. On 25 January 1801 Flinders, now appointed to lead the expedition, formally took command of refurbishing and equipping the obsolescent 334 ton sloop *Xenophon*, appropriately renamed *Investigator*, and was promoted to the rank of commander three weeks later on 16 February. The venture, however, was entirely directed by Banks who planned it as an extension of the *Endeavour* voyage. Flinders he saw as another James Cook, a highly competent officer who could be relied upon to navigate the ship safely around the continent and make whatever hydrographic charts and geographical discoveries were possible; the underlying purpose would be the scientific investigation and collection of natural history specimens, chiefly those of value to Kew. That plan was brought out quite explicitly at the end of the voyage where, in a draft introduction to Flinders' account, Banks wrote that the aim of settling and exploring Australia was 'to advance the progress of natural knowledge in various branches – [and] to open fresh fields for commerce, and new ports to seamen' (*Brabourne Papers*, 1886:23). Banks already had George Caley in Sydney collecting as his paid agent since 1798, but Caley was limited in his range of travel: the *Investigator*, it was intended, would traverse the entire coastline of the continent and complete the more limited charting of the

Endeavour voyage. The scientific team having been selected, on 1 March 1801 the Admiralty wrote to Flinders advising him of the appointment of Brown, Bauer, Westall, Good and Allen.

Flinders, as a naval officer, was entirely absorbed in preparations for the voyage; as his own writings record, he knew almost nothing of natural history and left those matters to Banks. He was, however, very thorough in organising the details from a naval aspect, having already planned the route the ship would take. Since Cook had so ably charted parts of the east coast of New South Wales, Flinders wrote to Banks on 29 April 1801 that he intended to reverse Cook's route by travelling outside the Reef where Cook had taken the inner route, 'but where no coast was seen by him it would be proper to come in upon it through the reefs' (*HR.NSW*:IV.352). Flinders later set out his purposes in exploring that section of the Reef very explicitly:

> From 16° [near Cape Tribulation], northwards to Cape York, an extensive chain of reefs had been found to lie at a considerable distance from the coast, [on the outside] of the islands . . . It was important to ascertain the limits of these vast bodies of coral, were it only on account of the ships employed in the whale fishery; but in view of future settlements within the tropicks, it was necessary to be known whether these reefs might form such a barrier to the coast, as to render it inaccessible from the eastward: if not, then the open parts were to be ascertained (Flinders 1814:I.cciii).

The *Investigator* sailed from England on 18 July 1801 via the Cape of Good Hope and reached Port Jackson on 9 May 1802. The Admiralty had instructed Flinders to retrace his path from Sydney back to the west coast, and then proceed to chart the poorly known north-west section of New Holland. The month of May, however, is the beginning of the southern winter, and the weather of the Southern Ocean can be very forbidding with furious gales and high seas. Flinders secured Hunter's approval to use the winter months to travel north to the Reef instead. From the Reef he could then sail through Torres Strait and continue charting the northern and western coastlines, finally returning with favourable following winds across the Southern Ocean to Port Jackson.

The stay at Sydney was brief and Flinders was keen to continue on his great expedition: not only must he traverse the Reef, he must also clear Torres Strait before the summer monsoon and cyclone season began, often as early as November. Some ten weeks after arriving in Sydney, on 22 July, 1802 the *Investigator* sailed north, with the small ship *Lady Nelson* as consort, along the coastline that Flinders had already charted in late 1789 in the *Norfolk*, reaching the Great Sandy Peninsula by 28 July. It was here, Flinders believed, that the southern limit of the Reef began under the surface.

For the ensuing twelve weeks Flinders investigated and charted the coastline very thoroughly, observing in his diary that his bearings were diverging considerably from those on Cook's charts. From the Sandy Cape northward, he recorded 'greater differences began to show themselves' (Flinders 1814:I.13). Cook, it will be recalled, did not have the benefit of the newly developed marine chronometer on his *Endeavour* voyage, and so his chart bearings were not entirely accurate. Flinders' task was to correct those, chiefly the longitude measurements and compass variations for the entire Reef and Torres Strait, and then to ascertain others for the rest of the continent. Flinders was well aware of his responsibility, recording in his journal that false readings might lead to 'rocks or shoals, as near even as half a mile, which might prove fatal' to vessels. The necessary changes to Cook's readings were explicable, he continued, since 'timekeepers were in their infancy when captain Cook sailed on his first voyage' (I.v,vii). Indeed, Flinders was forced to make numerous changes, and his extensive revised tables of longitude show major corrections of between 20½ and 58½ minutes (I.vii). (Measured at the equator these give distances between 24 and 68 nautical miles, and at the tropic, approximately 22 to 63 nm).

Allowing time for shore excursions for the scientific party, Flinders continued his meticulous charting of what is at any time a forbidding coastline with a maze of islands and inlets, although, fortunately, few dangerous shoals. In the area of his traverse, from Sandy Cape northwards to Cape Bowling Green (24°42'S to 19°18'S) the Reef is some distance offshore, unlike the northern section where Cook had his misfortunes. In sequence, Flinders investigated Shoalwater Bay, Broad Sound, Curtis Island, the Keppel group, Cape Clinton and Port Bowen (now Port Denison). Near Shoalwater Bay (22°S to 21°S) more difficult waters were encountered: from the fifth 'to the 14 in a continual labyrinth of Coral reef Shoals and Breakers with water at times as smooth as a mill pond at other times looking like a whirlpool and such an eddy as to wheel the Ship about during which time we lost an Anchor and the *Lady Nelson* lost one and broke another' recorded the gardener Peter Good in his autograph journal (Good 1802:95). Indeed, the difficulties continued to increase: the *Lady Nelson*, fitted with innovatory sliding keels to enable shallow water surveys, lost the main keel and was unable to 'Sail against the wind' as Good recorded. With much regret, since it now had but one anchor and no keel, Flinders had no choice other than to send Lieutenant Murray and his vessel back to Port Jackson: 'having settled every thing between the two vessels & exchanged some hands from their own choice about 8AM weighted & stood North. Brig [*Lady Nelson*] Stood South & soon lost sight of each other by her sent some letters for England &c fresh Breeze soon came in Sight of Breakers – in various directions' wrote Peter Good. The same day, 18 October, Good continued to record, in the region of Cape Bowling Green (near modern Townsville) where the Reef begins to approach the coastline, the *Investigator* was frequently 'obliged to change our Course almost every hour to avoid [the reefs]' (96).

Having now logged 500 miles inside the Reef – 'so extraordinary a barrier' – Flinders decided on 20 October to move through the complex of small reefs into the open sea and found a way at 18°45' × 148°5' (today, Flinders' Passage), Good recording that sailing 'before noon past between two reefs about a mile distant each – past noon a considerable swell & heavy sea which indicated we were clear of the Reefs' (96). In his own journal Flinders commented that the Swain Reefs which begin around 22°50'S were first 'discovered in 1797 by captain Campbell of the brig *Deptford*; and probably also with those further distant, which captain Swain of the *Eliza* fell in with the following year'. He noted that Swain was wise to move out into the open sea at that latitude, even though 'through a long and tortuous channel'. Otherwise, the next opening would be at Flinders' Passage, and so the extent of what he continued in his journal to call the 'barrier reefs' would be 'near 350 miles in a straight line; and in all this space, there seems no large opening' (Flinders 1814:I.102).

Taking the outer route, since Cook had already charted the inner up to Cook's Passage at 14°32', the *Investigator* continued blue water surveys, moving eastwards into the Coral Sea where Flinders charted and named the Eastern Fields (10°5' × 145°40'). Since these were in the latitude of Torres Strait he then set a course westbound, knowing that he would soon encounter the Reef again. What, though, he puzzled, is the relationship between the reefs of his charts and those of Cook? Neither navigator had travelled the full extent of the Reef. From the data they had both obtained, Flinders wrote that he

> assume[d] it as a great probability, that with the exception of this, and perhaps several small openings, our Barrier Reefs are connected with the Labyrinth of captain Cook; and that they reach to Torres' Strait and to New Guinea, in 9° south; or through 14° of latitude and 9° of longitude; which is not to be equalled in any other known part of the world (I.102).

NEW HOLLAND: MISSING LINKS IN NATURE'S CHAIN?

By the time of the *Investigator* expedition natural history had undergone many profound changes since the voyage of the *Endeavour*. The three major divisions – botany, zoology and geology – had been in rapid development, and were accompanied by intense, often bitter, debates over new hypotheses and explanations. Initially, the dominant theme had been the quest to reveal nothing less than the entire structure of the universe, understood as a Divine Design. By the first decades of the nineteenth century, however, the nature of the design – and the chain of creation – became increasingly questioned as the Linnaean framework was proving inadequate to accommodate the discoveries being made, both in botany and zoology. Geology, too, was exercising a greater influence, since it was developing at first into the new discipline of stratigraphy as technology allowed deeper mines to be dug, and cuttings were constructed for the burgeoning canal systems, and later the railways. Layers of earth previously unknown were revealed, and in the process, geology became deeply involved in debate over the puzzling nature of the fossils being found in ever greater numbers.

It had been in a similar, but less controversial, intellectual framework that the natural history of the *Endeavour* voyage had been conducted. When the *Investigator* was dispatched, the debate among scientists had intensified although it was the French who had taken the lead. The French, however, despite the purging of much religion from science during the Enlightenment of the eighteenth century, the profound anti-clericalism of the Revolution and the promotion of the cult of Reason, still held – although with growing doubt – belief in a *scala naturae* and the fixity of species. The Linnaean system, in default, remained in operation, and French scientists still sought to explicate, in complete detail, the chain of creation. The scientific significance of the *Endeavour* and then the *Investigator* voyages, for them, was to complete the overall mapping of world discoveries, to seek by all possible means the 'missing links' in nature's chain.

That, in part, explains much of the significance of their work on coral formation and the need to understand the nature and role of the polyp (discussed in chapter 11). When André de Peyssonnel and John Ellis in the eighteenth century finally confirmed that coral is constructed by animals in apparent plant form, they established the polyp as a significant 'link' in the continuous blending of nature from the most imperfect beings to the most perfect: mankind. In his *Contemplation de la Nature* in 1769, Charles Bonnet put it succinctly when he wrote that ignorance of 'the gap between the animal and the vegetable' is constantly being closed by science, demonstrated by the fact that 'the polyp has come to fill it and to demonstrate the admirable gradation there is between all beings' (cited, Lovejoy 1936:233).

By the end of the eighteenth century virtually all of Europe's biota had been brought into the Linnaean scheme, and significant inroads had been made into Africa, Asia and the Americas, while geology, too, was rapidly maturing into an autonomous science. Australia, as the last great landmass to become known to European science was, therefore, of profound significance: its plants, animals and rocks could complete the intellectual mapping of all creation. In particular, the puzzles set by fossils could well be explained. If all nature were complete, and in existence since Creation, as was believed, then those specimens known only in the Old World as fossils should be found alive in the New.

Scientific societies were now proliferating as knowledge advanced rapidly. Earlier, in 1788 a society devoted entirely to natural history had been founded, basing its charter on the quest initiated by Linnaeus, and taking its title from the Swedish version of his name, Linné, which gives the (seemingly anomalous) spelling of Linnean Society. Its Latin motto proclaimed its goal as *Naturae discere mores*: 'To know the ways of nature'. The first president was the wealthy gentleman botanist James Edward Smith, and Banks was one of the first four honorary fellows. Smith had already moved energetically into natural history: earlier, in 1784, he had bought the entire natural history collection of the deceased

Linnaeus. In 1805 he moved it to London, thereby providing a formidable axis between it and Banks' residence for the study of natural science, in competition with the Jardin and Muséum in Paris.

As specimens came back from Australia in both British and French ships they were described and catalogued in Linnaean taxonomy, although often their place in the chain of nature was difficult – even impossible – to determine. That difficulty was beautifully illustrated in one of Smith's captions in *A Specimen of the Botany of New Holland*, a fascicle of fifty-four pages and sixteen coloured plates, intended to be part of a much longer series that never eventuated. Attempting to describe *Ceratopetalum gummiferum* of the Cunoniaceae family (today, the very popular NSW Christmas Bush) sent to him by John White, Surgeon-General of New South Wales, he wrote one of the most celebrated passages in the early history of Australian botany:

> When a botanist first enters on the investigation of so remote a country as New Holland, he finds himself as it were in a new world. He can scarcely meet with any certain fixed points from whence to draw his analogies; and even those that appear most promising, are frequently in danger of misleading, instead of informing him. Whole tribes of plants, which at first sight seem familiar to his acquaintance, as occupying links in Nature's chain, in which he has been accustomed to depend, prove, in a nearer examination, total strangers, with other configurations, other oeconomy, and other qualities; not only all the species that present themselves are new, but most of the genera, and even natural orders. The plant before us justifies the above remarks. Its botanical characters are so new, we can scarcely tell to what tribes it is allied; and although, from the peculiar felicity of the LINNAEAN sexual system, founded on parts which every plant *must* have, we are at no loss to find its class and order (i.e. *Decandria monogyna*) in that which is an artificial system, we still scarcely know what genera are its natural allies (Smith 1793:Pl.3, p.9).

Equally puzzling were many of the Australian biota, especially the marsupials; when Cook saw his first kangaroo while surveying the Reef he described it as a cross between a greyhound and a hare. Both the possum and the kangaroo gave London scientists considerable difficulty. At first it was thought that they were closely related: 'some may choose to classify [the kangaroo] with the genus opossum' wrote George Shaw, the great taxonomic zoologist of the British Museum of Natural History, or, 'alternatively as a gigantic Jerboa'. He was certain however that 'we need not have the slightest hesitation in forming for the kangaroo a distinct genus' (Shaw & Smith 1793:39).

Even more enigmatic was the platypus. When the first specimen arrived in England it was considered a hoax and Shaw recorded that when he first set eyes on it 'I almost doubt[ed] the testimony of my own eyes . . . yet I confess that I can perceive no appearance of any deceptive preparation'. Examining the bill most carefully, he concluded that it came to 'seem perfectly real' and that, finally, 'nor can the most accurate examination of expert anatomists discover any deception in this particular' (Shaw 1800:I.1.232). Satisfied that it was genuine, he catalogued it at the lowest level of his taxonomic chain in the order *Bruta*, in with anteaters, terming it *Platypus anatinus*, although it later had to be renamed *Ornithorhyncus anatinus* ('Bird-billed duck') since, unknown to Shaw, *platypus* (Gk 'flat-footed') was already in use for a genus of beetle. In both England and its colony, the puzzle of Australian biota continued for years. When the Van Diemen's Land Scientific Society was founded in 1829 (defunct two years later), it took the platypus as its emblem with the alternative name of *Ornithorhyncus paradoxus*, and as its motto, to continue the double entendre, *Quocunque aspicias hic paradoxus erit* (literally, 'Whatever one examines here will seem an absurdity') but commonly translated, with a considerable sense of mischievous intent, that in Australia 'All things are queer and opposite' (Hoare 1969:198).

ROBERT BROWN: PLANT TAXONOMY IN A NEW WORLD

It was, then, into such a climate of scientific thought that Robert Brown (1773–1858) entered when he was first introduced to the riches of the Banksian and Linnaean collections in London, and this determined his thinking when setting out on the voyage of the *Investigator*. Benefiting from the *Endeavour* experience, the *Investigator* was equipped with plant cabins, along with all necessary supplies including collecting jars, transport cases for live plants, puncheons for dried specimens, and paper for pressings and scientific instruments.

While the *Investigator* completed Cook's earlier partial survey of the Reef from Breaksea Spit to Cape Bowling Green between 28 July and 20 October, Brown, Bauer and Good collected as energetically as they could along the Reef in the brief periods they were able to get ashore, while Westall made the necessary topographic coastline sketches, along with drawings of Aborigines and animals. Brown's expectations, after the riches of the southern regions of the continent, however, were not to be sustained. In fact, quite the opposite occurred since the Reef coastline was found to be very sparsely populated with botanical species. The Reef province did, however, provide some new finds, and when the dangerously deteriorated *Investigator* reached Timor on 31 March 1803 for temporary repairs and recuperation for the seriously ill crew, Brown sent a further report to Banks expressing disappointment that 'Our acquisitions in botany are much fewer than I had hop'd for in a country so completely new . . . the number of absolutely new species hardly amounting to 200'. Zoology had fared no better since 'we have not done much, nor do I think that much was to be done. Of quadrupeds we have met only with the kangaroo, the dog [*Canis familiaris dingo*], and *Didelphis obesula* [Porculine Opossum], of Shaw'. (Brown's description of the last fits the antechinus, possibly *A. flavipes*.) 'Our additions to ornithology', he continued, 'are exceedingly few'. Marine collections were almost nothing: 'I have not been able to attend to ichthyology, which would probably have afforded the greatest variety, and both insects and shells are few in number, and by no means beautiful or singular. Mineralogy continues as barren a field as ever' (*HR.NSW*:V.82–83).

Brown was clearly expecting far too much: 200 new species was an impressive achievement and the Reef littoral and the few continental islands visited yielded some very interesting discoveries, of which a selection, illustrated by Bauer, remain as artistic masterpieces. Significant for Reef science was Brown's classification of one family of mangroves as Rhizophoraceae. Right at the outset, in Broad Sound, Brown discovered what he thought was a unique tree, which, although rather insignificant in appearance, reaching a maximum of 15 metres and having but small white flowers, was, he believed, much more than a single species or even a single genus: it was, even higher in the taxonomic tree, a family on its own. Unique and defying easy classification – as he considered Flinders to be in the nautical world – he decided to honour the captain by naming it *Flindersia australis* (popularly known today as Australian Teak). Further revisions now classify it as a genus of the very populous *Rutaceae*, and *Flindersia* remains the generic for fifteen species altogether.

In the Cumberlands the scientists had a painfully enlightening experience with a cycad. Brown classified it as *Cycas media* – a very ancient species originating in Jurassic times – which was beautifully illustrated by Bauer, while at the same time, Good collected the seeds that grow in a large cluster in the centre for eating. For millennia Aborigines had collected cycad seeds or 'nuts' as a very rich source of food although extended experience has taught them to soak the seeds in water first in order to remove the powerful toxins which affect the central nervous system, fatally in some cases, a property unknown to the botanists at the time. Having found 'the fruit being both pleasant to the taste and sight' recorded Peter Good, 'I eat some as also Mr Brown and Bawer, on coming on board Mr Bawer and I were taken with violent reaching with sickness which continued with short intervals the greater part of the night' (Good 1802:104). Fortunately, no lasting harm was done.

Leaving the Reef, Flinders continued westbound to complete the remaining survey of the continent. The *Investigator*, however, was not a seaworthy craft: it was made available only because it was too decayed for naval service against the French. Despite having been re-sheathed in copper before the voyage, the strakes were rotten, and by the time it reached Torres Strait the pumps had to be employed continuously. The *Investigator* had to be careened urgently and so Flinders passed through the strait in only three days, 31 October to 3 November, and was eventually able to find a suitable location in the Gulf of Carpentaria near the Sweers Group of islands. Although the carpenters were able to effect temporary repairs, the ship had to return without further delay, if at all possible, to Sydney. By this time the crew were all ailing, chiefly from dietary deficiencies: the Reef had not provided them with sufficient nourishment, and some of their efforts were disastrous.

Good recorded on 5 October that 'some very large Shellfish one of which weighed 47 lb with the Shell and without the Shell 31 lb 2oz' was brought aboard (95); Flinders reported that 'many enormous cockles *chamas gigas* [today, the giant clam *Tridacna gigas*]... were taken on board the ship, stewed in the coppers; but they were too rank to be agreeable food' (Flinders 1814:I.88). The scientist, Robert Brown, was more direct in his account: 'they were servd out to the messes of the ships company but were almost universally dislikd & in some produced nausea & vomitting' (Mabberley 1985:101–02). By the time the ship reached Kupang in Timor on 31 May 1803 illness was increasing even faster; after a week's stopover for some rest, Flinders departed for Sydney, making a wide westerly sweep into the Indian Ocean, with continuing disaster. Reaching Cape Leeuwin a month later, five of the crew had died; on 31 May the industrious gardener Peter Good died of 'dysentery'; when they entered Port Jackson on 9 June 1803, four more were dying.

Flinders' career following his return to Port Jackson became a decade of continuing tragedy. Having recuperated somewhat in Sydney, he left for London on 10 August 1803, with his charts and journals, to report to the Admiralty and to raise the possibility of a further expedition to explore and survey the still inadequately charted north-west coast of the continent. Three vessels formed the convoy: Flinders was aboard the *Porpoise*, accompanied by the *Cato* and the *Bridgewater*, which, to avoid the dangers of the still poorly charted Reef, took a north-easterly course to the Coral Sea – which Flinders also named – on the outside route to Batavia, and thence across the Indian Ocean to Cape Town and then London. Barely a week into the voyage the *Porpoise*, under the command of another captain, and, despite the exercise of extreme caution, crashed into the weather side of an uncharted reef in the middle of the Coral Sea, situated some 175 nautical miles east of the outermost part of the Reef in the Swains Group, almost on the Tropic of Capricorn. It rapidly disintegrated in the pounding surf and to this day the location is known as Wreck Reef (22°14'S × 155°20'E).

The *Cato* was then also dashed on to the Reef in the heavy conditions: the *Bridgewater*, ahead and on the lee side, stood off to effect a rescue. Then, in a controversial action, it sailed off. One plausible explanation is that Captain Palmer would have experienced great difficulty manouevring his large East Indiaman ship, drawing 20 feet (6 metres), in a beat to windward around the extensive maze of reefs to be in position to rescue the survivors, and had a primary duty to secure the safety of his own ship (Ingleton 1986:241). Flinders, given the distress of the episode, accused him of desertion, but the truth will never be known. Some time after, the *Bridgewater* disappeared without trace.

As senior officer at Wreck Reef, Flinders took command and organised a salvage operation of supplies to the island on the platform reef using the ships' boats; then, in a ship's cutter with Captain Park of the *Cato*, and twelve seamen, set out for Sydney to organise a rescue for the eighty men left on the island. Six weeks later Flinders returned in the *Cumberland*, rescued the stranded mariners and continued north, yet again, for England. By that time, 10 November 1803, the *Cumberland* was leaking badly: after a stay in

Kupang for temporary repairs, Flinders set out across the Indian Ocean for Cape Town, the ship being pumped out continuously. Unaware that war between Britain and France had again broken out, Flinders decided to put into the French colony of Ile de France (today, Mauritius) for further caulking and other repairs. In a drawn-out confrontation, partly involving personal intransigence on both sides, Governor Charles Decaen refused to honour Flinders' safe passage passport, since it had been granted for the *Investigator* and not the *Cumberland,* and detained him as a prisoner of war for the next six and a half years. As a small courtesy Flinders was later allowed to move into a private house to work on his charts and papers and to send letters to London.

SCIENCE ASSESSED: BROWN AND BAUER IN ENGLAND 1805–1814

When Flinders left for London neither Brown nor Bauer wished to return at that time. For them, the riches of southern New South Wales and Van Diemen's Land remained alluring and so far, as Brown wrote, 'even botany has fallen short of my expectations' (Brown to Banks, 6 August 1803, *HR.NSW*:V.181). Brown persuaded Flinders and Governor King, less than a week before Flinders embarked on the *Porpoise* for England, of the value of remaining for another eighteen months to continue the botanical survey, and for Bauer to complete the artistic record. Brown was, however, very unhappy at the thought of sending his precious specimens of the Reef in Box T for *Littus intra Tropicum* ('The Tropical Coast') and the other collections in the *Porpoise* since it 'is so much crowded that she can take but a very small part of the collection of specimens, and even this must be put in the hold'. And, he added, ominously, 'She is, moreover, so wet a ship that I am afraid, small as it is, it may suffer very materially in the passage' (V.182). His forebodings were to be realised. When the *Porpoise* foundered on Wreck Reef just ten days after he wrote the above letter to Banks, all of his precious collection was lost – seeds, live plants, pressings – although he had been prudent enough to retain duplicates of the pressings. Sadly, as he advised Banks, it was the better set that went down with the *Porpoise*. Gallantly, Bauer exchanged some of the second set from superior examples in his own collection. To compound the loss, Allen lost most of his geological samples, while much of Westall's topographic and descriptive artwork either went under, or was severely water damaged. None the less, the Admiralty commissioned him to illustrate Flinders' *Voyage to Terra Australis* for which, in due course, he completed nine paintings for copperplate engraving.

Brown and Bauer left Sydney later, on 23 May 1805, aboard, of all vessels, the *Investigator*, and, perhaps to their great relief, arrived in Liverpool safely with their duplicate collections on 13 October 1805 while the ship continued on to the breakers' yards. They soon settled in. Brown moved into the Banks residence at Soho Square – where he remained till his death fifty-three years later – to take up appointment as the first salaried secretary of the Linnean Society and to begin work on the classification of the thousands of specimens he had brought back, while Bauer rented a house of his own. As a result of Banks' submission to the Admiralty, Bauer was commissioned to work his drawings into finished watercolours, and of the total of 2703 made in Australia and Norfolk Island, 300 were translated into superb and enduring works of both scientific description and art.

By 1809, the year before Flinders was able to return to England, Brown had completed a mammoth botanical task: writing to Banks on 2 June 1809 from his position in the Linnean Society in London, he advised, in considerable understatement, that he had described 1600 plants from his collection, of which 1300 were unpublished, and these contained 100 new genera (*HR.NSW*: VII.160). His descriptions, from a scientific viewpoint, are masterpieces of meticulous, thorough and comprehensive observation. Brown was particularly taken by the Protea family (Waratah, Grevillea, Hakea, Banksia, Dryandra, Lomatia, Stenocarpus, Telopea, Macadamia genera) since it is well represented in Australia and along the Reef littoral, and contributes much character to the floral landscape with 44 genera out of a world total of 75, and 900 of the world's 1500 species. In that same year, 1809, he read a paper to the Linnean Society 'On the natural orders of plants called Proteaceae' which the society published in their *Transactions* for 8 March 1810. His descriptions of Protea were brilliant since, in seeking a more natural taxonomy, he introduced a major innovation through the use of the microscope to examine pollen grains as a significant determining criterion.

In 1810, having moved into permanent residence in Soho Square, Brown brought all of his work together in his great publication entitled *Prodromus florae Novae Hollandiae et Insula Van Diemen* (*Introduction to the Flora of New Holland and the Island of Van Diemen*). Throughout his labours in composing the *Prodromus* Brown had wrestled with the difficulties, felt by all scientists at the time, of the restrictive taxonomy imposed by the 'great chain of being' and the artificiality of the Linnaean system. In the same period a major expression of dissent had been published in 1809 by Lamarck in his *Philosophie Zoologique*. Professor of 'Inferior Animals' (Invertebrates), at the Muséum d'Histoire Naturelle in Paris, Lamarck had become absorbed in the problems of classification and the fundamental question: are categories real? Late in the seventeenth century in his very influential *Essay Concerning the Human Understanding* John Locke had asserted the contrary position that 'the boundaries of species, whereby men sort them, are made by men' (III.§35) and that categories therefore are nothing more than human impositions on nature. Lamarck's *Zoological Philosophy*, subtitled 'An exposition with regard to the natural history of animals' had a profound effect on the thought of the time, and especially on Robert Brown who came to share many of the same ideas. 'No importance is now attached in France', Lamarck wrote, 'to those artificial systems which ignore the natural affinities among objects; for these systems give rise to divisions and classifications harmful to the progress of knowledge' (Lamarck 1809a:32).

The discoveries of this new world, already commented upon by Smith of the Linnean Society, had also puzzled Lamarck. Arguing that even if scientists become able to establish an order of nature (*ordre naturel*), 'the classes which we are obliged to establish in it will always be fundamentally artificial divisions'. The existing system, he continued, was completely inadequate to deal with continuing discoveries such as the platypus and the echidna (or 'spiny anteater') which alone in the class Mammalia lay eggs. 'Already the *Ornithorhyncus* and the *Echidna*' he wrote, 'seem to indicate the existence of animals intermediate between birds and mammals. How greatly natural science would profit if the vast region of Australia and many others were better known to us!' (23). Foreshadowing his revolutionary and controversial theory of transformism, Lamarck denied any natural great chain of being ('*échelle de la nature*') and asserted that '*la nature n'a réellement formé ni classes ni ordres, ni familles, ni genres, ni espèces constants, mais seulement des individus qui se succèdent les uns aux autres et qui resemblement à ceux qui les ont produits*' [in reality, nature has formed neither classes, orders, families, genera, nor invariable species, but only individuals which follow one another and resemble those from which they have been generated] (Lamarck 1809b:I.1,1907).

In the Preface we read his account of the difficulties encountered in classifying the flora of New Holland in which it was 'absolutely necessary to adopt a natural system of classification, for only in this way could I hope to avoid the more serious errors, particularly in forming those new genera for which New Holland is pre-eminent'. So, even though he followed de Jussieu, he had reservations that the system 'is often artificial and at times founded, it seems to me, on artificial principles'. The next sentence marks a great step forward in taxonomy which was to make Brown a major figure in his assertion that 'nature itself, by linking organic bodies after the manner of a network than of a chain, scarcely acknowledges such a succession' (*ipsa natura enim, corpora organica reticulatim potiùs quam catenatim connectens, talem vix agnoverit*) (Brown 1810:v). The innovative concept of a 'network' (*reticulum*) instead of a chain, along with Lamarck's concept of transformism, helped lay the groundwork for the evolution theory of Wallace and Darwin.

Bauer, meantime, was busy preparing his own *magnum opus* which he planned as a comparable volume entitled *Illustrationes florae Novae Hollandiae*. Brown reported to Banks in a letter of 2 June 1809 advising that Bauer wanted to publish as soon as possible in order to forestall the French who had also been collecting in New Holland. Proceeding very slowly due to his meticulous and highly accurate methods, Bauer misjudged publication dates badly and so decided to issue the work first in fascicles (small separate sections), planning, on completion, to bring them all together subsequently into a grand collected edition. By 1814 only three fascicles had appeared and his project, like Brown's *Prodromus* and Banks' *Florilegium*, lapsed into obscurity. Brown and Bauer were attempting incredibly ambitious projects in a time of great social dislocation due to the Napoleonic Wars which convulsed all of Europe and did not end until Wellington's final victory at Waterloo in Belgium on 18 June 1815.

A VOYAGE TO TERRA AUSTRALIS: THE JOURNAL PUBLISHED, 1814

In June 1809 the British captured the Ile de France and Flinders – by now seriously debilitated – was repatriated to England where he arrived on 24 October 1810. His health, in fact, had been failing for years, stemming from two gonorrheal infections contracted while serving on Bligh's *Providence* which evidently caused renal deterioration leading to chronic cystitis, kidney infection and calcification of the bladder. Reunited with his wife Ann, whom he had rarely seen, and although suffering badly, he began the massive task of editing his journal and charts as *A Voyage to Terra Australis*. When complete, it was illustrated with nine engravings by William Westall, followed by an 80 page supplement by Robert Brown entitled *Botany of Terra Australis*, and ten illustrative plates by Ferdinand Bauer.

In his journal Flinders devoted several pages to a description of the Reef, as charted by Campbell, Swain, Cook and himself: his own charting from 22°50'S led him to hypothesise that 'Break-sea Spit is a coral reef, and a connexion under water, between it and the barrier, seems not improbable' (Flinders 1814:I.101). In these pages, Flinders showed a flair for neologisms by renaming Cook's 'Labyrinth' as 'The Great Barrier Reefs'. The terms 'barrier reef' and 'barrier reefs', often capitalised, occur throughout his journal. On the large General Chart, along with sectional and regional charts as an Appendix, these reefs are labelled, in capitals, 'THE GREAT BARRIER REEFS', thereby providing a term which continues in use today, gives the region a distinct identity, and will stand for all time as a memorial to Flinders. His final description of the Reef that ends the relevant section of his journal, is as true today as when he wrote. Any commander who seeks to transit the Reef, he stated, must 'feel his nerves strong enough to thread the needle, as it is called', otherwise, 'I would strongly recommend him not to approach this part of New South Wales' (I.104).

The *Voyage* is a complex account of the exploration of the coastline, and for the Reef is exceptionally detailed and perceptive. Flinders, too, was taken by the strangeness of the new land, and as the *Investigator* sailed along the east coast among the coral formations he too became as impressed as Cook: for Flinders it was, in his carefully constructed account, 'a new creation' although 'imitative of the old'. His descriptions of corals, seen always from the deck of the ship, or one of its tenders, verge on the poetic and sustains the imagery of earlier times that coral formations were plants: 'We had wheat sheaves, mushrooms, stag horns, cabbage leaves, and a variety of other forms, glowing under water with vivid tints of every shade betwixt green, purple, brown and white; equalling in beauty and excelling in grandeur the most favourite *parterre* [arrangement] of the curious florist'. Yet, as a counterpoint, he strengthened the growing mystique of the Reef as a seductive danger in the closing comment that even 'whilst contemplating the richness of the scene, we could not long forget with what destruction it was pregnant' (Flinders 1814:II.88).

At the same time, Flinders showed a considerable awareness of the growing debate over the nature of coral reef formation, demonstrating his understanding that reefs were becoming more widely recognised as productions of minute animals. In a section describing the various formations of submerged platforms, rubble cays and vegetated cays, he knew, in some way not acknowledged, but possibly gained from Robert Brown, that 'animalcules erect their habitations upon the rising bank', and that the 'care taken to work perpendicularly in the early stages, would mark a surprising instinct in these diminutive creatures' (I.115). A more complete understanding of reef formation was well into the future, coming to occupy scientists in a quest that still continues.

Despite many disasters, and the tragic loss of lives, his voyage led to a major advance in geographical knowledge. The survey of the Gulf of Carpentaria eliminated the possibility of finding any waterway dividing New Holland from New South Wales: the continent was now confirmed as a single landmass. Again displaying a proclivity for neologisms, Flinders recorded that once those two regions 'were known to form one land, it seems best to have a general name applicable to the whole' and he regretted having to revert to the term Terra Australis. 'Had I permitted myself any innovation upon the original form', he continued, 'it would have been to convert it into AUSTRALIA; as being more agreeable to the ear, and an assimilation to the names of other great portions of the world' (Flinders 1814:I.iii). In fact, Flinders still slipped in the new word since the full title of the General Chart included in *A Voyage to Terra Australis* reads 'General Chart of Terra Australis or Australia', providing a stimulus for later action.

There was considerable debate in London over the best term to use since both New Holland and New South Wales were current and often used synonymously. Once Flinders had finally demonstrated that these refer to a single landmass, a suitable term had to be found. Some, including both Banks and Flinders, objected to New Holland since it applied only to Dutch discoveries in the west; yet, they were mindful of the fact that the term New South Wales did not give the Dutch their due credit. Banks, therefore, opted for Terra Australis, writing in a draft of an introduction to Flinders' later book that it would be 'unjust to the British nation, which has had so large a share in their discovery, that New Holland should embrace the whole, as it would be to the Dutch, that New South Wales should be so extended'; therefore, when speaking of New Holland and New South Wales together, 'the original Terra Australis has been judged the most proper' (Brabourne Papers 1886:24). Flinders, however, was obdurate. Back in London preparing his journal and charts for publication, he showed drafts to the Hydrography Office of the Admiralty which saw no problem with the term Australia, although he was advised that Captain Burney may well object to such a 'novelty'. Burney was in a position to do so. Having sailed on the *Resolution* with Cook on the second voyage, when retired to a desk in London he began his publication of the vast, five volume, standard *Chronological History of the Discoveries in the South Sea or the Pacific Ocean* (1803–17). Indeed, Burney – the unquestioned authority at the time – sided with Banks for the conservative position, and so, in deference, Flinders titled his publication *A Voyage to Terra Australis*. Even so, he used the word Australia on his general chart, and kept using it in his correspondence to

Banks. Flinders' achievement, then, was that of helping convert the adjective 'Australian' into the noun 'Australia' – an enduring accomplishment equal to that of putting together the disparate terms of 'barrier' and 'reefs' already in common use by chartmakers, into one powerfully evocative term, 'The Great Barrier Reef'.

The final tragedy of that last decade came in 1814. The 40 year old Flinders, ailing ever more seriously throughout the earlier months of that year, became bedridden, his days solaced by visits from, among others, his brother Samuel, a lieutenant who had also sailed on the *Investigator*, and botanist Robert Brown. Flinders' personal diary makes melancholy reading. On 27 February 1814 he wrote that he told his surgeon Hayes of 'all the symptoms of the complaint which alone ever troubles me, and which appears to be either stone or gravel in the bladder'. The disorder grew in intensity; on 1 May he noted that the pain was so intense he was having difficulty concentrating on correction of the proofs of his journal. On 29 May he wrote that he 'Rose at eleven with tolerable ease, but sitting brought on pain which lasted the whole day, and in the evening the urine became bloody. Took Uva-Ursi [patent medicine] and 3 bottles of Seltzerwater; voided a good deal of fresh mucus, and some crystals in it'. By Saturday 9 July the nights were getting worse, the pain increasing. That was his last entry (Flinders, Autograph Diary, Mitchell Library, CY 515).

When the printed copy of his *Voyage* – two volumes with a separate appendix of charts – arrived at his London home nine days later Flinders had already lapsed into a coma and never saw the finished work: he died the next day, 19 July 1814. His wife Ann, in a letter of the same day wrote that the autopsy showed his bladder to be 'in the most dreadful state of decay . . . literally torn to shreds by an incalculable number of small crystals . . . in his last few weeks . . . he looked full 70 years of age and worn to a skeleton' (Mitchell Library MSS Safe 1/57, Private Letter Book, Vol.III).

CHAPTER 5

THE REEF EXPLORED: EARLY SURVEYS, 1821–1844

EXPANSION OF THE COLONY: FIRST DECADES

The Napoleonic Wars had consumed most of Britain's energies throughout the early years of the nineteenth century, and – as the failure of the publishing projects of Banks, Brown and Bauer testify – the natural sciences now had a lower priority. The infant colony at Sydney received little attention from London and grew only slowly under a succession of naval governors – Arthur Phillip, John Hunter, Philip Gidley King and William Bligh – all of whom experienced considerable difficulties with the army officers and senior public officials. Both of those groups sought extensive privileges and large grants of land with the intention of recreating the social structure of the mother country, establishing themselves as a colonial aristocracy and using the transported convicts as a serving class.

All four governors sought to administer wisely and evenhandedly, but were hampered by lack of real legislative authority and were glad to leave the colony, especially Bligh who had done his best to correct the many abuses – trafficking in rum and the indiscipline of the New South Wales Corps – and to resist the growing demands of the privileged sector. When Lachlan Macquarie arrived in 1810 with his own trustworthy highland regiment the colony began a slow upward climb to stability and development, characterised by expansion of the settlement across the adjoining Sydney plain and exploration of the regions beyond, a task becoming particularly urgent since the water supply in Sydney Town was inadequate and unpredictable. Because the Sydney plain was hemmed in by harbours and rivers to the north and south, and by a forbidding mountain divide 50 kilometres to the west, while roads were primitive and horses and wheeled vehicles in short supply, considerable early exploration took place by sea. Already Bass and Flinders had made their major discoveries to the south, enabling Van Diemen's Land to be settled in 1803 at Hobart Town, which became an important base for the whalers and sealers, many of whom came from the far off Yankee ports of New Bedford and Nantucket in Massachusetts. Since Flinders had made considerable progress in determining the general features of the continent, the way was now open for further expansion.

In a long dispatch to Colonial Secretary Bathurst of 4 April 1817 Macquarie advised that he had sent Surveyor-General John Oxley and Assistant Surveyor George Evans across the Blue Mountains route to explore the west of the colony, and that 'With these two Gentlemen I have also Joined Mr. Alan Cunningham, One of the King's Botanists, who arrived in this Colony some few Months ago [20 December 1816] from Rio de Janeiro, to Prosecute his Botanical Researches, and to Collect rare Botanical Plants for His Majesty's Gardens at Kew' (*HRA*:IX.1.356). Oxley explored in a large clockwise loop from Sydney travelling west and then north-easterly until he reached the sea where he named Port Macquarie.

In the same month that Oxley was beginning his overland explorations, Earl Bathurst wrote to Macquarie on 8 February 1817, requesting him, as a matter of urgency, 'to explore, with as little delay as possible, that part of the Coast of New Holland which was not surveyed or examined by the late Captain Flinders' (*HRA*:IX.8.207). Better charts of the Reef waters were becoming a necessity, especially to verify and coordinate the random discoveries being made by various mariners in the region. In 1803 Ebenezer Bunker, master of the whaling ship *Albion*, reported finding a group of islands offshore from Bustard Bay, one of which had already been charted by Flinders with an entry in his Journal for 4 August 1802 which he confirmed in his *Voyage* with the comment that 'A cluster of low islands about fifteen leagues from the coast, was seen in the following year by Mr Bunker, commander of the *Albion*, south whaler' (Flinders 1814:II.14n.). By that time there was growing traffic in the area from whalers, and other ships travelling north, with a corresponding frequency of shipwrecks, of which the *Cato* and *Porpoise* were noteworthy, followed by the *Mersey* in Torres Strait in 1804, the whaler *Britannia* on Middleton Reef in 1806, the *Morning Star* in Torres Strait in 1814, the *Eliza* in 1815, and the *Lady Elliot* off Cardwell in 1816 (Loney 1993:30–31).

The need for better charting arose from the growing traffic between New South Wales and Asia, particularly India following the increasing British occupation of that subcontinent after Robert Clive, in the Battle of Plassey in 1757, established supremacy. There was a paramount need to find the safest and shortest way north. The route inside the Reef, known as the 'inner', had a forbidding reputation from the accounts of Cook and then Bligh and Flinders, not only from the hazards of the myriad small cays and platform reefs, but also from the need to anchor every night and so lose valuable sailing time. Yet the 'outer' had little more to recommend it given the dangers of submerged reefs in the Coral Sea such as Saumarez, Wreck Reef and numerous others, and the few safe passages through the northern section of the Reef such as Bligh's Channel and Cook's Providence Channel which allowed access to Torres Strait and then Timor, as the gateway to the East

Indies and Asia. Following Flinders' voyage through the southern section of the inner route, all vessels chose the outer way, even Flinders himself on his ill-fated first attempt to return to England aboard the *Porpoise*, accompanied by the *Cato*. Then, in 1815, the next charting of the inner route took place, rather adventitiously, by Lieutenant Charles Jeffreys (1782–1826).

AUSTRALIAN COASTLINE CHARTS COMPLETED: JEFFREYS AND KING 1815–1822

In 1810 Governor Macquarie had petitioned the Colonial Office to provide more ships for the expanding colony, since by that time it had extended as far south as Van Diemen's Land and north-east to Norfolk Island. In tardy response, the armed brig *Kangaroo* was finally dispatched in 1813 under command of Lieutenant Charles Jeffreys, a Royal Navy officer, arriving on 10 January 1814 after a seven month voyage. Almost immediately the seeds of conflict were sown: Macquarie considered the *Kangaroo* a colonial ship, Jeffreys held it to be a naval vessel and not under direct control of the military governor.

Consequently, relations between Macquarie and Jeffreys began rather badly, and grew steadily worse. Jeffreys was insubordinate, asserting his role as a naval commander and not a cargo carrier, although even in that latter capacity he had talent for inadequacy, being exceptionally dilatory in effecting the commands entrusted to him. Macquarie suspected Jeffreys of incompetence and timidity, and wrote to Colonial Secretary Bathurst on 16 March 1815 that he had 'every reason to believe that Lieut. Jeffreys is very Unequal to Such a Service, being said by his Own Officers to be a Very timid Seaman and ignorant of his Duties as such' (*HRA*:VIII.1.442).

Given the shortage of vessels, Macquarie was constrained to use the *Kangaroo* to carry a regimental detachment to Ceylon under the command of Jeffreys and in discharging that task Jeffreys achieved naval distinction by completing the charting of the still unsurveyed part of the Reef which neither Cook nor Flinders had seen. Leaving Port Jackson on 19 April 1815, he elected to take the inner route through the Reef, following Flinders' track as closely as possible. At the location where Flinders left the Reef at 18°45'S (opposite modern Ingham) Jeffreys continued along the still uncharted coast all the way to Torres Strait. During the journey, which turned out to be quite uneventful, he named some twenty topographic features, many being retained to this day, including the Howick Group, Cape Melville, Bathurst Bay, Pipon Island, Flinders Group, Clack Island, Princess Charlotte Bay and Cape Sidmouth. In addition, he confirmed that Bligh's 'Island of Direction' was in fact part of the mainland, and he revised it as Cape Direction. The voyage came to a safe anchorage in Ceylon where Jeffreys had 'the happiness to land the detachment, with their families, in a state of health, which from the variety of climates and changes of atmosphere passed through, could not have been hoped for' (text in Gill 1988:148–50). Jeffreys then took a leisurely period in revictualling the ship and did not return to Port Jackson until 5 February 1816, more than nine months later, to meet Macquarie's fury at such delay.

Giving little credit to his final charting of the northern section of the Reef, Macquarie vented his seething hostility towards Jeffreys in a letter to Bathurst of 18 March 1816, declaring that his behaviour in Ceylon had only served 'to Confirm Me in the Opinion that I had formerly expressed that Neither the Kangaroo nor her Commander are at all fit for, or calculated ever to render any important Service to this colony . . . [consequently] I have fully resolved on sending the Kangaroo to England in January next' unless contrary instructions come from the Colonial Office (*HRA*:IX.1.61). A year later, in a dispatch to Bathurst of 4 April 1817 Macquarie advised that he was sending the *Kangaroo* back to England since 'it has rendered [the colony] very little service . . . owing to the Inactivity and Negligence of her Commander', describing him as a 'Vain, Conceited, Ignorant, Young Man, and totally Unfitted for such a Command' (*HRA*:I.IX.359).

Jeffreys was sent back to England to be dismissed from the British navy – after which he emigrated to Van Diemen's Land where he took up a grant of land. Nevertheless he retains the distinction of being the first navigator to chart the remaining dangerous northern section of the Reef between 14°42'S and 12°32'S not traversed by Cook, Flinders or Bligh. Indeed, Jeffreys was praised by his naval contemporary Phillip Parker King for the charting skill with which he 'filled up the space between Endeavour River and Cape Direction which Captain Cook did not see' and such an achievement 'does him very great credit' (King 1827:I.228f. s.v. 'Sailing Directions'). Three years earlier a small 90 ton brig, the *Cyclops*, under command of a Captain Cripps, seems to have taken the inner route to Bengal, but no record or charts of the voyage exist, only a brief reference in Phillip Parker King's *Narrative* (1827) and an oblique reference to the *Cyclops*, but not to Cripps, in *Horsburgh's Indian Directory* (Gill 1998:fn.151). That now lost voyage may have been known to Jeffreys, but none the less, the achievement of Jeffreys in a much larger vessel and the chart he produced certainly warrant the praise given by King.

Whereas Jeffreys had never been sent to discharge duties as a naval hydrographer, Phillip Parker King (1791–1856), son of former governor Philip Gidley King, was chosen in 1817 explicitly for that purpose in order to map the vast unknown sections of the continent. Born on Norfolk Island, Phillip Parker King has the distinction of being the first Australian to survey the entire continental coastline in detail. In what was, and remains, the greatest single hydrographic feat in Australian maritime history, over the ensuing four and one half years, 22 December 1817 to 25 April 1822, King made four voyages from his base in Port Jackson. The first was via the western coast to survey Melville Island and Raffles Bay on the nearby Coburg Peninsula, as possible sites for a tropical garrison, returning by the same route. The remaining three were complete anticlockwise circumnavigations of the continent, in the course of which he traversed and charted the Reef along with making very competent geographical and scientific observations. These were supplemented throughout by the natural history collections and studies of Alan Cunningham, the King's Botanist, who accompanied him on all four voyages. When the charting of the Australian coastline was completed, King returned to England where he spent two years, 1822–24, writing up his discoveries in one of the great foundation documents of Australian maritime navigation, published in 1827 by the Admiralty in two comprehensive volumes as a *Narrative of a Survey of the Intertropical and Western Coasts of Australia performed between the years 1818 and 1822*. Included as an extensive Appendix, along with the detailed hydrographic data, was a careful catalogue of shoals and reefs, and 'Directions for a passage within the reefs and through Torres Strait', along with the scientific record of Cunningham's collection, written up by others in London – John Edward Gray for zoology, Robert Brown for botany and William Henry Fitton for geology – in a lengthy account organised into sections on mammals, birds, reptiles, fish, insects, zoophytes (marine animals), shells, geology, with the largest individual section being on botany (King 1827: II.408–629).

Earl Bathurst had instructed Macquarie on 8 February 1817, to 'place at the disposal of Lieutenant King either of the Colonial Vessels, which he may consider best suited to his purpose' (*HRA*:IX.1.207) the *Mermaid* being chosen. King's instructions from the Admiralty were 'to examine the hitherto unexplored Coast of New South Wales' (strictly, New Holland) in the west and northwest, especially the inlets and bays, in order 'to discover whether there be any river on that part of the coast likely to lead to an interior navigation into this great continent' and specifically to chart 'all parts of the coast which have not been laid down by Captain Flinders, M. de Freycinet, or preceding navigators'. In fact, the instructions were very extensive, in that King was ordered to observe everything of potential value including the topography, minerals, biota and the Aborigines, in particular, their languages. With respect to science he was to 'receive on board Mr. A. Cunningham, a botanist, now in New South Wales, who has received the orders of Sir Joseph Banks to attend you' and 'on all occasions of landing, you will give every facility to the botanist' (I.xxvii–xxxi). On 21 December 1817, Macquarie replied to Henry Goulburn, Undersecretary to Colonial Secretary Bathurst, informing him that Lieutenant Phillip Parker King and King's Botanist Alan Cunningham had departed to complete the charting of the still unknown stretches of the continental coastline. In sending his dispatch, Macquarie advised he had every expectation that these two very skilled explorers would 'be able to make very important additions to the Geographical knowledge already acquired of the Coasts of the Continent of *Australia*, which I hope will be the Name given to this Country in future, instead of the very erroneous and misapplied name, hitherto given it, of "New Holland", which properly speaking only applies to a part of this immense Continent' (*HRA*:IX.1.747). Flinders' name for the continent therefore became fixed for all time.

Alan Cunningham (1791–1839) proved to be an ideal appointment. His career began in 1808 as a clerk at Kew Gardens where he later met Robert Brown who recommended him to Banks as a botanical collector. Following that, Cunningham was sent to Brazil in 1814 where he worked for two years. Then in 1816 he was sent to New South Wales in Banks' employ as 'King's Botanist', a situation somewhat tense at the time since Maquarie already had Charles Frazer as Colonial Botanist. Cunningham, however, soon proved himself a valuable asset with the work he did on the Oxley expeditions into the interior. Macquarie went so far as to append an encomium to Earl Bathurst (5 September 1817) praising both Cunningham and Frazer: 'The extensive and Valuable Collections of Plants formed by Mr. A. Cunningham, and Mr. C. Frazer, Colonial Botanist', Macquarie wrote, 'will best evince to yr Excellency the Unwearied Industry and Zeal bestowed on the Collection & Preservation of them. In every respect, they merit the highest praise' (*HRA*:IX.1.484). When King set off on his four voyages of coastal exploration, Cunningham – who was amiable and well-liked – gladly volunteered his services for all expeditions, with the result that he made a significant contribution to the natural history of Australia, one that continued after the completion of the coastal surveys.

Sailing from Port Jackson on 8 May 1819 on the first of the surveys to begin in Reef waters, the initial task was to take Lieutenant John Oxley to Port Macquarie for terrestrial survey, and there King made the memorable record on 13 May 1819 of 'Several cedar-trees (*Cedrela toona*) of large growth . . . one of which, being measured, was found to be ten feet in diameter at the base' (King 1827:I.171). Continuing north and anticlockwise, the direction King chose for all of the three circumnavigating surveys, the coastline was charted ever more accurately and its natural history became supplemented in closer detail. Cunningham, following his mentor Robert Brown, classified all material according to de Jussieu's 'natural' taxonomy – which attempted to show genuine biological affinities – set out in *Genera plantarum secundum ordines naturalis disposita* ('Plant classification according to natural features'), commonly referred to as the '*Ords. nat.*' which was carried on board, along with Brown's *Prodromus* and Hawkesworth's version of Cook's *Journal*.

King's text is in the usual diary form and reveals that he moved rapidly through the Reef, chiefly because his brief, and his interest, was in the unknown north-western and western parts of the continent. His first opportunity to add to Reef charts came when, sailing 'across Hervey's Bay towards Bustard Bay [we] passed a small island that was discovered by the ship Lady Elliot in 1816, and that had not yet a place upon the chart of this part of the coast' (King 1827:I.180). From Lady Elliot Island north, the charts were revised where necessary, while Alan Cunningham, often joined by King who was interested in and knowledgeable about natural history, went ashore as occasion arose to collect. However, where Matthew Flinders and Robert Brown had already been they did not linger although they did spend some time collecting seeds in the Endeavour River area. Relatively little, however, was added to Reef knowledge and the natural history section of Reef discoveries in the scientific Appendix occupies only three pages (II.568–570) in King's *Narrative*, although the geological comments by Dr Fitton raise intriguing questions about current theories of terrestrial formation.

Most of the Reef content of the *Narrative* is of a generalised, geographical descriptive character, valuable in itself, especially that part dealing with the Aborigines. King made it clear in his *Narrative* that the primary aim of all four voyages was a hydrographic survey of the continent – set out in the tables in the Appendix – and that relatively little time was available for natural history collecting (II.410). Even so, Cunningham was still able to make sizeable additions to the biological knowledge of Australia and the Reef, and zoological specimens as well as botanical were collected for dispatch to London. Of the former, insects were the most numerous, and King records having collected 192 personally – beetles, phasmids, locusts, dragonflies, butterflies, cicadas, bugs, flies and spiders (the last were then included in Insecta) which were classified in Sydney by the newly arrived Colonial Secretary, Alexander Macleay. A number of other animals were collected including a flying fox, a dingo, a possum, a dolphin, 14 birds, nine reptiles and seven fish. Of marine specimens, 111 molluscs and 25 corals or 'zoophytes' described as 'Fungia, Tubipora musica, Porites, Madrepores, Gorgonia and Spongia', were collected and classified according to the scheme in Lamarck's *Histoire naturelle des animaux sans vertèbres* ('Invertebrate Zoology').

When the specimens arrived in London, the most exciting find, at least from Robert Brown's perspective, was a complete specimen with flowers and fruit of a Grass Tree which Brown had only been able to collect unfructified (that is, without the complex spore-producing features) in 1801 at King George's Sound. Since Brown was pioneering the microscopic taxonomy of plants from their reproductive parts, the addition of the fruit and flowers enabled a more accurate classification which he determined to be close to other Grass Trees of the genus *Xanthorrhoea* (now with 28 species) but not within that genus. This new specimen was found, to his great excitement, to be monotypic, a single species within its own genus. On account of its uniqueness, in a paper read to the Linnean Society of London on 1 and 15 November 1825, Brown named the genus *Kingia* to honour both Philip Gidley King who 'materially forwarded the objects of Captain Flinders' voyage' and 'my friend Captain [Phillip Parker] King, who on all occasions gave every assistance in his power to Alan Cunningham, the indefatigable botanist who accompanied him' (Brown, *Appendix* in King 1827:II.535). The monotypic species is *Kingia australis*.

GEOLOGICAL OBSERVATIONS: RISE OF CONTROVERSY

King's concern as a navigator was always directed to the very practical issue of the navigational hazards of 'numerous reefs, many as yet unknown' (King 1827:I.386). Although he may not have known the finer details of current theories, he was also certainly aware, from the section on geology in the Appendix of his *Narrative* written by Dr William Henry Fitton, president of the Geological Society of London, of growing speculation among geologists regarding the nature of coral reefs and the puzzling discovery of marine deposits in elevated positions inland. These had already been reported in numerous other locations around the globe, especially in South America, raising some of the liveliest issues of the times. Fitton dealt, then, with some of those that had already received prominence in the observations of François Péron, naturalist aboard Baudin's scientific voyage of 1800–1804 in the *Géographe*, concerning the discovery along the coast of New Holland and Timor of '*marine* shells . . . in cemented masses, at heights above the sea, to which no ordinary natural operations could have conveyed them'. Subterranean geological activity was becoming suspected as a major contributor and Péron inclined to the gradualist argument that the shells could not have been transported inland by 'any very turbulent operation' since they were found in a condition of 'perfect preservation'. Fitton described similar studies and cited a Dr Jack to the same effect that inland marine deposits 'must have been effected with little violence' and that 'the phenomena are in favour of an "*heaving up of the land*, by a force from beneath" '. (Fitton, in King 1827:I.592–93; Fitton's emphasis).

Although the pressing need at the time of King's voyages was to continue more accurate charting of the Reef, and all other Australian waters, at the same time scientific interest in the geological and biological character of coral reefs was accelerating as information continued to be gathered from exploratory voyages. The Great Barrier Reef was thus propelled into the forefront of European speculation on reef formation, and thereafter ships' captains were specifically instructed to make appropriate geological observations.

ACCURATE REEF CHARTS: *BEAGLE* SURVEY 1837–1844

Britain had emerged the victor from the Napoleonic Wars. Recognised as the world's pre-eminent naval power, it began to create a great empire, containing the major land masses of Canada, India, Australia and (in 1840) New Zealand, along with a very large number of island colonies in the Caribbean Sea and the Atlantic, Pacific and Indian Oceans. In the decades from 1815 until 1869, when the Suez Canal was opened, it was essential to protect the trade routes to Australia and India, and to that end a chain of naval bases was created. By the Treaty of Paris of 20 November 1815, Britain gained from France St Helena in the South Atlantic, as well as Mauritius and the Seychelles in the Indian Ocean. To strengthen the chain, barren Ascension Island in the south tropical region of the Atlantic was occupied and fortified, and the Cape of Good Hope wrested from the Dutch. The naval sequence to Asia was completed with the foundation of Singapore in 1819 and the ceding of Hong Kong from China in 1842. Sydney was the major terminus of that route and the passage through the Reef to the north became even more significant. To coordinate British naval power in the Indian Ocean region a vice-admiral's command, the East Indies Station, was established in Ceylon, occupied earlier in 1795, located at the naval base of Trincomalee, with a major shipyard at Bombay.

The primary need in charting the Reef stemmed from its continuing dangers, and loss of shipping; that is, to ensure the safety of the shorter and less turbulent inner route. By the time King had completed his surveys in 1822 eleven ships had been lost in Reef waters; in the ensuing decades up to 1861, some 148 further losses were officially recorded between Moreton Bay and Torres Strait (Loney 1993:26f.). Moreover, from the 1840s on, expansion of the colony was increasing as settlers pushed northward along the coast and for whom the most efficient transport was by sea. Most of the sinkings were of small vessels without major loss of life, although the wreck of the *Grimenza* on Bampton Reef on 4 July 1853 enroute from India to Peru with 800 coolie labourers on board resulted in at least 650 fatalities (68).

It was, of course, the dramatic and sensationalised wrecks that commanded public attention and fear of the Reef, and so intensified its awesome mystique. In the early nineteenth century the most dramatic, and widely popularised was the wreck of the *Stirling Castle* on the Swain Reefs in 1836. Accounts are notoriously unreliable, but it seems that eighteen survived and reached the mainland in two parties: eleven, including Captain Fraser and his pregnant wife Eliza, in a longboat, and seven others in a pinnace. The longboat landed on what is now Fraser Island – K'gari to its traditional owners – only to be met by hostile Aborigines. The captain was killed and the first mate Brown died of injuries, while some of the others escaped. Mrs Fraser was captured, stripped naked, and taken away. By that stage, it seems, the newborn baby had died. Some weeks later the survivors from the pinnace reached the Moreton Bay settlement and the authorities sent a search party that eventually rescued Eliza.

Her account of captivity was highly sensationalised in a number of lurid booklets published in Sydney, London, and even New York. The New York version, in fact, has a frontispiece steel engraving showing an Aborigine in calf-height moccasins, knee-length loincloth and a three-feathered headdress dragging Eliza – dressed in a voluminous crinoline – into a slab hut, and was entitled 'Narrative of the Capture, Sufferings, and Miraculous Escape of Mrs. Eliza Fraser'. The lengthy description goes on to state that she was held 'for several weeks in bondage and . . . compelled to take up her abode in a wigwam and become an adopted wife of one of the chiefs'. Other publications had highly over-written accounts with such phrases as 'The horrible barbarity of cannibals inflicted upon the captain's wife' and 'her unparalleled sufferings'; while the survivors endured 'dreadful slavery, cruel toil, and [had] excruciating tortures inflicted upon them' until finally they received 'deliverance from the savages' (Alexander 1971:*passim*; Williams 1982:34–43).

Whether such accounts had any direct influence upon the Admiralty is unknown but they certainly created enormous public interest. What is evident is that trade and naval supremacy demanded better knowledge of the Reef and all Australian waters. Following the work of King, four more hydrographic surveys were commissioned to complete the task: that of the *Beagle* in 1837–44, first under command of John Clements Wickham, who was succeeded by John Lort Stokes in 1840, the *Fly* under Francis Greenwood in 1842–6, the *Rattlesnake* under Owen Stanley in 1846–50 and the *Herald* under Henry Mangles Denham in 1852–61. Although primarily hydrographic voyages, all had a wider brief including natural science, and consequently – except for the *Beagle* – carried naturalists as part of their complement.

The *Beagle* was a small ship of 235 tons and 95 feet (29 m) in length: barely two-thirds the size of Cook's *Endeavour* of 368 tons and 106 feet. Launched in 1820 as a two-masted, 10-gun brig, it spent its first twelve years on survey work, mainly in South American waters; in 1831 it was rebuilt as a three-masted barque, and under the command of Robert Fitzroy spent the years 1831–36 on a world survey. It was on this voyage (described in chapter 11) that Charles Darwin, sailing as captain's companion, and in effect as supernumerary naturalist, began formulating his theories on both the formation of coral reefs and the origin of species.

Following that epochal voyage, the *Beagle* was again dispatched to Australia to continue survey work under the command of a Scot, John Clements Wickham (1798–1864). When Wickham retired from the naval service in 1840, he was succeeded by his first officer, John Lort Stokes (1812–85), under whom the *Beagle* survey continued the task of repeating the previous circumnavigation and refining all measurements and observations as much as possible. Again, the Reef was traversed very quickly, in less than three weeks (3 to 21 June 1841) Stokes recording that around Cape York 'We found little to add to Captain King's chart' (Stokes 1846:354). Although the official records, as always, were embargoed as commercially sensitive documents, at the end of the survey he was entrusted with the task of preparing the public document, *Discoveries in Australia*, which was published commercially in 1846.

Discoveries in Australia was a discursive account – as distinct from official hydrographic reports of soundings, compass bearings, magnetic deviations and other technical matters – and by its nature, appealed to a wider, educated public. The text covered all aspects of the survey, especially topographic descriptions of the coastline for a British public curious about that strange south land where so much was seemingly contrary to Europe. Included were less attractive accounts of the indigenes, especially the fierce Murray Islanders who lay in wait in their canoes to pillage those unfortunate ships that became wrecked in Torres Strait, such as the *Charles Eaton* which ran aground on the Charles Hardy Islands in 1834 where all of the crew were massacred except for two boys who were sold to another Murray Islander for a bunch of bananas. The islanders, Stokes reported, had excellent vision which 'resembles that of the carrion bird itself [South American condor] while their rapacity and recklessness of blood is fully equal to that of the lower animals' (Stokes 1846:I.359,364; Loney 1993:37).

Of greater significance to Reef science, however, were the geological observations. In the Admiralty Instructions, issued originally to Wickham, and fulfilled by Stokes on both surveys, was the specific requirement, given the lively issues raised in Fitton's analysis in King's *Narrative*, and the intense controversy surrounding geological evidence so far, to continue observations on the structure of coral reefs: specifically since:

> It has been suggested by some geologists that the coral insect, instead of raising its superstructure directly from the bottom of the sea, works only on the summits of submarine mountains, which have been projected upwards by volcanic action. They account, therefore, for the basin-like form so generally observed in coral islands, by supposing that they insist on the circular lip of extinct volcanic craters; and as much of your work will lie among islands and cays of coral formation, you should collect every fact which can throw any light on the subject (Stokes 1846:I.21).

On his return from the previous voyage of the *Beagle* between 1831 and 1836, Charles Darwin had commenced the natural history report required by the Admiralty and, as an additional personal endeavour, was attempting to draw together the masses of accumulating geological evidence into a more general theory of the formation of coral reefs. In May 1837, just before the 5 July departure of the *Beagle* from Plymouth on its next survey, Stokes, clearly interested in current geological theories, had attended a meeting in London of the Geological Society. There, he wrote in *Discoveries*,

> I had the pleasure of hearing a valuable paper by my friend Mr. Darwin, on the formation of coral islands, read at the Geological Society; my attention being thus awakened to the subject, the interest of this important paper was to me greatly enhanced by a series of queries, kindly furnished by Mr. Darwin, and drawn up with a view to confirm or invalidate his views, his purpose being to elicit truth from a combination of well attested facts, and by inducing the research of others to further the objects of science.

Since Darwin had not sailed along the Great Barrier Reef, and its geological structure would be important to his inquiries, he asked Stokes to make further observations on questions that Stokes quoted verbatim: 'Are there masses of coral or beds of shells some yards above high water mark, on the coast fronting the barrier reef?' (I.331).

In response, Stokes made what observations he could when the *Beagle* made its two brief traverses of Reef waters. The first real opportunity came at Cape Upstart (near modern Townsville) where Stokes 'found a flat nearly a quarter of a mile broad, in a quiet sheltered cove, within the cape, thickly strewed with dead coral and shells, forming, in fact, a perfect bed of them – a raised beach of twelve feet above high water mark' (I.332). Darwin's query related to that intractable geological and coral reef problem: how to explain the discovery of well-preserved marine remains in elevated inland locations. Darwin had wrestled with that enigma ever since he had read the observations of

Alexander von Humboldt on South American limestone strata, Caribbean coral reef outcrops and earth movements in his *Historical Account of a Voyage to Tropical Regions of the New World between 1799 and 1804.*

Geological discoveries were continuing to reveal more numerous examples of marine strata inland from the coast, and at elevations well above sea level that defied ready explanation. In *Discoveries* Stokes made the relevant observation that:

> Had it been on the seaward side of the Cape, I might have been readier to imagine that it could have been thrown up by the sea in its ordinary action, or when suddenly disturbed by an earthquake wave; but as the contrary is the case, it seemed impossible to come to any other conclusion, than that an upheaval had taken place (I.332–33).

That observation was amplified further in his significant comment that:

> The remarkable breaks in this singularly great extent of coral reefs, known as the Barrier of Australia, being in direction varying from W. to W.N.W., generally speaking N.W., leads me to believe that the upheaval by which the base of this huge coral building was formed, partakes of the general north-westerly direction, in which a large portion of the eastern world apparently emerged from the water (I.375).

In regard to other aspects of natural science – botany and zoology – very little was achieved, nor was it the brief of the *Beagle* to do so. Stokes was a marine surveyor but, since no naturalist was carried on board, unlike other naval surveys, he also acted as naturalist on occasions. It is not surprising, therefore, that few biological data were obtained. In the Instructions, in fact, the Admiralty commented that large natural history collections were neither possible nor expected, although some recording should be attempted wherever possible (I.24). Such collections are recorded by Stokes in the Appendix (pp.479–521) where the biological specimens are listed and described: numerous birds, six fish, lizards, one snake, and many beetles. The most interesting item is one not hitherto listed by previous naturalists: a crocodile, then designated *Crocodilus palustris* (Lat. *palus*, swamp or marsh), known today as the fierce salt water estuarine *Crocodylus porosus* (Gk *porosus*: 'rocklike', so, 'scaly humps').

JUKES AND MACGILLIVRAY: NATURALISTS ON THE *FLY* 1843–1845

While Stokes was still surveying in the *Beagle*, the corvette *Fly*, under the command of Francis Price Blackwood (1809–54), was enroute to Australia. Leaving Falmouth in April

1842 the *Fly* arrived at Port Jackson on 15 October the same year, where Blackwood was able to gain some first-hand knowledge of Australian waters from the recently retired Captain Wickham. The Sailing Orders to Blackwood were in similar vein to those issued to previous commanders, with an emphasis this time on finding secure passages from the outer route into inner Reef waters. Many ships to Asia, India and Britain choosing the Torres Strait began with the outer route and then sought an opening towards Torres Strait through which to enter the inner waters. Given the increasing fatalities, the Orders stated that it was imperative 'to have the Great Barrier Reef explored, and to have gaps surveyed, in order that some means may be devised for so marking the most eligible of these openings, that they may be recognised in due time, and passed through in comparative safety' (Jukes 1847:II.255). The *Fly*, therefore, was instructed to survey the entire Reef 'from Breaksea Spit to the shore of New Guinea', and in so doing, to collate the findings of Flinders, Bligh, King and others, to fix upon 'some comparatively safe channels' and to find sheltered anchorages, with suitable trees, for the repair of vessels and their spars (II.257).

The Admiralty provided this time for two naturalists. In that period the British government was in its most extreme phase of ecological imperialism, and the search for plants, directed by Kew Gardens, with the aim of expanding colonial agricultural settlements, was a continuing need. As for previous surveys, therefore, a botanist was also considered essential, to whom the captain was 'to give ... every facility to pursue those vocations whenever the business of the survey may permit' (II.260). Zoology still seemed a more 'abstract' scientific pursuit without the obvious economic advantages offered by botany.

Despite that intention, the botanist position was filled by John MacGillivray (1821–67), son of a famous ornithologist, Regius Professor William MacGillivray of Aberdeen. Before completing his medical degree at Edinburgh, MacGillivray received the patronage of the Earl of Derby, an enthusiast for natural science, as his personal zoologist. So, although a botanist was sought, a youthful zoologist went instead. MacGillivray proved a valuable acquisition to the ship's complement, and became one of the new colony's leading zoologists, especially in his specialist fields of ornithology and conchology. MacGillivray, in fact, was so good that he sailed on the next two surveys, those of the *Rattlesnake* (for which he wrote the official *Narrative of the Voyage*) and the *Herald*, although on the latter survey he became embroiled in a strange controversy that makes him one of the more fascinating and enigmatic characters in Reef science. Given the increasing evidence of inexplicable reef formations, particularly inland marine depositions, it was also decided to include a geologist. That position was offered to Joseph Beete Jukes (1811–69), a very competent professional geologist, and former student of the redoubtable Adam Sedgwick, professor of geology at Cambridge.

The *Fly*, accompanied by a smaller support vessel, the cutter *Bramble*, under command of Lieutenant Charles Yule, spent three years, from December 1842 to December 1845, in surveys of Australian waters, making two circumnavigations of the continent, via Torres Strait and the Gulf of Papua where the estuary of the Fly River was charted and named.

When the expedition returned to England, Jukes joined the geological survey of North Wales, and prepared the official account of the Australian work under the title *Narrative of the Surveying Voyage of H.M.S. Fly . . . during the years 1842–1846*, which was published in two volumes in 1847. The first descriptive volume was entirely written by Jukes; the second, containing the more technical and scientific data, has an excellent section on the geology of coral reefs by Jukes (311–47) while other parts were prepared by specialists to whom the data were supplied.

The major survey of the Reef took some eight months, commencing on 7 January 1843 when Jukes wrote, as his opening sentence,

> I landed for the first time in my life on a coral island. This was a little islet called the First Bunker's Island, in the northern part of the Capricorn Group, which is an assemblage of islets and reefs on the north-east coast of Australia, having the 152° of longitude, and the tropic of Capricorn passing through them (Jukes 1847:I.1).

Jukes, however, was mistaken. The first island in the Bunker group is Lady Elliot (24°6'S), first seen by Thomas Stewart, master of the *Lady Elliot*, in 1816. The island on which Jukes landed, in fact a coral cay and later named Lady Musgrave, was not exactly on the tropic (which passes through Heron Island at 23°26'.) but a little further north, at latitude 23°50'S.* Three days later, moving north, the party then landed on the already named One Tree Island where Jukes recorded his disappointment: 'The whole was very different from my preconceived notion of a coral reef, and I erroneously imagined it must be an exception to their general character. It looked simply like a half drowned mass of dirty brown limestone on which a few stunted corals had taken root' (Jukes 1847:I.5). Distance, however, often lends enchantment, and a few days later he conceded that 'although there is not much variety, there is considerable beauty in a small coral reef when viewed from a ship's mast-head at a short distance in clear weather' (I:10).

The first Reef voyage of the *Fly* concentrated on three major regions: the Capricorn Group from Fraser Island to the Whitsundays, from Rockingham Bay to the Endeavour River, and the far northern coast of Cape York. As the *Fly* progressed north Jukes was active in recording the geology of all of the sites visited, and demonstrated an intimate knowledge of coral types, now becoming essential as stratigraphy became fundamental to understanding geological processes. Numerous commentaries on the landscape – now being actively explored for settlement – occur throughout the *Narrative*, but almost nothing complimentary. In the region behind Cape Upstart, as a result of short forays inland, he 'concluded that beyond the mangroves there must be a considerable quantity of very good land (*for Australia*) within a slight distance of the sea' (I.87). His emphasis 'for Australia' again confirmed the common perception by most explorers since the first days of the settlement that Australia had nothing other than a dreary, monotonous and impoverished landscape. 'There is, I should imagine', he wrote, 'no coast in the world, of anything like the extent, so utterly destitute as that of Australia, of everything in the shape of fruits, vegetables, or any other edible, except limpets and oysters' (I.116). His views remained unchanged throughout the surveys although he did make one moderated statement, in a footnote, that:

> After twice circumnavigating Australia, and visiting all its colonies, especially those of the southern coast, I look back upon this tract between 22° and 20° [around modern Mackay], with still higher expectations than before, and certainly have never seen any part of Australia, *near the sea*, of equal fertility, or of nearly equally pleasant and agreeable aspect, or combining so many natural advantages (I.51n.).

The Reef itself, however, despite his earlier negative comments, came to be better appreciated. From his first opinion at Lady Elliot Island, that when he first saw the Reef he was 'rather disappointed . . . and had not seen much to admire' given the drab colours as seen from the deck of the ship, he later agreed, on closer, more informed inspection – and doubtless guided by the enthusiastic collector MacGillivray – that there was also a romantic quality to the beauty and dazzling Reef corals and fish which would undoubtedly have intrigued readers. When anchored one day in a 'sheltered nook' he saw the corals 'in full life and luxuriance':

> their colours were unrivalled – vivid greens, contrasting with more sober browns and yellows, mingled with rich shades of purple, from pale pink to deep blue. Bright red, yellow, and peach-coloured nulliporae clothed those masses that were dead, mingled with beautiful pearly flakes of eschara and retepora; the latter looking like ivory. In the branches of the corals, like birds among trees, floated many beautiful fish, radiant with metallic greens or crimsons, or fantastically banded with black and yellow stripes. . . . All these, seen through the clear crystal water, the ripple of which gave motion and quick play of light and shadow to the whole, formed a scene of the rarest beauty, and left nothing to be desired by the eye, either in elegance of form, or brilliancy and harmony of colouring (I.117–18).

A second, shorter voyage was made the following year when the *Fly* left Sydney in May to return to Raine Island in order to erect a navigation beacon to guide ships from the outer route safely into inner waters, thereby beginning the essential process of placing navigation aids in strategic locations. The Raine Island beacon was intended to help ships avoid the more treacherous Pandora passage 20 kilometres to the north, where the *Pandora*, returning with some of the captured mutineers from the mutiny on the *Bounty*, foundered and sank in 1791. Materials and twenty convict volunteers sailed in the *Prince George*, accompanying cutter to the *Fly*, and using coral beach rock cut on the site, along with timbers salvaged from another wreck, the *Martha Ridgway*, a cylindrical tower, 45 feet (13.7 m) high and 30 feet (9 m) in diameter, with walls 5 feet (1.5 m) thick, with inner floors and a canvas canopy on top, was constructed in four months, by mid September (Loney 1993:34). From then, until 18 December 1845 when the *Fly* sailed for Port Essington to return to England, they were engaged in further surveys of Torres Strait and the Gulf of Papua. The *Bramble* thereafter continued survey work in Torres Strait and the Gulf of Papua as tender to the next survey ship, the *Rattlesnake*.

The assiduous work of Jukes and the young MacGillivray was to make the first major contribution to the zoology of the Reef, especially in respect to sea snakes, crustaceans, sea stars, shells and the dugong. MacGillivray did not, however, do any of the describing: the carefully preserved specimens were taken to London to be studied by experts. John Edward Gray, Keeper of Zoology (and later Director) of the British Museum, was excited to discover a 'new genus of snakes'. Already the Aipysuridae had been distinguished from the Hydridae, but 'the transition', he wrote, 'was so abrupt, and it was to be expected, that there must exist some genus which had hitherto escaped the observation of naturalists which would shew a gradual approximation; such a genus has at length been discovered by Mr Jukes on Darnley Island' (in the north-east of Torres Strait). That transitional genus was designated Hypotrophis and the binomial, to honour Jukes, was designated *Hypotrophis jukesii*, Gray 1846 (Jukes 1847:II.332). Several taxonomic revisions have since been made and none of the above terms now apply.

Gray also, briefly, described the Asteridae – starfish – some 34 altogether, with an indication that a fuller description would appear in a future monograph. Gray criticised current taxonomy, and stated that the best descriptions would come from attending to 'the proportion given, and the difference in the shape of the arms and spines' (II.339). A second honour for Jukes was the naming of a new species of crustacean from Darnley Island as *Cymopolia jukesii*. MacGillivray had collected a large number of shells and these were sent for description to Adam White, a Fellow of the Linnean Society. Among these he identified 13 new species: 6 from the Reef and 7 from Port Essington. In addition, MacGillivray, following his ornithological interests, collected numerous birds, the larger ones by rifle, using fine bird shot. These were sent on to the British Museum and were later included in the stupendous seven volume, 600 coloured-plate publication by the great English naturalist John Gould (1804–81) as *Birds of Australia* (1840–48) which covered 600 species of which more than 300 were new. On the voyage of the *Fly*, and the subsequent voyage of the *Rattlesnake*, MacGillivray was to provide a number of the type specimens for Gould. Although he did not contribute directly to Jukes' *Narrative*,

---

* From Jukes' description and drawings the island was mistakenly identified later in the century by the great Reef scientist William Saville-Kent (1893:96) as Lady Elliot Island (24°6'S), and that confusion has been perpetuated in the literature (see Heatwole 1984:39). Given the similarity of the Bunker vegetated cays, and the primitive steel engraved sketches which accompanied the scientific narratives of the period, it is easy to understand Saville-Kent's error.

MacGillivray did publish four articles in the *Zoologist* for October 1846, of which two referred specifically to the Reef: 'An account of Raine's Islet, on the N.E. coast of New Holland' (IV.1846:1473–81) and an 'Ornithological excursion to the north coast of New Holland' (IV.1846:1481–84).

At Ince Bay near Cape Palmerston they came across the remains of a dugong on the beach that Aborigines had eaten. But, since the skeleton was broken up, it was useless for science. Fortunately, somewhat later at Port Essington they found a complete dugong that was scientifically quite valuable; its skeleton was prepared for transport back to Richard Owen, London's leading – and highly controversial – comparative anatomist, for further study. In the scientific report 'Notes on the characteristics of the skeleton of a dugong', from a careful examination of the dentition it was found to be different to the common *Halicore indicus* of India and the Red Sea, and was classified as a new species: *Halicore australis* (Gk *halys:* 'sea' + *kore*: 'maiden'; so, 'mermaid'). Interestingly, the Malay word for the animal, 'duyong' had already been transcribed and adopted in France by Buffon in 1765 as 'dugong', which was to give its modern binomial of *Dugong dugon*.

The rich store of new species arriving in Britain from these remarkable reefs quickened the interest of scientists in Britain who were eager to see the investigations extended, particularly into corals and associated reef biota. That opportunity was to come the following year when, given the need to continue the essential hydrographic surveys, the Admiralty dispatched a larger vessel, the three-masted frigate *Rattlesnake*.

CHAPTER 6

EARLY REEF CHARTS COMPLETED: 1846–1862

REEF SURVEYS OF THE *RATTLESNAKE*: 1847–1850

Under command of Captain Owen Stanley (1811–50) the *Rattlesnake* sailed from Plymouth on 11 December 1846 to continue the survey of Reef and New Guinea waters begun by the *Fly* and the *Bramble*. Aboard was John MacGillivray, on his second tour of duty, this time as official naturalist. He was accompanied as assistant surgeon by the 21-year-old and totally inexperienced Thomas Henry Huxley, who recorded that he had prepared a program of studying the 'corallines' and other marine species. Due to the untimely death of Captain Stanley in 1850 from some unidentified tropical illness contracted during the third northern cruise of 1849–50, MacGillivray later also assumed the task of writing the official report, *Narrative of the Voyage of HMS* Rattlesnake. In addition, the artist, Oswald Walters Brierly (1817–94), kept extensive observations in a series of diaries which enable further reconstruction of the voyage.

On 11 October 1847 Stanley sailed with the *Bramble* to improve the survey of the inner route through the northern Great Barrier Reef and find other reliable openings besides Raine Island. The first landing was to examine and report on Port Curtis (modern Gladstone) where a party of eighty-eight had attempted to found a new colony of 'North Australia' in January 1847 but had been forced to abandon the enterprise after a few months. Even so, it was advised that increased squatting expansion would require Port Curtis to be developed, not only to service the hinterland, but also as a coaling station and whaling port, while dugongs in the region, it was proposed, should be 'the object of a regular fishery on account of [their] valuable oil' (MacGillivray 1852a:I.48). Stanley continued to survey from Port Curtis north to Upstart Bay, and then, rounding into Cleveland

Bay, the present site of the port of Townsville, recorded that, for a future port, it was 'the worst possible place I ever saw for such a purpose' (I.46, 52). At the same time, MacGillivray was busy collecting various marine species, some 'curious fishes' and 'several rare and new crustacea' (I.58). Returning south the *Rattlesnake* left off the survey at 'Keppel's Island' on 24 December 1847 and sailed for Port Jackson. Arriving on 14 January 1848, Stanley learned that an expedition was being organised to explore the land adjoining the Reef from Rockingham Bay north to the tip of Cape York.

Since Port Essington, established as a military outpost in Arnhem Land in 1823, had proved too remote to provide protection for ships travelling through Reef waters, it was decided in 1846, that it had to be closed, and a base established on the eastern seaboard. A possibility had already been suggested from the earlier survey by the *Fly* and the *Bramble* in 1846 when Lieutenant Yule had reported that a good anchorage and base could be established in the channel between Albany Island (11°44'S) and the mainland, right at the tip of Cape York. MacGillivray ventured his own opinion in the official report – doubtless having discussed the situation with Captain Stanley and other officers – that Port Albany would offer a refuge for ships, act as a source of supplies and coal for vessels in transit, and help to control the always 'troublesome' natives and the 'little trade in tortoiseshell' (Hawksbill turtles). It would also make a good base both for military purposes and for the growing missionary activity in the tropical north (I.319–20).

In order, therefore, to further the exploration of that region – still virtually unknown apart from brief forays inland from ships – and possibly to establish a permanent settlement, the government in Sydney planned an expedition from Rockingham Bay, adjoining Hinchinbrook Island (18°S), to the tip of Cape York. The task was to be undertaken by Edmund Kennedy (1818–48) who was to have been deputy two years earlier to Thomas Mitchell in an expedition north-west from the Darling River to seek a route to Port Essington. That plan, however, was abandoned when news reached them that Ludwig Leichhardt had already succeeded in 1845. Captain Stanley agreed to provide an escort for Kennedy to Rockingham Bay, and so the *Rattlesnake* and *Bramble* remained at anchor in Sydney Harbour while the expedition was organised and loaded aboard its transport, the barque *Tam O' Shanter*. The three vessels sailed from Port Jackson on 29 April 1848 to travel via the inner route to the starting point of the expedition. Kennedy was accompanied by twelve other men and the Aborigine Jacky-Jacky, along with a large number of horses, sheep and supplies.

Arriving at Rockingham Bay on 23 May 1848, the crews of the *Rattlesnake* and *Bramble* helped unload the barque and then set off on their own major task of surveying Reef and New Guinea waters to the north, while MacGillivray began his seine trawling at Dunk Island, a little to the north of the bay. It was decided, however, that the *Bramble* would rendezvous with Kennedy's expedition at Princess Charlotte Bay three months later in early August. Kennedy never arrived and after waiting ten days the *Bramble* sailed on to meet up with the *Rattlesnake* on 21 August at the Pipon Islands. A small supply ship, the brigantine *John and Charlotte*, arrived at the Cape on 21 October to wait for Kennedy, and on 27 October the schooner *Ariel*, which had been chartered to maintain a watch, also arrived. Two months later, 22 December, the *Ariel* sighted a single figure on the shore, Jacky-Jacky, who alone had reached the goal. Kennedy had been speared by Aborigines not far inland, while illness and Aborigines accounted for ten other members of the expedition. The two survivors, who had cut across to the coast earlier, were later rescued by the *Ariel* at Cape Grenville (12°S).

From 25 May to 4 November 1848 the two vessels continued survey work, separating initially for the *Bramble* to chart Princess Charlotte Bay while the *Rattlesnake* went on to work in Cape York and Torres Strait waters. Joining up on 4 November at Booby Island, the *Bramble* continued surveys of Endeavour Strait and the Prince of Wales channel after which it was instructed to return to Sydney while the *Rattlesnake* went across the Gulf to Port Essington. Finding the small settlement in a deplorable condition with low morale and serious tropical illnesses among the remnant marine garrison, the *Rattlesnake* returned to Sydney where it arrived on 24 January 1849. So bad were conditions at Port Essington, and the adjacent settlement of Victoria, that plans were made to abandon it completely, and the vessel HMS *Maeander* under Captain Henry Keppel, senior naval officer in Australian waters, was sent to effect the final evacuation, completed on 30 November 1849. Of the 1847 garrison of six officers and 58 other ranks only four officers and 33 men remained to be taken aboard: thirteen had already died there and a further fourteen had been invalided out earlier (MacGillivray 1852a:I.136–39).

The third cruise of the *Rattlesnake* was outside Reef waters, chiefly along the southeastern coast of New Guinea and its extension as the Louisiade Archipelago, named by Bougainville in 1767 after Louis XV. Stopping at Cape York for fresh water, on the return journey to Sydney some five months later, MacGillivray reported yet another dramatic shipwreck similar to that of Eliza Fraser when 'on October 16th [1849] a startling incident occurred to break the monotony of our stay': 'in the afternoon some of our people on shore were surprised to see a young white woman come up to claim their protection from a party of natives from whom she had recently made her escape' (I.301). MacGillivray himself did not witness the incident directly: the main evidence comes from the diary of the artist–illustrator Oswald Brierly, now preserved in the Mitchell Library, Sydney.

The *Rattlesnake* and *Bramble* had anchored on 8 October off the cape in Evans Bay where they joined the brig *Sir John Byng* which, en route to Manila, was waiting with supplies for the two survey vessels. It was a week later, on the 16th, that Brierly, along with three others met up with Captain Levien of the *Byng*, one of his passengers, a Mr Aplin, and the surgeon of the *Bramble*, Dr Schloss, when they were told that a party of sailors and marines on the beach washing and drying their clothes had been approached by several Aborigines accompanying a badly sunburnt white woman, naked apart from a few leaves covering her pubic region, and blind in one eye from ophthalmia. Suddenly discomfited by her nakedness before white men, Brierly recorded, she was given two shirts, one which she put on as a blouse, the other wrapped around as a skirt. Taken aboard the ship she was given medical treatment since, at first, she seemed very vague, as if she were 'just waking up from a deep sleep' (Brierly, ed. Moore 1979:77). Some time later, as she recovered, she was also given calico, needles and thread to make a shift and underwear for herself.

It was learned that Barbara Thompson (MacGillivray spells it Thomson) was a native of Aberdeen and totally illiterate who, at the age of sixteen had eloped with and married William Thompson, owner–master of a small vessel the *America*, and left with him from Moreton Bay on a salvage operation to Torres Strait to retrieve a valuable cargo of whale oil from a wreck. Unfortunately, the *America* itself was wrecked on Prince of Wales Island, which the Aborigines called Muralag: all five men aboard were drowned, while Barbara was taken by natives who incorporated her into their small community to work with the other women. One man claimed her, from her white skin, as Gi'om, the ghost of his long-lost daughter, and from her own account, treated her kindly. Although between twenty and thirty ships passed through the channel every year, none stopped. When she heard that the men were talking of the two 'war canoes' that had arrived there recently, she persuaded them to take her to the cape, with the suggestion that the sailors would give them various kinds of presents. The stratagem worked: each received an axe and a knife, along with tobacco and ship's biscuits, which they had come to enjoy from previous ships.

The *Rattlesnake* remained at the cape for nine weeks that provided further opportunities for ethnographic study of the locals, extensively recorded in the works of both MacGillivray and Brierly, and for the natural science of the Reef region. Brierly makes only passing mention of Surgeon Huxley, but refers to MacGillivray's industry throughout his diaries, with a description of his characteristic technique: 'on October 25th, having gone ashore, MacGillivray was soon off, crouching and stealing along like an indian [Aborigine] after some new tern that he saw skimming along on the hillside' (80). MacGillivray continued his relentless search for natural history specimens while Brierly spent his days in topographic sketching and ethnography, using much of his time there, and on the voyage back to Sydney, to question Barbara Thompson as closely as he could regarding her knowledge of Aboriginal customs and behaviour. In her five years with the Kuaregs she had acquired extensive information regarding their culture which, recorded by Brierly, was to prove invaluable in understanding the life of that most southerly community in the Torres Strait during the era of pearling and bêche-de-mer collecting in the ensuing decades.

When the two vessels weighed anchor from Evans Bay on 1 December 1849 with a now very ill Captain Stanley aboard, the Aborigines became agitated on learning that Barbara, of whom they apparently had become very fond, would be leaving with the ship. On the voyage south Stanley continued to worsen and died on 15 March, five weeks after arrival in Port Jackson. Once back in Sydney Barbara Thompson was reunited with her parents who had migrated to Sydney in 1836 when she was a child, and it seems that she recovered much of her health and lived into her eighties (MacGillivray 1852a:301f.; Jack 1922:262). Brierly went on to become one of the greatest seascape painters of the period:

in 1874 he was appointed official maritime artist to Queen Victoria, and in 1886 was knighted for his services to art.

RATTLESNAKE NATURALISTS: MACGILLIVRAY AND HUXLEY

While the *Bramble* remained in Sydney, the *Rattlesnake* on 2 May 1850, carrying a crowded total of 230 persons including the Port Essington garrison, sailed for England. When it arrived MacGillivray, as ship's naturalist, was requested by the Admiralty to write the official public record, which he completed two years later under the customary style with a lengthy descriptive title: *Narrative of the Voyage of H.M.S. Rattlesnake, commanded by the Late Captain Owen Stanley, R.N., F.R.S. &c. during the years 1846–1850, including discoveries and surveys in New Guinea, the Louisiade Archipelago etc. to which is added the account of Mr. E. B. Kennedy's expedition for the exploration of the Cape York Peninsula*. Based on MacGillivray's detailed notes kept in a series of notebooks, along with assistance from Admiralty records, the two volumes were published in 1852.

MacGillivray was primarily a zoologist, and his reports of the three voyages mark a decisive turning point in Reef science, from a preoccupation with land-based botany to zoology: marine, terrestrial and avian. Having served aboard the *Fly* as a junior to Jukes, MacGillivray was now in his element as official naturalist, often irritating the self-confessed lazy and 'morose' Huxley who recorded in his diary (not discovered until a century later) that he was 'mortally sick of this wearisome monotonous cruise' (Huxley 1935:116,122).

In those days all marine research depended in the first instance on ship-based seine netting and dredging, sounding line collection (by retrieving sea bed materials such as sand and shells, stuck to the tallow inserts in the recessed bottom of the lead weight) and low tide lagoon and estuary collecting on foot, the major preoccupation being with the discovery, description and classification of hitherto unknown species. At the time, the taxonomic imperative was helping to move the scientific climate towards the concept of evolutionary development, proposed by Erasmus Darwin in the late eighteenth century, articulated by Lamarck in the early nineteenth century and carried forward by a number of thinkers, of whom the most influential was to be Charles Darwin, Erasmus' grandson. Even so, the achievement of MacGillivray was formidable and the two scientific appendixes of the *Narrative* – occupying 193 pages altogether, 74 in Volume 1, and 119 in Volume 2 – mark a great advance in knowledge of the marine life of the Great Barrier Reef. A measure of MacGillivray's competence can be gauged from the fact that during the course of the voyage he collected 85 species from 29 genera altogether; 54 of those species were of Polyzoa and 31 of 'zoophytes' (today, Cnidaria). When sent to George Rusk, a fellow of the Royal Society of London, for description, 78 species were found to be undescribed, which

Rusk recorded in his *Account* in the fourth Appendix to Volume 1 of the *Narrative* as 'the largest and most interesting of the kind ever brought into this country' (Rusk, 343f. in MacGillivray 1852a:I). Rusk classified the 'zoophytes' within the order Anthozoa hydroida.

MacGillivray's molluscs were sent to Edward Forbes of the Geological Survey who was considered one of the most promising marine scientists of the time in Britain, and one of Darwin's circle, who died, tragically, of kidney failure in 1854 just months after his appointment to the chair of natural history at Edinburgh. Ever thorough in his field work, MacGillivray had collected and labelled his specimens according to the stratum in which each was found, ranging from salt marshes through mud and mangroves, littoral zones down to 'laminarian regions', that is, the various layers of the sea (Lat. *lamina*: 'thin plate') from one fathom (1.8m) to 17 fathoms (31m). Forbes was highly complimentary and in describing 71 species of those received he wrote that the 'comparative paucity of undescribed species produced in the littoral zone and the large proportion of new or doubtful forms among those taken by the dredge', due to MacGillivray's industry, indicate that it could be conjectured that 'a rich harvest has yet to be reaped in the deeper regions of the southern seas' (Forbes, in MacGillivray 1852a:II.366).

Given his strong background in ornithology, gained from his father, MacGillivray was equally assiduous in collecting birds from north-eastern Australia and the Torres Strait, amounting altogether to 171 separate species. Since the voyage of the *Fly* when he collected birds – many of them new species, and in cases, genera – for description and illustration by John Gould in *Birds of Australia*, MacGillivray continued to correspond with Gould (recorded in MacGillivray's notebooks 1847–49) which led Gould to publish an article in the journal *Contributions to Ornithology* for 1848–52. Under the title 'A brief account of the researches in natural history of John M'Gillivray, Esq., the naturalist attached to H.M. Surveying Ship the *Rattlesnake*', Gould read that paper to the July 1850 meeting of the British Association for the Advancement of Science.

One of the more intriguing aspects of the voyage of the *Rattlesnake* was the role of Thomas Henry Huxley (1825–95) as assistant surgeon and amateur biologist. Unlike MacGillivray who came from an academic background, Huxley was the son of George Huxley, a country schoolmaster in rural Middlesex. Sponsored by his brother-in-law, Dr John Scott, in north London, he became an apprentice to John Cooke, and then to Scott, to study medicine. Not yet fully qualified, he joined the navy, and through some influence, received his first assignment to the *Rattlesnake* in 1846 when it was being prepared for the voyage to the Reef.

Very little is known of Huxley's activities throughout the three and a half years of the *Rattlesnake* survey: he is mentioned, almost always incidentally, in the records of both MacGillivray and Brierly, and then only in a supporting role. Neither makes any comment about individual or independent work by Huxley. Apparently unknown to all aboard, however, he kept a diary in which he recorded his thoughts throughout the voyage. Nearly a century later, 1933, it was found in family papers by his grandson, Julian, who himself was to gain fame as a distinguished biologist. Published in 1935, and edited by Julian Huxley, it provides some understanding of the formative scientific years of one of England's greatest biologists of the nineteenth century.

When the *Rattlesnake* sailed in 1846, Charles Darwin's epoch making and highly controversial book *The Structure and Distribution of Coral Reefs* had been in circulation for over four years and was creating one of the great geological controversies of the time. The issue was one that every captain of British naval survey ships had been instructed to investigate: how are coral atolls and coastal fringing reefs formed? Prevailing catastrophist beliefs held that atolls – in their classic form of a circular chain of islands with a central lagoon (Maldivian *atolu*: 'interior') – had been formed on the base of the ocean and had been lifted, in some mysterious way, by great explosive upheavals of the earth. That theory had been challenged in Charles Lyell's three-volume *Principles of Geology* (1830–33) where he argued that atolls were 'nothing more than the crests of submarine volcanoes, having the rims and bottoms of their craters overgrown by corals' (Lyell 1832:II.290). Darwin added to the controversy by hypothesising an alternative theory: that atolls had formed on top of slowly subsiding extinct volcanoes, while fringing reefs grew upwards from subsiding continental margins. But of that topical issue, Huxley's diary betrays no knowledge whatsoever.

The mystery is heightened by the fact that the young Huxley set off on the voyage to the Reef with the express intention of making a 'careful study of all matters relating to coral and corallines' (Huxley 1935:52), in the greatest region in the world for such a project. Yet, given such an unrivalled opportunity, he evidently made no such studies at all, and never, throughout his lifetime, published a single report on reef corals. Instead, he was preoccupied with other coelenterates, especially the medusae or scyphozoans (jellyfish), and hydrozoans (stinging corals, Portuguese man-o-war and similar colonial animals). Most of his specimens were actually collected by tow-netting on the way out from England, supplemented with others found in Reef waters. His published descriptions and skilful drawings reveal, however, that he had made a meticulous study and dissection of them through the microscope. It seems, in fact, that Huxley had spent all of his scientific time aboard the *Rattlesnake* at the microscope, as his engraved plates illustrate, even though he complained later of the lack of skill of the engravers in the printed works.

In Sydney Huxley met chief administrator Colonial Secretary Alexander Macleay, his son William Sharp Macleay (1792–1865) and his 28 year old nephew, William John (1820–91) who had arrived from Scotland after his father's death, in 1839. William John was particularly taken by Huxley's dramatic recounting of the adventures of biological collecting being done aboard the *Rattlesnake*. In March 1848, while in Sydney after the first cruise, Huxley showed a paper on the medusae to William Sharp, an enthusiastic naturalist who was attempting to develop a new taxonomy, (the 'Quinary System'), although one safely within the orthodoxy of 'certain limits as assigned by the creator' (Stanbury & Holland 1988:37). William Sharp Macleay, who succeeded his father as chairman of the boards of trustees of both the Australian Museum and Botanical Gardens when Alexander died in 1848, was so impressed with the quality of Huxley's paper that he took the responsibility of sending it off in Owen Stanley's dispatches to the Zoological Society of London while Huxley set off on the next cruise of the *Rattlesnake*. On arrival it was published in 1849 in the Philosophical Transactions of the Royal Society of London under the title 'On the anatomy and affinities of the family of the Medusae'. Indeed, this resulted in Huxley's election, a few months after he returned to England in 1850, as a Fellow of the Royal Society.

Huxley, in effect, had made a major contribution to marine science in advancing the taxonomy of invertebrates which at the time was still very confused. Demonstrating particular skill in dissection and classification, he had established that if affinity depends upon corresponding organs being homologous (similar structures) then, accordingly, medusae belong to the phylum Coelenterata, so named using their basic anatomy as a criterion of classification (Gk *koilos*: 'hollow' + *enteron*: 'gut'). Huxley's term coelenterate is now generally used to cover what some taxonomists consider two separate phyla, the Ctenophora, or comb jellies (Gk *kteis*: 'comb') and Cnidaria, those with stinging cells (Gk *knidé*: 'nettle'). Huxley's achievement was significant: he 'had provided scientists with a criterion on which to base the classification of the whole group [of coelenterates], which had previously been assumed to be related to the Radiata (of which the starfish is a member)' (di Gregorio 1984:7).

AUSTRALIA IN A NEW ERA: CHANGING INTERNATIONAL RELATIONS

By the time the *Rattlesnake* had finished its survey of Reef waters in 1850 the world situation had changed quite dramatically, and although Australia was seemingly remote from the main theatres of international events in Europe and North America, it was involved, none the less, both as a colonial outpost of British imperialism and in its own right as a new, developing society.

A major concern was the growing expansion of Russia, not only west into Europe, but also east into Pacific waters, especially after its victory ending the Napoleonic war with France. Fanned by alarmist statements in the press, the rapid spread in Australia of what Barrett (1988) has termed 'Russophobia', had a direct bearing on Australian preparations for defence, leading to calls to establish forts along the undefended coastline, including a northern base at Cape York. Russian territorial expansion had been occurring since its 1581 conquest of Siberia from the derelict Mongol empire. By 1732 it had secured a Pacific base on the Kamchatka peninsula, and by 1800 had occupied Alaska and established a string of trading posts down the coast of California.

As a result of earlier Australian agitation, and additional concern over a likely Russian occupation of New Zealand, at the time already being appropriated by British settlers, a reluctant Britain was forced to annex those islands in 1840. After the Crimean War of 1852–54, in which Australian troops joined Britain and France to prevent Russian expansion south to the Black Sea, serious alarm grew when the large Russian fleet, from its new Pacific port of Vladivostock, undertook active exploration to find trading and colonising opportunities in the Pacific.

Anxiety mounted even more in 1853 when, following continuing French annexation of islands throughout the Pacific, including Tahiti and the Marquesas, on 25 September, it occupied and declared sovereignty over New Caledonia, uncomfortably close to the Reef. Fear in the 1850s was heightened further by awareness that the population of the four eastern colonies of Australia was still very small: the census of 1 March 1851 counted 402 174 persons, excluding Aborigines. This was soon to change.

The discovery of gold, first in California in January 1848, then in New South Wales in 1851 and Victoria in 1852, brought a rapid influx of miners, opportunists and settlers. The population of Victoria, separated from New South Wales in 1851 as an independent colony, rose dramatically in a decade to 538 628 in 1861, a remarkable increase of 230 per cent, while the population of New South Wales likewise rose by 220 per cent. As traffic increased from North America, with all shipping needing to cross the Coral Sea, there was a sharp rise in shipwrecks both inside and outside the Reef. From 1841 to 1850 at least 40 losses were recorded, rising to 77 in the following decade, 1851–60 (Loney, 1993, *passim*). As a result, for both commercial and military purposes, the Admiralty was stimulated to survey more thoroughly the Reef and its difficult openings to the east, the inadequately charted route across the adjacent Coral Sea to North America and the dangerous passages through the reef-studded Torres Strait on the well used northern route to the East Indies, British India and the Gulf of Suez. If the faster outer route along the Reef were better charted, vessels could use it as far north as possible before passing through the Reef via the Great North-East Channel to Torres Strait, or taking an easterly course across the Coral Sea. The imperative was to provide accurate charts of these routes in order to reduce travel times from Australia.

CORAL SEA SURVEYS OF THE *HERALD*: 1853–1860

That intention was set out explicitly in an Admiralty directive of 17 May 1852 to Captain Henry Mangles Denham (1800–87) who had been appointed to make the new survey due to

> the rapidly increasing traffic between our Australian Colonies & the Western Coast of America, and moreover of the inadequate knowledge we possess of the intervening navigation among its insulated rocks & intricate clusters of islands which extend to the eastw'd of New Caledonia, & considering the great benefit that distant commerce & maritime enterprise would derive from a thoro' examination of the

Region, from having its dangers fully explored, and from having its harbors so charted & described (David 1995:Article 1, 434).

The ship chosen for that more accurate survey of the Coral Sea, and especially the still unknown and very dangerous outer, eastern edge of the Reef, was the *Herald*. Although a traditional three-masted sailing vessel, the *Herald* marked a new era in hydrography: its crew were much better trained and equipped with improved technology, including twelve chronometers in order to determine longitude with great precision. Another mark of the coming era was the inclusion of a camera for the use of the artist, James Glen Wilson. For the first time a photographic record of still subjects was made for posterity. Early cameras, of course, could not record moving subjects and we are fortunate for the historical record that Glen Wilson (as he was known) was a skilled landscape and portrait artist. His watercolours, pencil drawings and line and wash sketches of the two ships, anchorages, natives, and various plants and animals give an accurate and immediate sense of presence to the viewer. Another significant innovation was the choice of a tender: for the first time a steel-hulled sail-assisted paddlewheel steamer, the *Torch*, under the command of Lieutenant William Chimmo, was assigned to assist the *Herald*.

The Admiralty Instructions issued to Denham consist of seventeen articles which cover the usual technical concerns for accurate hydrographic observations, as well as for commercial information about the natural resources of the islands of the Coral Sea region, especially 'their physical character & their mineral & vegetable productions, for which a Naturalist well skilled in botany, zoology & geology has been attached to the *Herald*, & to him you are to afford every reasonable facility both in making & preserving his collections' (Appendix 1, 'Admiralty Instructions to Captain Denham', Art.4, David 1995:434–37).

Preparations for the cruise of the *Herald* were very thorough: in addition to a comprehensive inventory of survey equipment and necessary stores, careful attention was given to the personnel. Altogether seventeen officers, some with considerable previous survey experience, and 83 other ranks were assigned to the *Herald*, in addition to the nine officers and smaller crew of the *Torch*. The choice of Berthold Seemann, a Hanoverian working at Kew as naturalist, who had sailed in that capacity on the previous voyage of the *Herald* to the north Pacific under command of Captain Henry Kellett, had to be deferred due to the fact that he was still writing the official record, *Narrative of the Voyage of HMS Herald from 1845–1851* which was published in 1853. So, with the concurrence of William Hooker, director of the Kew Botanical Gardens, John MacGillivray was appointed, for his third tour of duty, as official, civilian naturalist, assisted by William Milne, a gardener from the Edinburgh Botanical Gardens.

The *Herald* arrived in Sydney on 18 February 1853 while the much slower *Torch* had been forced to separate and follow at its own pace. After an initial survey of Lord Howe Island a sequence of six further surveys was undertaken over the ensuing years, until 23 May 1860. Of these, two were in waters between Norfolk Island and Fiji to survey the still uncharted

Norfolk Island Ridge which stretches across to New Caledonia and the adjacent South Fiji Basin. The third survey was from Bass Strait to Shark Bay in Western Australia, an area also being considered as a penal settlement. These were followed by a further three cruises in the Coral Sea closer to the Reef, and in the southern part of the inner Reef. In both of those first two northern cruises to Norfolk Island, New Caledonia, Fiji and the Solomons, Denham was active in making soundings, compass bearings and magnetic deviations, and determining longitude while at the same time searching for a very large number of hazardous but elusive reefs that had been reported by the captains of commercial ships.

At the same time, while the *Herald* surveyed the waters of Norfolk Island, Aneityum in the New Hebrides (today Vanuatu), the Isle of Pines in New Caledonia and Fiji, MacGillivray was eternally active, and Milne, whose duties were largely those of collecting live plants and seeds for propagation at Kew, was equally busy. A sense of MacGillivray's eager anticipation for further natural history discoveries can be gained from his handwritten, and still unpublished journal (kept in the Admiralty Library, London), 'Voyage of HMS *Herald* under the command of Capt. H. Mangles Denham' (begun in 1852 and kept continuously thereafter: today catalogued by the Admiralty for the year 1852), where the entry for 3 September 1854 headed 'Fiji Islands' reads 'At length we have reached the promised land, after a remarkably pleasant voyage of 795 days from Plymouth' (MacGillivray 1852b).

The year 1855 was not a happy one for MacGillivray since he became embroiled in one of the most unpleasant controversies in Australian natural history, one that still has not been, and apparently never can be, explained satisfactorily. The story, as best it can be reconstructed is that on 25 April 1855 MacGillivray was examined by a Court of Inquiry aboard the *Herald* in Sydney Harbour and then dismissed from his position as naturalist (David 1995:161f.). Reconstruction of events comes chiefly from correspondence between the Admiralty hydrographer Francis Beaufort and Sydney Fremantle – senior naval officer in Australia before the Admiralty established the independent Australia Station on 25 March 1859 under Commodore William Loring – and among Denham, John Washington (who succeeded Beaufort) and William Hooker, as recorded by David (161). No independent defence by MacGillivray seems to exist. Essential to any really fair assessment of the dismissal, however, is the very curious – even suspicious – fact that none of the Admiralty records of the court proceedings now exist and that much of the situation seems reminiscent of the conflict between Governor Macquarie and Lieutenant Jeffreys, a clash of personalities.

Whatever the reasons or justification for the dismissal of MacGillivray, Australian Reef science was deprived of further achievements of one of the nation's really great naturalists, a much merited accolade that runs through the literature to this day. Natural science work on the *Herald* therefore came to an abrupt end, and although the two surgeons, Rayner and Macdonald, were asked to do what they could in addition to their normal medical duties, nothing significant in natural science ever eventuated from the six voyages of the *Herald* – neither from the ensuing three cruises over the years 1858–60, nor even from the earlier three. Neither officer had either the scientific knowledge or literary skill of MacGillivray, which is why no 'narrative' of the work of the *Herald* was ever written. Commander David has provided an excellent account from a hydrographic, nautical perspective, but we still lack a comprehensive account of the natural history of the voyage of the *Herald*.

From 1858 to 1860 Denham took the *Herald* on three more cruises, all to chart the Coral Sea, in a region bounded by the Swains in the south, New Caledonia in the east and the isolated Willis Islands in the north. The primary aim of all three cruises was to correct the numerous conflicting locations given by previous survey reports and sightings, chiefly by commercial vessels, but also by naval ships, chiefly British and French. Denham's task was to make a continuous survey of the Bampton and Chesterfield Reefs – a dangerous, poorly charted longitudinal stretch of some 200 nautical miles (370 km) lying between New Caledonia and Townsville – after which he was to continue to Wreck Island, Cato Island, Moreton Bay and back to Port Jackson.

Four months later, on 11 April 1859, Denham sailed with the *Herald* on its second Coral Sea cruise, lasting six months, which involved him in two successive sweeps of the sea, separated by a return to the inner Reef waters to replenish water supplies on the Percy Isles, to the north of Broad Sound. Working his way south, Denham continued soundings and bearings, and attempted to reconcile variant chartings of the coral reefs that stretched westwards towards the Great Barrier Reef itself. Eventually he reached 'the charted position of the outer edge of the Great Barrier Reef without any signs of it... [and continuing his search] towards the outer margin of the reef, which was eventually sighted 47 miles south of its charted position'. 'Even today', David continued, 'the outer edge of the Great Barrier Reef in this vicinity is still not accurately defined' (373–74). With supplies running low and half the crew suffering from scurvy, some twenty seriously ill, Denham set the ship on a southerly course for Lady Elliot, Breaksea Spit and Moreton Island and then Port Jackson which he reached on 14 October 1859.

A little over three months later, on 21 January 1860, with the crew having recuperated from the rest period in Sydney, with a plentiful supply of fresh oranges to ward off scurvy, the *Herald* again sailed north to make its third and final survey of the Coral Sea, with the particular intention of finding the exact location of the Bougainville Reefs and the Diane Banks, lying between 14 and 18 degrees south, north of Lihou and Mellish Reefs and due east of Cape Grafton on the mainland. Although the venture was inconclusive, observations were made throughout the region and as far north as the Willis Island group that Denham believed, mistakenly, had been named by Bougainville himself in 1768. Denham was unable to find those reefs, and after further routine charting, headed south via Mellish Reef, through the Coral Sea and finally anchored in Port Jackson after a final five-month survey on 23 May 1860.

From a purely hydrographic point of view the survey of the *Herald* was an outstanding success. Denham had begun the voyage with an Admiralty warning that 'the nature of this expedition is somewhat more extensive than that which we have usually assigned to Naval Surveys' (435) and there is no doubt that he performed his tasks exceptionally well, proving to be an outstanding hydrographer. His achievement, and that of a highly dedicated crew, has been well described by a former Australian naval hydrographer, Commander Geoffrey Ingleton, as a new era in surveying: 'gone were the explorer–surveyors of earlier days' who were now replaced by 'keen young surveyors, more scientifically trained . . . progenitors of the present-day charts of the Australian coast'. In those years Denham had produced 200 sheets of charts and drawings, noted 163 geographical positions, made 2419 compass variations afloat and 191 ashore, had contoured 700 nautical miles (1300 km) and erased 23 falsely charted shoals (Ingleton 1944:77).

Unfortunately, very little was published at the end of the voyage. The Royal Geographical Society printed a short memoir in 1862 with a three page account of the general nature of the voyage and four pages listing the surveys, adding an addendum of 20 items of 'miscellanea' concerning the scientific results. Had MacGillivray been retained we could now well have a detailed scientific *Narrative*. The Admiralty, however, was very pleased with the work of Denham. Lord Paget, Secretary to the Admiralty, was instructed:

> to convey to you their Lord'ps especial thanks for the useful work done in the *Herald* in the region known by the name of the Coral Sea, in continuance of the labour of Flinders, Ashmore, King, Blackwood & Owen Stanley, whereby the passages up to & through Torres Strait have been cleared of imaginary dangers & the limits defined of the Outer and Inner Routes which the mariner may navigate in safety (cited in David 1995:432).

Probably the best tribute to the work of all the Reef surveyors – Flinders, Bligh, King, Jeffreys, Wickham, Stokes, Blackwood, Stanley, Denham – came in a statement by Captain George Henry Richards, Admiralty Hydrographer, to a meeting of the Royal Geographical Society in London on 11 April 1864 when he commented that even

> 'twenty years ago this inner passage from Cape York to Moreton Bay was the most intricate in the world; now it was as easy to navigate as the English Channel. Silently but steadily this great work had been progressing during those long years, and it was undeniably one of the most gigantic and splendid undertakings ever carried out by any nation (*Proceedings*, Royal Geographical Society, VIII.4, 1863–64, 119–20).

Denham was promoted to Rear Admiral in 1864, and retired in England in 1866. On the retired list he was later promoted to Admiral, and died in London in 1877. So ended a remarkable era in the survey of the Great Barrier Reef.

CHAPTER 7

THE REEF AS A MARITIME HIGHWAY: COLONY OF QUEENSLAND, 1859–1900

Throughout the first century of European involvement, the Great Barrier Reef, in addition to its daunting hydrographic challenges, and its significance for science – fascinating as it was to biologist and geologist alike – from its sheer magnitude exerted an even more powerful and determining influence on the social and economic history of Queensland. From the beginning of settlement, Queensland was the only Australian colony whose development was almost entirely constrained by a seemingly impassable barrier of complex reefs, with only the port of Brisbane directly accessible to the open sea. Reef waters, however, well before Queensland's separation from New South Wales in December 1859, were exploited from the earliest days. Whaling and guano mining, along with the later expansion of pearling, bêche-de-mering and fishing, came to assume economic importance in Queensland second only to the agricultural and pastoral industries. Although in the later nineteenth century the best possible interpretation was to be placed on its existence as the 'Grand Canal' or 'a safe and secure harbour' – today known as the 'lagoon' – the Reef was actually a serious impediment to development. As settlers moved along the restricted coastal lowlands – the only lands suitable for widescale permanent occupation – its effects on agriculture, grazing and industry were to often create intractable problems.

From the time of the First Fleet of 1788 there had been a continuing expansion of the original colony of New South Wales beyond the Sydney plain. As pressure for land continued unabated, opportunists moved outwards – north, west and south – taking up land as squatters, a process the government in Sydney was powerless to prevent. The only remedy was to legalise land seizure. In 1836 Governor Bourke introduced the *Squatting Act* whereby illegal occupation was approved with the payment of an annual licence fee of

£10, irrespective of the area taken up. So began a movement into the north of New South Wales, as the entire eastern region of the continent was then called: a search for harbours, town sites and routes of transport. At the same time, its natural history began to focus also on suitable commercial crops for agriculture, and animals for pasture.

The first exploring expedition beyond Sydney by Surveyor-General John Oxley and assistant Alan Cunningham in 1817 had returned with a very dismal picture of the far western regions of the colony. Port Macquarie, which Oxley reached later, by contrast, was quite promising and settlers began to move north along the coast where they found the valuable pastoral and cedar lands of the northern rivers region (of today's New South Wales) with its two great rivers, the Clarence and the Richmond, known as the 'District of Clarence River'. Surveys further north of those two great rivers were less encouraging. Five years later, Oxley was dispatched from Sydney aboard the cutter *Mermaid* on 23 October 1823 to report on the potentialities of a remote convict settlement for intractables either at Port Curtis (modern Gladstone) or further north at Port Bowen (named by Cook after his naval friend Captain James Bowen, and subsequently renamed Port Clinton to avoid confusion when the town of Bowen was founded in 1861 with its harbour of Port Denison). Due to threatening weather, Oxley went only to Port Curtis and his lengthy report was totally negative, advising Colonial Secretary Goulburn on 10 January 1824 that 'Having Viewed and examined with the most anxious attention every point that appeared in any degree eligible for the site of a Settlement... in my opinion Port Curtis and its vicinity does not afford Such a Site; and I do not think that any convict Establishment could be formed there' with any economic viability (*HRA*: I.xi:218).

Having rejected Port Curtis, Oxley sailed south to the still unexplored Moreton Bay where he landed at Red Cliff (today Redcliffe) and subsequently found a large river to the south which he followed as far upstream as possible and, in the same report, advised Goulburn that 'out of respect to His Excellency the Governor... [it] was now honoured with the name of *Brisbane River*' (*HRA*: I.xi:222; Oxley's emphasis). On its banks he recommended a site for settlement that was to become the city of Brisbane. From 1824 to 1840 the District of Moreton Bay was primarily a convict settlement; by the 1830s, however, squatters had begun to move into the region. Although transportation ceased in 1840, the convicts had already been removed by 1839 when Governor Gipps opened the 'District of Moreton' to free settlers. In 1842 land sales began.

Partly in response to such squatter movements, Governor Bourke, who arrived in 1831, had begun to consider as early as 1838 a free settlement and perhaps a new colony in this region even though it had relatively few settlers. The official census listed 1337 persons in the Moreton Bay district in 1828; rising to 2525 in the next census of 1846. The census of 1851 recorded only 8525 in Moreton Bay and the squatting districts, a marked contrast to the populations in 1851 of some 260 000 in New South Wales, around 190 000 in Victoria and 70 000 in Van Diemen's Land (renamed Tasmania when it became a separate colony in 1856) (Knibbs 1911:42,49).

Despite the small numbers at the time, the idea of separation of the Moreton Bay district from New South Wales began to be raised throughout the 1840s. The administrative centre of Sydney was a long way off: at least a week from Brisbane by sailing vessel, even longer from the inland settlements of Darling Downs. The British government under the new prime minister, Lord John Russell, who succeeded Peel in 1846, was supportive of the idea of a new colony, and one was proclaimed in May of that year, although no further action was taken at the time. News of that proposal and the possibility that it could involve the use of slave labour created considerable consternation among the anti-transportation groups of whom a major leader was the Reverend Doctor John Dunmore Lang. Having visited Brazil on three occasions – Rio in 1823, and Pernambuco (today Recife) in 1839 and 1846 – and, from witnessing there the appalling atrocities inflicted upon the indigenous peoples and black African slaves by both landowners and Portuguese Catholic clergy, Lang was reinforced more strongly in his Presbyterian hostility to Catholicism and became a vigorous opponent of slavery.

His central proposal was the promotion of cotton farming under humane conditions, as that fibre was becoming dominant in the textile industry, to counter the appalling abuses of the slave-based systems in the south of the United States and Brazil, which at the time supplied the factories of Lancashire. Unlike the system depending on the abject African slaves he saw in Brazil, he believed that the free colony, which he suggested should be named Cooksland, could produce easily greater yields due to 'the superior intelligence and energy of the capitalist or employer of labour' (Lang 1847:175). The significance of his proposals lies in his idea of the extensive cultivation of much of the Reef coastline: in a combination of 'at least a thousand miles of land in longitudinal extent and of various breadth, of first quality for the cultivation of cotton and sugar . . . *having a river-frontage directly accessible to steam navigation*' (178; Lang's emphasis). All settlement, he wrote, must be on 'the banks of principal navigable streams' (232).

Lang went even further to promote his ideas. Preempting any Westminster approval, he formed the Cooksland Colonization Company and, subjected to much official disapproval, even hostility, the first of his three chartered vessels, the *Fortitude*, arrived in Moreton Bay on 20 January 1849 with 245 unauthorised migrants aboard. Magistrate Wickham – former captain of the *Beagle* – had been instructed to deny them any courtesies or privileges and they were detained in uncomfortable circumstances in a valley 2 kilometres north-east of the main township. To this day Fortitude Valley, now part of the city of Brisbane, records that episode.

The Legislative Council in Sydney, anxious among other things to retain the valuable northern District of Clarence River, which included the fertile Richmond and Tweed river valleys, attempted to found a new centre to take precedence over the nascent Brisbane in, of all places, Port Curtis, with the hope that, when separation occurred, the capital would be as far north of Sydney as possible. Throughout the 1850s tension increased between Sydney and the two northern districts, Moreton Bay and Darling Downs, which only came to a temporary halt when Queen Victoria signed Letters Patent on 6 June 1859 to establish the new colony. Six months later, after the decision had been conveyed to Governor Denison (the incipient telegraph system had not yet reached Australia), Sir George Ferguson Bowen arrived in Brisbane and was sworn in as the first Governor on 10 December 1859. Brushing aside suggestions for a name tendered to her, however, having personally chosen 'Victoria' for the Port Phillip District in 1851, the Queen announced that she would call the new colony 'Queensland'. As Governor Bowen tactfully phrased this further display of arrogance in his Address in Reply to his welcome to Brisbane on 12 December 1859, 'Queensland was entirely the happy thought and inspiration of Her Majesty Herself' (Bowen 1889:I.90). Thereby a new colony was created.

QUEENSLAND EXPLORATION AND EXPANSION: 1859–1870

In the decades leading up to separation the quest to open up grazing lands through the great interior of the continent continued unabated and the unknown regions to the north of the Downs beckoned. In 1845 the intrepid adventurer Ludwig Leichhardt, encouraged by the colony's Legislative Council, had completed a marathon overland trek from the Darling Downs to the Gulf of Carpentaria, then across Arnhem Land where he reached the settlement of Victoria at Port Essington.

The Reef region, however, was relatively free from rapid occupation. It remained a fearsome place, having acquired a negative image from sensationalised accounts of shipping disasters and a justly heightened apprehension of attacks from Aborigines, confirmed in 1848 by the tragic failure of Kennedy to reach Cape York overland from Rockingham Bay. Already, in 1847, for example, Jukes wrote in his *Narrative*, basing his opinion on experience of the Reef littoral, that there is 'no coast in the world, of anything like the extent, so utterly destitute as that of Australia' (Jukes 1847:51n.). In 1852 John MacGillivray, during the survey cruise of the *Rattlesnake*, recorded Captain Stanley's observation that Cleveland Bay (today, Townsville), as a location for a settlement and a harbour 'is the worst possible place I ever saw for such a purpose' (MacGillivray 1852a:I.46). While the naturalists were delighted with the many finds of new species of plants and animals, the surveyors who described the Reef coastline were unenthusiastic in respect to new settlements.

Following the explorations of Leichhardt and Mitchell into the interior of the north there was relatively little activity until 1859 when impending news of separation created a flurry of activity. In February a syndicate of wealthy pastoralists and Sydney investors financed an expedition, under the leadership of George Elphinstone Dalrymple, which sailed to Rockhampton and began searching for suitable grazing lands in the Burdekin Basin, already reported upon favourably by Leichhardt. In July two brothers, Christopher

Lizard Island, named in 1770 by James Cook after the large sand monitors seen there. The Australian Museum research Station is to the right of the beach.
Photograph: Andrew Elliott.

Dunk Island, leased in 1900 by Edmund Banfield. Brammo Bay lies left of the sand spit in the foreground.
Photograph: Leon Zell.

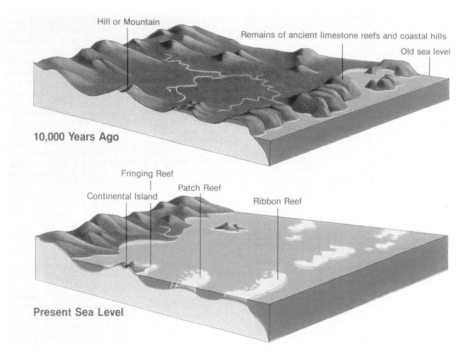

Origins of the Reef. Coral reef formation on lands submerged during rising sea levels after the last Ice Age.
Diagram courtesy of Geoff Kelly.

Coral ecosystems: the brilliant teeming life of reef communities depends on coral outcrops for survival.
Photograph: Courtesy of Great Barrier Reef Marine Park Authority.

The Low Isles. Backed by its mangrove cay and reef flats the small lighthouse cay was the site of the 1928–29 British Expedition research station.
Photograph: Peter Isdale.

Heron Island, a wooded cay, largely a national park, with a tourist resort and research station established in 1952.
Photograph: Leon Zell.

Green turtle excavating a nest in foreshore dunes on Heron Island to lay up to 120 eggs before returning to the sea after covering them with sand. Photograph: Julie Jones.

Crown of Thorns starfish that has completely destroyed a large acropora plate coral formation by sucking out all of the polyps. Photograph: Bill Legg.

Silt discharge into Reef waters. The Barron River discharging silt from inland agricultural clearing and coastal cane farming into Reef waters. Photograph: Andrew Elliott.

Dugong and her calf graze in shallow coastal waters on seagrass meadows, now threatened by development. Photograph: Ben Cropp.

and John Allingham, set out north from the New England tablelands with a large mob of sheep to the grazing district west of the Darling Downs known as the Maranoa, followed by a similar expedition by John Mackay. Then, barely three weeks before Bowen took office the Legislative Council of New South Wales, under the governorship of William Denison, who was vigorously opposed to separation, proclaimed two new grazing districts of Mitchell and Kennedy open for settlement. Kennedy encompassed virtually all of the Burdekin Basin, Mitchell adjoined its western boundary and extended to the edge of the arid regions of the far interior. The land rush began.

Governor Bowen was very enthusiastic concerning his task, and perhaps a little too eager in his forward-looking optimism when he wrote to Sir Edward Bulwer Lytton, Minister for Colonies, on 6 March 1860, less than four months into his tenure, that:

> fresh bands of pastoral settlers, driving their thousands of cattle, sheep, and horses before them, are fast pushing out into the wilderness; and it is confidently expected that, in the course of the next five years, there will be a chain of stations from Moreton Bay to the Gulf of Carpentaria; and thence a line of steamers to Singapore, opening up commerce with India and China, and with the Dutch and Spanish colonies of the Eastern Archipelago (Bowen 1889:I.107).

Continuing to write enthusiastic dispatches to London, barely a year after taking office he wrote on 4 December 1860 to Henry Clinton, Duke of Newcastle and Secretary of State in London, that:

> there is something sublime in the steady silent flow of pastoral occupation over north-eastern Australia ... [which] resembles the rise of a tide or some other operation of nature, rather than the work of man ... at the close of every year we find that the margin of Christianity and civilisation has been pushed forward by some 200 miles (I.193).

A contentious issue, however, was soon to arise.

The Letters Patent authorising the separation of Queensland from New South Wales followed the original instructions of 1787 to Governor Arthur Phillip whose jurisdiction was 'from the northern cape or extremity of the coast called Cape York, in the latitude of 10°37' South to the southern extremity ... in the latitude of 43°39' South' with the very ambiguous phrase 'including all the islands adjacent in the Pacific Ocean'. At the time, the implicit intention (although no examples were listed) seems to have been to include those islands immediately offshore, including the nearby Torres Strait group of continental islands, that is, Prince of Wales (indigenous Murulag), Horn (Nurupai), Thursday (Waibene), and the numerous smaller islands including Albany, Goode and Booby, since the latitude of Cape York in fact, is 10°41' whereas the demarcation given of 10°37' is slightly to the north and runs through both Booby and Horn islands. Governor Bowen believed that his jurisdiction encompassed the landmass of Queensland with its immediately contiguous islands, perhaps those able to be reached by small day boats, and he sought clarification. To his astonishment he was advised by Colonial Secretary Newcastle that it extended only to the conventional three mile limit. The Torres Strait islands, in fact, were outside his authority and remained under the nominal control of New South Wales until 1872 when Queensland's maritime boundaries were extended in all directions by 60 nautical miles.

COASTAL SETTLEMENT OF THE REEF: TOWNS, PORTS, RAILWAYS

Within his first year, assisted by Colonial Secretary Robert Herbert, Bowen set about the task of developing the colony as rapidly as possible. In addition to opening the interior of the new colony to settlement, he gave equal attention to the Reef waters as the main axis of communication by encouraging further settlements along the coast. Two months before separation Captain Henry Sinclair, seeking an ephemeral promised reward for the discovery of a good harbour on the coast, had sailed the *Santa Barbara* north and avoiding the difficult bar at Rockhampton, on 16 October 1859, entered an excellent, previously undiscovered harbour to the north which he named after Governor Denison. The following year the new government of Queensland sent the ketch *Spitfire* to search for other harbours and report on Sinclair's claims for Port Denison. Finding them accurate, on 12 April 1861, now having a governor of their own, the Colony of Queensland proclaimed its small settlement of Bowen as the first official town on the Reef coast, situated midway between Brisbane and Cape York. Neither Gladstone, declared a Port of Entry in 1859, nor Rockhampton, settled in 1857, were yet officially designated as townships.

As a necessary complement to coastal development Governor Bowen returned to earlier concerns for a defensive northern outpost on Cape York, citing a number of major considerations: maritime safety, control of hostile aboriginal tribes, defence of the colony, and increasing trade with Asia and India, all drawn from MacGillivray's *Rattlesnake* narrative (MacGillivray 1852a:I.319–20). Maritime safety, initially, was uppermost as a concern as ship losses continued to mount: in the decade 1831–41 at least 13 were recorded, rising to 40 from 1841–51 and nearly doubling to 77 between 1861 and 1871 (Loney 1993, *passim*). Despite the fact that traffic through the Strait had been increasing rapidly most ships still took the more dangerous outside route and entered Torres Strait at one of the hazardous Reef openings at the treacherous Raine Island or others further north. In his narrative of the *Beagle* voyage Stokes warned mariners of the risks in attempting Reef entry in that region around 11°50'S. Records had been made there of 'a strong northerly current outside the reef, in some instances nearly three knots' and such a course, in preference to the safer, if slower inner passage due to overnight anchoring, was 'attended with much greater risks . . . the anxieties and danger of the outer passage, which, short as it is, has doubtless sprinkled grey hairs over many a seaman's head' (Stokes 1846:I.373). In contrast, Stanley, in his survey by the *Rattlesnake*, had referred to the 'tranquillity and security' of the inner route while Jukes had also argued that the inner route was the 'most practical route to and from Australia to India' (Jukes 1847:I.303f.). None the less, whether by the inner or outer route, Torres Strait continued to be used by shipowners, especially for the tea trade with China which was becoming a major commodity import to the colony, despite ships being lost and the insurance premiums levied by the southern companies based in Sydney, Melbourne and Hobart. In 1854, with losses mounting, the managing committee of Melbourne Underwriters endorsed all insurance policies with the stipulation 'Not warranted for Torres Strait' and imposed a 2.5 per cent surcharge (Foley 1982:17–18).

In order to minimise inherent dangers, Bowen ordered HMS *Pioneer* under command of Captain Robinson to complete the survey of adjacent waters, especially those to the west in the vicinity of Booby Island. Robinson was quite enthusiastic about the natural advantages of the region, from a maritime point of view, and in a report of 2 July 1863 to Bowen he cited the opinion of Commodore Burnett that 'the Great Barrier Reef, hitherto regarded by seamen as a bugbear, was one of Queensland's greatest blessings, being a natural breakwater to the South Pacific Ocean, and making the intervening sea from the Percy Islands to Cape York *one great and secure harbour*' (Robinson 1864:117; his emphasis).

Aboriginal hostility was another major factor in seeking a northern base. It is difficult to generalise about the numerous indigenous tribes: many were quite friendly, as the treatment of Barbara Thompson illustrates, but others were engaged in determined resistance to European occupation. Some massacres had received wide publicity and become part of white prejudice to all blacks, of which those of the crews of the *Charles Eaton* in 1834 and the *Stirling Castle* in 1836 were highly sensationalised. An anonymous account of the former, believed to be by Charlotte Barton, entitled *A Mother's Offering to Her Children*, was published in Sydney in 1841, allegedly to warn children of the dangers from Aborigines. In August 1834 seventeen survivors of the *Charles Eaton*, which was wrecked off the Sir Charles Hardy Isles, were massacred and decapitated, as she wrote, by 'treacherous savages' in Torres Strait as they made for Timor and their skulls, found on Aureed Island, were used as totem decorations by Murray Islanders. Survivors in the second longboat reached Amboyna and raised the alarm.

The episode received wide publicity in the *Sydney Herald* (30 June 1836) and there were several other accounts in 1836 and 1837 published at the same time the story of Eliza Fraser of the *Stirling Castle* was being promoted in equally gruesome detail. At the very moment of separation yet another grisly massacre was being widely publicised, this time of the crew of the *Sapphire* which was wrecked attempting the Raine Island entry. The crew took to two separate longboats; 18 survivors in one boat were massacred on Hammond Island (Keriri) in the Thursday Island group.

An official presence at the Cape was clearly needed, and seeking approval for the foundation of a northern base, Bowen sent a report to the Duke of Newcastle on 9 December 1861 in which he advised that from

> a naval and military point of view, a post at or near Cape York would be most valuable, and its importance is daily increasing with the augmentation of the commerce passing by this route, especially since the establishment of a French colony and naval station at New Caledonia . . . [A] small armed steamer with a light draught of water . . . would command the whole of the commerce between the South Pacific and the Indian Oceans.

Further, with the *Sapphire* episode in mind, he recommended that a gunboat would be useful to both patrol the waters and 'might complete the surveys and charts of Torres Strait and of the Great Barrier Reef' (Bowen 1889:I.215). In the interests of trade Bowen went so far as to suggest that a base there could well become a second Singapore, although concerns were expressed at the time when it was asked by one unidentified member at a meeting of the Royal Geographical Society in London, what 'produce were the steamers to get there? Cape York would only be of advantage as a port of call to British ships passing through Torres Straits' (*Proceedings*, Royal Geographical Society, VIII.4, 11 April 1864,120).

The site finally chosen, on the mainland opposite Albany Island, was named Somerset after the First Lord of the Admiralty and, in subsequent communications to the Secretary of State for Colonies, Governor Bowen expanded on its advantages. At the time, only the harbours of Brisbane and Bowen, connected by the new technology of steam navigation, were functioning ports for the produce of the interior. The base at Somerset would allow the extension of steam navigation since a coaling station could be established there. Along with another port at Rockingham Bay (the future Cardwell) a connection could then be effected with Timor where the Dutch government had already established a fortnightly steamer service to Europe. Further reflection indicated that Albany Island was unsuitable and an alternative on the nearby mainland was finally chosen; in July 1864 HMS *Salamander* and the cargo ship *Golden Eagle* set out from Brisbane with the first supplies to establish the base. John Jardine, magistrate in Rockhampton, was appointed resident official (that is, magistrate and police inspector) and his sons Frank and Alick travelled overland through Cape York driving their herd of cattle to join him, massacring Aborigines in their path. When they arrived Frank boasted of the eighty notches he had cut on the butt of his rifle (Sharp 1992:65). Despite the move from Albany to the mainland the location was not suitable, especially since by 1870 Torres Strait was developing as a region of commercial exploitation of the pearlshell and bêche-de-mer industries. In 1877 Somerset was abandoned – becoming derelict – in favour of a new base on Thursday Island (Waibene) which has remained the administrative centre of the Torres Strait region. Within a decade the settlement pattern of the Reef coast was virtually complete: eight towns in addition to Ipswich and Brisbane – Maryborough, Bundaberg, Gladstone, Rockhampton, Mackay, Bowen, Townsville and Cardwell – were in their early stages; in the 1870s Cairns, Cooktown and Port Douglas were also established.

The original stimulus to growth of the coastal towns was to service the pastoral districts of the interior, chiefly as ports for an effective maritime transport system in the absence of serviceable roads across the steep mountains. In the nineteenth century the great obstacle to development of land transport, and therefore of the Reef towns, was the rugged Great Divide mountain range which runs close to the coastline, reaching the sea in a number of places – between Cairns and Townsville, at Mackay, and near Gladstone – and this meant that roads to the coast from the pastoral districts were constrained to follow the few suitable routes. The Divide, moreover, prevents the formation of great coastal rivers, although explorers of the 1830s and 1840s still held hopes of finding another Mississippi or Amazon. Only two medium river systems exist, the Burdekin, discovered by Leichhardt, and the Dawson–Fitzroy that discharges into Keppel Bay, near modern Rockhampton. In both cases access was denied to early river shipping by the mudflat-mangrove delta of the Burdekin in Upstart Bay south of modern Townsville and the rocky bar at the mouth of the Fitzroy.

All the settlements struggled in the early decades to build their population base, and, since convict transportation had ceased in 1840, many farmers were looking to various kinds of indentured labour, particularly from India and China. In a curious blend of liberal thought and racism, Bowen supported the idea of cotton and sugar as staple crops, produced by free settlers without any slavery at all, although, given the fact that the Queensland coast was, lapsing into Spanish, a '*tierra caliente*, or hot region' he saw no difficulty with importing Indian coolies for menial labour (Bowen 1889:I.126). In a discussion by the Royal Geographical Society in London in 1862 of problems with the labour shortage in Queensland, and the difficulties of a climate 'too hot for Europeans to inhabit and multiply in', the idea was taken a step further. One proposal was 'to get the Chinese to do the labour . . . there would be two distinct races; the Chinese would be the helots [serfs of ancient Sparta] and we would be the masters' (*Proceedings*, Royal Geographical Society, VIII.4, 11 April 1864,120).

Equally, however, there was a growing resistance by philanthropic and church groups to any kind of non-European workforce. Labour-intensive cotton production was therefore impracticable and, consequently, white settlers turned to sugarcane. By the 1870s it had become a major crop, originally centred in the Mackay region, spreading south to Bundaberg and Maryborough and in subsequent decades as far north as Cairns. Even so, labour shortages, and the demanding physical effort in the days before mechanisation, caused many of the larger landholders to seek imported workers who were captured, throughout the final decades of the century, from the Pacific islands, and whose presence was to create a period of major social and political turbulence.

Tropical agriculture, however, seemed a major psychological obstacle to immigrants from Britain, Ireland and Northern Europe. Many farming subdivisions remained unsold and so, when a renewed push for white migrants occurred in the 1880s, the colony had to confront its prejudices. As early as the first days of separation Colonial Secretary Herbert resisted southern European migration – Italians in particular – not on racist grounds, but on those of religion: southern Europeans would increase the Catholic population. Interestingly, Governor Bowen was not opposed to it. A classical scholar, he was formerly rector of the Ionian University on Corfu in the Adriatic Sea, where most of the locals were either Italian from the days when it was part of the Venetian Empire (before French and then British occupation), or else Greek. Moreover, his wife, Roma Diamantina – whose name commemorates a city and a river in Queensland – was Italian, daughter of Count Candiano di Roma, President of the Senate of the Ionian Islands. Rather curiously, James Quinn, first Catholic bishop of Queensland, was also opposed to Italian migration, especially of Italian priests, giving as his reason the belief that his flock of Irish Catholics could lose their faith (Douglass 1995:34–5). Perhaps that said something of the character of Irish Catholicism.

Premier McIlwraith was also hostile to the idea of southern European migrants but Griffith, who succeeded him in 1883, responded positively to representations from the Catholic priest Paolo Fortini – Superior of Augustinian missionaries in north Queensland – to accept Calabrian migrants. Later, the idea took hold that Mediterranean Europeans with their olive complexions could handle the tropic sun more easily. Even so, ingrained attitudes could not be put aside so lightly, and Anglo-Celtic prejudices continued to exaggerate the impact of southern European – Iberian, French and Italian – migration, with the Italians being the major target of opposition. In fact, very few Italians migrated to Queensland in the nineteenth century, despite public perceptions to the contrary.

The chief stimulus to population growth, however, was the discovery of gold in payable quantities. The first finds in Queensland were at Canoona, north-west of Rockhampton in 1857, but the alluvial beds soon petered out and the great influx of prospectors left almost as fast as they arrived. Then a major strike that was to last for some decades occurred inland from a settlement at Cleveland Bay. Already at that location, John Melton Black, a manager for Captain Robert Towns, a Sydney merchant engaged in numerous trading activities including the sandalwood and bêche-de-mer trade, had established a boiling-down works from the abattoirs processing animals from the pastoral inland and so the settlement received its tautonymous designation as Townsville, which was declared a port of entry in 1865. It began its development in 1868 when gold was found inland at Ravenswood in 1868 and at Charters Towers in 1872 when gold was also discovered in payable quantities further to the north along the Palmer River, a tributary of the Mitchell that empties into the Gulf of Carpentaria.

Prospectors soon found their way to the Endeavour River – scene of Cook's disaster in 1770 – as the nearest place to trek inland over the formidable sandstone ranges – and before long Cook's Town was filled with tents and developed rapidly as Europeans flooded in. Then came the Chinese, who first arrived in 1874, at times in specially chartered ships from Hong Kong, chiefly to seek their fortunes in the Palmer goldfields. Their arrival was strongly resented, mainly because of their disproportionately large numbers and a general hostility to Asians as a lesser race of humans, exemplified in Governor Bowen's suggestion of importing them as labouring serfs. A recent documented record gives the Chinese resident population of Queensland in 1881 as 11 253; for 1891 as 8522; and for 1901 to be 8448 (John A. Moses, cited Douglass 1995:18).

By 1874 Cook's Town had 94 hotels, a large number of shops and services to meet the needs of the arrivals, and had become a thriving port. Meanwhile ships brought in, not only hopeful miners, but also great quantities of building and other infrastructure materials. Renamed Cooktown on 1 June 1874 it was declared a municipality in 1876 and, for a time, became a flourishing settlement – in fact, the third port of Queensland after Brisbane and Townsville.

Throughout the later decades of the nineteenth century the coastal towns of the Reef became locked in adversarial conflict for development and dominance, each seeking to develop port facilities to attract the maximum coastal trade, the only means of passenger and cargo transport along an immense eastern coastline of approximately 2400 kilometres (1500 miles) from Point Danger at 28°S to Cape York at 11°S. The final arbiter in deciding port dominance came from the role of the railways when construction across the colony began to accelerate in the 1880s.

Along the Reef coast some railway building was commenced in the 1880s by the Colonial Sugar Refining Company to carry cane wagons along narrow gauge tracks (42 inches; 1.067 m) from outlying independent farms to their mills in Bundaberg, Mackay and Innisfail. At the same time, other private venture companies and even shires began building small branch lines from their regional centres, chiefly Maryborough, Gladstone, Bowen and Ayr. The government also moved to begin extending the railway north along the Reef coast in the same period, and by linking this with private lines a continuous track from Brisbane to Rockhampton was completed by 1892. Following passage of the 1910 *North Coast Railway Act* further construction began, rather piecemeal, to connect other centres: Proserpine to Townsville in 1913 and from Cairns southwards to Daradgee in 1919 where the South Johnstone River delayed further construction for a period. The great difficulties were negotiating coastal ranges and bridging rivers, which presented formidable engineering problems, given the technology of the times. By 1923 those difficulties had been overcome and a continuous line linked Brisbane with Townsville. A year later, with the completion of the bridge over the South Johnstone, the railway terminated at Cairns. In 1929 the line was officially designated the 'Sunshine Route' and on 27 May 1935 the first purpose-built 'Sunshine Express' passenger train entered service.

REGULATING THE REEF: COASTAL SHIPPING

For all of the colonial period of the nineteenth century, and the early decades of the twentieth, since there was no land route available the Reef was virtually a maritime highway along which most of Queensland's local traffic travelled. Even the steel rails, rolling stock and associated equipment to build the railway were carried in ships, in an instance of ultimate irony since, when the rail network was finally completed in 1924, shipping companies lost a considerable part of their trade. None the less, in the early years of the colony ships played a major role in the development of Queensland, and especially in the structure of the port, harbour and railhead system.

Before separation, the major shipping lines were based in Sydney and Hobart, chiefly to provide commercial and passenger links with England. Ship owners used their vessels to carry cargoes in both directions: wool, wheat, cattle and other primary produce to England (and horses to India for the army); inward cargoes of manufactured goods essential to the developing colonies. For local traffic, in contrast, a new development occurred with the introduction of the first steam-assisted sailing ships in 1839 when the Hunter River Steam Navigation Company, based in the coal mining town of Newcastle, used nine paddlewheelers on the eastern coast of Australia. Such vessels, however, proved unsuitable for the rolling seas in open waters and by 1869 were relegated to placid inland rivers. Expanding its services as the colony grew, the Hunter River Company was restructured in 1851 as the Australasian Steam Navigation Company (ASN) and rapidly expanded its fleet in strong competition with several other lines. By 1860 in the first year of Queensland's colonial period the ASN began sailing regularly to Port Curtis and Rockhampton with intermediate stops in the New South Wales northern rivers region to take on cargoes, including the valuable cedar, from the Richmond and Clarence regions.

By the 1860s steam engines and the screw propeller began to come into use but a difficulty remained in that the load of fuel required was disproportionately large. By the end of the century, however, as efficiency increased, screw-driven powered vessels came to dominate. For bulky cargoes, chiefly wool, timber and grain, sailing ships continued to be used into the early twentieth century, although improving technology led to hulls, masts and spars being constructed of steel, and winches being mechanically powered. When the Suez Canal was opened in 1869 Queensland pressed a case for the inner Reef route as the fastest and most economical way to India and England.

By that time the government in Brisbane sought to distance itself even further from southern dominance by Sydney and Melbourne. To that end it signed a mail contract – avidly sought by shipping lines as an attractive subsidy – with an overseas company, the British India Steam Navigation Company (BISN) which began in April 1881. The new company was required to use the Torres Strait route to Singapore where cargoes were transferred to other lines such as the Peninsular and Orient (P&O) for onward carriage to England. By 1880 refrigerated ships came into use allowing direct transport of frozen meat to England via the Suez Canal, giving a considerable boost to the grazing industry. Sugar too, was exported directly from Queensland ports in direct competition to that grown in Mauritius. The ASN felt even greater pressure as other lines came into competition.

Creating and subsidising shipping lines did not guarantee Queensland a secure export industry since there remained considerable prejudice among ship owners and masters against the Torres Strait route, both via the inner and outer passages, due to the very high probability of loss. Despite the brilliant work of a century of hydrographic charting of the main channels by the British navy, risks were still considerable, and shipwrecks continued to increase since many hazards remained undiscovered and therefore uncharted. In the days of timber-hulled sailing vessels, which were necessarily slow and difficult to manoeuvre in confined waters, especially against a lee shore, there were many accidents and fatalities along the entire length of the Reef, as well as in the open waters south of the Reef to Moreton Bay. The greatest threats were in the outer Reef waters of the Coral Sea where, despite Denham's improved charts, Middleton Reef and the Chesterfield group continued to claim a large number of ships, while the inner route, especially north of Cooktown, was equally hazardous.

As best as can be estimated – since many shipwreck locations could not be found due to the simple fact that vessels failed to reach their destinations – in the first decade of separation the Queensland coast (Brisbane to Torres Strait) claimed 88 ships of which 28 were north of Cooktown; the following decade 1871–80 of the 102 recorded losses, 27 were in the northern section; in 1881–90 the number was 36 out of a staggering 141; in 1891–1900 24 out of 108 (Loney 1993:47–105 *passim*). The disasters which most gripped the public imagination in that period were the tragic 1890 sinking of the supposedly safe steel-hulled steamship *Quetta* in the Adolphus channel between Cape York and Mount Adolphus Island, and the horrendous Bathurst Bay cyclone (inside Cape Melville) of 1899, when 59 pearling luggers and 17 other vessels were totally destroyed.

The task facing the new government was formidable given that, in 1859, it inherited 3236 miles of coastline – from Point Danger south of Brisbane to Normanton in the Gulf, and after 1872 to beyond the Wellesley Islands – much of it poorly charted, with the ports at the time in flood-prone river estuaries, and with the immense, dangerous Reef enclosing most of the eastern side of the colony. In contrast, New South Wales had its great harbour of Port Jackson and relatively few hazards with which to contend. The indispensable mariner's guide, *Pugh's Almanac*, for example, in the 1864 edition continued to warn mariners never to trust the lead line for soundings since 'most of the reefs and coral patches . . . spring up so abruptly . . . that the lead frequently gives no warning of their vicinity before a vessel approaches too near to avoid them'. In order to control the great problems of navigation in a colony whose population was spread thinly along much of the coastline, for whom the sea lanes and the Reef route was its main artery, the government in 1862 passed the *Marine Board Act* to begin the process of regulating maritime traffic. The Marine Board was to have a Portmaster and four other members whose duties, among many, were to superintend construction and maintenance of wharves, harbours, lights, port fees and the licensing of pilots for the numerous difficult entries which required good local knowledge (Davenport 1986: 52f.).

As early as 1859, in fact, Queensland had begun to take responsibility for its own coastline and had appointed naval lieutenant George Poynter Heath (1830–1921) as marine surveyor for the colony. Just twenty-nine years of age, the young Heath was already a veteran of Reef navigation. Having joined the navy as a 15 year old cadet his first assignment was to the *Rattlesnake* for its three cruises under Owen Stanley, 1846–50, and then, for the ensuing five years he served on *Calliope* and *Fantome*, two other naval ships conducting Reef surveys in the period 1850–55. Back in England he spent the following years, before returning to Moreton Bay, as an Admiralty draftsman helping draw charts of the surveys of the *Rattlesnake*. On 13 January 1862 Heath was appointed Portmaster for Brisbane and in 1869 Chairman of the Marine Board. For twenty-eight years, until his retirement on 30 June 1890, he worked tirelessly at the task of improving the ports and harbours of the entire Queensland coast and most especially at one of his greatest contributions: increasing the safety of the inner route.

The Reef itself, however, was not always the culprit: there were numerous instances of incompetent masters, and of unsuitable vessels being sailed well beyond their design limits. Moreover, there were few navigation aids on the Reef. To deal with that problem, and the considerable reluctance of owners and masters to use the inner reef route, the Queensland parliament in May 1864 established a 'Select Committee On the Rivers and Harbors of the Colony' to inquire into the condition and improvement of the harbours and shipping lanes of the entire coast, and to seek remediation of the very great dangers, including the provision of adequate navigation aids, including lighthouses, lightships and channel markers. In addition, the committee recommended that the newly built telegraph lines be extended to the lighthouses in order to afford 'the advantages of meteorological and shipping reports, and a greater facility for rendering assistance in cases of distress' (QPD, 31 August 1864, 1201).

The committee summed up the problem as a commercial issue for Queensland stressing the necessity to deal with the 'serious task of so lighting what is called the Inner Passage within the Barrier Reef, that not only the trade to our own rapidly increasing ports may be protected, but that much of the trade with India, China, and other countries to the North of this Continent may be diverted from the Western to the Eastern line of Passage' (*Journal*, Q.Legis.Cncl.VII, 1864, Paper 37). Ships taking the outer route, if they survived the hazards, simply sailed through Torres Strait and on to Coupang (today Kupang), on the southern tip of Timor, and points north and west: the government wanted to encourage use of the inner route in order to entice ships to carry cargoes from Queensland ports.

In 1865, to consider further navigation safety, a second inquiry by the Select Committee on Steam Navigation was conducted. For the first time the need for competent pilots was raised. Lieutenant Heath was called as an expert witness and was unequivocal in advocating navigational beacons and a pilot service. Drawing from years of experience, he advised that the inner route could be made much safer and more attractive to shipowners, and 'with beacons and a skilful pilot it might not be necessary to anchor more than two or three nights' (cited Foley 1982:23). The issue of navigational lights came up in February 1873 at the Intercolonial Conference held in Sydney where the agenda listed four major concerns: the condition of existing lights, the location of new lights, the efficiency and economy of navigational management, and the equitable sharing of costs among the colonies. In regard to the last item, it was decided that each colony would maintain its own mainland lights, but the costs of those offshore – which materially affected Queensland – would be shared.

Throughout his years of service as Portmaster of Brisbane and Chairman of the Marine Board of Queensland (1862–90), Heath, who became known as the 'father of Queensland lighthouses' worked at the task of marking the coastline, and, in particular, the northern inner passage, travelling to every installation to inspect the work and make follow-up observations. In 1857 a lighthouse had been built – the only one in Queensland of stone – on the northern tip of Moreton Island to guide shipping into the bay. Following separation, the first lighthouse erected by the new colonial government was constructed of prefabricated cast iron plates and assembled on the site in 1868 at Bustard Head; the second, also of cast iron components, came two years later at Sandy Cape. Those were the only iron lighthouses; the remainder were built – usually on concrete foundations – of timber framing and heavy gauge galvanised sheet-iron cladding. In steadily increasing numbers they appeared in the most suitable, and often dangerous locations: Bustard Head in 1868; Sandy Cape in 1870; Lady Elliot in 1873; Cape Capricorn in 1875; North Reef and Low Isles 1878; Dent Island and Cape Cleveland 1879; Double Island Point 1884; Pine Island 1885; Goode Island and Cooktown 1886; Booby 1890; Pipon 1891. By the end of 1894 (with the addition of Archer Point and Rocky Islet) there were 17 primary sea coast lights operating, 5 lightships, 8 secondary coast lights and 106 river and harbour lights, with an additional 4 in the Gulf to mark the Norman river entry, as listed in the official navigational guide compiled by the nautical surveyor E. A. Cullen (1895).

Heath's final year, unfortunately, was clouded by a controversy that was never finally settled, created by the tragic sinking of the British India Company steamer the *Quetta*. The sensational press reporting of the incident created public outrage when it was stated that the ship, bound for India via Torres Strait, struck an uncharted pinnacle of rock in deep water in Mt Adolphus Channel on 28 February 1890 at 9.15 p.m. on a calm, moonlit night. The ship sank within three minutes with no time to launch most of the lifeboats: 133

persons died and 158 survived. Receiving extensive press coverage, with numerous illustrated engravings of the vessel sinking bow first, the outcry was directed largely at the Marine Board which was accused of having failed to chart a hazard reported six years earlier by the master of the *Thales* when it 'struck an underwater obstruction . . . in the vicinity of Cambridge Point (Adolphus Island)' (see Foley 1990:107f.).

The cause of the disaster was never determined. Accusations in the press were levelled at Heath who denied the *Quetta* had hit the same rock as the *Thales*. An official inquiry was then announced for 2 April 1890. Conflicting evidence came forward about the grounding of the *Thales* and Heath, as a witness, was somewhat evasive. His most significant comment was that it was 'too speculative to say that if a search had been made for the *Thales* rock the *Quetta* rock would have been found' (111). The official report of the Marine Board of Inquiry, 26 April 1890, concluded, not very convincingly, that the location identified as the *Thales* rock 'is evidently an error' by the master of the *Thales* and 'from the evidence solicited in the course of this inquiry it appears more than probable that the *Thales* struck no unknown rock but grounded on the outer fringe of [another] reef' (*Report*, Appendix 5; collected in Foley op. cit. 160–62).

Despite that major tragedy, the achievement of Heath by the time he retired with the naval rank of commander, was impressive. Heath had worked assiduously to improve the safety of the Reef, and his efforts are vindicated by the statistics. In 1895, five years after his retirement, 6157 vessels (overseas and intercolonial) docked in Queensland ports (Davenport 1986:257) and while but 17 ships were lost on the entire coast, the majority around Moreton Bay, only four were wrecked along the inner route (Loney 1993:97–99). Even so, the task was not finished and further lighthouses and markers needed to be installed by the Queensland government until, in 1915, the Commonwealth assumed the task of building and maintaining all lighthouses and navigational aids for the entire Australian coastline.

Efforts were also made to meet another major need: establishment of a pilot service through the Inner Route (now so termed as a designated waterway), and the Torres Strait. The *Queensland Navigation Act* of 1876 was the first move towards licensing of Reef pilots (as distinct from local harbour pilots). However, although it made provision for such control, nothing specific was implemented until 26 May 1884 when the Marine Board gazetted 'Regulations for Pilot Service, Torres Strait', setting out ten provisions. Significantly, however, while the behaviour of pilots was prescribed, pilotage was not made compulsory for shipowners. It was simply left to the good sense of shipowners to determine whether it would be in their own best interests to pay the scheduled fee, set out in the 1884 Regulations: £15 through Torres Strait, £25 from Torres Strait to Cooktown, £30 from Torres Strait to Moreton Bay, a situation almost unchanged until the end of the twentieth century.

In just four decades, the colony of Queensland had made remarkable progress in the development of a viable economy and construction of an extensive infrastructure of communications: roads, railways and designated shipping lanes along with towns, harbours and ports. Dominating it all was the Great Barrier Reef which, in the words of Captain Robinson to Governor Bowen in 1863, provided 'one great and secure harbour'. The Reef, moreover, lies mainly within the geographical limits of one political entity, the State of Queensland, and it alone in the world has such a remarkable maritime boundary. Towards the end of the colonial era, however, the Reef had become more than a source of excitement to scientists who had marvelled at their discoveries of a separate creation; it would also be perceived as a dangerous wilderness to be subdued and a resource for unthinking exploitation.

CHAPTER 8

FROM NATURAL HISTORY TO SCIENCE, 1850–1900: VOYAGES OF THE *CHALLENGER* AND THE *CHEVERT*

CABINET COLLECTORS AND MUSEUMS: DARWINISM RESISTED

The final surveying of the Reef by the *Herald* and the subsequent development of a port and a more secure navigation structure by Heath and the Marine Board marked the end of one era and the beginning of a new phase. The work of the naturalists, however, had made no comparable advances. Apart from the few scientists who were carried aboard the British naval survey vessels, whose findings were taken back to England to be integrated into the maturing disciplines of biology and geology, little progress occurred in the colony. Natural history in Australia, in contrast, in its formative years had been the preserve of a minority of conservative and wealthy gentlemen amateurs and cabinet collectors. Although membership in the naturalist societies was limited to men, a few talented women also made contributions. Their efforts, however, were almost entirely limited to botany, and then chiefly as illustrators, and some of their work has enduring distinction, including that of the sisters Harriet and Helena Scott; and Ellis Rowan, whose paintings of Queensland flora are excellent examples of botanical illustration. The one woman to achieve distinction in both botany and zoology was the German collector Konkordie Amalie Dietrich (1821–91) who travelled throughout Queensland on behalf of the Godeffroy Museum of Hamburg. Unfortunately, all of her specimens left the country (Moyal 1986).

As the terms imply, most gentry collectors in those decades were neither scientifically trained nor interested in the growing movement for the practical and experimental study of nature as a product of evolution. On the contrary, they were generally devout Christians who sought to reveal the beauty and extent of divine design in their collections, and to that end they employed promising young men to do the actual collecting, some of whom were to advance natural science significantly, and in some cases, controversially. The challenge of understanding the 'separate creation', in fact, had stimulated the foundation of Australia's first naturalist society, when a small group of affluent men formed the Philosophical Society of Australasia in 1821 as a forum for discussing the natural history of the country. After lapsing a little over a year later, it was revived in 1850 as the Philosophical Society of New South Wales and in 1866 became the Royal Society of New South Wales.

The idea of building large collections, housed in impressive glass-panelled cedar and mahogany cabinets, had been generated in the seventeenth century. Some of these were extremely comprehensive, such as that of Sir Hans Sloane which formed the initial collection of the British Museum of Natural History, and that of Linnaeus acquired by James Edward Smith and passed on to the Linnean Society of London, as well as the extensive collection of shells of the Duchess of Portland and the great botanical collection of Sir Joseph Banks. In Australia the 'cabinet' – in fact a large number of display cases – of Colonial Secretary Alexander Macleay was unrivalled, its entomological collection second only to that of the British Museum. After the death of Alexander Macleay the collection was inherited by his son William Sharp Macleay in 1867, and in turn by nephew William John Macleay who offered it in 1873, under the terms of the original bequest by Alexander, to the University of Sydney as the foundation of its natural history collection. In 1888 it was transferred to the new purpose-built Macleay Museum in the university grounds where it remains to this day.

In the same decades, and from the same motives, the movement to found a public museum of natural history in Sydney began, also under the stimulus of Alexander Macleay. Originally part of the Botanical Gardens, its small zoological collection was moved in 1849 to its present location in College Street (leaving botany to the Gardens), under the name Australian Museum, a territorial claim that still discomforts other State museums. Incorporated as a State instrumentality under the *Museum Act* of 1853, prepared by William Sharp Macleay (1792–1865), by the end of the century it had grown to five departments: zoology, palaeontology, mineralogy, ethnology and numismatics. It was dominated at first by Christian fundamentalist Alexander Macleay as chairman of the Board of Trustees and in turn by William Sharp Macleay until the latter retired in 1862 due to ill health, when the succession went to William John Macleay (1820–91).

Probably the best illustration of the fundamentalist belief in divine design and the great chain of being that guided the early naturalists, and continued to be championed by the Macleays, was given in a series of public lectures in Melbourne in 1869–70 by Frederick McCoy, foundation professor of natural history in the university, who explained to his audience that the significance of Australian endeavours to date lay in finding evidence to complete the picture. The visible world we experience, he declared,

> the whole of the vegetables, the whole of the animals, are part of one, great, complete, universal, perfect plan, which was conceived by the Almighty, in the beginning while as yet there were none of them, and that all of the separate parts were brought into existence at His own different times, following laws some part of which we may dimly perceive.

As those missing 'separate parts' or 'creations' (such as Australia) were discovered, so they could be fitted into the overall design. At first there were many gaps, he declared, but, as exploration and collections continue, when 'you go to look at some other country you find many of the creatures that were wanted to fill your gaps, to make up the perfect sequence' which will lead you to 'find that many of them follow in such exact succession that admiration is excited at the beauty and continuity of the chain' (McCoy 1870:23–32).

McCoy, as a bitter opponent of the new theory of evolution, was trying to counter the pernicious doctrines set out by Darwin in *The Origin of Species by Means of Natural Selection* of 1859 that were creating a furore in both scientific and religious circles in Europe. 'There is no authority', McCoy declaimed in a public lecture, 'either in Scripture or science, for the passage from a low creation to a higher one' (McCoy 1869:10).

Despite strenuous efforts by fundamentalists to conserve the inerrancy of Scripture, scientists continued to experience difficulties in ordering nature and discovering the underlying design. Following Linnaeus there had been further efforts to replace his artificial taxonomy with a more effective 'natural' system that linked life forms to each other by similarities, 'affinities' – thereby reflecting the seamless continuity of nature. After the de Jussieu dynasty at the Jardin des Plantes in Paris had attempted the task, without success, others sought a solution. In Australia, William Sharp Macleay took up the challenge and developed a 'quinary' natural system built upon numerous clusters of five contiguous circles, each circle containing related species; where the circles touched each other the points of contact indicated affinities of structure and function. The system seemed promising, from a grand design point of view, since Macleay, like both his father and cousin, was an uncompromising creationist. It was, however, inherently faulty and it, like other such attempts, was swept aside by the powerful impact of Darwin's new theory.

In *Origin of Species* Darwin dealt ruthlessly with fundamentalist beliefs by asking the question:

> But what is meant by this system? ... many naturalists think that ... the Natural System ... reveals the plan of the Creator; but unless it be specified whether order in time or space, or what else is meant by the plan of the Creator, it seems to me that nothing is added to our knowledge ... I believe that ... propinquity of descent – the

only known cause of the similarity of organic beings – is the bond, hidden as it is by various degrees of modification, which is partially revealed to us by our classifications (Darwin 1859:413).

Humans could no longer be considered the divinely created pinnacle of the *scala naturae* but had taken their position over deep geological time, Darwin argued, through slowly acting processes of natural selection. Time, space, descent: those were the key concepts that enthused the minds of the growing generation of 'scientists' – a neologism of William Whewell in 1840 to distinguish between cabinet collectors and the new experimentalists – who were convinced that life has a progressive character and were seeking an explanatory mechanism. Since 'our classifications are often plainly influenced by chains of affinities' Darwin continued, 'community of descent is the hidden bond which naturalists have been unconsciously seeking, and not some unknown plan of creation' (421). Darwin had provided the explanatory mechanism.

By the 1870s the amassing of large collections was beginning to be derided as a mere hobby – 'species hunting' – irrelevant to the purpose of science. Indeed, 'natural history' was being replaced by the word 'science', with its connotations of the careful collection of observed data, induction and testing of hypotheses, leading either to verification or falsification.

One of the first persons to bring the new approach to Australia had been Gerard Krefft soon after his appointment as Curator of the Australian Museum in 1861 by William John Macleay, who was evidently unaware of Krefft's evolutionary sympathies. The misalliance soon became increasingly obvious as Krefft began reorganising the Museum on lines which he described in a lecture at the Museum in August 1868 on 'The Improvements effected in Modern Museums in Europe and Australia' (Strahan et al. 1979:30). There he cited the British Museum, under the directorship of Dr John Edward Gray, in which displays were simplified, and arranged according to classes, orders, families and genera, with a selection of each species from the genera as the world's best. Krefft was critical of crowded cases that only confused visitors and, repelled by the arrays of cabinets in Sydney holding 'horrible mounted impossibilities', moved towards making 'living collections' (30).

After the death of William Sharp Macleay in 1867, his cousin William John continued as a powerful influence in the Museum and was determined to remove the heretic from the citadel of creationism. In concert with his cousin (once-removed), naval Captain Arthur Onslow – who had earlier sailed on the *Herald* surveys of the Reef as a lieutenant under Henry Denham – Macleay began an underhand campaign to dismiss Krefft, which, on trumped up and patently false charges, succeeded in 1874. Those two, Krefft wrote to the editor of *Nature* on 20 March 1874, 'were known to be my bitterest enemies'. Onslow in particular, he stated, was 'a firm adherent to the ancient doctrine of the "creation according to Moses", a man who detests Darwin, Huxley and Haeckel, and abhors me still more' (Krefft correspondence: AMS 48).

Matters came to a head when a pretext for Krefft's dismissal came: a gold nugget was reported stolen from a display case and a charge of managerial negligence was made. Krefft, in continued correspondence with Darwin, kept him informed of the situation and asked for help, writing on 22 October 1874 that the real reason for his expulsion was, in his own words to the editor of *Nature*, because 'I have tried to make people understand what the theory of evolution really is' (AMS 48). Darwin, by then a seriously ill man himself, was able only to commiserate, writing on 6 December 1876 that although 'I am sorry to hear of your troubles it is impossible for me to organise a subscription. But I beg leave to enclose a postal order for 5£ which I should subscribe if there be a subscription' (Mitchell Library MS.5828). But Krefft was doomed. In one of the less honourable episodes in the Museum's history Krefft refused to go quietly and on 29 September 1874, still seated in his office, was carried out bodily by two burly hired prizefighters in his director's chair and deposited in College Street, on the footpath outside. Although Krefft subsequently won a suit against the Museum, he was never reinstated. However, his memory is sustained in a satirical ceremony, still maintained in the Museum, of carrying out the retiring director, most recently Frank Talbot in 1975 and Des Griffin in 1999, in his chair.

The most striking illustration of William John Macleay's attitude to natural history is revealed in his evidence given to the Select Committee of the Legislative Assembly of New South Wales inquiring into the Australian Museum in 1874. Questioned on the purpose of extensive collections he stated, in reply to Question 2398, that 'the more complete and perfect the collection is, the more valuable it will be'. Pressed to explain further by being asked whether the purpose of the Museum would be 'better served by showing the development of insects, their habits, mode of feeding, and kind of food, than by simply putting the insects in cases and affixing scientific names to them' he assumed, in reply, that the question meant 'by breeding the animal – having it in a live state?' The chairman responded with a statement suggesting a more dynamic role for museums: 'No, by showing the transformations, and the kinds of plants upon which the different kinds of insects feed'. Macleay denied any knowledge of Krefft instructing an employee 'to make up a series showing the development of an insect and the plant it fed upon', with the peremptory reply that 'I have never heard of it' (Q.2401). He seemed totally lost and missed the entire point of the growing interest in basing collections and museums on ecological (if not evolutionary) principles. Or, perhaps, he was deliberately appearing to be obtuse due to the concurrent fracas at the Museum over the dismissal of curator Krefft.

OCEANOGRAPHY AND POLITICS: *CHALLENGER* AND *CHEVERT* VOYAGES, 1872–1876

A major disadvantage suffered by the small number of experimentally minded scientists in Australia throughout the nineteenth century was their remoteness from the main centres of Europe and the consequent delay in learning of new developments, along with the difficulty of implementing innovations in a society where governments, dominated by cabinet collectors, were reluctant to provide funding and the general awareness of the value of science was low. In Europe, by contrast, marine science was becoming quite important, particularly due to its economic potential in the fishing industry, especially in Britain now seeking to maintain its commercial pre-eminence as a manufacturing and trading nation. In 1870 the British Association for the Advancement of Science, usually known as the British Association, or the BA, at its meeting in Liverpool formed a committee 'for the purpose of promoting the foundation of zoological stations in different parts of the world' (Groeben 1985:4f.). By 1870 European dynamics were changing rapidly as Germany, created out of the various newly federated states by Bismarck, became yet another rival, along with the United States and Sweden, to the established colonial powers, all of whom were seeking to extend their commercial reach.

In 1871 the British Admiralty, therefore, was receptive to the idea that a wide ranging survey of the world's oceans, which included the northern waters of the Great Barrier Reef, should be undertaken to gain the best information available. At the practical level, for example, since the first undersea telegraphic cable had been laid from England to France in 1850, and the first successful transatlantic cable in 1866, knowledge of the structure and composition of the sea floor and the chemistry of its waters would be of inestimable value, providing strategic advantages as well. At the same time the Royal Society was expressing considerable interest from a purely scientific viewpoint. With the cooperation of the Admiralty the project was conceived of a worldwide survey of the oceans: in effect, the world's first oceanographic expedition.

The corvette HMS *Challenger* was refitted as a surveying and dredging vessel, specifically to undertake four main tasks: to study the physical conditions of the sea; analysis of its chemical composition; the topography and nature of the ocean floor and its deposits; and the distribution of organic life down through the water column. Under the command of Captain George Nares, *Challenger* sailed from Portsmouth on 7 December 1872, carrying a complement of civilian scientists led by a Scot, Regius Professor of Natural History at the University of Edinburgh, Charles Wyville Thomson (1830–82) who, as an enthusiast for deep sea research while at Queen's University in Belfast, had come into prominence in 1872 with his highly acclaimed book *The Depths of the Sea*. The scientists included Henry Nottidge Moseley (1844–91), medical practitioner by training but naturalist by disposition and achievement, along with John Murray (1841–1914), born in Canada of Scottish parents, who studied medicine at Edinburgh but did not take a degree.

Murray's interests were really in the sea: already in 1868 he had sailed to the Arctic aboard a whaler where he collected marine organisms and made physical observations of the chemistry and behaviour of ocean waters. A major limitation, not appreciated at the time, was the failure to include a geologist in the survey.

Beginning with a year-long traverse of the Atlantic, *Challenger* then surveyed eastbound from the Cape of Good Hope, arriving in Sydney in April 1874 where the officers and scientists were entertained by the gentry in a series of functions through April and May. William John Macleay – at the time still embroiled in the contretemps with Krefft – was lavish in his hospitality. It seems, during a dinner party in his residence, the elegant Elizabeth Bay House, that Thomson suggested to Macleay, who already had been enthused from reading *The Depths of the Sea* earlier that year, that he might consider making a survey himself to the virtually unknown New Guinea since the *Challenger* was limited in time and Captain Nares had been instructed to sail west through Torres Strait to the Dutch East Indies and then north to Japan.

Leaving Sydney, the *Challenger* went first to New Zealand and then to Cape York. Taking the outer route it entered northern Reef waters through Raine Island Entrance and spent some time at Somerset. Moseley, who did more of the natural history work than any other on the survey, has left the most comprehensive account of the biological studies on both Raine and Albany islands, first having thoroughly briefed himself by reading Jukes' 1847 *Narrative* of the voyage of HMS *Fly*. He made botanical notes on grasses and flowering plants, and the large number of turtle shells on Raine (one of the Reef's largest green turtle rookeries), supposing – mistakenly – that turtles came there to die. Most of all, however, he was particularly taken by seabird life, recording numerous species and observing that 'They are in such numbers as to darken the air as they fly overhead' (Moseley 1880:300). Moseley also took the major part in writing the natural history section of the 1885 *Report* which, although nominally under the names of Thomson and Murray, was written by four of the scientists.

The scientific results of the voyage again contain numerous observations on birds, both land and sea, as well as marine invertebrates of many kinds, seaweeds and grasses, and the sand of the shores; in the Torres Strait region they found 'a new genus *Moseleya* [believed to be] a new subfamily of Astraeidae, Moseleyinae' (Thomson & Murray 1885:I.544). In addition, considerable observations were made on the Aborigines of Cape York, including some of the appalling habits introduced by white settlers: 'Aborigines seemed degraded, *Cymbium* shells used to hold water were smashed, replaced by square gin bottles, of which there were plenty lying about the camp, brought from the settlement'. They were, in addition, 'heavily into smoking tobacco' (537f.). From Somerset the *Challenger* went on to Japan, then in turn to Hawaii and Tahiti and by way of the Chilean coast around Cape Horn and, making a second traverse of the Atlantic, reached England on 24 May 1876.

At the conclusion of that epic of oceanography, the *Challenger* had travelled 68 890 nautical miles and from 362 stations had made 133 dredgings and 492 deep soundings, some as far down as 8183 metres (26 850 feet) (Thomson 1877:I.xvi). To plumb such depths, the hand line was obviously out of the question. By then, in response to the needs of cable laying enterprises, mechanical apparatus had been developed by commercial companies, and patented Baillie sounding machines with steam-driven winches and huge spools of hemp rope were employed aboard the *Challenger*.

The data collected eventually filled fifty volumes while a seriously ailing Thomson managed to publish a two volume popular account under the title *The Voyage of the 'Challenger'* in 1877. Tragically, Thomson died at fifty-two just six years after the voyage ended, and the task of seeing the fifty volumes of scientific findings by numerous authors to completion in 1895 was undertaken by John Murray at the specially created Challenger Expedition Commission in the University of Edinburgh, which thereby became a centre of world oceanography. After the voyage Moseley was appointed professor of human and comparative anatomy at Oxford University where he wrote his own account of the voyage under the title *Notes by a Naturalist during the Voyage of HMS* Challenger, published in 1880. Sadly, he too became a tragic figure, dying of bronchitis, exacerbated by overwork, at the untimely age of forty-seven. Murray, however, who went on to live a full life, made a number of major contributions to the study of the chemistry and geology of the sea bed and the remains of marine organisms, publishing in 1891, jointly with Abbé Alphonse Renard of Ghent, one of the scientists of the voyage, the volume on *Deep Sea Deposits*.

The visit of the *Challenger* to Sydney had a major impact on marine science in Australia since it brought together a large number of men interested in natural history, and provided a window to a wider world of scientific debate. For William John Macleay the greatest matter of interest was the prospect of emulating the *Challenger*, albeit on a more modest scale, and this fired his imagination. New Guinea – as suggested by Wyville Thomson with whom he discussed collecting prospects – was still largely unknown in Australia, apart from a few coastal villages. It was becoming noticed, however, by various European powers as possibly another Africa to be carved up and occupied. When John Moresby returned to Sydney from his two surveys of Reef and New Guinea waters in the *Basilisk* in 1871 and 1873, he warned at the time, and later recorded in his *Discoveries and Surveys in New Guinea*, that 'Russian, French and Italian travellers were now exploring this island, the possession of which must in the future be a necessity for Australia because of its near vicinity and its strategic and relative geographic position, and it is feared that these [European] efforts might lead to a foreign occupation in time' (Moresby 1876:168). In similar vein he stressed the significance to Australian defence of the northern region dominating 'the Torres Strait route, the transit of Queensland mails, and our newly-discovered route for Australian trade to China, commanding the rich and increasing pearl-shell fisheries . . . and also the bêche-de-mer fishery' (207).

By 1874 annexation was in the air: a meeting to discuss the exploration of New Guinea was reported in the *Town and Country Journal* on 10 January (p. 47) and the liberal British Colonial Secretary Lord Carnarvon sought an expression of Australian interest. Always concerned about foreign naval ships coming close to Australian shores, Sydney became alarmed. Premier John Robertson drew up a cabinet minute urging Britain to annex New Guinea; Moresby went so far as to take possession of Hayter Island off the eastern tip of Papua on 24 April 1873, since he believed that a naval base there would command all the waters of the northern Reef (208). The conservative newspaper *Sydney Morning Herald* on 20 May 1875 took up the idea and encouraged a public protest meeting to prompt the Westminster government to annex New Guinea as it had done the year before in the case of Fiji.

In that climate, some time in 1874, Macleay decided to mount his Great Barrier Reef and New Guinea expedition: the prospect of such a daring project as yet another indication in the colony of British progress in science, had germinated in his mind. His express intention – later to be intensely, even scathingly, debated and questioned both in the press and political circles – was that of collecting biological specimens throughout the Reef, across Torres Strait and up the Fly River. It would be, in effect, an Australian equivalent to the *Challenger*: the first from the fledgling nation to an unknown region; not, however, plumbing the depths of the oceans but traversing the great Reef and the mysterious interior of an island still unexplored by Europeans. The tantalising references to the Fly River in Jukes' *Narrative* and the accounts of fabulous birds and animals in Moresby's *Discoveries* intrigued Macleay who believed that the Fly could lead, like another Amazon, to the interior. Already, in conjunction with Arthur Onslow, Macleay had been instrumental in founding a natural history society on the model of the Linnean Society of London, which held its inaugural meeting on 30 October 1874 with Macleay elected president.

At the end of 1874 Macleay made contact with Captain Charles Edwards, former partner of Robert Towns, with twenty-five years experience in Pacific and Reef waters where he was engaged in the bêche-de-mer trade, who agreed to help find a suitable vessel and to sail as master. In Sydney Harbour they inspected, and subsequently purchased, a former French naval barque of 30 metres and 318 tonnes, built in 1860, named the *Chevert*. Macleay was very active in promoting the voyage with numerous press releases and public meetings and claimed to have been deluged with hundreds of requests from prospective travellers. The actual complement was smaller: Edwards was in command of a crew of twenty; with Macleay were Arthur Onslow, George Masters and the visiting American surgeon W. H. James as specimen preparers, and four zoological and two botanical collectors. Since there were few coaling stations along the Reef, and none in New Guinea, a sailing vessel was mandatory. However, for inshore work, and travel up rivers, where sailing craft cannot go, Macleay had a 10.6 m (35 foot) wood-fuelled steam launch constructed which was towed behind, under the care of Lawrence Hargrave (1850–1915), a 25 year old engineer and son of Judge John Fletcher Hargrave, Macleay's neighbour. Hargrave, later, was to achieve fame as an aviation pioneer.

Macleay promoted his expedition tirelessly, addressing the New South Wales parliament on the project which held a banquet in his honour on 3 May 1875. On 18 May the day of departure from Sydney's major shipping terminal, Circular Quay, there was a farewelling party of 200. Guests included Premier John Robertson, Macleay's son-in-law and Colonial Secretary Edward Deas-Thomson, Catholic Archbishop Vaughan, proprietor of the *Sydney Morning Herald* John Fairfax, visiting English author Anthony Trollope and his immigrant son, most of the cabinet and many members of both houses of parliament, who gathered for an astonishing ten hour banquet aboard the paddlewheeler *Coonanbarra* to the music of a brass band (*Chevert* archives, Macleay Museum; Goode 1977:102f.).

The *Chevert* travelled rapidly through Reef waters, now that these had become better charted and marked by navigation beacons, reaching Cape York in a mere 25 days, after recording collecting stops at eight island locations: off Townsville, and then Palm, North Barnard, Fitzroy, Low Isles, Barrow, Claremont Group, and Forbes, all visits having to be so brief that it seems they were tokenistic. Their brevity lends support to the belief that Macleay was not really interested in collecting on the Reef at all, but was preoccupied with getting to New Guinea as soon as possible in search of unknown exotic specimens. Two weeks were spent in the Cape York region and then, on 26 June the *Chevert* sailed for New Guinea, collecting in Torres Strait along the way, until they reached Saibai and the Katau River eight days later.

At that stage of the exploration a mystery exists based on very partisan accounts by Macleay on the one hand, and his arch opponent, Charles Macfarlane, of the London Missionary Society (LMS), on the other. Tension between missionaries and traders was very high in the period, in large part because the missionaries, and Macfarlane in particular, who imposed flogging punishments on islanders who were convicted of adultery or fornication, 'had been accustomed to running [their] missions as petty theocracies, sometimes in the absence of secular government' (Beckett 1987:41). Hostility between the two was ignited when Macfarlane, accompanied by his wife, visited the *Chevert* at Darnley Island in the mission launch *Ellengowan*, supposedly to bring replenishment supplies that Macleay had requested earlier. Learning that no supplies had been brought, Macleay refused to meet them, a snub of the greatest magnitude in colonial society at the time and especially in such a remote location. Macfarlane was determined on revenge.

There is no doubt that once in the Gulf of Papua Macleay wanted to traverse the Fly but Captain Edwards refused. Macfarlane, however, spread the story that Edwards had 'lost his nerve' although it is known that he was quite ill at the time. Macleay recorded another version in his diary: that Edwards was 'most averse so long as the strong southeast trade wind continued ... to approach such a dangerous coast' (cited Goode 1977:109). Whatever the reason, the attempt was not made and the ship then headed east across the gulf to Yule Island as a compromise where collecting continued. It then retraced its course to the Torres Strait islands of the Reef region for more collecting. By August both Onslow and Macleay, who was suffering from fever and the heat quite seriously, were ill. Macleay was also deeply troubled by the continuing invective coming from Macfarlane which reached a rancorous level later when he wrote, supposedly confidentially, to

Dr Mullens of the LMS in London that 'the expedition is a complete failure . . . Captain Onslow RN has quarrelled with Mr Macleay and has left the ship. He [Onslow] and leading members of the expedition . . . say they are ashamed to return to Sydney' (120–1).

On his return to Sydney Macleay found himself embroiled in vituperative argument, inflamed by Macfarlane's bitter attacks of his purposes in mounting such a costly expedition. Macfarlane criticised the pomp and ceremony of the departure, and contrasted it with his belief that the voyage was disorganised, disagreement-ridden, unachieving, and that Edwards was rebellious and timid (Macmillan 1957:133). In self defence, Macleay composed a letter to the *Sydney Morning Herald* on 20 September which was not published – rather strangely – for another three weeks, on 11 October 1875. Meantime, the *Town and Country Journal* which before the voyage had given Macleay extensive favourable coverage, printed a highly critical article on 9 October stating 'it is the general opinion that the enterprise has resulted in total failure'. In the letter sent to the *Sydney Morning Herald* Macleay had asserted that the voyage was not an exploring expedition to New Guinea but one of natural history collecting.

His explanation was rejected by the newspaper in a lengthy editorial which clearly expected that Macleay should have produced evidence to justify the annexation of New Guinea in order to maintain Australia's interests in Torres Strait and to forestall foreign powers. The relentless criticism commented that 'there is an impression that both science and commerce have much to gain, and nothing to lose, by an immediate occupation of New Guinea'. Continuing, the editorial chided Macleay for the 'smallness of the results of the expedition' which 'is on many accounts to be greatly regretted . . . the voyage of the *Chevert* has not proved a success'. In fact, the *Herald* (whose proprietor Fairfax had been at the farewell banquet) accused Macleay of 'not doing much more than going to New Guinea and coming back again'. Taking up Macfarlane's criticisms of the pompous farewell, it made the sardonic observation that 'if Mr Macleay had slipped away noiselessly, he might have returned without comment. But the public, having been led to expect something, and having paid in advance in the shape of praise, is perhaps unreasonably disappointed'. It concluded with the damning comment that 'The voyage of the *Chevert* would have been remembered with greater interest, satisfaction and pride, if the promises at the beginning had been less, and the performance at the end had been somewhat more' (*Sydney Morning Herald* 14 October 1875).

Whatever the real intention, Macleay, still driven by the cabinet collecting mentality, stoutly defended his expedition and delivered a lengthy account to the Linnean Society at a meeting eleven days after the *Herald* article on 25 October 1875 which was published in the next issue of the *Proceedings*. To justify the collecting claim, he itemised the material gathered as 82 cases containing 1000 birds, 800 fish, 'numerous marine molluscs' and 112 bags of shells. No specimens of fish were brought back since the trawl nets became entangled in the reefs, and in any case, in what may have been meant as a humorous aside, 'the natives would never part with anything edible' (*Proceedings*, Linnean Society of NSW,

November 1875, 35). To house that vast array he built two museums in the grounds of Elizabeth Bay House, the larger one measuring 35 by 10.6 metres (115 × 35 feet) and employed a carpenter for an entire year to construct the display cabinets.

Clearly Macleay was an inveterate collector, but a question remains: why was the collecting done? Some was accomplished, of course, in the original traverse north through the Reef, but a suspicion remains. Having failed to convince Edwards to travel up the Fly, did Macleay then salvage the trip by turning to collecting in Torres Strait for two months to justify the expedition? We can never be sure.

So ended, rather ignominiously, the first marine exploration of the Reef and Torres Strait mounted by a private citizen in Australia. It is, however, too judgmental to look back. Many scientific ventures fail to achieve their express purposes and at the same time others, in the process, make fortuitous discoveries that open fresh fields for investigation. In those years the voyage of the *Chevert* did just that: it brought before the Australian public a quickened interest in the Reef and waters to the north, and although his mindset was already obsolete, Macleay stimulated an opposition that was beginning to move science in different directions.

QUEENSLAND MUSEUM: MARINE SCIENCE LAGS, 1859–1880

The situation in the 1870s and 1880s was no better in Queensland where there was absolutely no interest in marine science, much less the Reef. The Queensland Philosophical Society had been founded at the time of separation in 1859 and as it acquired collections it was allowed to house them in the derelict Windmill building – standing today as a heritage structure – on Wickham Terrace overlooking the city and river to the south. From the meagre surviving evidence it seems the collections were mainly geological. In 1879 a Museum was constructed in the city to be replaced in 1891 by a larger, impressive and florid 'Victorian Revival' building which remained in use until 1984 when the Museum moved again to its present commodious and functional quarters on the south bank of the Brisbane River.

Like the Australian Museum, the Queensland Museum in its foundation years reflected the approach of cabinet collecting. Few staff were appointed in the early years and Charles Coxen, one of the founders of the Philosophical Society, became first curator in an honorary capacity. The first salaried appointment was Karl Staiger (1833–88), a German trained at the Stuttgart Polytechnic who came primarily as government analytical chemist, although his Museum appointment was necessarily part time and his official designation was 'custodian', not curator. Staiger failed to become full time curator when William Haswell was appointed in November 1879.

The first serious efforts in Queensland on the scientific study of the Reef were initiated by a vigorous young Scot from Edinburgh, William Aitcheson Haswell (1854–1925). A very bright protégé of the great Edinburgh luminaries Wyville Thomson, Archibald

Giekie and the famous surgeon Lord Lister, the 25 year old Haswell had earned his MA in zoology (winning an astonishing seven university medals) with a particular interest in marine invertebrates. Appointed curator in December 1879 and expressing a strong interest in Reef research, Haswell threw himself almost immediately into the task of reorganising the Museum which meant, in his own words, recorded in a series of letters to Director Edward Ramsay of the Australian Museum, that he was 'getting it tidied up' (8 February 1880: Mitchell Library MSS 1589/2–3).

Irritated to discover that he had been badly contracted in that his salary was a niggardly £200 per annum, Haswell began negotiations with John Murtagh Macrossan, the responsible government minister, for an increase. Due to the 'financial condition of the colony' it was denied, although Macrossan advised that he would get 'a proper salary as soon as they could afford it'. Haswell was not mollified and expressed his mistrust in a letter to Ramsay on 26 May 1880, that his low status was because Macrossan regarded 'the whole scientific department as a superfluity'.

Despite his poor treatment by the government, Haswell was the first scientist in Queensland to display an interest in the marine biology of the Reef, undertaking field expeditions to the Bowen region – Queens Beach, Stone Island and Holbourne Island. He received, however, neither encouragement nor support from the government and by August his working conditions had deteriorated to the extent that he was forced to consider resignation. On 16 August he wrote again to Ramsay stating that the conflict with Macrossan and the trustees 'is almost enough to make one give up the place in disgust – if it were not at the same time extremely ludicrous'. For his part, Ramsay was keen to get Haswell to Sydney where marine research in Sydney Harbour was being actively promoted and a biological station was being considered. Ramsay found Haswell a private position as a collector for William John Macleay who, despite the *Chevert* fiasco, remained the dominant trustee of the Australian Museum. In response to Ramsay's telegram with an offer of employment in Sydney, Haswell wrote on 16 November 1880 that 'I shall be unable to resist the temptation of dredging and shall say good-bye to Brisbane . . . one stagnates here in a surprising manner' (Mitchell Library MSS 1589/2–3).

By the end of the year Haswell was in Sydney in the temporary employ of Macleay. The following year, 1881, he travelled aboard HMS *Alert* in northern Reef waters and in 1882 was appointed to a demonstratorship in comparative anatomy, physiology and histology at the University of Sydney. Haswell's rise was meteoric: in 1883 he was acting curator of the Australian Museum where he plunged into the task, begun by Gerard Krefft, of modernising the Museum by boldly relabelling and reclassifying many exhibits to reflect the new trends in scientific thought overseas. In 1884 he was appointed lecturer in zoology at the University of Sydney, and on 1 March 1890 he became Challis professor of biology in succession to the foundation professor of natural history, William Stephens. The change of title also marked a new direction in Australian science, particularly since Haswell, apparently unknown to Macleay, was an uncompromising Darwinian.

Queensland's loss was Sydney's gain: marine science, and especially Reef studies, passed to Sydney, both at the Museum and the university, where Haswell went on to become the leading invertebrate scientist in the nation, while marine science stagnated in Queensland – in fact it scarcely existed – for another thirty years.

SCIENTIFIC CHANGE 1888–1900: DARWINISM ESTABLISHED

By the 1880s the era of McCoy, the Macleays and their ilk was at an end. In 1888 the government astronomer of New South Wales, Henry Chamberlain Russell, gave his Presidential Address to the initial meeting of the Australasian Association for the Advancement of Science (AAAS) in which he criticised, in courteous fashion, the original Philosophical Society of Australasia as 'more of a friendly association . . . meeting over a cup of coffee and a biscuit' (Russell 1888:1.4). The new AAAS,* a national body 'assembling the workers in each branch of science from all the colonies' was urged to get on with serious investigation and research and so advance the task of 'keeping alive the dying embers of the first attempt [by the PSA] to plant science in this part of the world' (1.8,3).

The impetus to founding the AAAS which Russell outlined in his address was seen as a natural extension to Australia of the parent body in England, the British Association for the Advancement of Science which had been founded in 1831, largely in response to a devastating criticism of the Royal Society by Charles Babbage in his 1830 publication, *Reflections on the Decline of Science in England and Some of its Causes*. Babbage, professor of mathematics at Cambridge, and brilliant inventor of several calculating machines, one of which incorporated many of the principles of the modern computer, was profoundly disturbed that the scientific advances on the continent of Europe were well ahead of England.

At the time, England had only the two universities of Oxford and Cambridge, along with a small college in London (in contrast to Scotland with four universities) which were still dominated by studies of the classical era of Greek and Latin antiquities. The Royal Society, Babbage argued, had degenerated into a club of autocratic peers, knights and gentlemen, who were not pursuing science but simply using the accolade 'FRS' as a social honorific. Babbage argued for the abolition of cronyism and the Society's complete reform, with sustained government funding, as on the continent, to get the society going again. His mind always attuned to practical affairs, he stressed the need for intimate links between science, technology and commerce. The following year, on 27 September 1831, 39 scientific societies in England and 353 representatives convened in the city of York to revitalise the British Association by giving 'a stronger impulse and more scientific direction to scientific inquiry'. By 1835 it had become organised into seven sections, with botany and zoology labelled Section D. When that model was proposed in Australia the same structure was adopted. To the initial meeting of the AAAS, held in Sydney in September 1888, all 38 known societies were invited to send representatives and 28 responded, sending 820 persons.

The foundation address to the AAAS by Russell gives a good survey of the development of scientific activity in Australia in the early formative period. He described the Society's aim as 'assembling the workers in each branch of science from all the colonies ... [to promote] the exchange of ideas upon the same subject, and the pleasure which that affords'. Further on he urged that

> One of our first duties will be to work up all the facts known in every branch of Australasian science ... all those branches of science which are more immediately connected with the material advancement of the colonies ... Every worker knows how necessary it is to have all facts in clear and orderly arrangement, as a preliminary and necessary step to any safe advance (1.8,11).

No comments were made on the increasing number of controversies in science, especially in biology – or on the inclusion of women. The aim was more generally pitched at 'bringing to the front many men amongst us now scattered through the country who have the ability and genius, and who are willing to take up some of the subjects which require investigation' (11).

Russell's hopes were soon realised. All seven sections – mathematics and physics, chemistry and mineralogy, geology and geography, zoology and botany, medical science, statistics, mechanical science – were mobilised. At the conference in Christchurch three years later Haswell, president of Section D, delivered a ringing endorsement of the new direction in which biology was heading. His paper, 'Recent Biological Theories' declared a vigorous research program for the future and swept away any possibility for a continuation of the cabinet collecting era. In his opening paragraph Haswell set out his manifesto: the development in the life sciences of 'new theories or new modifications of old theories that have found the light in the course of the last year or two'. In a devastating broadside at the preceding era, he commented that previously 'the word "theory" was almost *anathema* to the vast majority of students of plant and animal life ... The naturalist of the old school went plodding along, accumulating his descriptions of species and his records of remarkable and interesting facts, without much thought of theoretical explanation'. Then followed his stark declaration: due 'mainly to the influence of Darwin's writings ... a very important change has come over biological research ... in the nature of an illumination, and the illuminating influence has been theory, and more especially the theories of descent and modification by natural selection' (Haswell 1891:173–74).

Haswell then proposed four main areas for the future of biological research: 1) the ultimate constitution of organised bodies, 2) the mechanisms by which the living world has become evolved, 3) theories of the genealogy of the animal and vegetable kingdoms as a whole, and 4) theories which seek to explain individual phenomena in terms of evolution (174). All these would be geared to the solution of 'the causal connection between the utility of a part and its development that constitutes the main feature of Darwin's theory' (175). The research focus was to be brought directly to Darwin's 'conclusion... that Nature selects advantageous variations'. But, Haswell continued, 'how do variations arise?' He examined and dismissed the rival theories: Lamarck's transformism by transmission of acquired characteristics, August Weismann's concept of transmission in 'germ-plasma', Hugo de Vries' proposal of 'pangenesis' – once supported by Darwin – whereby every cell in the body carries a self-copying 'gemmule', in every case raising logical objections to their conclusions. Haswell himself offered no solutions, which is not surprising since Mendel's work of the period 1856–68 was not rediscovered and published until 1900, a decade after Haswell's address to the AAAS was delivered. He did, however, set biology in Australia in a new, irreversible direction.

The central theme of Haswell's paper of 1891 was developed at length in the massive two volume *Text-book of Zoology* (1897), written jointly with Jeffery Parker at the University of Otago, which served as a major textbook for the following fifty years. The outstanding characteristic of Haswell's otherwise inspiring paper is the narrowing of biological science exhibited in the positivist stance adopted. Divine design and cosmic purpose are totally absent; concern is focused directly upon solving problems of the development of material phenomena. In Section XV, headed 'The Philosophy of Zoology', written by Haswell, Darwin's theory is presented as self-evident fact, requiring no qualification. 'The animal and plant life of the globe', he wrote, 'has come to be as it is now by a process of *evolution* which has been going on continuously from an early period in the history of the earth to the present time. The plant and animal worlds, have been *evolved* by a gradual process of development, in the course of which the higher forms have originated from the lower' (Parker & Haswell 1897:II.608; Haswell's emphasis). No debate: no alternative possibilities. The history of biology itself was presented in a masterly survey from the time of John Ray in the seventeenth century up to Wallace, Darwin and Weismann in the late nineteenth as a progressive triumph to certitude. The climax of that ascent came, Haswell believed – ignoring the comparable theories of Lamarck, and Wallace in his time – in the *Origin of Species*: 'It is to Charles Darwin that we owe the most thorough and consistent explanation of evolution that has hitherto been put forward – the explanation known as the theory of Natural Selection' (II.613).

While the theoretical aspects of science at the end of the century were being articulated with the new impetus to seek explanations of change solely in terms of natural selection, field work was still far from satisfactory. That problem was carefully delineated by Joseph Fletcher in his presidential address of 1900 to Section D of AAAS. A geologist by training,

---

* The association included New Zealand, hence 'Australasia', although in response to continuing objections from across the Tasman it was renamed ANZAAS in 1930.

and librarian of the Linnean Society, Fletcher delivered a paper 'On the rise and early progress of our knowledge of the Australian fauna'.

The central thrust of his argument was, in comparing the two branches of natural science, that botany had progressed reasonably well, having been given a great start from the work of Robert Brown and Ferdinand Mueller. Zoology, in contrast, had lagged well behind. Botany, of course, had been supported by the ecological imperialism of Kew Gardens, whereas zoology was left to 'private, non-responsible collectors and dealers – to anybody who chose to do it. The collecting was consequently haphazard and miscellaneous in character, not exhaustive and representative. Collections made by private individuals found their way into privately-owned museums' (Fletcher 1900:78). Although Australia had been colonised for more than a century, he wrote, 'only two great groups have ever been properly collected' and he cited the birds of John Gould and the insects of the Macleays (83). Like his contemporaries, Fletcher criticised the non-critical cabinet collecting mentality, 'the study of the fauna [being] for too long synthetical when it should have been analytical' (100). He made a strong plea for the strengthening of public museums in order to eliminate sporadic collecting and scattered collections which gave rise to concern that 'some of the large invertebrate groups ever being completely worked up becomes more and more remote' (100). Fletcher's summary of the condition of zoology at the end of the nineteenth century was that of a 'disastrous heritage'; moreover, as the twentieth was about to begin, the task had become that of dealing with 'a heritage which needs to be put into liquidation, with a view to reconstruction' (101).

That is precisely what happened as science replaced natural history; as collecting and ordering came to be superseded by a narrow empiricism of an increasingly positivist and presentist cast, concentrating upon reductionist studies of phenomena. The future of science became one of seeking to understand the mechanisms of nature and thereby its control and domination. Whatever the limitations of natural history of the previous era, it did seek to place nature in a context of meaning and purpose: in the new twentieth century that was to disappear. Scientists no longer studied the history or broad context of their discipline: the demands of presentism, the urgent pressure to reveal structures and mechanisms made science no longer a broadly educative pursuit, but changed it into the handmaiden of industry, economics and technology. Natural science – and Reef science was no exception – went willingly into commercial applications, and scientists became salaried employees of corporations, or of institutions, becoming cautious of critical debate that might alienate their funding sources.

CHAPTER 9

EXPLOITATION AND RESOURCE RAIDING: 1860–1890

As Queensland's colonial decades progressed the Reef was also developing into a region of marine industries, based chiefly on four products: bêche-de-mer, tortoiseshell, turtles and pearling. All found a ready market in Asia: bêche-de-mer for Chinese cuisine, green turtles for meat and soup, tortoiseshell and pearling for overseas manufacture in the luxury jewellery trade. For over a century the dominant ethos was that of the open frontier of an unlimited resource potential, simply there for the taking, a process designated as 'resource raiding' (Ganter 1994). So, as cabinet collectors became obsolete and natural history evolved into science, one major concern became directed towards discovering and exploiting economic products to sustain the increasing numbers of settlers who moved along the Reef coastline. That process, however, was slow to develop since in Reef waters there was, at first, no understanding of limited resources. The abundance of the marine environment was taken for granted, even though the dangers of such practices were already evident in the almost extinct sandalwood trade. Only when resources were close to extinction was scientific inquiry brought to bear.

One of the early European exploitative activities in tropical Pacific waters, along with the extensive spice trade in the Dutch East Indies, was the logging of sandalwood that had a ready market in China, where various species – red sandalwood (*Santalum album*) being the most prized – were used for building temples, houses and reliquaries. From the heartwood, close to the roots, highly aromatic oil is distilled, used in the manufacture of incense sticks for Buddhist ceremonies and the fragrant perfume, 'attar of sandalwood'. A number of tropical Pacific islands had rainforest stands containing sandalwood, especially Fiji, New Caledonia, the New Hebrides, the Marquesas and Hawaii. Given that the tree had to be completely dug out by the roots in order to distil the aromatic oil in the base of the trunk it was known that supplies would necessarily run out. As Western intrusion into the Indo-Pacific region intensified, and sandalwood became increasingly valuable, logging brought it close to total extinction. By 1815, following the boom years of 1808–10 when Sydney traders were involved, all of Fiji's accessible sandalwood had gone and by 1825 all the Hawaiian stands. In the decades of the 1840s to 1850s New Caledonia and the New Hebrides were stripped bare (Shineberg 1967).

As supplies diminished the sandalwood traders, among whom Robert Towns was to become a major figure, supplemented their cargoes by harvesting and processing another readily available resource, bêche-de-mer, to fill the holds of their vessels. Called a 'fish' in the trade, it is found abundantly in tropical reef waters, and has a history of commercial use spanning several centuries at least. The term is of Portuguese origin, *bicho da mar*, literally 'sea worm', which was corrupted by English speakers around 1814 into the quasi-French 'biche-de-mer' which evolved into *bêche-de-mer*. The animal, known more generally as a sea cucumber or sea slug, due to its appearance, is an echinoderm – in the same phylum Echinodermata as starfish and sea urchins – varying in length between 15 cm and, in some species, up to a metre, with sedentary habits, and 126 species are found in northern Australian waters. In the nineteenth century, however, only six species were considered edible and commercially valuable of which the most highly prized was the 'teat-fish' (*Holothuria mammifera*), so named, with Victorian delicacy, because it had a 'series of mammiform excrescences developed along each side' (Saville-Kent 1893:56). Taxonomic revision today names it *H. nobilis* (Cannon & Silver 1986:49). It was collected by swimming divers in depths up to 10 metres and, due to its poor keeping qualities, had to be processed almost immediately. The 'fish' was slit lengthwise, gutted, held open by small stick skewers and boiled for anything between one and ten hours and then dried rapidly, either on wire racks in the sun or else in smoke houses fuelled by mangrove timber (Jukes 1847:I.359–61). It was then ready for shipment to Asian markets.

The earliest evidence of the bêche-de-mer trade suggests that it was a commercial activity in the early eighteenth century in the Dutch East Indies where a report of 1754 mentions 'dried jelly fish' (Macknight 1976:95). At that time, and perhaps for centuries, fishermen from Macassar in the Celebes (today, Ujung Pandang in Sulawesi) sailed every year across the Timor Sea to collect and process 'trepang' as they called it in the waters of north-western Australia, ranging from sites near modern Darwin around to the east into the Gulf of Carpentaria to Groote Eylandt and the Pellew Group, an extensive littoral region which they called Maregê. Staying for around six months, to take advantage of the monsoon winds in both directions, and establishing regular relationships with coastal Aboriginal communities, the Macassans then took the processed fish home for onward transport to various Chinese destinations. The Malay name of trepang has remained in use for the north-western fisheries: in contrast, in Pacific regions, including Torres Strait and the Reef, the term bêche-de-mer is used, as it was in all official Queensland documents and also, after 1901, in all Commonwealth reports.

The exact time when sandalwood traders came to the Reef is still uncertain, but Jukes' references suggest the 1840s. With an intimate knowledge of the Asian markets where sandalwood was sold, those traders would have been familiar with the profits to be gained from bêche-de-mer, and the Reef was an inviting region for exploitation. Sometime in the 1840s, for example, it is known that Martin McKenzie was trading between Sydney and the Dutch East Indies and had established a fishery in the Coral Sea and along the Queensland coast. Around 1864 bêche-de-mer fishing had reached Torres Strait which became the main region for the industry, and was to have a pronounced impact on the lives of Torres Strait Islanders (Mullins 1992b:23; and 1995.55f.). There is no evidence of the Macassans in the Torres Strait (Macknight 1976:36).

Throughout the Reef regions, once collected bêche-de-mer were then taken to shore stations for processing. The most graphic account of declining resources survives in the tragic story of Mary Watson of Lizard Island whose husband had to travel longer distances away as local stocks were depleted, leaving her on the island with their baby and two Chinese servants. The Watsons were apparently unaware that Lizard Island, known as Jiigurru by the Dingaal people, was a sacred initiation site. The Dingaals could not speak English nor could Mary speak their language, and when the Aborigines apparently came across to request them to leave, she could not understand the import of their message. So, on 27 September 1881 while her husband was away, a group of Aborigines attacked, as they later said, to chase away the white intruders. In the following days they returned, killing Ah Leong and wounding Ah Sam. Mary Watson kept them at bay with a rifle but made plans to escape with her infant son Ferrier and Ah Sam by loading some provisions into the iron tank used for boiling the bêche-de-mer. For over a week the tank floated north, Mary Watson keeping a diary of events; around 8 October all three died of thirst and exposure where the tank came aground in the Howick Group. Her diary, a simple notebook, is preserved today in the Oxley Library in Brisbane, and the iron tank (some 1.5 m square) with its two paddles is in the Museum of Tropical Queensland at Townsville. Mary Watson's funeral in Cooktown on 29 January 1882 saw the entire town turn out for a ceremonial parade through the town to the Church of England for the service. At the graveside there was an outpouring of grief and tributes; numerous Chinese attended to farewell Ah Sam who was interred at the same time, with burning candles, incense and a fusillade of firecrackers (Pike 1979:68).

At the time of Mary Watson's death the bêche-de-mer industry was becoming well-organised: boats were licensed by the government and statistics recorded. For the decade 1880–89 the Queensland Commissioner of Fisheries reported in 1889 that annual tonnages of exports varied from a low of 160 in 1880 to a high of 255 in 1882, the average being 208.6 tons while over one hundred boats were registered (Saville-Kent 1893:231).

As an additional commodity, bêche-de-mer traders collected so-called tortoiseshell from the natives, and to keep the trade going, usually bartered with addictive substances –

tobacco, alcohol, sugar – in return for the carapaces. Only one of the seven species of marine turtle was used in the trade: the hawksbill, *Eretmochelys imbricata*. Unlike other species of turtles the scutes of the hawksbill are translucent with a range of very attractive amber-hued colours from pale cream through to a dark brown. The naturally glossy surface and the material strength of the shell made it very suitable for manufacture into toiletry articles such as combs and backs of hairbrushes, as well as inlays of various kinds and, as spectacles became more commonly used, for the more expensive frames.

Like bêche-de-mer, the tortoiseshell trade has a long history but it never became a major contributor to the Queensland economy, in large part because there was no monitoring of activities since collecting ships simply sailed directly on to Asian ports. From the sparse data available, in the period 1880–90 the average annual value registered in government records was approximately £400 per annum. In comparison, for the same period, annual wool exports averaged around £2 000 000 and pearlshell £70 000 (Saville-Kent 1893:322).

The green turtle (*Chelonia mydas*), like the hawksbill, was also a minor item of trade from Reef and Torres Strait waters, not for its shell but its flesh which was used for both meat and further processing into a soup that had high demand as a gourmet dish. Green turtles are easily collected since the females come ashore annually, between November and February, to lay their clutches of eggs in the sand dunes, usually numbering from 90 up to 150. Small, white and spherical and around 30 mm in diameter (golf-ball size), the eggs were always sought by coastal Aborigines for food, along with the turtle itself which, although agile in the sea, moves slowly on land and is easy to catch. Since the adults weigh anything between 100 and 200 kilograms it usually takes two strong persons to roll them sideways on to their backs where they lie unable to right themselves and can be left to die. Again, like the hawksbill, statistics for the colonial period are elusive but, since males never come ashore and mate always at sea, and are therefore harder to catch than the females, it seems likely that as hunting intensified so the breeding populations of females became increasingly reduced as the twentieth century progressed. In many tropical reef regions today they are on the endangered list.

PEARLS AND PEARLSHELL: RESOURCE DEPLETION

Human fascination with, and desire for, pearls goes back to our earliest records. Our most extended classical description occurs in two books of Pliny's *Naturalis historia* from the 1st century AD. Men face the hazards of collecting pearls, he wrote, despite the dangers of the ever present sharks, since 'wealthy women must have them dangling from their ears, a disease for which there is no cure' (IX.110).

The pearl itself is a natural product, resulting from a foreign irritant, such as sand or grit, being covered by successive coatings of nacre (Persian *nàkara*: pearlshell) which over a period of time build up to a spherical or ovoid form. Despite its long history of fascination for humans its only merit comes from its relative rarity. Marine biologist William Dakin summed pearls up as 'nothing more than abnormal products, calcareous pathological bodies . . . some perverted growth' (Dakin 1913:53,92).

Pearls, although they had acquired their romantic cachet in antiquity, were always secondary in economic value to the shell itself in Queensland's early decades of exploitation when pearlshelling became the dominant industry. The shell of the so-called pearl oyster (it is more closely related to the mussel in the class Bivalvia) was also used in antiquity for decorative purposes, its attraction coming from the nacreous lining to each half of the shell. As the oyster grows it secretes a fluid form of calcium carbonate ($CaCO_3$) to extend the margins of its shell that precipitates into crystals of aragonite. The resultant crystalline lining, nacre, is transparent but, due to its mineral character, the crystals act as miniature prisms and both reflect and refract light, creating the rainbow patterns that prove so attractive to the human eye. Like tortoiseshell, pearlshell – known also as 'mother of pearl' – is very strong with good mechanical workable properties and was used extensively for jewellery, decorative inlay work, as well as knife handles, knobs, and buttons and buckles in the higher grades of clothing.

In some unrecorded time the peoples of the Pacific region collected oystershell and began fashioning it into decorative ornaments. Europeans first saw it on Pacific Islanders, commented on by Jukes when he saw that 'men on Darnley Island wore round their neck circular or crescents of mother of pearl' (Jukes 1847:I.171). The great Cambridge anthropologist Alfred Haddon, during his famous expedition to the Torres Strait in 1888, also observed the same practice and recorded in the *Journal of the Anthropological Institute* (1890:XIX.339) that the 'native name mai, or mari, [was] made into a crescent breast ornament *danga-mari*, literally tooth of pearl shell'. In the early years of the settlement at Port Jackson pearlshell was being imported and by 1810 it was a well established trade, upon which Governor Bligh imposed customs duties. In 1814 the *Queen Charlotte* brought in the largest load to that time of 70 tons (Branagan 1996:5).

Pearl oysters in Reef waters occur mainly in Torres Strait, where they were discovered somewhat accidentally in 1869 by Captain William Banner, master of the brig *Julia Percy*. Two of the major traders in sandalwood and bêche-de-mer in the 1850s were the commercial rivals James Paddon and Robert Towns, both of whom were concerned to sustain profitable trading for their shipping fleets as the sandalwood supplies began to falter. In the desperate search for new stands to exploit Paddon sent Banner in the *Julia Percy* in 1859 to seek fresh stands of timber along the north Queensland coast and the Torres Strait Islands, and suitable beds for harvesting bêche-de-mer. Paddon died, however, in 1861 before he could act on Banner's discoveries and it was his partner, Charles Edwards, who began to exploit the new bêche-de-mer fields in the Torres Strait when he established the first island-based station in 1864 on Darnley Island (indigenous Erub) in the far north-east of the strait (Mullins 1992b:22,23). While working in the central part of the strait Banner came across the first known pearl oyster beds on Warrior Reef. Scarcely had the location become known than an era of relentless exploitation began.

Like bêche-de-mer, pearl oysters were initially gathered by swimming divers from two-masted luggers and in the first year five were on the scene. Within five years they increased to 17 luggers and 40 other vessels, many using technology which had recently arrived in the form of the helmeted diving suit, maintained by two deck hands taking alternate shifts to pump air down manually through a hose, although in time they were replaced by mechanical air pumps. Two years later, in 1877, the government agent Henry Chester reported to the Queensland parliament that 'There are now 16 firms engaged in the fishery employing 109 vessels and boats, 700 natives and 50 Europeans, 63 boats are fitted up with diving apparatus and the amount of capital invested about £40,000' (QPD, V&P 1877, III:1123–24). In order to monitor the growing industry more effectively the administrative centre was moved to Thursday Island in 1878, where it remains to this day.

The species which yielded the finest shell for manufacturing were the two varieties known as gold lip and silver lip, later described and named taxonomically by Lyster Jameson in 1901 as *Pinctada maxima*. The first specimens collected were huge, around 10 to 12 inches across (25 to 30 cm) although, as harvesting intensity increased the size of the shells correspondingly decreased. A second, smaller and less desirable species was also harvested, the black lip, described by Linnaeus in 1758 as *Pinctada margaritifera*, the specific in Greek meaning 'pearl bearing'.

Oystershells are not found in dense colonies, but must be searched for on the sea bed where they attach themselves in the early stages to a rocky substrate with fibrous threads, known as a byssus, which is secreted from a gland inside the animal. At first they were found in water so shallow that swimming divers could collect them by hand; by the 1890s when depletion became noticeable the average depth for collection was between 40 and 50 feet (12–15 m) although helmet divers were known to descend to around 120 feet (40 m) which is, at any time, a very dangerous level where the water pressure reaches 5 atmospheres. To prevent the lungs collapsing under pressure divers breathe air compressed to the same level at which they work, which causes nitrogen to accumulate in body tissues. Return to the surface requires very careful management and lengthy periods of decompression to allow dissolved nitrogen to slowly dissipate from the diver's body in order to prevent damage, termed barotrauma, in which nitrogen bubbles from the highly compressed air form during too rapid an ascent and accumulate in the blood vessels creating blockages or embolisms. These, known as the 'bends', cause the diver to contract in agony, leading to paralysis, which can result in death, as happened all too frequently.

The usual collecting technique by helmet divers, who formed the majority by 1890, was by drifting along under the lugger with the flow of the tide, attached by a safety line, stopping to collect shells in baskets as they were found. In 1893 it was recorded that 'a fairly remunerative quantity of shell for a boat to bring in, as the result of one month's work, is from 600 to 700 pairs, which, consisting of, reckoned as, 3-lb. Shell, would represent but little short of a ton in weight'. In the same context it was recorded that under the most favourable conditions 'as many as from 1,200 to 1,800 pairs may be obtained' and some owners encouraged higher productivity by offering 'a bonus for all shell collected numbering over 1,000 pairs' (Saville-Kent 1893:206). Since the luggers worked every day in the season it is readily estimated that the daily average was between 20 and 24 complete shells, each oyster making what the trade called a 'pair', each pair weighing between 3 and 4 pounds (1.4–1.8 kg).

The shells were opened aboard the ship and the animals discarded into large drums after which they were searched for pearls which customarily belonged to the diver as his perquisite, although in some cases the owner claimed them. No record of the pearls discovered was ever kept and therefore the value obtained in the market remains unknown. Drying and cleaning shells, at first, was carried out at shore stations on various islands, although later schooners were used as floating stations for that purpose. Such floating stations became a serious problem since, if foreign owned, they were usually beyond the reach of the law; wasteful practices of discarding undersized shells could not be monitored, and the clandestine export of shells directly to Asia meant that Queensland customs duties were being avoided with a loss of revenue. Once cleaned, the shells were packed into standard sized boxes for shipment overseas mainly to England, Germany and the United States for the next manufacturing stage. There was virtually no attempt at value-added processing in Australia in the nineteenth century.

A CLIMATE 'UNFIT FOR EUROPEANS': SLAVE TRADING

No sooner had the bêche-de-mer and pearling rush begun than abuses became evident, particularly in relation to the treatment of Aborigines, Torres Strait Islanders, and later, the indentured labourers brought in from Pacific islands. Such abuses were but part of a much greater problem throughout Queensland and the south west Pacific. A considerable literature confirms the appalling treatment of indigenous peoples in that region during the nineteenth century, and also confirms native reprisals against Europeans. Frontier racism was the norm for early resource raiders, with murders and massacres on both sides reported regularly in the press. At the same time, there was a strong current of liberal, humane belief that indigenous peoples were to be accorded respect, especially in Pacific island regions where there were well organised tribal societies with effective internal government. The government in Westminster was concerned to moderate the many abuses by British island traders: difficulties came from the limitation of resources available to prevent transgressions, as well as the extensive ocean regions involved.

In Queensland, of all the colonies, the call for non-European workers was strongest, in large part due to a chronic labour shortage throughout the nineteenth century and the difficulties of agricultural work in a hot, enervating climate, but primarily from an ingrained belief in white superiority. Despite the pleas of those such as Dunmore Lang for all immigration to be restricted to white settlers, and the energetic recruitment campaigns by the Queensland government, the generally favoured solution to the labour shortage was the importation of coloured workers. Aborigines were considered inadequate due to their understandable reluctance to work as forced labour on their own lands; other subject races were canvassed and their use was advocated in a report of 1868 by naval surgeon Alexander Rattray who gave an extended analysis of his observations of Cape York as a region for white settlement.

Rattray had sailed aboard HMS *Salamander* when in August 1864 it carried the first men for the foundation of Somerset. From his experiences he recorded that the 'four months of the hot rainy season are both less pleasant and healthy, and although the young and vigorous may withstand, perhaps for some years, the debilitating influence even of this season, various complaints are apt to occur . . . such as rheumatism both acute and chronic, while even the hale feel languid and listless'. His conclusion was quite definite: 'when longer tried and better known, [northern Queensland] will be found . . . unsuited and even dangerous for prolonged residence of Europeans, and especially unfit for open-air work in the sultry sun' (Rattray 1868:409, 411). His solution was generally accepted: to encourage the 'immigration of Chinese, Malays, New Hebride[an]s, and other South Sea Islanders accustomed to the solar heat and exposure without causing inconvenience or running any risk to health; by whom outdoor work may be done' (408).

The importation of menial workers into Queensland first came in 1863 when Robert Towns brought 67 Pacific islanders to his property 'Townsvale', a little to the south of Brisbane, to work as indentured labourers in his canefields. Although the venture was unsuccessful, the precedent had been set and for the ensuing forty-five years the so-called 'kanakas' remained a heavily oppressed labour force in the Queensland agricultural sector. At the moment of federation the new Commonwealth government was to draft the *Immigration Restriction Act* – the first item on its cabinet agenda – implementing what became known as the 'White Australia Policy'. Part of that initiative was to repatriate all the Pacific islanders, a process that was finally completed on 31 July 1908 when, apart from a few who chose to remain, the last of the indentured labourers left for their home islands.

Abuses were rife in the first years and one of the major catalysts for regulation came in January 1868 when the recruiting ship *Syren* under the command of Captain McEachern docked in Brisbane. There was a public outcry, led by missionaries and the Anti-Slavery Society, when it was learned that twelve islanders had died on the voyage and another twelve dead were still on board (Docker 1970:52). The Colonial Office in Westminster was fully aware of the problem, as was the government in Brisbane, and later the same year the first attempt at some kind of regulation was attempted with the passage of the *Polynesian Labourers Act* (31 Vic. c.47) which required licences to bring islander workers into the colony. Even the title of the Act demonstrates the lack of official awareness at the time of the significant differences between the cultures of the more sophisticated Polynesians of the central Pacific, already well-accustomed to European ways, and the more primitive and belligerent Melanesians of the western regions.

Despite the term *kanaka*, an Hawaiian word meaning 'young man', the recruiting grounds were not in Polynesia but much closer in Melanesia, chiefly New Caledonia, the New Hebrides and the Solomons. Of the many surviving records of the recruitment methods, known more vulgarly as 'blackbirding', one of the most disturbing is an account by W. B. Churchward which, although recording events in the 1880s, is illustrative of a practice of much longer standing. The technique was to lure the islanders aboard the ships on the pretext of offering them enticing trade goods: once they became friendly 'a lot of them were persuaded to go down below' when 'the hatches were clapped on' and at the same time the crew fired volleys to sink the canoes moored alongside to prevent escape. Those on the deck were grabbed and tied up, but others escaped by leaping overboard. Churchward recorded that 'one big fellow broke loose . . . jumped over and . . . began swimming at an awful rate . . . [and], as he passed me, I shot him dead, all for nothing' (Churchward 1888:62–63). The manacled captives, 42 in all, were landed surreptitiously at night on the beach of a coastal town where they went into virtual slavery.

Traders wanted both secure exploitation grounds in Torres Strait and a cheap, docile labour force the lack of which, from a European point of view, came to a head very early in the pearling era in 1869 when an event occurred that was magnified into a major episode, the *Sperwer* incident. The crew of the *Sperwer* had been engaged in bêche-de-mer fishing near Prince of Wales Island (Murulag) when their camp was attacked by Torres Strait Islanders: the captain, his son and three others were killed, the ship stripped of metal parts and then set alight and adrift. In 1870 a punitive expedition was mounted by the temporary government resident administrator in Torres Strait, Henry Chester, aboard HMS *Blanche*. Three leaders of the tragedy were captured and shot by summary execution. The intense racism in the white mind is revealed in Chester's account of the massacre where he justified his actions – taken with neither government knowledge nor approval – as just retribution for

> a long series of massacres of defenceless people perpetrated with absolute impunity [which] had accustomed these miserable savages to regard Europeans as an easy prey, and until last year they have congregated every south-east season at the Prince of Wales group in readiness to take advantage of any disaster that might occur to shipping (collected in Prideaux 1988:100).

Previously, in 1866, Governor Bowen, recognising that missionaries could well exercise a moderating influence upon the often violent behaviour of settlers and traders, requested the Society for the Propagation of the Gospel in Foreign Parts (SPG) to send some to Somerset; the following year Frederick Jagg and William Kennett arrived where they set up a school and organised church services. Following customary missionary practice, to ensure attendance, they handed out food, totally unrelated to the indigenous culture – rice, sugar and biscuits – at the end of each day (Sharp 1992:46). The extent of conversion, measured against the rewards for attendance, however, has been questioned since they left after barely two years. There were no missionaries in the region for the following four years, in which period Chester wrought his savage reprisal for the *Sperwer* incident, until the more zealously evangelical London Missionary Society (LMS) sent Archibald Murray and Samuel McFarlane with some teachers for the school.

A major problem for Queensland was the lack of jurisdiction outside its borders which stopped at the three mile limit, a problem brought home very effectively when it was learned that the British Crown had reserved title to the offshore islands and, acting on a commission of 10 June 1868, New South Wales had given a guano extraction lease on Raine Island to a Sydney merchant. New South Wales, in fact, still claimed all the islands off the Queensland coast. Queensland immediately sought to make the Great Barrier Reef islands part of its own territory; Governor Normanby negotiated successfully with the Colonial Office to extend the boundaries of the colony out to 60 miles around the entire coastline. Already in 1862 Queensland's western terrestrial boundary had been moved further into the Gulf from 141°E to 138°E longitude; the extension of its boundary 60 miles out to sea by Proclamation and Deed of Transfer of 22 August 1872 brought the Wellesley Islands in the Gulf of Carpentaria – on which New South Wales had earlier issued pastoral leases – within Queensland's jurisdiction.

At the same time, to strengthen the provisions of the 1868 *Polynesian Laborers Act*, in 1872 the Queensland parliament passed the *Pacific Islanders Protection Act* (35 & 36 Vic. c.19), known colloquially as the 'Kidnapping Act', which provided, in Section 3, that 'No British vessel [shall] carry native labourers unless the master has given a bond and received a licence'. Offences included such actions as 'decoys a native', 'carries away, confines or detains any such native', 'ships, embarks or receives . . . natives without [their] consent', and so on through a lengthy number of offences. Penalties for breaches went so far as to prescribe the seizure of the offending vessel. The Act, however, was difficult to enforce, particularly since the waters of the region were high seas. Three years later in 1875 the Act was amended (38 & 39 Vic.c.51) to assist naval policing of the region with a requirement on all masters for any labourers carried 'for the purpose of carrying on any fishery, industry or occupation in connexion with the said vessel . . . to be entered properly in the ship's log'.

In the decade of the 1870s, following passage of the 1872 'Kidnapping Act', there was a period of confusion and conflict that had a significant impact on the pearling industry. The original intention of the Act was to prevent the forced introduction of Pacific islanders into the sugar industry; yet at the same time it created problems in Torres Strait. The 1872 extension of Queensland's boundaries seaward by 60 miles had the effect of cutting off the northern third of the strait since the distance from Cape York to Saibai Island off the New Guinea coast is a little over 93 miles. Consequently the islands of the north and north-east were seen as part of Pacific waters unprotected by the Act, while those of the south, chiefly the Prince of Wales group, were part of Queensland. The net effect was that there was no protection for the northern islanders who were subjected to numerous abuses by traders.

Even further problems remained: all of the Reef south of Townsville, chiefly the Swains, was outside the boundary, as well as the northern half of Torres Strait and most of the Gulf of Carpentaria. Torres Strait in particular was a very sensitive zone: pearling and bêche-de-mering were proving to be valuable industries while at the same time, as other European powers were showing interest in the region, the trade route had to be protected. Further, abuses of, and by, the indigenous populations had to be subjected, as far as possible, to the rule of law. One particularly irritating point was that much of the extractive industry was directed from Sydney where politically powerful groups were financing the pearling luggers. In an effort to secure full control by the government in Brisbane, the Queensland Parliament, by Letters Patent of 11 October 1878 issued in Britain, in June the following year passed the *Coast Islands Act of 1879* (43 Vic. No.1). The attached schedule defined the new range of Queensland's territory to be all islands within a line which encompassed the Swains, Torres Strait up to the New Guinea coastline, and the whole of the Gulf waters. Most significantly, however, Queensland did not lay claim to the waters of the zone as territorial waters (Cumbrae-Stewart 1930:18).

By the 1870s Britain was becoming a reluctant imperialist as the empire became overextended and financially draining. Already in 1875 it had been forced to annex Fiji in order to control the growing lawless white population. To exercise the rule of law and good government throughout the whole of the region, therefore, it created the High Commission for the Western Pacific with headquarters in Suva. At the same time the intention was to involve the Australian colonies in sharing costs and supervision. Even though Queensland had already passed the *Pacific Islanders Protection Act* of 1872 and the *Coast Islands Act* of 1879 whereby the islands had 'become part of the said Colony and subject to the laws therein', abuses continued. Further legislation was required to protect immigrant workers: the *Pacific Islanders Protection Act* of 1880, the *Pearl Shell and Bêche-de-Mer Fishery Act* of 1881 and the *Aborigines Protection and Prevention of the Sale of Opium Act* of 1898.

There has been unresolved controversy over the years on the extent of abuses of non-European workers in frontier Queensland. In the agricultural sector, chiefly in the sugar industry that by the mid 1880s had become a major revenue earner, demands for field workers saw forced labour and brutal coercion continuing until the beginning of the twentieth century. In marked contrast, in the maritime industries, while there is a vast amount of irrefutable evidence of atrocities, it does seem that on balance, as the industry became better organised and regulated, the abuses were moderated greatly. The British Royal Navy had the task of enforcing the law, and maintained a number of ships for that purpose on the Australia Station.

By the early 1890s the coloured labour issue in the pearlshelling industry, in marked contrast to the agricultural sector, was being brought progressively under control. Several recent informed authorities have argued very cogently that the islanders were increasingly able to organise themselves and to negotiate better conditions. In a thorough and scholarly study of Torres Strait in the period 1864 to 1897 by Steve Mullins – an experienced Torres Strait diver and fisherman before becoming an academic and historian – while the abuses of the early years are documented carefully he too makes it clear that the islanders continued to become better organised, especially as they absorbed the teachings of the missionaries (Mullins 1995:125f.). The missionaries, in fact, had begun to exert a significant influence on the social structure of Torres Strait; one of them, Frederick Walker of the LMS, went so far as to help the islanders of Mabuig (Jervis) buy a lugger and pursue an 'alternative production strategy', although, sadly, the venture failed (Ganter 1994:69).

ADVENT OF THE JAPANESE

The race controversy in Reef maritime industries became even more complex in the two final decades of the nineteenth century as increasing numbers of Japanese divers entered the workforce. Following the opening up of Japan to the Western world in 1853 and the deliberate program of Westernisation that commenced in 1868 under the Emperor Meiji Tenno, unlike other countries from which the maritime workers came – Malaya, Java, Macassar, Philippines, Pacific Islands – Japan had rapidly become a powerful nation with a highly developed technology and a strongly projected international political presence. With few natural resources and a large population, in order to enter the Western industrial economy based upon rapid production and capital accumulation, Japan was forced to become an overseas trading nation with a focus on high quality value added products. The pearling industry was one area into which it moved quite swiftly.

By the 1880s the first Japanese divers were being hired to work in the Torres Strait; after 1890 their numbers grew rapidly and they soon developed diving skills of a very high order. In the 1890s the Japanese started to become the dominant ethnic community, and in the administrative centre of Thursday Island they increased from under 100 in 1892 to 700 in 1894, at which time they outnumbered Europeans (Sissons 1979:9–27; Ganter 1994:103). Over the first major twenty-year period of Japanese activity in Reef pearling, from 1895 to 1914, the percentage of their divers rose to a marked extent. From the *Reports* of the Inspector of Pearl-Shell and Bêche-de-Mer Fisheries on Thursday Island

(Table 3 in Ganter, 1994:103) Japanese divers in 1895 had reached 49.5 per cent of the total, by 1898 their percentages were 64, by 1902, 75. From 1906 to the outbreak of World War I their numbers rose rapidly, remaining between 97 and 99 per cent from 1908 to 1914. Percentages alone are inadequate and must be measured against the absolute numbers of divers which reveal even greater dominance of the Japanese. Non-Japanese divers however, continued to decline, from 122 in 1895 to just four in 1914.

The advent of the Japanese, however, was not without controversy and considerable resistance. By 1890 the Japanese had gone well beyond the role of contract diving and had begun to buy their own luggers. By 1897 the thousand Japanese residents of the administrative centre at Thursday Island was double that of the Europeans while one third of the pearling fleet of 231 licensed boats was in Japanese hands (Ganter 1994:105–6). Australian hostility increased markedly: in the same period the Chinese were being deliberately excluded and their numbers waned rapidly, especially since there was a sustained program of race hatred and immigration restriction. Concern over Japanese inroads into the pearl industry had now come to the fore, especially as the first signs of depletion became evident in the early 1890s. In 1883 the British parliament had created the Federal Council of Australasia as a forum to express common interests, to enable legislation for internal matters under British law, and to explore the possibility of federation. At its first meeting in Hobart in 1886 the subject of the depletion of the pearling beds of Torres Strait and the north-western shelf of Western Australia was on the agenda, the outcome of which was that in 1888 Queensland, for its part, enacted the *Pearl Shell and Bêche-de-Mer Fisheries (Extra-Territorial) Act* (51 Vic.No.1). That Act was coterminous with the Coastal Islands Act of 1879: the same wording in the Schedule was employed, except that 'islands' was altered to 'waters', along with the important addition of the words, missing in the 1879 Act, 'and thence by that meridian southerly to the shore of Queensland'.

The circumstances allowing the Act came from a British landmark judgment of 1876 known as the Keyn Case (R. vs Keyn [1876] 2 Ex.D.63) in which Chief Justice Cockburn ruled that Britain had no legal power over foreign nationals in its territorial waters, that is, within the conventional three mile limit. Accordingly, in 1878 the Westminster parliament enacted the *Territorial Waters Jurisdiction Act* to escape from the bind created in the Keyn Case by giving the imperial government power to control the actions of foreign nationals in British territorial waters, which was automatically extended to the colonies. Even so, the Act did not vest rights to the commercial products of the sea bed in the Crown, which was to complicate matters enormously for Queensland. Since the colonies did not have what in law is termed an 'international personality' they could not, normally, enter into treaties with other nations. All colonial legislation was derived from British law, a situation not changed until the twentieth century when the Statute of Westminster of 1931 gave independent international status to those former colonies to be known henceforth as members of the British Commonwealth of Nations. There was, however, a provision for the more advanced colonies (including Queensland) not to be automatically bound by international treaties between Britain and other nations: rather, they were 'invited' to make 'voluntary acceptance'.

The extra-territorial Act of 1888, however, only gave Queensland power over vessels sailing under the British flag on the high seas, and foreign vessels only when they were within the three mile territorial limit surrounding each island within the colony's jurisdiction. Lord Carnarvon of the Colonial Office, in a dispatch of 5 April 1875, had made the position abundantly clear that Queensland had no rights over the extra-territorial sea bed. Consequently, in one case of alleged poaching, the court ruled that oysters could not be stolen outside the three mile limit because they were, in law, *ferae naturae*, a legalism meaning 'wild animals', which are unable to be stolen until they are first brought into possession (Jameson 1912:426). Japanese pearling ships were quite free to ignore the provisions of the Act. Queensland responded by an alternative means: preventing Japanese from owning vessels; in effect, to keep divers and crews – essential to the industry – as paid labour aboard Australian owned vessels which were subject to the extra-territorial Act.

In that period Britain was acting in concert with Japan to resist Russian expansion in the north Pacific and on 16 July 1894 they signed the Anglo-Japanese Commercial Treaty which the colonies were invited to adopt. All except Queensland spurned the treaty and increased their opposition to further Japanese migration, fearing the growth of what was perceived as another Asian enclave which would replace the Chinese one which had just been successfully expelled. Queensland, acting from its own self-interest, exercised the right of 'voluntary adherence' and continued to admit Japanese divers and support staff to the pearling waters.

The colony, in fact, had little choice; it had no power to abrogate the treaty of 1894 since it was between the United Kingdom and Japan. However, in an attempt to regulate Japanese pearling, acting under advice from Britain (itself responding to diplomatic pressure being exerted by Japan), Queensland passed a face-saving amendment in 1898 (63 Vic. c.3) to the *Pearl-Shell and Bêche-de-Mer Fishery Act* of 1881 to prevent, not Japanese, but 'aliens' from owning or registering boats. None the less, as was generally recognised, the clear intention of the Act was to curtail Japanese exploitation even though the continued viability of the pearling industry required their expertise. When the first federal government passed the *Immigration Restriction Act* of 1901, persons engaged in the pearl shell and bêche-de-mer industries were exempted from its provisions, and until the outbreak of the Pacific War on 7 December 1941, Japanese workers remained active in Torres Strait.

CHAPTER 10

FOR MAXIMUM YIELD: REEF BIOLOGY

Throughout the three pearling decades of the late nineteenth century the major problem confronting the industry was continuing fluctuations in supply as beds became exhausted by the depredations of resource raiding, for which the Japanese had become the prime targets of abuse. The problem, however, was much more systemic. Ship owners wanted the best possible return from each venture and to that end developed the bonus incentive for higher yields described earlier. Following the first serious decline in the late 1870s the industry was stimulated when new fields near Badu, but also stretching to New Guinea, known as the 'Old Ground', were discovered in 1881. Again these were raided and by 1885 were so depleted that many of the luggers, and their schooner 'mother ships', left for new fields off Broome in Western Australia. As the larger shells had virtually disappeared, smaller and smaller sizes were harvested, down to 5 inches internal nacre measurement, which made them 'practically valueless [since] . . . no workable mother-of-pearl or nacre is left' (Saville-Kent 1890a:4).

Queensland had attempted its first regulation in 1881 with the *Pearl-Shell and Bêche-de-Mer Fishery Act*, the same year that New South Wales, concerned over depletion of its table fish stocks, established the Macleay Royal Commission that recommended controls. Both colonies found that legislation was easy, enforcement difficult. In the pearling grounds, consequently, further decline continued as both quantities, and quality, fell. To the growing number of professionally trained scientists it had become evident by the end of the nineteenth century, and the first decades of the twentieth, that a scientific knowledge of the biology of natural resources and their environmental needs was mandatory for better management. Even so, the primary concern was not conservation but ensuring maximum yield for commercial profit; the temporary closure of the pearling beds was intended only to allow sufficient regeneration to enable extraction to continue unabated.

REEF SCIENCE: ACHIEVEMENTS OF SAVILLE-KENT

To that end William Saville-Kent (1845–1908), an oyster specialist, was appointed Commissioner of Fisheries by the Queensland government in March 1889 for a three year term. Saville-Kent was one of the new generation of applied scientists who were coming to replace the gentlemen amateurs and cabinet collectors of earlier decades. Born in Devon, he enrolled in 1867–68 at King's College, London, where he studied microscopy and invertebrate marine life, becoming a keen Darwinian under the stimulating influence of Thomas Henry Huxley, and then moved on to employment in the British Museum in the period 1869–73.

In July 1884 Saville-Kent arrived in Hobart as Inspector of Fisheries to work on the valuable table oyster industry and the acclimatisation of salmon (*Salmo salar*). In 1889, following a return to England, Saville-Kent accepted an invitation by the Queensland government to report on serious problems affecting the Moreton Bay oyster beds. Fortuitously, he was subsequently invited to join the cruise of HMS *Myrmidon* which was leaving Brisbane via the Inner Route to Western Australia to survey the waters of the north-western shelf. Like every other traveller, Saville-Kent became enamoured of the Reef and accepted an offer by Premier Thomas McIlwraith for a three year contract as Commissioner of Fisheries.

During the first year he made a major survey of the pearl and bêche-de-mer resources of Torres Strait, sending his critical Report to parliament in 1890. Thorough in his collection of data, he stressed to the government the great 'importance of the pearl-shell fishery [which] . . . demands every attention and encouragement in the direction of its judicious conservation, and scientific development' (Saville-Kent 1890a:2). The second section, *Suggested Regulations and Concessions*, contains his carefully considered recommendations. Since it is 'only too evident that shell was being taken of such small size and immature condition as to very seriously injure the future prospects of the pearl-shell fishery . . . a standard size . . . demands consideration', he advised. He recommended that a 'minimum standard therefore of eight inches across the nacre or mother-of-pearl for trimmed shell would in my opinion represent the most appropriate gauge for adoption' (4). In order to conserve the species, and allow time for regeneration, he advised closing certain areas, and, anticipating objections from the less reputable sector of the industry, commented that correction of the 'exhausted condition' of the grounds is endorsed 'by several of the most prominent members of the shelling trade' (5).

In his final, third section he made it clear that very little was known of the life cycle of the pearlshell oyster and that the future viability of the industry required not only conservation measures but also experimental inquiry into controlled aquaria cultivation and, subsequently, re-seeding of depleted areas. The government responded and in 1891 amended the *Pearl-Shell and Bêche-de-Mer Act* to incorporate his minimum size recommendations of a 6 inch (150 mm) internal measurement of nacre. Even so, decline from overfishing, due to illegal collecting of undersized shells, continued while industry complaints about the minimum size were so effective in influencing the government that it reduced the legal minimum to 5 inches in 1896.

As a result of Saville-Kent's recommendations George Bennett was appointed inspector of pearlshell and bêche-de-mer fisheries, with the 25 foot (7.6 m) *Alert* as patrol boat. He began a program of vigorous enforcement, including prosecuting offenders for both illegal fishing and the ill-treatment of natives. In 1897 Bennett took the extreme, but necessary, step of totally closing the Darnley Deeps where most of the unprecedented number of 22 divers – 7.2 per cent of the diving force – had died that year from the bends, mostly in dangerous depths of 40 metres or more.

Such measures were now far too late: none the less, in 1898 Bennett recommended the closure of Endeavour Strait, along with other areas, for a period of some years, and for the appointment of a consultant marine biologist to make a thorough scientific report on the region's sustainability. At the same time Bennett sought to better regulate abuses by bringing all pearling divers and crew under the provisions of the British *Merchant Shipping Act* of 1894, which required Articles of Agreement and more explicit terms of labour contract – although this, in practice, proved cumbersome to enforce (Davenport 1986:339). Bennett's report reached the government on 9 February 1901, indicating that

> the pearlshelling industry in Torres Strait has reached a critical stage. From the time when the earliest shellers each with a schooner and a few open boats, picked up shell practically at low water mark, there has been a gradual evolution in skill, in appliances, and in organisation, to meet the incredible difficulty of obtaining shell. Year by year the shell bearing area has been pushed further and further from the shore and year by year the boats have ventured further afield (342).

By 1905 the depletion had become so serious, and enforcement so stringent, that half the pearling fleet left for the Aru Islands in the Moluccas, the most easterly group in the Dutch East Indies, where, beyond the reach of Australian law, they sought fresh grounds to exploit. Alarmed, the Queensland government requested Saville-Kent, at the time working for the Lever Pacific Plantations company in the Solomons, to conduct yet another inquiry into the ailing industry. In a short four page report, tabled in the Queensland parliament on 12 October 1905, Saville-Kent's opening paragraph was quite forthright: 'I am unable to avoid the decision that their unsatisfactory state has been mainly brought about through the over-depletion of the natural shellbeds'. Referring to the Aru migration in passing – with an oblique suggestion of the damage to State revenues – he asserted that even natural regeneration by closing depleted grounds would take 'some considerable time', in effect, 'the lapse of many years'. The best response, in his view, was 'the establishment of a certain number of Government pearlshell-breeding reserves' along with 'an experimental cultivation laboratory'. In his opinion, that would be the only way out: otherwise, 'I fail to see a way by which the depleted Torres Strait fisheries can be restored to profitable productiveness within any reasonable time' (Saville-Kent 1905:3–4). Despite his unequivocal evidence, the government failed to act on the report and the idea of scientific study and cultivation was ignored.

In his brief Queensland tenure Saville-Kent was remarkably active on a number of fronts. In 1890, when elected president of the Royal Society of Queensland, he gave an historically significant presidential address on 20 November 1890. His lecture, with the famous marine research stations of Naples and Plymouth obviously in his mind, developed at length a single topic: the need for government funded scientific research in marine biology for economic purposes in a new university and a 'well appointed zoological station and a marine biological laboratory' (Saville-Kent 1890b:22). Although Sydney, Melbourne, Adelaide and Hobart had already founded universities (1852, 1853, 1876, 1890 respectively) in Queensland, without a university, there was considerable indifference, and even resistance, although the possibility was being raised at the time by a few progressives. Saville-Kent stressed that 'a science of biology' was essential for a colony heavily dependent on 'animal and vegetable organisms of direct utility to man . . . in order to keep abreast of the times'. It was vital, he stressed, 'to conduct every operation on a thoroughly scientific basis', citing the fact that 'many European and American universities [already have] special seaside laboratories' (18). Another of his major points was the need to recognise that

> the peculiar advantages possessed by Queensland are associated with the circumstances that she possesses a line of sea-board stretching far up into the tropics, and embracing the world-famous Great Barrier coral reef, and the many islands entering into the composition of the Torres Strait Archipelago. This extensive area is rich beyond imagination, in the production of a marine fauna redundant with forms possessing both an economic and a scientific value (22).

His suggested location for a biological station was Thursday Island due to its strategic location, both in terms of communication, and as the centre of the valuable fishing grounds.

In addition to the commercial benefits to be gained from marine science, he proposed also the study of 'the formation and growth of coral reefs' of which 'next to nothing is as yet known' (35). Apart from the general scientific knowledge to be gained, there was also the 'practical point of view' that once the rate of growth of coral reefs could be measured, then waters currently 'declared safe for navigation . . . [which] may now contain many hidden dangers arising from the upward growth of isolated or accumulated coral masses' could continue to be charted accurately (37). A fundamental requirement, he stressed, was 'the establishment of a zoological station or biological laboratory at Thursday Island . . . the mainstay of [which] should, no doubt, be the Queensland University'. In conclusion he urged the support of 'every Australasian scientific society, and more especially that of the Australian Association for the Advancement of Science' (it was, in fact 'Australasian'), founded just two years earlier in 1888 (41–42). There was no political response and the foundation of a university had to wait another twenty years, until 1910.

Despite the disastrous situation in the pearling beds described in his Report of 1895 – which he reiterated in 1905 – Saville-Kent was buoyantly optimistic over the riches to be garnered from the Reef. That theme was developed at length in his magnum opus of 1893, *The Great Barrier Reef of Australia* with the revealing subtitle: *its products and potentialities*. That impressive book, dedicated to Samuel Walker Griffith, Premier of Queensland in the year of its publication, divides into two roughly equal sections. The first three chapters provide a survey of the 48 monochrome photographs illustrating the coral reef flats and various types of corals followed by a comparative survey of opposing geological theories. The similar theories of Charles Darwin and James Dana (an American whose field researches supported Darwin) are compared to those of John Murray, of the *Challenger* voyage. Darwin and Dana hypothesised that reefs were built on subsiding landforms whereas Murray argued that reefs were built from the sea floor upward on accumulated debris. Saville-Kent's allegiance to Darwin is strikingly evident: although he dealt as impartially as he could with the two main opposing theories, those of Darwin and Murray. His sympathies were obvious when he wrote that Darwin's *Structure and Distribution of Coral Reefs* (of 1842) 'laid the foundation of the imperishable reputation of that distinguished naturalist' whose text employs a 'highly logical and sagacious line of reasoning' (Saville-Kent 1893:73,81). A comprehensive account of the geomorphology of the Reef, and a fourth chapter on corals, coral animals and current taxonomy conclude Part 1. The ensuing four chapters in the second part of the book cover, in sequence, pearl and pearlshell fisheries; bêche-de-mer; table oysters; and food and fancy fishes. The large-format work was illustrated with sixteen plates of Reef biota: anemones, corals, bêche-de-mer, molluscs, oysters and fish, all hand-painted by Saville-Kent himself in striking accuracy and detail and reproduced by what was then called the 'chromo' process. It marked a revolutionary approach to introducing the general public to the incredible beauty and variety of Reef life.

In the final, ninth chapter, headed 'Potentialities', Saville-Kent discussed a wide range of products which he believed could be extracted from the Reef. The opening sentence reveals his exploitative stance: 'Many and various are the potentialities of the Great Barrier Reef of Australia. They are associated, practically, with every one of the several fishing industries described in the preceding chapters, and include numerous subjects hitherto untouched' (311). 'So little of the available fish fauna are turned to account' to be harvested, he wrote, including the abundant shark. Trawling, however, he dismissed as neither possible nor profitable, 'especially among the intricate channels of the Barrier'. The greatest potential, he claimed, warming to his specialty, would remain in pearling, although by that time he was aware that the future of pearls themselves would come from artificial cultivation. At that time true spherical pearls had already been successfully developed by the Japanese, beginning with the achievement in 1892 of Mikimoto Kokichi to whom Australian patents were granted in July 1914.

Further products were suggested for extraction, especially the green turtle which he noted 'abounds throughout [the Reef] and breeds extensively on the sandy shores of the coral-cays and islets. Except, however, for local consumption, and for the export of a limited number to Sydney and Melbourne, little or nothing is done with this valuable commercial article'. To secure and maintain economically viable quantities he suggested 'large turtle-breeding ponds and lagoons, that might equal in importance the celebrated establishment in the island of Ascension' (321). Sponges, too, he recommended for investigation, of the bathroom toiletry kind (*Euspongia officinalis*), at that time imported from the Greek Mediterranean islands, as well as shells and coral specimens for household decorations.

The final suggestion was one closest to his heart: 'a biological or zoological station' would be mandatory, preferably at Thursday Island. With the movement to federation of the six colonies foreshadowed by Henry Parkes in his draft proposal of 1889, Saville-Kent, looking ahead, proposed that it should be 'essentially a federal Australian institution, and would look for the main means of its foundation and maintenance to Australian corporate support, and Australian private liberality'. Pointedly noting that already mineralogy, forestry and agriculture were well endowed, he proposed that the one remaining area of neglect should be remedied so that 'the science of Marine Biology should not long be disappointed of her hope' (334). In fact, the discipline had to wait another seventy years, mostly in the face of government obduracy, born of ignorance of the significance of the marine environment to a nation entirely surrounded by water, isolated in a vast southern ocean, and almost entirely dependent on navigation across the sea for its commercial livelihood.

Saville-Kent's work in Australia, and particularly on the Reef, was followed intently by scientists and amateurs alike. In 1907 the first ever illustrated lantern slide lecture on the Reef was held at a Brisbane meeting of the Royal Geographical Society of Australasia, presided over by Vice-President Arthur Morgan. Made in London, the slides came from Saville-Kent's photographs of 15 fringing reefs, 15 seascapes and 24 of various biota (anemones, bêche-de-mer). What must have been a highlight of the evening were the 16 'coloured chromo plates of anemones, corals, oysters, and quaintly marked fish' all taken from Saville-Kent's own skilful illustrations.

Equally interesting would have been the discussion on the intriguing and still controversial aspect of Great Barrier Reef geology and formation. Dr Guppy presented the rival theories to Darwin, proposed by Murray and Agassiz and reported in the *Queensland Geographical Journal*, on the vexed issue of elevation, commenting that 'This fascinating area of animated nature can no doubt originate in a submerged portion of the mainland, cut off during a prolonged period of subsidence, affecting most of the coast-line'. Guppy then suggested that the evidence leads to the inference that slow subsidence followed, which affects most of the coast, encroachment by the sea and masses of rock carried from inland into the waters by torrential rains. The islands, he argued, were older rock formations, and the Chillagoe caves, 120 kilometres inland over the Great Divide from Cairns, were a 'typical example of an ancient submarine structure, where the old coral formation had been developed'. Yet, after this final analysis he was by no means dogmatic and ended his talk with the observation that 'the subject of reef-formation was one requiring further consideration' (Proceedings, *Queensland Geographical Journal* 1907–1908:87–90).

COLLAPSE OF THE PEARLING INDUSTRY: 1906–1916

By 1906 the pearlshelling situation had deteriorated to such an alarming extent – as reported acidly by a strident press – that the government, under Premier Robert Philp who had a wide range of commercial interests in the area, was forced into action, authorising a 'Royal Commission to Inquire into the Working of the Pearl-Shell and Bêche-de-Mer Industries' that issued its Report in 1908. The Commission was chaired by Captain John Mackay who succeeded Thomas Almond as Portmaster on 1 September 1902. The inquiry was very comprehensive, calling 78 witnesses, representing boat owners, shellers, divers and government officials. Only one was Japanese, T. Kashiwagi, President of the Thursday Island Japanese Club, along with M. Shiojima as his interpreter. Significantly, no scientists were called, simply because there were none available with any relevant expertise: Saville-Kent had given his evidence in the 1905 Report and had returned to England where he died in 1908, the same year that the Report was tabled in parliament.

The inquiry found, not surprisingly, that the pearl oyster beds were severely depleted. The general tenor of evidence was overwhelmingly in favour of white ownership of vessels and that, in time, only white divers should be licensed. The Report recommended a limit to the number of vessels, restricting their size to 25 tons to prevent them being used as floating stations, with a maximum of five to be owned by any one company. In addition, certain severely affected areas should be closed permanently.

A dawning realisation of the need for sustainability by means of knowledge and sound management comes through some of the responses. Given the evidence of many witnesses of the need for serious research, as advised by Saville-Kent, the Report recommended that

> a thoroughly competent Marine Biologist [should] be stationed in Torres Strait ... a man of the highest attainments in marine biology will be required ... he should be furnished with every requisite, including sufficient staff, a well-equipped

laboratory, an area upon which to conduct practical experiments in cultivation, and a vessel or vessels for the collection of bivalves to stock his propagation area (Mackay 1908:LX.58).

Research, however, was not the major concern: rather, it was economics and the race issue. Despite the fact that the Australian Senate, acting in terms of the constitutional powers in Section 51 of the *Commonwealth of Australia Constitution Act*, resolved on 30 November 1905 'that no coloured alien should be admitted for the pearlshelling industry other than those required to fill vacancies as divers or crews' no effective restrictions were placed on Japanese entry (CSIRO archives, Series 1:68/186). In the year of the Royal Commission Report, 1908, the fact that 172 of the 174 licensed divers were Japanese undoubtedly led to the emphatic recommendation in Section 61 that Europeans should be in control: 'Steps should be taken to reduce this vast preponderance of aliens, and to have this outpost guarded by a hardy population of loyal and patriotic Australians'. Four final recommendations were given, which appear in Section 77: 1) no new vessels be licensed unless such boats were built in the Commonwealth and by British subjects, and 2) all aliens on licensed vessels to be registered under the Pearl-Shell and Bêche-de-Mer Fishery Act (Lxxvii.116); 3) all vessels before being licensed, to be measured and registered in Queensland (Lxxvii.118); and finally, 4) the owners of licensed vessels should furnish a quarterly return showing the weight and value of pearlshell, pearls, bêche-de-mer, turtle and turtleshell obtained by each vessel owned by them (Lxxvii.135).

As in the past, nothing was done and the pearling industry crisis worsened, in large part because it was caught up in the wider issue of national politics. When the colonies federated in 1901 there was no distinctive Australian identity: people still thought in colonial terms. Both sides of politics, the conservative Liberal Party (renamed National Party after 1917) as well as Labor were both strong supporters of the White Australia policy, a position vigorously promoted in Queensland by the militant workers and the Australian Workers Union (AWU, founded 1904) who were determined to keep jobs for white Australians. In contrast, the small but politically powerful minority of sugar producers and the owners of large shelling fleets alike saw the need for a low paid coloured labour force to maintain productivity and profits. The sugar industry was bought off in 1906 by a subsidy on sugar produced by white workers, but no such remedy was available in the shelling sector. Despite the 1905 Senate resolution the State government was unable to act at all decisively.

In the first decade of federation the party political system was emerging, and in 1910 the Labor Party won a decisive victory securing 43 of the 75 seats in the national parliament. It then moved to find a sound basis for the now seemingly intractable pearlshelling crisis by appointing its own Commonwealth Royal Commission on the Pearlshelling Industry under the chairmanship of Senator Frederick Bamford. The commission submitted a Progress Report on 9 September 1913 and a Final Report on 17 July 1916, published as Commonwealth Parliamentary Papers 54 and 56 for 1913, and 326 for

1914–17. Despite the findings of the Bamford Commission, which were similar to those of the Mackay Commission of 1908, there was still a major disparity between evidence and opinion. On 25 November 1916 the Commonwealth Statistician's Office released a table to the Advisory Council on Science and Industry headed 'Pearl, Pearlshell and Bêche-de-Mer Fisheries in the Commonwealth, 1907 to 1915' covering Queensland, Western Australia and the Northern Territory. The figures for the Territory were very small: an average of 201 men were employed over the nine year period while the value of pearlshell was correspondingly low, with an annual average of £10 464, the lowest return of £6135 being in 1915. Western Australia maintained its employment rate throughout of an average 2500 men and average earnings from 1907 to 1914 (no figures for 1915 due to World War I) of £202 000 per annum. Since no serious depletion problem was experienced in that period in Western Australia, normal practices were continued there.

In comparison, Queensland, where the pearlshell and bêche-de-mer problem was most serious, the figures reveal a consistent and disastrous downward trend. Average annual employment of 1331 fell to 844 in 1915, tonnage of shell from 472 to 112 in 1915, value obtained from an average of £75 500 to just £18 512 in 1915. In contrast, bêche-de-mer showed an improvement from an average annual value of £21 637 to £39 918 in 1915, possibly reflecting boat owners harvesting it as an income alternative in face of declining shell returns (CSIRO archives, Series 1:82/1).

The Bamford Commonwealth Commission accepted the argument that white men could not be trained as efficient divers and thereby acceded to the demands of the boat owners that Japanese divers continue to be accepted. A major problem remained in that there was still no scientific knowledge of pearl oyster biology, nor were there any scientists or facilities for training them, even though serious concerns had been raised by Saville-Kent back in 1890 and again in 1905 and had featured in the Mackay Report of 1908. For their part, the Japanese avoided restrictions that excluded them from boat ownership through the practice of 'dummying' – well developed by squatters in the rural sector – by nominally registering their boats in the names of Australian stooges, and continued their relentless stripping of resources. As the pearl oysters diminished, beginning around 1912, they began to collect a reasonable, if poorer substitute in the trochus shell. Moving south of the strait down the Inner Reef below Cairns as far as Mackay, the resource raiders collected everything possible in addition to pearlshell: trochus, bêche-de-mer, copra, and green and hawksbill turtles. In a visit to Cairns in 1914 Dr Hamlyn Harris, director of the Queensland Museum, commenting on Japanese plundering, stated that 'it is aggravating [to see] many foreigners coming here day after day taking away these resources' (*Cairns Post* 26 April 1914, p. 8).

As the pearlshell supply continued to diminish in the early 1900s a quest for alternative sources commenced and it was discovered that the previously ignored trochus or 'top' shell was a possibility. Late in the first decade of the century a trial shipment was sent to Japan and found suitable for the main need of garment buttons; a first shipment of 35.7 tons was consigned in 1912 (Davenport 1986:347). The same year the Queensland government proclaimed trochus subject to the provisions of the *Pearl-Shell and Bêche-de-Mer Fishery Act*.

The trochus animal is a gastropod in the class that includes all snails, slugs and the familiar seashore periwinkles. Classified in the smaller family of Trochidae, there are many genera and hundreds of species. The Trochidae generally build a conical, spiral shell and conceal their opening with a horny plate known as an operculum (Latin for 'lid'). Within the genus *Trochus* are many species, the largest known as *Trochus niloticus* or the 'commercial trochus', which was first classified in the sixteenth century by the great systematist Ulisse Aldrovandi, professor of natural history at the University of Bologna. In the mistaken belief that they were endemic to the River Nile he designated them, in the Latin scholarship of the day, 'Niloticus', which remains its specific term. They were given their full binomial by Linnaeus in 1767 who set the style that all scientific names had to be in Greek or Latin (or at least seem latinate) and since they resembled a child's play toy he took the Latin for top, *trochus*, (from the Greek *trochos*, 'anything of a round shape'). To this day they are known in popular literature as top shells.

The trochus shell is found widely distributed throughout the shallower Indo-Pacific waters. All trochus build their shells out of calcium carbonate, like pearls on a conchiolin foundation, which thereby constitute a nacreous lining giving the same mother-of-pearl iridescence. Only *Trochus niloticus* was found, however, to have workable qualities, and even then, it is inferior to true pearlshell since in time it discolours and delaminates. Moreover, unlike pearlshell that is fully translucent, trochus buttons always have reddish brown streaks on the underside which, in addition to their poorer lasting qualities, relegated them to the cheaper kinds of clothing. Growing to a maximum size across the base to 150 mm, at that stage the shells usually show signs of abrasion and wormholes; the most workable size for the trade was found to be within the range 75–100 mm. Button blanks were cut, as for the pearlshell, by a small circular saw drill down the spirals to the base.

Pearl beds received a temporary respite in 1914 due to the hostilities of World War I that turned the Indo-Pacific region into a war zone. Soon after the cessation of hostilities in 1918 Japan resumed pearling, to the continuing irritation of most Australians in the region. After a decade or so many of their ships withdrew to Palau, which, along with Okinawa in the Ryukuyus, served as their major shelling bases. Four of the major Japanese companies formed the Palau Pearlers Association and, using a fleet of over a hundred diesel-powered luggers, supported by mother ships which brought all supplies down, and carried the shell back, continued to exploit the resources of the Reef ruthlessly. With no attempt at conservation, by the end of the 1930s the industry was again at the point of collapse. Hostilities throughout the 1920s and 1930s between the Japanese and Australians reached dangerous levels of armed conflict, especially when the Japanese established clandestine shore stations on the Reef coast to poach copra (edible coconut fruit within the shell) and its outer husk used for matting, along with bêche-de-mer, turtles and trochus shell. Following ugly confrontations, instances were recorded of the Japanese setting fire to the Australian copra and husk drying sheds (Bartlett 1954:280–3). Only the outbreak of the Pacific War in December 1941 gave the various resources yet another temporary respite. Unfortunately for the industry, but fortunately for trochus, the era of plastics was at hand, and before long button manufacturing was within the province of synthetic materials. Trochus disappeared, bêche-de-mer became a minor product, and pearlshell a specialty item for the highest grades of designer clothing and jewellery.

CHARLES HEDLEY'S 'MARINE BIOLOGICAL ECONOMICS'

Research into the economic utilisation of the Reef at a national level came rather fortuitously, as a consequence of the outbreak of World War I, under the initial stimulus of Charles Hedley. In the first year of hostilities the British government was horrified to discover that many necessary war materials were manufactured in Germany, including such essentials as zinc, magnetos, optical glass and tungsten steel. More bizarrely, since Germany had become the world leader in dyestuffs, even the khaki dye for British military uniforms was manufactured there. Hurriedly, the Westminster government set up a Coordinating Science Research Council to remedy those deficiencies, and requested the dominions to do likewise. In January 1916 Australian Prime Minister William Morris Hughes established a temporary Advisory Council on Science and Industry, (ACSI), later to be formally established by the *Institute of Science and Industry Act*, No.22 of 1920, under the directorship of the highly respected George Knibbs, Commonwealth Statistician, and restructured in 1926 as the Commonwealth Council for Science and Industrial Research (CSIR), forerunner to the Commonwealth Scientific and Industrial Research Organisation (CSIRO) of 1949. The function of the ACSI was to confer with the States on cooperating with science and industry to deal with wartime deficiencies and Gerald Lightfoot, from the Bureau of Statistics, was appointed secretary. With the Commonwealth matching funds equally with those provided from the private sector, a range of pressing issues were undertaken, dealing with zinc processing, cattle tick infestation of export beef, prickly pear invasion of grazing lands, brown coal energy research, aluminium alloys and chemical production (Currie & Graham 1966). All, however, were terrestrial.

After Saville-Kent, the most indefatigable proponent of research on the extraction of Reef resources was Charles Hedley (1863–1926) who was the first to seek ACSI assistance for Reef research. Hedley, in fact, became one of the greatest of Reef scientists. Although employed for most of his working life in the Australian Museum in Sydney, Hedley was more active in Reef research and field exploration in Queensland throughout the first decades of the twentieth century than any other person. Born in Yorkshire, son of Canon Thomas Hedley, a fellow of Trinity College in Cambridge, the young Charles was an asthmatic who migrated to New Zealand in search of a milder climate in 1881 and then to

Queensland in 1882. A self-taught naturalist with an interest in molluscs, by 1889 he was employed in the Queensland Museum, and he had published an astonishing nine papers in the two years of 1888 and 1889, most on land snails, in the *Proceedings* of the Royal Society of Queensland. In 1890 he was invited to travel to New Guinea with an expedition to the Louisiade Archipelago, where he spent his time collecting land snails in the Milne Bay region. An outcome of this was a stream of further papers in major journals, most in the *Proceedings* of the Linnean Society of New South Wales.

His reputation growing rapidly, by 1891, like Haswell in the same year, Hedley moved to Sydney, in his case to the more congenial Australian Museum, as an assistant in the conchological department. By 1896 he was appointed official conchologist, in 1908 Assistant Curator, and in 1920 Acting Director. With a wide range of interests, including the geological issues of coral reef formation, he was sent by the Museum in 1896 as an assistant to help in the Royal Society of London expedition to the atoll of Funafuti to test Darwin's reef formation hypothesis. Throughout those years Hedley not only became Australia's pre-eminent conchologist, and equally recognised in malacology (study of the animal within the shell), but had also established an even greater reputation as a taxonomist. His papers in the field, over his active lifetime of nearly thirty years, run to an incredible list, compiled by the the Museum's Tom Iredale and appended to the panegyric obituary in 1936 in the *Proceedings* of the Linnean Society of New South Wales by Charles Anderson, at the time Director of the Australian Museum (Anderson 1936).

Since the Australian Museum funding was minimal – Hedley, in fact, undertook many field expeditions at his own expense – he also sent off a request to the ACSI for financial support for research into the 'marine biological economics of tropical Australia', specifically, the Reef. Hedley had requested research into the major industries at the time, namely pearling, bêche-de-mering, turtle and dugong processing and trochus cultivation, as well as the possibility of harvesting bathroom sponges. On 26 September 1916 he wrote to Lightfoot stressing the urgency of the need for research, and the depredations of the resource raiders, both Australian and Japanese. The Reef, he wrote, 'is an area which has in the past furnished subsistence to many and large fortunes to a few. The harvest of it has been reaped with the least knowledge of the subject as a savage might reap it in haste and greed. In most cases the area has been heavily over-fished and is now too depleted to be worth much attention'. While 'the School of Tropical Medicine at Townsville might afford facilities for at least a preliminary campaign on the life history of the Pearl Oyster' as well as the bêche-de-mer and trochus, in the longer term the better way would be to obtain 'a science graduate from the Queensland University ... to pursue such investigations'. Moreover, in the short term a small grant would enable Edmund Banfield to 'obtain a quantity of Trochus and plant them on a beach at Dunk Island' as a preliminary field trial (CSIRO archives, Series I.82/1).

As a consequence of separate representations from Haswell and Hedley, in November 1916 the ACSI decided to establish an investigating committee on 'Biological Economics' with Hedley as chairman, and four other members: Alan McCulloch of the Australian Museum, Professor William Dakin of Sydney University – an expert on coastal ecosystems – along with Dr Hamlyn-Harris of the Queensland Museum, and Edmund Banfield, journalist and self-styled 'beachcomber' of Dunk Island. Hamlyn-Harris, unfortunately, resigned a year later due to ill health and was replaced by Harvey Johnston, professor of biology at the recently founded University of Queensland. Their task was to look at six possible products to be obtained from the Reef: pearling, bêche-de-mer, trochus, sponges, posidonia (sea grass) fibre, and cultured pearls. When ACSI secretary Lightfoot sought advice from Captain Vassal Forrester, Portmaster of Brisbane, regarding the so far undeveloped industries of sponge and trochus fishing, Forrester's assistant George Hamilton informed Lightfoot on 18 April 1917 that 'production of sponges in this state is an undeveloped industry, although there is evidently a considerable growth of this marine product in our tropical waters . . . though doubtless of commercial value there has been so far no serious attempt to fish for same'. However, 'On the other hand, the collection of trochus shell has of recent years developed into a considerable industry, and large quantities of the shell have been gathered and exported, principally to Japan, during the last two years'. Echoing the earlier complaints of Hamlyn-Harris of 1914, Hamilton noted that 'boats which were principally employed in the collection of pearl shell have been diverted to this purpose, the venue of their operations being the Great Barrier Reef extending from Torres Strait as far south as Mackay' (CSIRO archives, Series I.82/0/2).

The following year Hedley produced the committee's first report which, not surprisingly, recommended that pearl oysters, bêche-de-mer and trochus all needed systematic study of their life cycles over sustained periods. However, as Lightfoot wrote in his subsequent report to the ACSI, the work to 'be carried out would have involved the employment of specially qualified biologists, and for the reason that the Advisory Council's resources were of a very limited and temporary nature no action was taken to carry out the research work' (Lightfoot 1926:2). Posidonia fibre was considered uneconomical, and cultured pearls were not possible since the Japanese had already taken out Australian patents which would have precluded any action except 'by arrangement with the patentee' (4). Both trochus and sponges, however, were considered worth pursuing and Hedley was given a grant of £15 to enable him to conduct field trials with the cooperation of Banfield at Dunk Island.

Hedley had been actively gathering every item of information he could concerning trochus as a possible substitute for the pearlshell for button manufacture and a large number of research reports were sent to Lighfoot. In 1917 Hedley submitted his findings to the ACSI, basically in the form of an article which he had published at the same time in the *Australian Zoologist* under the title 'The economics of *Trochus niloticus*'. His report highlighted the fact that 'during the past six years [1910–16] an active request for Trochus by button makers has sprung up, advancing from £20 to £30 per ton' and that, by 1916, 950 tons had been exported from Queensland. In addition, there was the possibility that the animal itself, in addition to the shell, might have an economic value since 'in China smoked Trochus is esteemed a dainty' when diced and used as the base of a soup, that information having been supplied by Mr E. J. Banfield who 'tried this Trochus soup and reports it as very palatable' (Hedley 1917; also CSIRO archives, Series I.82/2). Unfortunately, the trochus experiments were not successful as Hedley wrote to Lightfoot on 10 December 1917. Despite the grant of £15, Banfield was unable to obtain supplies of live trochus due to 'the approach of the hurricane season, [so that] fishing in his district has come to an end' (CSIRO archives, Series I.82/2). Although Hedley continued to publish on the issue of trochus, and maintained a voluminous correspondence with persons throughout the South Pacific region, no trochus industry, managed on a scientifically controlled basis, ever eventuated.

Undeterred, Hedley turned his attention to the remaining possibility of commercial sponge fishing. Already on 18 October 1917 he had written to a Mr Durham, in Sydney, a trader for whom Hedley had arranged to collect samples from the Cooktown coast for assessment. Encouraged by that commercial interest, ACSI early in 1918 agreed to provide funding for the wages of a diver and the chartering of a suitable boat to obtain such samples. Probably the kindest account of the sponge project from then on – as it emerges for the first time from the original records stored in CSIRO archives under Black Mountain in Canberra – would be, at best, to describe it as a comedy of errors.

Since virtually all commercially valuable bathroom sponges, along with those used by painters and decorators for the then fashionable stippling of walls, came from the Greek Mediterranean islands, obtained by helmeted divers, Hedley decided that he would need to obtain the services of an experienced Greek sponge diver. Inquiries from a Greek shipping agent in Sydney led Hedley to Mr Michelides, a Greek commercial attaché in Perth, who in turn recommended Nicholas Kezalias, a former sponge diver from the Mediterranean island of Kastellorizo, living at the time in Fremantle. Kezalias, however, while he could speak English, could neither read nor write it, and so was dependent upon translators. It was decided to appoint Kezalias on a four month contract at £30 per month, expenses for boat hire in Cooktown and combination rail and boat travel to Cooktown for a survey of the sponge beds in Reef waters (AM Archives: AMS 272/1/10). After a marathon rail journey from Perth to Rockhampton, and then by steamer to Cooktown, Kezalias arrived on 30 April 1918 where he reported to the harbourmaster and was introduced to the enthusiastic members of the Chamber of Commerce.

On arrival, Kezalias found that, due to the war, the navy had requisitioned all serviceable diving suits and the rubberised fabric of one provided for him had perished, leaving it, in effect, unusable. To compound problems, Kezalias discovered that the cost of hiring a suitable boat would be an expensive £4 per day. A flurry of telegrams between Cooktown and Hedley in Sydney then ensued, the harbourmaster initiating it on 2 May with news that a functioning diving suit was required, along with a boat and a crew of six whites, and requesting authorisation to proceed. Hedley replied the next day with a telegram noteworthy for its brevity: 'Please arrange preliminary excursion on cheaper scale and inform'. Hedley later suggested employing Aborigines at £2.10.0 a day even though 'they require constant supervision' to which the harbourmaster agreed on 8 May. Hedley then arranged for Alan McCulloch (1885–1925), his colleague at the Australian Museum, to travel to Cooktown to take charge of the proceedings, while Kezalias was instructed to await his arrival. Tired of waiting, nine days later, on 17 May Kezalias decided to travel back to Sydney to explain the difficulties in person, despite the fact that Captain Byrne, the harbourmaster, had warned him that 'he was under engagement and bound to obey instructions'. Arriving on 20 June, McCulloch was stunned to learn that while he was proceeding north, Kezalias had travelled south, all the way to Melbourne, unbeknown to Hedley, which in the absence of any documentary evidence, was to the offices of the ACSI. A frustrated McCulloch then initiated a second flurry of telegrams to Hedley, recorded on bright red paper to indicate their urgency, but to no avail: Kezalias had fled (AM Archives: AMS 272/1/10).

Not all of the documentation seems to have survived, but the final word on the specific project of commercial sponge fishing is possibly that of Hedley to Lightfoot on 18 May in which he stated that 'I have to make the sad announcement that, despite all our efforts, this sponge investigation has come to an abrupt and disagreeable conclusion. This ensues from the sudden and unexplained, even unannounced, retirement of Kezalias from his engagement'. Following further justification of the venture, Hedley concluded with transcripts of the numerous telegrams.

Two weeks later, however, from Melbourne on 31 May, Kezalias wrote to Lightfoot, through an amanuensis, his justification that he had been waiting in Cooktown for a fortnight and 'could go no further'. He reiterated his desire to be 'willing to do my utmost and prove to the satisfaction of the department the existence of paying sponges on the coast of Australia and only ask to be provided with the outfit required, [with] which I can work and also engage a Greek sponge assistant in Sydney'. His explanation seems to have assuaged Lightfoot who accepted that the course of events was indeed unfortunate and that both the ACSI and the Queensland governments had to share responsibility. He recommended to the Executive Council of ACSI that Kezalias be paid the agreed salary and expenses: while having 'fully discussed the matter with Mr C. Hedley ... under the circumstances it appears best that the work should be abandoned for the present and that Kezalias should return to Fremantle' (CSIRO archives, Series I.82/3). So ended the first ventures into Commonwealth government support for the economic exploitation of the natural resources of the Reef.

PART TWO

A New Era in Reef Awareness:

From Early Scientific Investigation To Conservation and Heritage

CHAPTER 11

ORIGIN AND STRUCTURE OF CORAL REEFS: FROM FORSTER TO DARWIN

Throughout the early decades of the nineteenth century when British naval ships were actively surveying the waters of the Great Barrier Reef, the new science of geology was becoming increasingly important in the search for an understanding of the origin and structure of coral reefs. In particular, growing evidence of marine deposits on land, especially in elevated strata, was attracting much speculative attention and rapidly initiated a new direction of investigation in the long history of inquiry into the nature of coral.

MYSTERY OF CORAL: THE ANCIENT QUEST

Already, a formidable record of theories and explanations concerning coral had developed over several millennia: in terms of recorded observations, from the time of Aristotle during the reign of Alexander the Great in the fourth century BC. The word 'coral' itself first appears in the work of Theophrastus (*c*.372–*c*.285 BC) who succeeded Aristotle as director of the Lyceum. In his treatise on minerals, *Peri lithon* ('On Stones'), Theophrastus described a kind of rock by the term *kouralion*, later transliterated into Latin as *curalium*, then, centuries later, into Italian as *corallo*, and French as *corail*, with cognates in all European languages. Only a fragment of *Peri lithon* survives today, but we find there the first mention of coral in a brief description: 'coral is similar to a stone, is shaped like a root, and found in the sea' (VII.39). Equally interesting are two lines following where he mentions the 'rock like Indian reed' (*o indikos kalamos apolelithomenos*) which, 'when it turns to stone is very similar to coral'. His Indian reed, of course, is actually the familiar organpipe coral, *Tubipora musica*. Most of the early studies, however, were not on the 'hard' reef-building scleractinians (Gk *skleros*: 'hard, unyielding'), but on the widely traded soft coral of which the most famous was the precious red species, *Corallium rubrum* found in the Mediterranean and the Red Sea.

In stages no longer known, the conception of coral as a stone changed to that of a plant. Only a few references occur in the succeeding Latin literature, chiefly in the works of the poet Ovid (43 BC – AD 17) and the great Roman author Pliny (Gaius Plinius Secundus, AD 23/24–79). Two brief lines appear in Ovid's famous *Metamorphoses*, a lengthy poem dealing with changes in nature, where we read in the final book that 'coral, once a soft plant under water hardens in contact with air' (*sic et curalium quo primum contigit auras tempore durescit: mollis fuit herba sub undis*), (XV.417–18). Pliny went much further in his huge encyclopedia *Naturalis historia* ('Natural History') with an entire entry on coral. Describing its characteristics, obviously referring to the red coral collected for the jewellery trade, he stated that 'it is like a shrub, coloured green with white berries which, once taken out of water immediately hardens and turns red' so that it must be cut quickly by a sharp knife. 'That', he informed his readers, 'is why it is called coral' implying that 'coral' came from the Greek verb *keiro* 'to cut' (XXXII.xi.20–21). Like so many of Pliny's stories, however, it was a fanciful comment and modern philology has determined that coral, in the Greek form *kouralion*, is of obscure Semitic origin.

Nearly two millennia later, a landmark publication *Traité du corail* ('Treatise on Coral') by the French scientist André de Peyssonnel (1694–1759) was read to the Royal Society of London on 7 May 1752. After studying both soft corals and the hard, reef-building species, he finally established that coral formations were created by minute animals, stating that what earlier investigators 'took for flowers were truly insects' (*Philosophical Transactions*) (Vol.47, 1751–52, 445–69) – 'insects' at the time being a general term for many invertebrates. In the same period the word 'zoophyte' began to reappear in a number of writings. This term, often attributed to Aristotle, was probably a neologism of Isidore, Bishop of Seville who, around AD 600 composed the famous *Etymologies*, an encyclopedic dictionary of terms. The word actually was a useful hybrid since its Greek provenance from *zôon* ('animal') and *phytos* ('plant') dealt nicely with the ambiguity. Once revived in later years of controversy, it continued to be employed, even when it was accepted that the coral polyp is an animal. Remarkably, it was not confirmed until the later decades of the twentieth century that all reef-building corals (also known more generally as 'hermatypic': Gk *herma*, reef; *typou*, building) depend on a symbiosis with a multitude of tiny algae in their tissues, known as zooxanthellae, which provide an essential nutritional exchange, enabling polyps to create the hard calcareous formations of coral reefs. With that discovery, 'zooxanthellate' has been introduced as a modern descriptor for hermatypic corals (see Veron 1995:97). This has prompted a suggestion of 'coralgal' as a more appropriate term for coral reefs. Perhaps zoophyte, after all, came close to a correct designation.

Peyssonnel's conclusive demonstration that coral is formed by animal agencies led to a phase change in coral research: an entire new taxonomy could now be constructed and this began in the work of John Ellis, using the microscope that had been developed in the previous century. In 1755 he published the first major synthesis of marine organism research under the title *An Essay towards a Natural History of the Corallines and other Marine Productions of the like Kind*, essentially a taxonomy of corallines collected in the waters off Dublin and the nearby island of Anglesea off the coast of Wales. The term corallines was used at the time to describe not only corals but a wide range of colonial animals including sea-mats, sponges and sea-pens, many of which also have a concentration of lime in their bodies, giving a strengthening cohesion. Since such investigations were only then beginning, Ellis had a huge task ahead in attempting to bring order and clarity into the masses of findings that were accumulating.

His central theoretical argument, set out in the Introduction, stressed the importance of microscopical study which 'seemed to indicate [corallines] being more of an animal, than vegetable Nature'. Indeed, he claimed that his own microscopic observations showed that 'the Subjects themselves . . . which had hitherto been considered by Naturalists, as Marine Vegetable, were in Reality of Animal Production' (Ellis 1755:vii). His taxonomy was based upon the external structure of the corallite (cup of an individual polyp) and microscopic analysis of internal morphology: shape, texture, colour, secretions. A continuing stream of observations followed, which were published in the *Philosophical Transactions of the Royal Society of London*. His crowning achievements were the two papers 'On the animal nature of *Zoophytes*, called *Corallina*', read to the Royal Society on 9 July 1767, and '*Actinia sociata*, or *Clustered Animal Flower*' of 12 November 1767.

Ellis then set himself to catalogue the huge accumulation of specimens that continued to arrive, especially from the West Indies – for which he had been appointed King's Agent – where hermatypic corals are dominant. He was also in close contact with Linnaeus in Sweden, who had begun a taxonomy of all living creatures in his *Systema naturae* of 1736, which was to become the standard reference work, to be continually revised as further species were described. Linnaeus, like all other scientists of the period, used Aristotle's concept of an infinite ladder of nature, *scala naturae*, as the framework on which to construct his taxonomy. Aristotle's system began with the lifeless, then ascended through plants – from those with no apparent sensitivity and in sessile form – to those which seemed animal, and others that clearly were. The great taxonomic problem, Aristotle commented, especially for life in the sea, is that gradations are so fine no clear lines of demarcation can be found. 'The progressive changes of nature are so small and imperceptible' he wrote, 'it is impossible to draw a boundary and determine their category' and so we 'are at a loss to know whether they are animals or plants' (*diaporeseien an tis poteron zoon e phyton estin*) (*Historia animalium* VII.588b).

Linnaeus had first classified corals as stony plants (lithophytes) but, due to the work of Ellis, with whom he had regular and warm correspondence, he was persuaded to reclassify some as animals in the 1758 tenth definitive edition of *Systema naturae*. Even so, Linnaeus remained uncertain, and clung to a belief, as he wrote to Ellis on 16 September 1761, that soft corals had a plant stem from which animals were generated by a still unknown metamorphic process 'granted by the Creator'. Soon after, following Ellis in choosing the term *Actinia sociata* ('clustering starlike flower') for one species of soft coral, he classified them collectively, as they remain to this day, as 'anthozoans' (Gk *anthos*: flower + *zôon*: animal). By the early 1770s Ellis had enlarged his original *Essay* into a huge catalogue of every species of coral then known, hard and soft, having collected and begun to describe sixteen genera, containing 279 species.

In those years, however, Ellis' health had begun to fail and on 15 October 1776 the great systematist died, his magnum opus still unfinished. The task of preparing it for publication was undertaken by the distinguished botanist Daniel Solander who had gained first-hand knowledge of coral reefs during the voyage of the *Endeavour* in 1770. Solander barely completed it before his own untimely death in 1782, aged only forty-nine. Ellis' life work, *The Natural History of many curious and Uncommon Zoophytes, arranged and described by D. C. Solander*, appeared four years later.

ANOTHER MYSTERY: FORMATION OF CORAL REEFS

The tentative acceptance by Linnaeus of polyps as animals in his *Systema naturae* of 1758, and the great work of Ellis in 1786, ended a major phase in the quest to solve the mystery of coral formation by minute polyps. Almost completely unexamined, however, had been the question of how reefs themselves are formed. How had such vast assemblages been constructed in clear, deep blue water, far from land? What held them together in often turbulent oceans? How can they rise from the vast expanses of the oceans to lie, just below water level, fearful hazards to navigation: as 'hard and ponderous as a stone' and making 'a most terrible surf breaking mountains high'?

An early hypothesis came in 1778 when the German naturalist Johann Reinhold Forster (1729–98), who had sailed on Cook's second Pacific voyage of 1772–75 aboard the *Resolution*, recorded his observations on coral reefs in considerable detail in *Theory of the Formation of Isles*. Forster was a great naturalist and his *Resolution* journal contains valuable insights into atoll formation. When describing the 'formation of Isles' near Tahiti, he asked,

> if the question be put, how it comes that the *Madrepores* (hermatypic corals) form such circular or oval ridges of rocks; it seems to me that they do it by instinct, to shelter themselves the better against the Impetuosity & constance of the SW winds; so that within the ridge there is always a fine calm Bason, where they feel nothing

of the Effects of the most blowing weather (Forster, 15 August 1773; Hoare 1982:324;cf.494).

That observation was developed further in his official journal, prepared for the Admiralty on his return to London and published in 1778 under the title *Observations Made during a Voyage around the World*. In Section 4, 'Theory of the Formation of Isles', Forster expanded his concept of polyp 'instinct' whereby they 'endeavour to stretch only a ledge, within which is a lagoon, which is certainly entirely screened against the power [of the ocean and winds]'. This, he continued, 'seems to me the most probable cause of THE ORIGIN of all THE TROPICAL LOW ISLES, over the whole South-sea' (Forster 1778:150–51; Forster's capitals). Despite the apparent simplicity of his explanation, Forster had distinguished a fundamental feature of atolls, namely that the formation of a circular structure enables the coral colonies to resist the 'rage and power of the ocean'. Many other questions, however, remained unanswered. Given the enormous depth of surrounding waters, unable to be sounded by the technology of the times, how had polyps established themselves in the first place? Upon what foundations had they erected their limestone structures? What was the nature of that 'instinct' by which 'the animalcules forming these reefs . . . shelter their habitation from the impetuosity of the winds'? (151).

During the voyage of the *Resolution* Cook became aware of the growing speculation that reefs had been formed by minute 'insectes', and were neither rocks nor petrified plants. Not only atolls, but also continental islands with puzzling elevated fringing relict reefs were investigated, and in his journal for June 1774 – anticipating debates over a role in reef formation for sea level change, not resolved until the 1930s – he asked

> If these Coral rockes were first formed in the Sea by animals, how came they thrown up, to such a height? Has this Island been raised by an Earth quake or has the sea receded from it? Some philosophers [that is, scientists] have attempted to account for the formation of low isles such as are in this Sea, but I do not know of any thing has been said of high Islands or such as I have been speaking of (Cook 1774:438).

Gradually, evidence was being adduced as exploring voyages brought together pieces of the puzzle, such as the findings of Adelbert von Chamisso, the naturalist who sailed on the first voyage in 1815–18 of the Russian ship *Rurik* under Otto von Kotzebue around the Pacific from Kamchatka to Alaska, California and then to the Hawaiian, Marshall and Mariana groups lying between the north tropic and the equator. In an account entitled 'On the Coral Islands' included as an appendix to Kotzebue's narrative of the voyage Chamisso made two important observations. Firstly, in contradistinction to Forster, he pointed out that corals thrive best in windward, turbulent reef fronts, stating that 'the larger species of corals, which form blocks measuring several fathoms in thickness seem to prefer the more violent surf on the external edge of the reef', a point amplified further on, that the windward 'side of the reef, exposed to the unremitting fury of the ocean, should first rise above the element that created it'. His second observation attempted to explain why atolls appear in wide expanses of turbulent oceans, almost out of nowhere. Because, he reasoned, 'the corals have founded their buildings on shoals in the sea; or to speak more correctly, on the tops of mountains lying under the water', and that, even further, variation in magnitude and distribution of atoll clusters 'probably depends on the size of the submarine mountain tops, on which their basis is founded' (Chamisso 1821:III.331, 334).

When the *Rurik* docked at Portsmouth in 1818 on the way home to St Petersburg, Chamisso took the opportunity to travel to London where he met the ageing Joseph Banks, along with resident botanist Robert Brown, as well as Georges Cuvier from the Muséum National d'Histoire Naturelle in Paris, who was then visiting Banks. Back in Russia, Kotzebue published his findings in German three years later in three volumes, under the general title *Reise am die Welt* ('Voyage around the World'), with Chamisso writing much of the third volume, which became available to English geologists in a translation the same year, 1821.

A second essential element was contributed at much the same time by Jean René Quoy and Joseph Paul Gaimard, naturalists aboard the French corvette *L'Uranie* on its Pacific voyage of 1817–20, chiefly to the Mariana and Hawaiian Island groups. Presenting their findings in a joint paper of 1823 entitled *Mémoire sur l'Accroissement des polypes lithophytes considéré géologiquement* ('Geological aspects of coral formation') they argued from their extensive examination of reefs in Pacific tropical waters that it would be a mistake to ascribe all atoll formation to polyps alone, growing up from the ocean floor. Rather, they believed that coral reefs are surface features that 'have as a base the same element, the same minerals which concur to form all the known islands and continents . . . that [in effect] they build their dwellings on the submarine rocks, enveloping them entirely, or in part, but properly speaking they do not form them'. Thus, 'all these reefs', they conclude, 'are, in our opinion, platforms arising from the conformation of the primitive surface' (Quoy & Gaimard 1823:273, 290).

Neither the *Rurik* nor the *Freycinet* visited the Great Barrier Reef where only British ships had surveyed. From Flinders' apt description the Reef is not an atoll formation but, quite literally a 'great barrier' that parallells the coastline for a thousand miles, at varying distances from the coast. What, it was being asked, are the processes that form barrier reefs, given that the polyps are of the same species as those found on atolls? Answers to such questions were being sought in the new geological and similar societies being established on the model of the Geological Society of London, founded in 1807. At their meetings papers were being presented which attempted to draw together the increasing volume of findings from the survey voyages of various nations, particularly the French, English and Russian, including the geological section written by William Fitton in the 1827 *Narrative* of the Great Barrier Reef survey by Phillip Parker King. Although the findings of coastal geology seemed unrelated to atoll formation, they were part of the growing accumulation of evidence on both the coral reef question and the broader issues involved in establishing a general theory of geology.

A commentary on those issues came from the Prussian naturalist Alexander von Humboldt (1769–1859). Accompanied by the French botanist Aimé Bonpland, he travelled throughout the Caribbean and adjacent regions of central and South America from 1799 to 1804 making the first geological survey of those lands. Paying particular attention to the limestone strata, he observed their similarity with both the Jura limestone of Europe and the rock of which the coral Cayman Islands were composed. He saw this as relevant to the question of coral reef origins, in view of the theory of the distinguished geologist Leopold von Buch – that the whole Jura formation consists of elevated ancient coral reefs, located at a distance from mountain chains. Humboldt himself had been involved in identifying the Jura formation (index stratum for the Jurassic Period), naming it after the Jura mountains in France. Now, from his concept of the earth crust as an integrated complex, subject to immense structural change over the millennia, he sought to explain coral reef formation in terms of global processes, recording his conclusions in his *Personal Narrative of Travels* (Paris 1814–25) a work that inspired Darwin's life. Well aware of current hypotheses on reef formation he doubted the belief that atolls, rising from great depths, had been built up entirely by coral polyps, and was unsure whether 'rocks formed by polypi still living are found at great depth below this fragmentary rock of coral'. He did suspect that 'those huge masses which are said to rise from the abyss of the Pacific to the surface of the water . . . had some primitive or volcanic rock for a basis, to which they adhere at small depths' (Humboldt 1852:III.186).

LYELL'S SOLUTION OF 1832

What, then, was the origin of the 'platforms' of Quoy and Gaimard? What could have caused a 'heaving up of the land'? How could '*marine* shells . . . in cemented masses, at heights above the sea, to which no ordinary natural operations could have conveyed them' as described by Péron, be accounted for? How, in effect, are the curious atolls and barrier reefs related to the structure of the earth? In 1832 a solution was proposed by Charles Lyell in Volume 2 of his great three volume *Principles of Geology* published successively in 1830, 1832 and 1833, in which the subtitle revealed its uniformitarian approach: 'being an attempt to explain the former changes of the earth's surface by reference to causes now in operation', a theory first advanced in 1795 by James Hutton in *Theory of the Earth*.

Throughout that period the Biblically-based belief that the earth had been created 6000 years earlier according to the account given in Genesis, and that subsequent physical changes had been caused by various catastrophes such as floods and earthquakes, was becoming challenged by an alternative 'uniformitarian' theory, initiated by a Scottish gentleman-farmer, James Hutton (1726–97). In Hutton's day Edinburgh was the intellectual capital of Britain, centre of the 'Scottish Enlightenment' which parallelled the French movement, where the Scottish school of geology under his stimulus began to assert, from discernible evidence and consequent inductive conclusions, that the earth can be observed in continuous process of cyclical change. Hutton presented his views in a paper to the Royal Society of Edinburgh in 1788 in which he argued that the earth had been formed by the slow, never-ending and completely mechanical processes of inner forces and external weathering. He ended with the famous assertion that in seeking to understand the processes that have formed the earth, 'we find no vestige of a beginning, no prospect of an end' (Hutton 1788:I.304). Hutton reprinted that paper, along with others that elaborated the gradualist argument, in his epochal two volume book *Theory of the Earth with Proofs and Illustrations* of 1795.

Hutton's initial insight, however, as he acknowledged in his book, was not based on extensive fieldwork but was generated out of an hypothesis, drawn from the observations of others, and a giant speculative leap: 'I just saw it, and no more, at Petershead and Aberdeen, but that was all the granite I had ever seen when I wrote my Theory of the Earth. I have, since that time, seen it in different places; because I went on purpose to examine it' (Hutton 1795; cited in Gould 1987:72). Deluges and divine catastrophes were rejected as significant agents of change and the Biblical concept that the earth and all of nature came into existence only 6000 years previously was now challenged by the revolutionary conception of time reaching back through uncountable aeons.

Born in the year of Hutton's death, Charles Lyell (1797–1875), also a Scot from Edinburgh, was an uncompromising follower of Hutton. Indeed, Hutton's theories were developed in Lyell's great three volume *Principles of Geology* of 1830–33 which came to dominate geological thought throughout much of the century, although not without considerable dissent. After Lyell's death his sister-in-law Katherine collected, edited and published his papers in 1881. Here we find the most succinct summary of Lyell's work: in his own words, an attempt to demonstrate that a proper understanding of geological processes comes from the fundamental premise that 'neither more nor less than that *no causes whatever* have from earliest times to the present, ever acted, but those *now acting*; and that they never acted with different degrees of energy from which they now exert' (Lyell 1881:I.234; Charles Lyell's emphasis). The accumulating evidence of marine deposits – shells, ancient coral reefs, fossilised animals – discovered in strata sometimes hundreds, even thousands, of metres above current sea levels, along with massive unconformities of convoluted strata, often with intrusions of other very different rocks, was being revealed by engineering excavations in mountainous regions, especially as the industrial revolution accelerated the pace of canal and railway construction. This led him to argue that the crust of the earth, over the aeons, had been imperceptibly, but relentlessly, subjected to alternating periods of inner forces of elevation and subsidence.

Like Hutton, ignoring Bacon's rule that scientific theory has to come from direct observation, Lyell, who never saw a coral reef, gathered evidence from many who had voyaged in tropical waters – Kotzebue, King, Quoy and Gaimard, Beechey, de la Beche, von Buch and Maclure are mentioned specifically – and brought them together in Volume 2 into a sweeping synthesis. Coral atolls, he concluded, are formed by polyps, by an infinitely slow process, on the summits of submerged volcanoes on the ocean floor. 'The volcanic isles of the Pacific', he wrote, 'shoot up ten or fifteen thousand feet above the level of the ocean. These islands bear evident marks of having been produced by successive volcanic eruptions; and coral reefs are sometimes found on the volcanic soil, reaching for some distance from the sea-shore into the interior'. Pressing home his uniformitarian argument, Lyell dismissed any objection to the time required for such structures to be created, 'on the ground of the slowness of the operations of lithogenous polyps' (Lyell 1832:II.288). Two pages further on he drew his conclusions: 'The circular or oval forms of the numerous coral isles of the Pacific, with the lagoons in their centre, naturally suggest the idea that they are nothing more than the crests of submarine volcanos, having the rims and bottoms of their craters overgrown by corals' (II.290). In explaining archipelagoes, and similar distributions of reefs, Lyell attributed them to 'the ejection of volcanic ashes and sand . . . [which] may serve as a foundation for another' to the extent that, in the Pacific, 'they present the appearance of troops marching upon the surface of the ocean' (II.295).

But, he continued, why is it that 'there should be so immense an area in eastern Oceania, studded with minute islands, without one single spot where there is a wider extent of land than belongs to such islands as Otaheite [Tahiti], Owhyhee [Hawaii], and a few others, which either have been, or are still the seats of active volcanoes'? The answer he provided was that 'the amount of subsidence by earthquakes exceeds in that quarter of the globe at present the elevation due to the same cause' (II.296). The uniformitarian argument was advanced as demonstrative proof: active volcanoes certainly are brief catastrophic events, indicative of the release of subterranean energy, but they remain, none the less, mere transitory epiphenomena upon the continuing elevation and subsidence of the earth's crust. The Pacific, he reasoned, has simply sunk as the underlying forces have been released, and in corresponding motion (anticipating the theory of isostasy developed in the 1850s), he suggested that the nearby Andes had slowly risen out of the sea, taking their marine depositions with them.

In effect, Lyell confirmed the earlier observations of Forster, Chamisso, Quoy and Gaimard. His further development was to argue that atolls were formed on the summits of subterranean volcanoes that had emerged from the ocean floor, then '*gradually* elevated by earthquakes' (II.292; Lyell's emphasis). Periods of continuing elevation and subsidence would follow, and, when the water was shallow enough to allow polyp growth, coral colonies would build upon the detritus of previous formations. As examples of that alternating process Lyell cited the Maldive and Laccadive archipelagoes in the Indian Ocean, the Great Barrier Reef and the Rowley Shoals in north-western Australia surveyed by King, and some thirty-two of the reefs examined by Captain Frederick Beechey, commander of HMS *Blossom* which made a Pacific survey in the years 1825–28 with, among other tasks, orders to find the *Bounty* mutineers. The puzzle of the horseshoe shape of most atolls he attributed, as did earlier explorers, to the formation of inner rainwater lagoons which, upon flowing out thereby killed the polyps in their path, and to periods of 'alternate elevation and depression . . . [which] might produce still greater inequality in the two sides . . . while the action of the breakers contributes to raise the windward barrier' (II.294).

Lyell had made a determined attempt to fit field observations into his theory and solve the coral reef puzzle. Five years later, in 1837, it was challenged by a new explanation, developed by the 28 year old Charles Darwin, gentleman naturalist and neophyte geologist, recently returned from a five-year voyage around the world aboard HMS *Beagle*. Darwin then prepared a paper for the Geological Society of London 'On certain areas of elevation and subsidence in the Pacific and Indian Oceans, as deduced from the study of coral formations' which contradicted Lyell's theory of the formation of atolls. Having heard that Lyell had read a draft, Darwin wrote, 28 May 1837, to his erstwhile mentor at Cambridge, John Henslow, that 'On Wednesday I am going to read a short account of my views of the whole affair [of reef origins and formation], and Lyell I believe intends giving up the crater doctrine' (*Correspondence of Charles Darwin*, ed. Burkhardt et al.: II.21 & 22nn.3, 4. Hereafter cited as *CCD*). Lyell had indeed seen the draft four days earlier, on 24 May 1837, and was mortified, but gracious, having already written to his friend, the astronomer John Herschel, that 'I am very full of Darwin's new theory of Coral Islands, and have urged Whewell to make him read it at the next meeting. I must give up my volcanic crater theory for ever, though it costs me a pang at first, for it accounted for so much' (Lyell 1881:2.12). What, then, was the theory presented by Darwin, and what effects did it have on coral reef science, and geology generally? How did it affect the exploration of the Great Barrier Reef?

CHARLES DARWIN AND THE VOYAGE OF THE BEAGLE: 1831–1836

Of the several survey voyages of the *Beagle*, including that to the Great Barrier Reef of 1837–44 under the command of Wickham and then Stokes, it was the first in 1831–36 which captured both scientific attention and the public imagination and has remained deservedly prominent ever since. Central to this interest was Darwin's shipboard diary, revised and published in 1839 in the official *Journal of Researches* and a second, somewhat changed edition of 1845 under the same title, both editions at times being reissued under the misleading title of 'Diary'. The voyage of the *Beagle* also became the subject of highly romanticised accounts, by none more so than that by Darwin himself in his retrospective autobiography of 1876, written in his sixty-seventh year. Indeed, so thoroughly has the voyage of the *Beagle* been recounted it is only necessary here to give the essential details, and to point out that from his travel experiences, the data he collected, his relentless theorising, and voracious reading of most of the germane scientific literature available (much of it carried aboard the ship, at least 116 volumes) Darwin was able to publish two works of equal merit and scientific controversy that have endured to the present day. In geology, his much disputed account of *The Structure and Distribution of Coral Reefs* appeared in 1842, and in 1859, seventeen years later, came his monumental and even more controversial theory of evolution by natural selection in *The Origin of Species*, based in part on his study of the varieties of finches on the several islands of the Galapagos.

The historical background to the voyage of the *Beagle* was not very dramatic. Rather, it was part of the general program of the Admiralty to survey trade routes in that period when Britain was accelerating its colonial reach in search of new markets for its growing range of manufactured products, and regions for the supply of raw materials. Following the period of revolutions in the Spanish and Portuguese colonies, resulting in declarations of independence and the formation of Rio de la Plata in 1816 (expanded into Argentina in 1825) and Brazil in 1822, Britain was keen to develop trade with those emergent nations, now freed from the monopolistic control imposed by their former masters. Following his brilliant survey of the Australian coastline 1817–22, Phillip Parker King was sent in command of the *Adventure*, accompanied by the smaller *Beagle* under Robert Fitzroy, to survey South American waters in the years 1826–30. The Admiralty decided to continue the project in 1831 by sending Fitzroy in command of the *Beagle*, with Darwin on board, to survey all of South America's eastern Atlantic coast from Bahia (today Salvador) to Cape Horn, designated as the British navy's 'South American Station'.

Fitzroy was instructed to work mainly south of the Rio de la Plata (Silver River, often mistranslated as River Plate) where 'the more hopeless and forbidding any long line of coast may be, the more precious becomes the discovery of a port which affords safe anchorage and wholesome refreshments' (Fitzroy 1839: *Admiralty Instructions*, coll. in Browne & Neve 1989:386). Once completed, Fitzroy was then instructed to continue across the Pacific, using the incredible number of twenty-two chronometers carried on board, to determine even more precisely the longitudes of various places along the Pacific coast of South America, then those of Galapagos, Tahiti, Port Jackson and King George Sound; followed by those of Cape Town, St Helena and Ascension on the way home. As well, 'perhaps . . . if circumstances are favourable [he] might look at the Keeling Islands and settle their position' (393). For five years Fitzroy did exactly that, and in that time, with some 18 months aboard, and 39 months ashore, mostly in South America while the *Beagle* went off on separate surveys, Darwin was free to make his numerous geological and biological expeditions that provided data for his two great and contentious publications.

When the *Beagle* left Plymouth on 27 December 1831 Darwin was barely two months short of his twenty-second birthday; even so, by that time he had already demonstrated prodigious talent in a number of branches of natural science. Born on 12 February 1809, son of Robert Darwin, a wealthy medical practitioner, he was educated in the exclusive

Shrewsbury school of Dr Butler where, inducted into the mandatory classical education of the privileged classes, he claimed later to have forgotten everything. For Darwin natural history was everything that occupied his mind and activities. After a short apprenticeship to his father he was sent to the famous Edinburgh medical school, 1825–27, where he was repelled by the primitive surgical procedures of the period, and instead spent much of his time in the company of Robert Grant, a medical doctor who had moved into marine biology, with a particular interest in sponges.

Recognising the inevitable, Robert Darwin recalled Charles from Edinburgh and consigned him to Cambridge to study for the Anglican ministry. At Cambridge the 18 year old soon found others with an interest in natural science, including his cousin, William Darwin Fox, and in particular, John Stevens Henslow, professor of botany, and formerly of mineralogy, who urged the young Darwin to attend the lectures in geology of Professor Adam Sedgwick. Darwin proved to be a very apt student of geology, at the time a growing interest of more curious clerics as the Biblical account of Genesis was increasingly questioned. Indeed, the famous Royal Commission into the privileged private grammar schools in that period expressed alarm at the teaching of science then being introduced to complement the curriculum previously dominated by Greek and Latin. In particular, it warned, 'few boys could study geology without a violent disturbance to their religious beliefs' (Royal Commission 1864:III.xxi.4750).

While at Cambridge, ostensibly preparing for Anglican holy orders, but really following his passion for nature, and geology in particular, Darwin, fortuitously, was invited to sail aboard the *Beagle* on its voyage of 1831. Aware of the length and difficulty of the proposed two year voyage, and doubtless mindful of the requirements for a captain to eat alone in his cabin from officers and crew, which may have driven a distracted Stokes, former commander of the *Beagle*, to suicide in Patagonian waters, Fitzroy sought a congenial companion. His request was relayed to Cambridge and Professor Henslow, in turn, contacted Darwin, advising him to accept. As Henslow phrased it in a letter to Darwin of 24 August 1831, 'Capt. F. wants a man . . . more as a companion than a mere collector & would not take any one however good a Naturalist who was not recommended to him likewise as a *gentleman*' (CCD I:128–29; Henslow's emphasis). A very nice class distinction. The position of naturalist had been filled by the surgeon Robert McCormick although his tenure ended early when he left the ship at Rio to return to England.

Already well read both in the biology of marine invertebrates and contemporary geological theories and controversies, Darwin prepared carefully for the planned two-year voyage, taking a collection of books and a microscope made by the pre-eminent London firm of Robert Bancks & Son who had made the famous instruments used by Robert Brown, with which he later described 'Brownian movement' in cells. Brown himself, in fact, had advised the young Darwin to secure an instrument made by Bancks with the then excellent magnification power of around $\times 160$. The library aboard was also impressive. Of the known 116 volumes aboard, available for the use of the officers under strict borrowing conditions, 52 were travel books by previous explorers, 16 were on biology, 14 on geology, supplemented by atlases, dictionaries and other works. Altogether, 66 were in English; 43 in French; the remaining seven in Spanish, German and Russian. In addition Darwin had a personal copy of Volume I of Lyell's *Principles of Geology* (1830) (Volume II of 1832 was waiting at Montevideo and Volume III of 1833 at Valparaiso). Between them Fitzroy and Darwin packed their books into the tiny 10 by 11 foot (3 by 3.3 m) poop cabin where Darwin also had his bunk, and a work table which he shared with Mate and Assistant Surveyor John Lort Stokes and Midshipman Philip Gidley King.

Reconstruction of Darwin's experiences aboard the *Beagle* comes from a variety of sources: in the first instance from an autograph diary of 751 pages written aboard the ship from 18 small pocket notebooks in which he pencilled observations from his field trips throughout the voyage. The diary, in turn, was used back in London, along with a subsequent notebook of jottings made in 1837–39, now designated Notebook A, to prepare the official report published under his name in 1839 as Volume III of the *Narrative of the Surveying Voyages of His Majesty's Ships* Adventure *and* Beagle with the sub-title *Journal of Researches . . . 1832–1836*. Interestingly, the autograph diary itself was not published until 1933 when his granddaughter Nora Barlow used it to issue the first critical edition.

Similarly, we have an early version of his autobiography of 1876, originally edited by his son Francis in 1887 with a number of passages excised (6000 words) which the family considered too sensitive to reveal. In 1958 Nora Barlow published a fully restored edition which gives intriguing insights into Darwin's recollections of the genesis of his ideas on coral reef formation and his views on scientific method that guided his theorising. In addition to correcting his idiosyncratic spelling and syntax, she also restored the previously suppressed passages, mainly those concerning Darwin's religious beliefs as he progressed from orthodoxy to agnosticism (or possibly atheism), of how he 'had gradually come . . . to see the Old Testament [as a] . . . manifestly false history of the world', its God revealed as a 'revengeful tyrant', and therefore 'no more to be trusted than the sacred books of the Hindoos, or the beliefs of any barbarian' (Darwin 1876:85).

Once the *Beagle* had docked back in Falmouth on 2 October 1836 – the 'two years' having stretched out to five – Darwin began a process of writing and publication that continued unabated throughout his life. His geological theories on the origin and structure of coral reefs were to remain contentious and under attack long after his death, not finally settled until the era of atomic weapons testing on Enewetak in 1952.

FORMATION OF BARRIER REEFS AND ATOLLS: 'SO DEDUCTIVE A THEORY'

For the first three and a half years the *Beagle* surveyed both east and west coasts of South America with Darwin travelling as a 'supernumerary', that is, a civilian paying his own way and that of his servant Syms Covington. Not subject to ship's discipline, they spent most of their time ashore in personally funded (that is, by his wealthy and indulgent father) field excursions covering a very wide spectrum: plants, insects, fossil bones, marine depositions and geological specimens. Those excursions often went far inland, lasting for months and for which he hired guides, supplies and pack animals. Already, less than a month out of Plymouth and six weeks before first setting foot on South American soil at Bahia, he had determined on making geology his main pursuit. Carrying his copy of Lyell's first volume, which he 'studied attentively', while the ship was still in mid-Atlantic at 'St Jago' (Santiago), the main island of the Cape Verde group, Darwin wrote, forty-five years later in his *Autobiography*, at that time it 'first dawned on me that I might perhaps write a book on the geology of the countries visited, and this made me thrill with delight' (Darwin 1876:77, 81). At Rio, three months later, Darwin wrote to Henslow on 18 May 1832 that 'Geology & the invertebrate animals will be my chief object of pursuit through the whole voyage' (CCD I:237).

For the next three years Darwin explored the coastline and hinterland of South America, guided by the first two volumes of Lyell's *Principles of Geology*. Writing to his cousin Fox from Lima in August 1835 he enthused that 'I am becoming a zealous disciple of Mr Lyell's views as known in his admirable book'; then, with a tyro's bravado reinforced by his own 'Geologizing in S. America' confided that 'I am tempted to carry parts to a greater extent, even than he does'. The next line points to his naiveté at the time: 'Geology is a capital science to begin, as it requires nothing but a little reading, thinking & hammering' (CCD I:460). In the process of following Lyell, Darwin became ever more committed to the concepts of uniformitarianism and continental subsidence and uplift.

Wherever Darwin went he saw raised sedimentary strata and collected the embedded organic remains, chiefly marine shells, to support the uplift theory. He found the most dramatic evidence in marine deposits inland from Valparaiso in the Peuquene range where 'at the height of 13,210 feet, and above it, the black-clay slate contained numerous marine remains . . . which formerly were crawling about the bottom of the sea, now being elevated nearly 14,000 feet [4200 m] above its level' (Darwin 1839b:245). Equally interesting is his report on the examination of the 'step-formed terraces of shingle' running up a valley at 'Guasco' (Huasco) for some 37 miles (60 km) which 'Mr Lyell concluded . . . must have been formed by the sea during the gradual rising of the land' and which contained, not only marine shells but also 'the teeth of a gigantic shark, closely allied to, or identical with the *Carcharias Megalodon* of ancient Europe' (261).

The *Beagle* left South America in September 1835, sailing to the Galapagos where Darwin made extensive observations that were to provide significant material for his controversial theory of evolution by natural selection in *Origin of Species*. The next landfall was Tahiti where, on 20 October, he saw his first coral atoll. On 19 December the ship arrived at New Zealand and on 12 January 1836 it anchored in Port Jackson where Darwin was to spend the next eighteen days in Sydney and its hinterland Blue Mountains, travelling as far inland as the settlement at Bathurst which he found most surprising, particularly its distinctive vegetation and dry landscape. Even more strange were the fauna and he commented that he 'had the good fortune to see several of the famous Platypus, or *Ornithorhyncus paradoxus* ... diving and playing about the surface of the water'. In contrast to the rest of the world, he recorded, 'An unbeliever in every thing beyond his own reason might exclaim, "Two distinct Creators must have been at work"' (324–25).

DARWIN'S SUBSIDENCE THEORY: 1839–1842

After a little over two weeks in Sydney the *Beagle* sailed south for Hobart, and, although the voyage had stretched out much longer than planned, Fitzroy decided to head north to the Keeling (today, Cocos) group of coral islands in the Indian Ocean, as suggested in his *Instructions*, arriving there on 1 April. Over the twelve day sojourn, Darwin, in the company of Fitzroy, was able to make his first extensive examination of a coral atoll, recording his findings in detail in chapter 22 of the 1839 edition of his *Journal*. After surveying all previous theories he presented his own conclusions under the entry for 12 April 1836, which stands as his original extended explanation of the structure and formation of coral reefs:

> The theory which I offer, is, simply, that as land with the attached reefs subsides very gradually from subterranean causes, the coral-building polypi soon raise again their solid masses to the level of the water: but not so with the land; each inch lost is irreclaimably gone; as the whole gradually sinks, the water gains foot by foot on the shore, till the last and highest peak is finally submerged (Darwin 1839b:345).

Following that assertion, Darwin then extrapolated it to apply to what he termed fringing and barrier reefs. They were created by the same process of subsidence, he argued, with the enclosed waters becoming turbid and 'injurious to all zoophytes', and because – evidently following Chamisso – they only 'flourish on the outer edge amidst the breakers of the open sea'. In the same context, although he never saw it, he made his first specific mention of 'The great Barrier which fronts the NE coast of Australia ... [as] described by Flinders ... [which is] probably the grandest and most extraordinary reef now existing in any part of the world' (345).

In the *Journal*, the only source available to the public at this time, Darwin rearranged his observations and discoveries into topics of common interest, often drawing together quite separate excursions into the one chapter, although the *Journal* was apparently organised as a diary with a sequence of dates. That practice has a major significance in trying to understand the development of his theory of the formation of atolls and coral reefs: it is virtually impossible to determine exactly how he reached it. Our only clue comes from his retrospective *Autobiography* in which he made the oft-quoted statement that opens by neatly undermining Bacon's dogmatic prescription for inductive science that direct observation must come before theory: 'No other work of mine was begun in so deductive a spirit as this; for the whole theory was thought out on the west coast of South America before I had seen a true coral reef. I had only therefore to verify and extend my views by a careful examination of living reefs'. He continued in the same passage to state that his observations in South America 'necessarily led me to reflect much on the effects of subsidence, and it was easy to replace in imagination the continued deposition of sediment by the upward growth of coral. *To do this was to form my theory of the formation of barrier-reefs and atolls*' (Darwin 1876:98–99; emphasis added).

But it is not clear how his divergent theory was 'thought out', especially since he wrote to Fox from Lima (CCD I:460) in July 1835 that he was 'a zealous disciple of Mr Lyell'. In the 'extensive notes on geological observations that survive from the time CD spent on the west coast of South America' the editors of his *Correspondence* comment that 'No statement of the theory that could be described as "thought out" has been found' (I:567, *et seqq*.). It seems more than coincidence that Darwin's claim to a 'deductive' theory before he had ever seen a reef is closely parallel to that of Hutton back in 1795 whose theory of the earth came from a similar speculative leap: is it possible that Darwin borrowed the concept of 'speculative leap' directly? Just as Hutton went on to compile the evidence to support his theory, so it seems that Darwin did likewise: he collected a vast amount of data from all available sources to fit in with his revolutionary hypothesis.

After the *Beagle* docked at Falmouth, Darwin went to Shrewsbury, and then to meet Henslow in Cambridge where he renewed his contacts of five years earlier. A few weeks later, in London, he met Richard Owen and also Charles Lyell, with whom he formed a lifelong friendship. In March 1837 Darwin moved to London where he began writing the *Journal* based upon his diary and associated notebooks, along with recollections of events. It was then that Lyell urged Darwin to present his new theory of coral reefs to meetings of the Geological Society. In response, Darwin sent a short paper on 'Proofs of recent elevation on the coast of Chile' which was read on 4 January 1837. A second, dealing specifically with his theory of reef formation, was sent a few months later on 31 May. Drawn from his *Journal* entry for 12 April 1836 and written on Keeling, it was entitled 'On certain areas of elevation and subsidence in the Pacific and Indian Oceans, as deduced from the study of coral formations', and was published the following year in the *Proceedings*.

In that second paper he reiterated his *Journal* comment which presented his famous classification, still used today, of three kinds of reefs: lagoon (or atoll), fringing, and barrier, making the point that the 'Barrier reef, running for nearly 1000 miles parallel to the North-east coast of Australia, and including a wide and deep arm of the sea . . . is the grandest and most extraordinary coral formation in the world' (Darwin 1838:552). Departing from Lyell's volcanic crater theory, which asserted that volcanoes came from underwater quakes and that, when extinct, polyps subsequently built upon them, in his short paper (which he did not read personally) – basically a three-page summary that was later expanded into a chapter of his *Journal* – Darwin's position held that volcanoes do not erupt from under the water. Surprisingly, he seemed unaware of the eruption and emergence of a volcano from the Mediterranean between Sicily and North Africa on 13 July 1831 which created a sensation in the press, four months before the *Beagle* sailed from Plymouth on 23 November. For five months it remained as a small island, claimed by Sicily, England and France (the latter two planting flags of possession on it) until it subsided permanently on 28 December.

Certainly that was a rare event: Darwin did not mention it. Rather, from his extensive analysis of all previous findings, including his own, he argued that, on the contrary, elevated regions of volcanic activity, having become extinct, had later subsided, taking former volcanoes down with them. As a compensatory mechanism, subsidence of much of the Pacific had pushed the earth's crust up to form the Andes, raising oceanic sediments with them, thereby explaining the numerous marine findings high in the Peruvian and Chilean mountains. All along, like Lyell, Darwin had been anticipating the concept of isostasy, writing in his diary in April 1836 he was 'inclined to believe that the level of the ground was constantly oscillating up and down, [and that] . . . the amount of subsidence had been equal to that of elevation' (Darwin 1839b:354). Within that frame of reference, he proposed in his paper of 31 May 1837, which had been the substance of chapter 22 of his published *Journal*, that having become established on extinct volcanoes, 'as the land with the attached reefs subsides very gradually from the action of subterranean causes, the coral building polypi soon again raise their solid masses to the level of the water' (Darwin 1838:552–53). That was the position Darwin held to the end of his life, even though he continued to become embroiled in scientific controversy. Yet again, curiously, despite his intense interest in invertebrates gained from Grant's teaching at Edinburgh, Darwin never wrote on the role of the minuscule polyps in creating such immense geological structures. All of those issues, however, were to be taken up by others in the next decade, and the Great Barrier Reef was to become a major locale for their examination.

In 1842 came the publication of Darwin's long-awaited *The Structure and Distribution of Coral Reefs* which developed in greater detail the theories set out in his paper of 1838 and the *Journal* of 1839, which immediately initiated a century of controversy from its basic uniformitarian assumption that the crust of earth is in process of continuous, imperceptible oscillation. The monograph itself is an impressive achievement. For the first time in the history of world science it set out to bring together all available data from explorers on coral reefs and in six chapters elaborated the subsidence theory comprehensively. No new ideas were advanced but, in a most deliberate manner Darwin examined, and refuted all objections raised so far. What is significant, having a direct bearing on future research on the Great Barrier Reef, is his exploration of the causes of the formation of barrier reefs, with specific mention of the Reef itself throughout, gained chiefly from the observations he had requested John Lort Stokes to make on the *Beagle* survey in 1837. In particular he dealt with the very difficult question of the 'barrier-reefs of Australia and New Caledonia [which] deserve a separate notice from their great dimensions . . . The Australian barrier extends, with few interruptions for nearly a thousand miles' (Darwin 1842:46).

One very puzzling feature to explain was the contentious issue of vertical drop-offs on the seaward side that could, conjecturally, 'plummet' to 'a depth of 1200 feet [365 m]'. As Darwin noted, from this 'we must conclude that the vertical thickness of these barrier coral reefs is very great' which, in turn, raises 'a great *apparent* difficulty – how were the basal parts of these barrier-reefs formed?' (47,48; Darwin's emphasis). He admitted, in the case of barrier reefs, and the Great Barrier Reef in particular, that it is 'almost too preposterous to mention . . . that they rest on enormous submarine craters' given that the 'great reef which fronts the coast of Australia has been supposed, but without any special facts, to rest on the edge of a submarine precipice, extending parallel to the shore' (50,90).

In chapter 5 he worked meticulously through his subsidence explanation yet again, pursuing the question 'On what foundations, then, have these reefs and islets of coral been constructed?' Granted that the 'many widely-scattered atolls must . . . rest on rocky bases . . . we are compelled to believe that the bases of many atolls . . . were brought into the requisite position or level . . . through movements in the earth's crust' and since this 'could not have been effected by elevation . . . they must, of necessity have subsided into it, and this at once solves every difficulty'! (92–94). So, 'if the shore of a continent fringed by a reef had subsided, a great barrier-reef, like that on the N.E. coast of Australia, would have necessarily resulted . . . [and] continued subsidence of a great barrier-reef of this kind' would most likely have continued to develop 'into a chain of separate atolls' (102). Darwin then moved to deal with 'a formidable objection to my theory', namely, the 'vast amount of subsidence necessary to have submerged every mountain' (114). His answer was simple: given the immense seabed elevation found in the Andes 'no reason can be assigned why subsidences should not have occurred in some parts of the earth's crust on as great a scale both in extent and amount as those of elevation' (114).

His final sixth chapter, accompanied by a coloured map of the Indian and Pacific Oceans, discussed in detail the geomorphology of that enormous expanse of water in which he plotted all three kinds of reefs as then known, colour-coding them in red (fringing), bright blue (lagoon and atoll) and pale blue (barrier). From the data available he was able to demonstrate that fringing reefs (those attached to the land) are often in active volcanic areas that are either geologically stationary or else in process of elevation, while lagoonal and barrier reefs, free of volcanic eruptions are in deeper waters and areas of subsidence. To support the theory of elevation in volcanic areas he pointed out that 'on fringed coasts . . . the presence of upraised marine bodies of a recent epoch plainly show, that these coasts . . . have generally been elevated' (147). His concluding paragraph to the main body of text contains a final summary 'derived from a study of coral-formations': 'We there see vast areas rising, with volcanic matter every now and then bursting forth through the vents or fissures with which they are traversed. We see other wide spaces slowly sinking without any volcanic outbursts' which present 'a magnificent and harmonious picture of the movements, which the crust of the earth has within a late period undergone' (148). Throughout his life Darwin never lost his quest for a grand, wide-ranging vision of the unity of all nature.

The controversy continued throughout the nineteenth century as sides were taken for and against. In support of Darwin was Beete Jukes who, in his 1847 geological study of the Great Barrier Reef made during the survey of the *Fly* in the years 1843–45, had arrived at the conclusion that it was not so much 'a true barrier' as 'a long submarine buttress, or curtain, along the north-eastern coast of Australia, rising in general from a very great depth' (Jukes 1847:I.321, 332). Reef building corals he stated, given the limited knowledge of the time, 'seem to belong only to the existing order of things, and not to have lived during any of the secondary or tertiary periods. The present is, so to speak, the *coral reef age of the globe*' (Jukes 1847:I.343: Jukes' emphasis). Jukes went further: he endorsed Darwin's book with high praise, writing that, 'after seeing much of the Great Barrier reefs, and reflecting much upon them, and trying if it were possible by any means to evade the conclusions to which Mr Darwin has come, I cannot help adding that his hypothesis is perfectly satisfactory to my mind, and rises beyond a mere hypothesis into a true theory of coral reefs' (347).

Once Darwin's theories were published, Humboldt gave strong support. Following his earlier observations on coral reef theories in the *Personal Narrative of Travels*, Humboldt gave a more detailed review of recent research in the 1849 edition of his celebrated essays on natural history, *Views of Nature* (English translation 1850), where he surveyed the history of reef theories from Forster, Chamisso, Péron, Quoy and Gaimard, Flinders, Lütke, Beechey, Darwin, D'Urville to Moresby and Powell. Concluding that the explanations of Quoy and Gaimard which guided Darwin were heading in the right direction – that coral reefs did not grow from great depths, as Cook and Forster believed, but from submerged platforms closer to the surface – he remained sceptical of the theory 'that Atolls . . . owe their origins to submarine volcanic craters' since some have diameters up to 60 miles (~100 km). Darwin's theory of gradual subsidence, however, was a persuasive solution: as an island mountain gradually subsides, the coral polyps of its fringing reef continue to grow upwards, 'forming first a reef encircling the island at a distance', and then when the enclosed island subsides below the surface, an atoll. Humboldt paid tribute to that achievement, commenting that 'Charles Darwin has with great ingenuity developed the genetic connection between shore-reefs, island-encircling reefs, and lagoon islands [atolls]': all mark the prominent points of submerged lands, indicating the former topography of the area (Humboldt 1850:261). No greater imprimatur could have come at the time.

THE PROBLEM SOLVED?

Darwin's belief that he had solved the 'coral reef problem' as it came to be called, continued to be challenged, however, chiefly by Carl Semper and John Murray, and, even more trenchantly, by Alexander Agassiz who came into the debate in the final few years of Darwin's life and continued to badger Darwin with evidence seeking to refute his theory. By that time, ailing seriously from his unidentified illness, Darwin, prematurely aged and weary of the immense volume of controversy his writings in geology and natural history had generated, avoided being drawn into further argument. The end came on 19 April 1882. Despite much establishment opposition to honouring the author of the then horrific theory of evolution and the refusal of Queen Victoria and Prime Minister Gladstone to attend, the agnostic Darwin was buried in Westminster Abbey – Britain's most sacred sepulchre – carrying controversy to his grave.

CHAPTER 12

DARWIN'S LEGACY: CORAL REEF CONTROVERSY 1863–1923

DARWIN'S OPPONENTS: SEMPER AND MURRAY, 1863–1923

Debate over Darwin's claimed solution to the problem of coral reef formation continued to gather momentum for more than a century. First ignited by Carl Semper, it was fanned by John Murray, and then erupted into what became a rather captious confrontation in Darwin's declining years, and even after his death, with the aggressive Alexander Agassiz, the most vocal among a number of dissentients.

In the years 1857–65 the naturalist and explorer Carl Gottfried Semper (1832–93), having graduated in natural science from the University of Würzburg, travelled throughout the Spanish Philippines, spending a year in 1862 on Pelelui and the other five major atolls in the Pelew, or Palau, group (now Belau) a little to the east of the island of Mindanao. Semper had read Darwin's 1842 volume and in 1863 sent to a German zoological journal a short twelve-page *Reisebericht* (Travel Report), in which he took issue with Darwin's central hypothesis of subsidence as the primary determinant of reef formation, asserting that the irregular configuration of the Pelew island chain with areas of both elevation and possible subsidence created serious problems for Darwin's theory.

On his return to Würzburg, Semper joined the university staff and in 1869 was appointed director of its zoological institute. In 1877 he was invited to Boston to present his researches to the Lowell Institute, a philanthropic foundation that sponsored lectures by distinguished persons. Those lectures were published in Leipzig in 1880 under the title *Die natürlichen Existenzbedingungen der Thiere*, and translated into English the same year as *The Natural Conditions of Existence as they affect Animal Life*. Essentially a textbook in animal ecology, chapters 7 and 8 are devoted to the formation of coral reefs, as the natural habitat of polyps and the entire range of associated marine life. Here Semper resumed his attack on the geological coral reef theory of Darwin, asserting that his own researches on 'the reefs of the Pelew Islands . . . present insurmountable difficulties to Darwin's hypothesis' (Semper 1880:233). Repeating at length his earlier Travel Report, he noted that the chain of islands exhibited gross irregularities: in the south, Pelelui atoll was elevated while the atolls in the north were submerged at high tide. In his view the formation revealed varying degrees of volcanic upheaval on which corals subsequently grew, with evidence of more geologically recent elevation on eastern sides. In the north where there were true atolls, he argued – obviously following earlier theories of atolls beginning to form on solid submarine foundations – that reefs had grown upwards in suitable waters around 45–60 metres deep.

Semper was adamant there had been no subsidence of those formerly volcanic structures. If Darwin had supposed subsidence to be uniform, he asked, how did it happen that 'in the north only isolated atolls, in the middle barrier reefs, and in the south only fringing reefs have been formed; and why, farther south still at Ngaur, all reef structure should have almost ceased'? . . . 'Hence we must in the first place conclude that subsidence alone has not here sufficed to produce the forms of the northern reef of this archipelago'. However, Semper acknowledged the possibility that the variations could have been due to tilting of the underlying strata platform, commenting that 'by a sort of tilt the northern section has subsided most, the middle but little and the south not at all, [and so] all the observed phenomena might be accounted for, or at least apparently explained' (Semper 1880:252–57).

Even before the publication of Semper's major study of animal life, Darwin had moved to answer criticisms with a second revised edition of *Coral Reefs* in 1874. In an appendix under the heading 'Pelew Islands', he responded to the objections raised in Semper's Travel Report of 1863. Darwin agreed that in the case of Palau subsidence may not have been the dominant factor. Later he acknowledged that view explicitly in a letter of 2 October 1879 in reply to Semper's letter of September 1879. Darwin then continued:

> I always foresaw that a bank at the proper depth beneath the surface would give rise to a reef that could not be distinguished from an atoll, formed during subsidence. I must still adhere to my opinion that the atolls and barrier reefs in the middle of the Pacific and Indian Oceans indicate subsidence.

Exceptions in particular cases, he maintained, did not invalidate his general theory for the greater mass of formations in Indo-Pacific waters: 'It would be a strange fact if there has not been subsidence of the beds of the great oceans, and if this has not affected the forms of the coral reefs'. Gently chiding Semper for attempting to generalise about the formation of coral reefs from just one group of atolls, Darwin pointed out, correctly, in the Preface to his second edition, that in many respects Semper had actually accepted many of his own arguments and had, in effect, simply confirmed the earlier theory of Chamisso that argued for submerged foundations for coral reef formation (Darwin 1874:viii).

On 5 April 1880 John Murray, back in Edinburgh editing the huge volume of data gained in the oceanographic voyage of the *Challenger*, added his voice to the chorus against Darwin in a paper entitled 'On the structure and origin of coral reefs and islands'. The opening sentence itself was confrontational, stating that Darwin's contentious study had 'been universally accepted by scientific men' (Murray 1880:505). Murray then attacked that alleged consensus by stating that in the Pelew group Professor Semper 'experienced great difficulties in applying Darwin's theories'. The aim of his paper, he continued, was to demonstrate, from the evidence gathered in the survey of the *Challenger*, that 'there were other agencies at work on the tropical oceanic regions by which submarine elevations can be built up from very great depths so as to form a foundation for coral reefs' and that 'the chief features of coral reefs and islands can be accounted for without calling in the aid of great and general subsidences' (506). Throughout the short paper, which depended heavily on the ideas of Semper, Murray presented data taken from the voyage of the *Challenger* giving details of the extensive number of species collected both in the tow nets and from soundings down through the water column, along with a discussion of the solvent action of carbonic acid in creating sufficient calcium to allow coral formation.

Murray proposed two alternative agencies to account for reef formation. Firstly, emergent volcanoes which, once extinct, had been degraded below the surface by wave action, thereby creating a large body of surrounding rock debris which formed a suitable platform. Secondly, deposition on that platform of the skeletal and shell remains of the billions of microscopic plankton – foraminifera, radiolaria, diatoms and so on – which accumulated faster than the carbonic acid (H_2CO_3) present in sea water could dissolve them. Once 'the accumulation of the dead silicious and calcareous shells' had formed a foundation, numerous species of marine organisms became established, and 'Eventually coral-forming species attach themselves to such banks, and then commences the formation of Coral Atolls', to the extent that 'it is in a high degree probable that the majority of atolls are seated on banks formed in this manner'. His argument was quite the opposite of Darwin's: 'it is a much more natural view', Murray argued, 'to regard these atolls and submerged banks as originally volcanoes reaching to various heights beneath the sea, which have subsequently been built up towards the surface by accumulations of organic sediment and the growth of coral on their summits' (513).

Barrier reefs received particular mention, including the Great Barrier Reef, their formation being attributed to coral growth having 'commenced close to the shore and [then] extended seawards, first on a foundation composed of the volcanic detritus of the island, and afterwards on a talus composed of coral debris, and the shells and skeletons of surface organisms'. The great width of the cays in the Australian Reef he attributed to the numerous openings, created by the solvent action of sea water, which admitted larger quantities of ocean water and consequently more nutrition for the polyps. In his dredging near Raine Island, to a depth of 155 fathoms (283 metres), Murray reported that the sea floor was composed of 'a coral sand which was, I estimate, more than two-thirds made up of shells of surface animals' (515). His conclusion was quite unequivocal: although Darwin had proposed subsidence to account for the formation of reefs and atolls, 'it has been shown that these were produced by other causes, – by the vigorous growth of the corals where the most nourishment was to be had, and their death, solution and disintegration by the action of sea-water and currents' which also carried out large quantities of coral debris, thereby extending the reef edge outwards (516).

Darwin was quite unimpressed and in a letter to Alexander Agassiz, written in 1881, less than a year before his death, he roundly criticised Murray's views, stating that he could 'not understand Mr. Murray, who admits that small calcareous organisms are dissolved by the carbonic acid in the water at great depths, and that coral reefs, etc., etc., are likewise dissolved near the surface, but that this does not occur at intermediate depths' (Agassiz 1913:283). Agassiz replied on 19 May 1881 in agreement: 'This part of Murray's argument seems to me untenable' (285). There was, at the time, no serious response from scientists and Murray's alternative theory seemed destined for oblivion. Yet, Murray was a great oceanographer and his theory of reef formation on submerged and eroded volcanic bases was strangely anticipatory of the discovery in the 1950s by the American oceanographer Harry Hess (1906–69) of submerged flattened seamount platforms which he named guyots (in honour of the nineteenth century geologist Arnold Guyot). It is now generally believed that guyots could well be the bases for some reef formations and that the accumulating mass of foraminifera deposits on continental shelves actually contribute significantly to subsidence.

ALEXANDER AGASSIZ: PURSUING A SOLUTION, 1881–1910

The most sustained and hostile criticism of Darwin's reef formation theory, however, came from Alexander Agassiz who was totally opposed to any form of subsidence explanation and not only disputed with Darwin in his final ailing years, but even after Darwin's death in 1882 continued to attack subsequent investigators who sought to verify the Darwinian hypothesis of coral reef formation.

As a child, during the 1840s Alexander Agassiz (1835–1910) had witnessed a period of intense hostility between his father and Darwin. The conflict began in 1839 when Darwin proposed a solution to the mystery of the four 'parallel roads' on the sides of the valley of Glen Roy in Scotland. Around the valley four distinct shorelines from previous geological episodes were plainly evident. Drawing from his experiences in South America where he had observed similar formations, Darwin inferred that the parallel roads had been formed at a time when the valley was under the sea, and in successive elevations, as a compensatory mechanism for subsidence elsewhere, the roads were cut by wave action during the stable periods. Darwin presented his theory to the Royal Society of London on 7 February

1839 and his paper was published in their *Philosophical Transactions* the following year (1839c:I.39–81).

At the same time, Swiss geologist Jean Louis Agassiz (1807–73) had been investigating the strange and inexplicable phenomenon of 'erratic boulders'. For many decades geologists had been attempting to account for great rocks which could be found across Europe composed of materials totally different to the surrounding strata. Mariners had long been accustomed to seeing large rocks floating on icebergs in high latitudes, having been carried down fiords by glaciers which, on calving as icebergs in the spring, carried them to sea, in some cases eventually melting and depositing the boulders on the shore. But how could they be explained inland – even one in Shrewsbury in midland England which Darwin saw as a child – or even more puzzling, perched high on the mountains in the Swiss Alps at elevations of 5000 feet (1500 m)? As early as 1787 the Swiss pastor Bernard Kuhn suspected that some kind of glacial action, common in Switzerland, had shifted the rocks. Later the botanist Karl Schimper suggested glacial action in a colder phase of the earth's climate and coined the term *Eiszeit*: 'Ice Age'. At a meeting of the Swiss Society of Natural Science on 24 July 1837 Agassiz put the term *Eiszeit* into currency with his theory that the transport of erratic boulders, and the gouging out of great valleys through the European countryside, had been effected by great glacial flows covering all of Europe.

Although scorned, chiefly by catastrophists, Agassiz presented his theory in September 1838 to the Freiberg meeting of the Association of German Naturalists. He then published it privately in 1840 as *Études sur les Glaciers* ('Studies of Glaciers'), and repeated it at the British Association for the Advancement of Science meeting in September 1840 at Glasgow, and again in November to the Geological Society of London, in which he argued that the great ice sheets covering Europe were also responsible for trapping and freezing the woolly mammoths in Siberia. And, it was suggested, the action of melting glaciers when each ice age ended had gouged the roads of Glen Roy. Darwin was implacably opposed to the glacial explanation and clung tenaciously to his theory of uplift for twenty years.

Agassiz', theory, however, stimulated the lively mind of Charles Maclaren, amateur geologist and editor of the principal Edinburgh newspaper, *The Scotsman*. In his review of the startling glacial theory, Maclaren presented calculations that the extent and thickness of the great ice sheets, reaching from 'the 35th parallel to the north pole', and about one mile (1600 m) thick, would be 'enough to reach to the summits of Jura' and 'the abstraction of the water necessary to form the said coat of ice would depress the ocean about 800 feet [243 m]' (Maclaren 1841:346–65). Here came an indication that coral reef formations could have been influenced by sea level changes. Almost seventy years earlier, on his second voyage in command of *Resolution* and *Adventure*, even Cook asked, as noted, in his journal, when sailing past the elevated relict coral reefs on the Friendly Isles (Tonga), 'Has this Island been raised by an Earth quake *or has the Sea receded from it?*' (Cook 1774:438, emphasis added). Darwin did not mention Cook's observations in *The Structure and Distribution of Coral Reefs*. When Maclaren's observations were published, at the time that Darwin was writing, the implications for Darwin's theory were serious.

Following that article, in a review of Darwin's *The Structure and Distribution of Coral Reefs* the following year, 1842, and concerned that Darwin had not considered sea level changes, Maclaren reviewed it critically in *The Scotsman* later the same year (29 October and 8 November) raising serious objections. He published an abridged review the following year in the *Edinburgh New Philosophical Journal* and entered into correspondence with Darwin who became profoundly irritated with Maclaren's criticisms. Darwin chose to ignore his objections and never considered sea level changes a factor in coral reef formation.

Not until 1861, when other surveys confirmed glacial action, did Darwin admit defeat. Recalling the course of the dispute in his *Autobiography* he wrote that the Glen Roy episode 'was a great failure and I am ashamed of it. Having been deeply impressed with what I had seen of the elevation of the land in S. America, I attributed the parallel lines to the action of the sea; but I had to give up this view when Agassiz propounded his glacier-lake theory' (Darwin 1867, ed. Barlow 1958:84). It was, he wrote, my 'greatest blunder' (*CCD* IX:247).

Alexander Agassiz had been born in Switzerland and arrived in the United States in 1847 when Louis took up an appointment as professor of natural history at Harvard University. In the ensuing years, as the conflict between Darwin and Louis Agassiz continued, so the hostility became transmitted from father to son. Alexander completed two degrees, one in natural history in 1857, the second in engineering in 1862. He went on to superintend a copper mine in Michigan in 1867 and through astute investment, along with enlightened management, acquired considerable wealth in what became the greatest copper mine in the world. His affluence enabled him to become one of the great public benefactors of the period, endowing the famous Museum of Comparative Zoology at Harvard, of which he became curator, and to charter ships for his numerous coral reef surveys.

Alexander Agassiz' lifelong fascination with reefs had been initiated by his father's work in 1852 when the United States Coast Survey sought information on the coral reefs off Florida and requested Louis to undertake an investigation to determine their origin and formation. The elder Agassiz, pointedly in conflict with Darwin – even opposing the theory of evolution by natural selection in his Harvard lectures – reported that there was no evidence of either subsidence or elevation to be found and that all coral reefs there had started on the rocky shores and extended outwards by growing on the rubble created by weathering: as he put it, 'on the subsequent accumulation of their products' (cited Dexter 1972:490).

Following the death of his father Agassiz continued research into coral reefs and in the years 1877–80, as a guest of the government, joined the Caribbean cruises of the United States Coast Survey ship *Blake* in the Gulf Stream, along the Florida reefs to their termination in the Tortugas, and to the Alacran Reefs off the north-eastern coast of Yucatan. The results were published in 1888 as *Three Cruises of the USCGS Steamer Blake*. As a consequence of that period Agassiz – who, at the time had never travelled into Pacific atoll regions – initiated a vigorous correspondence with Darwin disputing the subsidence theory.

Agassiz sent a sequence of letters to Darwin, that of 16 April 1881 being the most challenging – although written in a polite and scholarly style – stating that he had evidence 'to explain . . . the formation of reefs'. In the Tortugas he found 'no signs of elevation' but rather that 'the immense plateau which forms the base upon which the Peninsula of Florida is formed, was built up by the débris of animal remains, – Mollusks, Corals, Echinoderms, etc. (after it had originally reached a certain depth in the ocean), until it reached the proper height for corals to flourish' (Agassiz 1913:282). A very weary Darwin replied, 5 May 1881, as courteously as he could, that he had never disputed that particular cases, such as those adduced by Semper, and others, were possible. 'I have expressly said', Darwin wrote, 'that a bank at the proper depth would give rise to an atoll which could not be distinguished from one formed during subsidence' – therefore including, by implication, the formations studied by Agassiz. But, Darwin continued, 'I can hardly believe in the former presence of as many banks (there having been no subsidence) as there are atolls in the great oceans, within a reasonable depth, on which minute oceanic organisms could have accumulated to the thickness of many hundred feet' (Darwin 1908:283).

Tired of the controversy – possibly in the terminal stages of his illness – Darwin acknowledged the ideas of Agassiz and replied that 'If I am wrong, the sooner I am knocked on the head the better'. To reach a final solution to more than half a century of controversy he added the famous words – with Agassiz in mind – 'I wish that some doubly rich millionaire would take it into his head to have borings made in some of the Pacific and Indian atolls, and bring home cores for slicing to a depth of 500 or 600 feet' (Darwin 1887:283). If Darwin's theory were correct, then cores would reveal successive layers of coral that had continued upward growth as the landforms subsided, resting upon a foundation of solid rock. The correspondence of 1881 failed to develop any further: less than a year later Darwin was dead.

The dispute by then was more than scientific: it had assumed a serious political dimension in that British science, a mainstay of its maritime and economic hegemony of the world, was being brought into question and close to international ridicule with such denigration of their national icon. At that time Darwin and Murray had produced the pre-eminent reef theories and their open conflict was causing international attention. The entrenched bitterness of the debate, however, was evident when in 1890 Agassiz directed a derisive attack on Darwin when he commented in a letter of 27 January to Murray stating that Darwin's only experience of an atoll was at Keeling on the voyage of the *Beagle*. Although the ship anchored there for eleven days, 1–11 April 1836, and Darwin was ashore all the time with Fitzroy making investigations, Agassiz dismissed him with the contemptuous (and inaccurate) comment: '5 days on a reef at utmost and then nothing but examination of charts!! to build up a theory, absurd'.

FUNAFUTI SUBSIDENCE THEORY TESTS, 1896–1898

In the decades from 1890 to 1920 the entire Pacific became an arena of increasing trade, colonisation and international rivalry as a number of major nations – mainly the United States, Germany, Japan, France and Britain – staked out territories for annexation and economic exploitation. Those powers sent scientific expeditions to study the biology and geology of the entire ocean and Darwin's hypothesis in particular stimulated a period of intense investigation from the time of the *Challenger* survey (1872–76) to the mid-twentieth century. It had become abundantly clear to the British scientific establishment by 1892 that a face-saving operation had to be mounted and Darwin's hypothesis confirmed or denied once and for all. The only way was to follow Darwin's plea for a boring operation to find the foundation on which atolls and other reefs were built.

First moves had already been initiated informally in 1891 within the British Association for the Advancement of Science (the BA) between assistant secretary the Reverend George Bonney, professor of geology at University College in London, and William Sollas, an Englishman, professor of geology and mineralogy in Dublin. By 1894 Bonney had approached the Royal Society of London and the British Admiralty for support: the former offered a financial grant of £800, and the navy an obsolete vessel, HMS *Porpoise*. In 1895 a Coral Reef Committee of the BA, which included John Murray, had been convened and approached the Admiralty for ideas on a possible site. Admiral Wharton suggested the atoll of Funafuti (8°31'S × 179°13'E) in the British protectorate of the Ellice Islands (today Tuvalu) with Sydney as the Australian base. Sollas, who was appointed leader of the expedition at a meeting of the BA in September 1896, then sought advice from the president of the Royal Society of New South Wales, Anderson Stuart, professor of physiology at the University of Sydney, who used his considerable political influence to secure essential support from the New South Wales government which included the use of a diamond drill rig and operating crew.

Supervision of the operation at the Sydney end was undertaken by the professor of geology at the Sydney University, Tannatt William Edgeworth David (1858–1934). The Royal Society invited Stuart and David to join the expedition, but since it was on a very tight budget, at their own expense. Both declined and the invitation was finally passed to Charles Hedley, the next most senior member of the staff of the Australian Museum who, with independent means, was always ready to fund his own expenses in coral reef research and accepted immediately. Hedley was instructed by Museum director Robert Etheridge to assist in the drilling work and 'to devote your time when not occupied in the main work of the Expedition to collecting, preserving and making notes of all suitable specimens and material that may come your way' (Australian Museum Letterbook, 29 April 1896). Hedley did exactly that and produced a Museum Memoir in two parts on 'The atoll of Funafuti: zoology, botany, ethnology and general structure' (1896a) and a 'General account of the atoll of Funafuti' (1896b).

On 21 May 1896 a site was selected and boring commenced on 2 June. After sixteen days, having reached a depth of only 105 feet (32 m) through porous rock and sand, which flowed in 'faster than it could be pumped out' – along with 'the frequent failure of the machinery' – it was decided to abandon the site and seek another location (Sollas 1904:2). The drill was moved and a second attempt commenced on 3 July. The results were no better: 'very little "core" was obtained', water continued to flood in and 'the coral rock through which the bit advanced was highly cavernous'. Unable to go beyond 72 feet (22 m) after eleven days, Sollas recorded, that 'Baffled in our endeavours, and with no other part of the island offering more hopeful prospects of success, we had no alternative but to abandon the undertaking, and on July 30 were taken from the island in the *Penguin*, and returned to Fiji' (5,6). Sollas reported that none of the material brought up contained 'coral débris'; most of it consisted of 'calcareous algae' and 'large foraminifera': these, 'and the abundant growth of corals and calcareous algae, such as *Halimeda*, lead to the belief that the lagoon is slowly filling up' (5,6).

On his return to Sydney, as the only Australian member of the expedition, Hedley submitted his results to Etheridge, and it was this seventy-three page report to the Museum – 'General account of the atoll of Funafuti' – that was to embroil Hedley – quite innocently – in a serious rift with the Royal Society over the protocol of prior publication. Etheridge, apparently unwittingly, authorised its publication as a section of a Museum Memoir on the results of the Funafuti expedition. However, when a copy of Hedley's contribution reached London, the Royal Society was outraged – officially, as secretary Foster complained, because the Society had funded the operation and had rights to first publication. A bitter correspondence ensued which was only slightly moderated when it was learned that Sollas himself had been negligent in not advising the Australian Museum of the protocol (Rodgers & Cantrell 1988:275). Relationships remained strained for some time and it took all of Edgeworth David's diplomatic skill to effect an harmonious working relationship for the two ensuing expeditions.

In large part, however, the conflict erupted because Hedley ventured an opinion on the formation and structure of coral reefs from his observations at Funafuti in which he compared the rival theories of Darwin and Murray. 'Darwin's theory of coral reefs', he wrote,

> as opposed to Murray's is favoured by these facts: Firstly, soundings show the atoll to be planted not on a bank but on a cone; secondly, they also show it girdled by a precipitous submarine cliff, explicable only on the subsidence theory; thirdly, our

observations and experience of residents agree that the lagoon is filling up, whereas Murray demands its excavation (Hedley 1896b:18).

That was exactly the conclusion Sollas was to publish in the official report of 1904 which Hedley had unknowingly pre-empted. One further outcome of the rift, recorded long after Hedley's death by the Australian Museum's conchologist and historian Tom Iredale, was that, as a result of an understanding reached with the Royal Society, the Museum refused permission for Hedley to publish his more extensive, detailed views on coral reef formation (Iredale 1969:28). Faced with that disaster, English endeavours ceased.

David believed that Australia had an obligation to make a second attempt to solve the Darwin question. Between them, he and Stuart raised sufficient funds for this from the government, several benefactors and learned societies, including the Royal Society of London who responded generously. With strong government approval and support the second expedition sailed from Sydney on 3 June 1897 and arrived in Funafuti, via Fiji, on 16 June. Drilling commenced on 2 July, and by 5 September boring had reached a depth of 557 feet (170 m) when David had to return to his university duties in Sydney. George Sweet, a geologist from Melbourne, remained as leader and on 21 September drilling recommenced, finally reaching 698 feet (213 m). In his official *Narrative* David was able to report a more successful outcome: 'The depth for which Darwin had asked that a coral atoll should be bored with a view to testing his theory (500–600 feet) had been exceeded by reaching 698 feet, and a core had been obtained as good as could be expected considering the friable and fragmental character of much of the rock' (David 1904:54).

David pondered the results and concluded that although 'it was evident ... that the base of the coral-reef rock had not yet been attained' the samples retrieved indicated that, in addition to the 'calcareous rock chiefly composed of foraminifera, *Lithothamnion*, and *Halimeda*, as well as reef-forming corals' and since most of the machinery had been left at the bore site, it was considered desirable that the main bore could profitably be deepened (54). Consequently, with approval of all bodies concerned, David arranged for his student Albert Finckh to lead yet a third expedition to Funafuti. Arriving on 20 June 1898, after surmounting the numerous logistical difficulties of transporting considerable quantities of equipment, supplies and workers to the site, along with the, at the time, formidable tasks of lighterage – unloading 25 tons of gear and 40 tons of coal for the steam-driven drilling rig, and establishing camp on a tropical shore – drilling recommenced at the previous bore site. By 7 September a depth of 973 feet (296 m) had been reached and by 11 October extended down to 1114 feet (340 m). Still boring successfully, Finckh was obliged to stop only because the supply of diamonds for the drilling head had been exhausted.

So ended the third and final British attempt to solve the riddle of the reefs that Darwin claimed to have solved. In the concluding paragraph to his *Narrative* David at least was able to record that 'We have at all events succeeded in carrying out the wish of Darwin for a core from a depth of 500 to 600 feet in a coral atoll in the Pacific Ocean'. In addition to much valuable information on the geology and biology of the atoll, David expressed the hope that the work of the three expeditions would 'enable zoologists and geologists to lay more surely the foundations of our knowledge of the origin and growth of coral atolls, and so make our work of some use to science' (60).

VOYAGES OF AGASSIZ: GREAT BARRIER REEF TO FIJI, 1896

Meantime, Agassiz had been engaged in his independent efforts to solve the coral reef problem. An indefatigable reef explorer, his high level government connections had enabled him to join yet more official cruises. In 1891 he travelled aboard the *Albatross* on its Western Pacific survey from the Gulf of California in Mexico south to the Galapagos just below the equator. Then in 1892 he studied the reefs of the Bahamas and in 1894 chartered a ship to examine the reefs of Bermuda and Florida yet again. By that time, despite his extensive experience in the Caribbean and northern Pacific waters, he remained ignorant of the vast stretches of the Pacific and its extensive reef formations across which Darwin had travelled and formulated his reef theory. Agassiz had also read Saville-Kent's sumptuous book of 1893 on the Great Barrier Reef and noted that although both the two major theories of Darwin and Murray had been given balanced treatment in it, final approval was accorded to Darwin's subsidence theory. The time had come to remedy that deficiency. Rejected in his approach to join the Funafuti expeditions that the Royal Society wished to keep an all-British project, Agassiz was not deterred. As a doubly rich millionaire he decided to mount one of his own by surveying the Great Barrier Reef personally, and by drilling on an atoll in search of evidence to refute Darwin.

He planned his voyage to the South Pacific meticulously for several months, and for advice on the best season to travel he cabled Saville-Kent who advised a mid-year voyage, although Wharton of the British Admiralty warned him it was the time of the strong southeast trade winds when weather could be unpredictably stormy. Preferring Saville-Kent's advice, in 1896 Agassiz chartered a small cargo steamer, the *Croydon*, and made extensive and thorough preparations for its provisioning and replenishment supplies to be ready along the planned route from Brisbane to Cairns. 'We carried', he recorded in his account of the expedition, *A Visit to the Great Barrier Reef of Australia*, 'a complete photographic apparatus and an extensive outfit for pelagic fishing in the way of surface as well as of deep-sea Tanner nets, with the usual apparatus for sounding in moderate depths. All this, beside the necessary appliances for preserving the collections, was forwarded to Sydney early in the winter' (Agassiz 1898:103–104). Travelling to Brisbane by commercial passenger ship from San Francisco, he took with him two members of the Harvard Museum staff, William Woodworth and Alfred G. Mayor, the latter to achieve future distinction as director of the department of marine biology in the Carnegie Institution of Washington.

The record of the expedition is contained in two major documents: Agassiz's own 145 page account of 1898 which he published through the Museum, and his correspondence collected and edited by his son George Agassiz in *Letters and Correspondence of Alexander Agassiz* of 1913 (in which the *Croydon* is given as *Croyden*).

The season could not have been worse: from the time of the ship's departure on 16 April 1896 there was an uninterrupted sequence of storms and heavy seas. Heading north, bad weather prevented them from landing on Elliot Island and the rest of the Bunker group to the near north. Steaming on, the ship reached Townsville on 22 April where Agassiz was able to examine the reefs of Palm Island before reaching Cairns on the 26th in continuing foul weather where the ship remained 'at anchor doing absolutely nothing' (Agassiz 1913:315). Leaving Cairns on 5 May for Cooktown, from there the ship headed north-east to the edge of the Reef at Lark Passage (15°7'S) where, in a few days of fine weather, a moment of exultation occurred: 'The coral reefs here are superb, and I had no conception from the West Indian reefs of what a reef can be'. In the same letter to John Murray he also stated that 'There I began to have my eyes opened, and to get an explanation of the formation of the coral reef flats' (317).

By that time, due to 'the boisterous weather' for most of the voyage, he decided to terminate the expedition. 'I have never been connected with a greater fizzle' he wrote to Murray on 16 May: since leaving Brisbane

> we have had just five days of good working weather . . . [why Saville-Kent] advised me to come here during the time of the trades I cannot understand. Wharton warned me about the trades, and I knew little enough of trades to know that when they blow very little can be done. Yet when I cabled Kent he reaffirmed his opinion, and got Wharton to agree with him to boot (Agassiz 1898:317).

Despite the fiasco, Agassiz was determined to put the best possible construction on the expedition: 'I have, however, seen enough of the reef to satisfy myself of its mode of formation, and I fancy the subsidence people will not have much ground for support. It is very much like the Florida reef, only on an immense scale' (318). Denying that any subsidence could have occurred, 'it seems to me that what I have seen of the Great Barrier Reef of Australia, leads to no such conclusion, – and it certainly is not the case in Florida. On the contrary, the present condition of the Great Barrier Reef can be satisfactorily explained by the mere action of erosion and denudation' (127). Basing his argument on the numerous continental islands that rise from its waters, Agassiz asserted that the whole region had been raised in post-Cretaceous times (after 65 million years ago) 'and has since then been exposed to the most extensive erosion and denudation' (127). He adduced a formidable account of evidence revealing an extensive grasp of the literature from Flinders, Jukes, Rattray and Saville-Kent, along with a number of terrestrial geographers, and an array of articles from the journals of the Royal Geographical Society of Australasia.

Agassiz took particular issue with the interpretations of Jukes during the voyage of the *Fly* (1841–45), chiefly because Jukes had sided with Darwin, whose arguments he examined critically. Agassiz then argued that although Jukes had correctly analysed 'the main features of the Great Barrier Reef, and of its relations to the mainland and intervening islands', none the less, he was 'led to what seem to me erroneous conclusions' (136). The error of Jukes, he claimed, like those of Saville-Kent and others who had followed Darwin, was because 'He assumes, as we do, that the Australian coast at one time was just within the line of the present Barrier Reef; but it seems to me that the causes given by Jukes for the formation of the Barrier Reef are equally well explained by erosion and denudation' (139). He amplified that view in another passage where he accounted for the Reef in categoric terms: it was upon

> the upper part of submarine slopes, of a former geological period, but modified by the erosion and denudation up to recent times, that during the present epoch corals have obtained a footing and built up the Great Barrier Reef of Australia. Thus, instead of Jukes' tremendous buttress of coral, there should be but a comparatively thin veneer of coral rock overlying the denuded land (Agassiz 1913:320).

There, for the time being, he rested his case.

Agassiz' pursuit of Darwin had now developed into a compulsive obsession. Having returned to Massachusetts from the disastrous Great Barrier Reef expedition, he immediately set about making yet another attempt to prove Darwin in error. America's leading coral reef scientist, James Dana, before he died in April 1895, suggested to Agassiz that the Fiji group might be a place to investigate, and Agassiz decided to conduct his own boring expedition. Having the choice of Fiji confirmed by Wharton in London, Agassiz spent the early months of 1897 making plans. He chartered the *Yaralla*, a 500-ton, 200-foot waterline steamer and re-hired the crew that had accompanied him on the *Croydon*. As a highly experienced mining engineer, he himself engaged a skilled operator from the Diamond Drill Company along with one of its drilling rigs, powered by a kerosene motor to avoid the need for a steam engine which had been used on Funafuti, since atolls do not always have a reliable fresh water supply. Writing of his plans from Cambridge, Massachusetts, on 16 May 1897 to John Murray, with whom he maintained a continuing correspondence, he stated that he was 'prepared to go to 350 feet', adding that 'I shall put a hole in an elevated reef and in the edge of an atoll if I can find solid ground anywhere to start' (323).

Benefiting from his Great Barrier Reef experiences, he decided to explore Fiji at the end of the year, arriving in Suva on 6 November 1897 where he found the *Yaralla*, the equipment, and the hired workers awaiting him. Leaving two days later, for nearly a month the *Yaralla* cruised around Fiji waters within which are situated the two main continental islands of Viti Levu and Vanua Levu, along with a dozen smaller rocky islands, the whole being surrounded on the northern side by a full semicircle of atolls and barrier reefs some 800 kilometres (500 miles) in length.

The results put Agassiz into a paroxysm of joy: 'Hurrah!' he wrote to Murray on 3 December 1897, 'I have been and gone and done it, as we say in Yankee slang. We have come in from nearly a month's trip round the islands of the Fijis, and a more interesting trip I have never made . . . it looks to *me* as if I had got hold of the problem of deep [lagoons of] atolls, and of the history of the coral reefs of the group'. Agassiz had found large regions of elevated limestone that led him to assert that not 'a single atoll in the Fijis has been formed by subsidence! – Darwin and Dana to the contrary notwithstanding. This is eminently a region of elevation at least to *eight hundred feet!*' (329; Agassiz' emphasis). That was the estimate already arrived at by Maclaren in 1843 but Agassiz revealed no evidence of that important article appearing in the *Edinburgh New Philosophical Journal* and, like Darwin, never considered sea level change as a contributing factor in coral reef formation. At that time, a letter from Edgeworth David had arrived in which David had informed him that the cores from the second Funafuti expedition earlier in July to September that year had found only 40 feet of true reef in the entire depth. That information led him to continue in the same letter to Murray that his report to the Funafuti scientists will 'give them a dose of what they do not expect, and the theory of subsidence will, I think, be dead as a doornail and subside forever hereafter' (330).

The time had come to test the theory and the drill was then set up on an atoll on the north-east of the group at Wailangilala. Agassiz went on to continue his zoological collecting while the drilling team went down through 80 feet of elevated limestone and stopped. That was as far as the drill ever went: Agassiz concluded that nothing further could be established simply from cores. In an astonishing turnaround he decided that 'it would be foolish to go on boring here' and that further inquiry should be made directly on 'the face and slopes of elevated reefs and [where possible to] study their composition *in situ* on a large scale and not from a core' (331).

Agassiz was beyond himself with euphoria:

> Here I have been at work now nearly six weeks with only a couple of days bad weather, and I have been most successful! . . . I have learned more about the coral reefs during the past month than in all my previous expeditions, and think that I now understand the causes which have brought about the existing state of things (in coral reef ways) in the Fijis.

Still sensitive to the rebuff from the Royal Society and the BA in rejecting his attempts to join the Funafuti expeditions, he went on, in a letter of 15 December to his close friend Wolcott Gibbs, professor of applied science at Harvard University, that 'our English friends who are howling for joy at the results of the boring in Funafuti will be greatly surprised then they hear from me' (332–33). When David heard from Agassiz he responded in partial agreement that the boring at Funafuti was inconclusive, which led Agassiz to comment that 'we are still as far as ever from having a general theory of the formation of coral reefs' (343). In a letter of 29 April 1899 he looked back on his discoveries in Fiji as hammering home 'the last nail in the coffin of subsidence'.

Agassiz had seen enough to convince himself of the success of his mission. After a few more weeks zoological collecting he terminated the expedition and left Fiji aboard a commercial passenger liner for the United States on 13 January 1898. From that time forward – the last decade of his life – Agassiz never remained satisfied that he had put the Darwin theory finally to rest. He continued to travel throughout the Pacific, leaving San Francisco in 1902 for an inspection of hundreds of islands all the way to Japan. He went on to see the Maldives in the Indian Ocean and reported to Murray that it was his most successful expedition ever where he 'learned more at the Maldives about atolls than in all my past experience in the Pacific and elsewhere. I should never have forgiven myself had I not seen the Maldives with my own eyes and formed my own opinion of what they mean', that is, they were constructed on elevated platforms and not on subsiding landforms. Pursuing Darwin – now dead for twenty-five years – to the end, he added another burst of invective: what Darwin wrote, he commented in a letter of 18 January 1902 to John Murray, was 'Such a lot of twaddle – it's all wrong what Darwin has said, and the charts ought to have shown him that he was talking nonsense' (394).

In addition, Agassiz continued his other marine observations, including dredging in the abyssal zones of the Pacific between San Francisco and the Marquesas aboard the American survey vessel *Albatross* in 1899–1900, and again on its survey from Peru to Easter Island. After a cruise on the Great Lakes of Africa in 1908 his explorations were coming to a close: he died in his sleep aboard the luxury liner *Adriatic* en route to England in 1910, aged seventy-five, probably the most fitting end he could have wished. In that final decade of his life, however, others were taking up the coral reef quest in greater numbers, and despite his best efforts neither he nor Darwin was yet accepted as the final arbiter on that vexed question.

GEOLOGY AND BIOLOGY: CORAL RESEARCH EXPANDED, 1902–1920

By the time Agassiz sailed in 1902 to study hundreds of reefs in the western Pacific island regions, the controversy between his theory and that of Darwin had ramified throughout the biological and geological sciences with increasing numbers of investigators involved. Specific Australian interest in the geological aspects of the Great Barrier Reef was developing slowly in that period, again from Sydney where Charles Hedley and his lifetime friend Ernest Clayton Andrews (1870–1948) of the New South Wales geological survey, and later government geologist, made a survey in 1901 of the middle section of the Reef around the Whitsundays, centred on 20°S to 21°S. In 1902 Andrews published their findings, one of the first Australian geological studies of the Reef, in the *Proceedings* of the Linnean Society of New South Wales. Andrews had focused on the significant feature of the wide submarine continental shelf that extends from north to south of the entire eastern coastline. He conjectured that originally there had been a fringing reef on the northern sections of the coast, in suitable warm waters, broken in places by outflowing rivers. In Pleistocene times (after 1 million years ago) he wrote, there had been a period of subsidence which allowed water to cover much of the former land leading in turn to water-caused erosion, 'during which a uniform coast and smooth offshore bottom had been formed'. He suggested that

> the sinking of this uniform area allowed the sea to trespass far over the old coast sands into the ranges and corals . . . [which then] proceeded in the clear waters of the shelf margin, now removed far seaward, to invest the whole width of the smooth offshore deposits with their masses, and establish themselves as the Barrier reef.

A moderate uplift followed in the next geological phase, Andrews believed, producing the elevated strata observed by Jukes. Andrews believed that the pre-Pleistocene river estuaries, once drowned during subsidence, probably accounted for the maritime passages in the outer Reef (Andrews 1902:177).

Interest in all aspects of science was mounting in that period and Australians were keen to be involved in the wider arena. Already the BA had begun holding meetings in overseas British regions. In response to Australian approaches, it was decided that its eighty-fourth meeting would be held in Adelaide and Melbourne in August 1914, and a large number of delegates left Britain in late June and early July, along with a number of distinguished foreign scientists. While they were travelling aboard ship Germany declared war on Russia on 1 August and on France two days later, thereby leading Europe into World War I. Consequently, scientific fieldwork by the belligerent powers, especially in the Indian and Pacific Oceans, ceased until 1920.

MAYOR'S ECOLOGICAL SURVEYS: TORRES STRAIT AND SAMOA, 1913–1920

The United States attempted to maintain a policy of 'strict neutrality' and did not enter the conflict until 6 April 1917, which allowed American reef scientists to continue undisturbed. The expeditions of Agassiz had stimulated considerable American interest and as early as 1913 a series of major studies was commenced by the Department of Marine Biology in the Carnegie Institution of Washington, the first of which began, of all places, on the Great Barrier Reef. That study was one of the most notable: it became the first intensive investigation of a single Reef location over a continued period and became the paradigm model for reef research, particularly in Australia, for several decades to come.

Previously, much science of the Reef had been effected either from naval surveys, where scientists were constrained to collect data while continually on the move, or from the few individual studies by Saville-Kent, Haswell, Taylor, Andrews and Hedley. More was known of the biology and geology of Funafuti than any Australian coral formation and, consequently, very little had been achieved to establish a comprehensive understanding of both the biota and geological processes in the Great Barrier Reef itself.

In 1913, however, the Carnegie Institution expedition to the Reef arrived at Thursday Island under the leadership of Alfred Goldsborough Mayor (formerly Mayer, which accounts for variant citations in the literature) who had been to the region as an assistant on Agassiz's disastrous expedition on the *Croydon*. Mayor, now director of the Department of Marine Biology, in company with Frank Potts of Trinity Hall, Cambridge, had planned to make a comprehensive study of the total ecology of a coral reef. Arrangements were made with the scientific help of Etheridge and Hedley of the Australian Museum, and the ready cooperation of Governor of Queensland Sir William Macgregor, Premier Digby Denham and senior officials and missionaries of Thursday Island. The American visit was considered a major international event: Hubert Murray, Lieutenant-Governor and Administrator of Papua, invited the group to be his guests in Port Moresby.

Finding Thursday Island unsuitable due to heavy silting at the time, the project was moved to Murray Island (indigenous Mer) at the extreme northern end of the Reef (8°S × 144°E) where research was conducted for the month of October. Murray Island – Mayor used its indigenous name, spelling it Maer, not the modern Mer – is a volcanic structure 2.6 kilometres in length (1.6 miles) and 1.6 km (1 mile) across at its widest place, lying on its main axis south-west to north-east. In the southern centre of its symmetrically elliptical shape a small extinct volcano rises to 122 metres (400 feet), ringed at a distance by an elliptical arc to the south-west of several ash-filled minor craters. On the north-eastern end it is surrounded for half its circumference by a reef flat around 600 metres (2000 feet) wide on which the studies were conducted. Throughout October that year the small team pursued a thoroughgoing investigation which was published under the title *Ecology of the Murray Island Coral Reef* (1918). In using the concept of 'ecology', possibly for the first time specifically in the literature, Mayor indicated that he sought to investigate and record a wide range of biota in their habitat and analyse what he considered to be the major interactions which led to the formation and continuing sustainability of a coral reef.

The Carnegie team concentrated on several aspects of the ecology of Murray Island: a geological survey to test the rival hypotheses of Darwin and his followers on the one hand, and Murray and Agassiz on the other; and, as their main purpose, an extensive biological survey. In common with most other investigators up to that time, they were primarily zoologists and the focus of their interests was on the corals themselves: range of species, distribution, and the environmental parameters of air and water temperature, water chemistry and circulation. The geological observations were made by Frank Potts, using the theories of Darwin, Dana, Daly, Semper and Murray as the basis for comparison. In addition, they were also concerned to test the hypothesis of Andrews.

The geological conclusions favoured the Darwin hypothesis. The Murray–Agassiz 'solution theory' was considered to be faulty, and 'of the explanations so far proposed for living coral reefs' the development of Darwin's theory in the subsequent studies by Andrews, Reginald Daly and Thomas Wayland Vaughan (who had worked on the Tortugas) was believed 'to furnish the best working hypothesis' (Mayer 1918:14). Moreover, they concluded that 'The entire visible reef belongs to the recent period subsequent to the cessation of volcanic activity and has evidently grown seaward over its own talus' (16). Exercising caution, however, their report stressed that 'the subject of the formation of barrier reefs and atolls is far from being settled in the minds of its students' (12).

Most of the effort, as mentioned, was expended in the biological work and for the month of October the various species were mapped *in situ*, while daily measurements were made at stations across reef transects, of air and sea water temperatures, tidal movements and sea water circulation, the chemical composition of sea water and rate of solution of calcium carbonate. In addition, laboratory experiments were conducted on the effects on various species of coral polyps of sunlight deprivation, changes in water temperature, silting, dilution of seawater by fresh water, and drying out.

Although neither Murray nor Semper was mentioned by name in the laboratory experiments, it is clear that one aspect of the program was to test their hypotheses regarding the solution of calcium in sea water and the influence of tidal action in forming reefs. Murray's theory fared badly: from tests conducted by Mayor in the laboratory on pieces of *Cassis* shells (small, commonly-found 'helmet shells') they concluded that 'it would take at least 1,000,000 years to dissolve a layer one fathom [1.8 m] thick' (Mayer 1918:42) – that is, the calcium carbonate thickness of many established reefs. Semper's theory fared no better when observations revealed that 'Scouring by currents and disintegration are . . . of limited efficacy in the deepening of lagoons in the Murray Islands' (43).

Observations on corals revealed that species were distributed in zones (miniature isopan generic contours) across the reef flats to determine whether zonation was due to a gradient of changes in air and water temperature. Experiments were conducted in glass aquaria by manipulating changes to salinity of the water and silt cover which, not surprisingly, led to conclusions that corals thrive best in clean sea water. One of the most interesting observations, from a present day viewpoint, was that concerning the abundance of species and density on the outer reef fronts and their progressive decline in number towards the shore. Part of those observations was to estimate the rate of growth and, by a remarkable coincidence, an ingenious method was available in one instance. Using a photograph from Saville-Kent's magnum opus of 1893 (Plate II, facing p. 6) of a large coral head of the *Symphyllia* genus on a reef flat on Thursday Island as a reference datum, they were able to locate the same head 23 years later. Saville-Kent had measured its diameter as 30 inches (76 cm); Mayor found it to have increased to 74 inches in diameter (1.9 m), thereby concluding growth to have been at 'the rate of 1.88 inches [48.75 mm] per annum' (18).

Since the 1880s, especially with the rapidly improving compound microscope, increasing attention had been directed to the anatomy and physiology of the coral polyp itself. Once its animal nature had been definitively established by the work of the great pioneers Peyssonnel and Ellis, the way had been cleared for closer inspection of its role in reef formation. As coral polyps were believed to be exclusively carnivores and observations showed that the available zooplankton (animal food) supply was adequate in all photic zones, an explanation for variations was sought. One of the intriguing questions at the time was whether there might be some relationship between polyps and the tiny single-celled organisms discovered in their fleshy tissues and regarded at first as an infestation, possibly of animal origin. With that likelihood in mind they were named zooxanthellae, from the Greek zôon ('animal') and xanthellos (diminutive form of 'yellow'). In earlier days of investigation it had also been thought that polyps might eat vegetable matter. Since the organisms did not seem to affect the polyp, it was then assumed that there might even be some kind of symbiotic relationship with these 'infesting zooxanthellae' in which neither animal nor plant had a deleterious effect upon the other, but that perhaps one, or both, gained some benefit (28).

Mayor's studies on Mer reached four main conclusions: that the ecology of the reef, defined in terms of distribution of species, was dependent upon temperature, silting, moving water and other mechanical effects, and, with a fine Darwinian touch, 'the struggle for existence' (44).

Four years later, Mayor repeated the Murray Island survey on the volcanic island of Tutuila in American Samoa, site of the naval base at Pago Pago. Altogether three separate expeditions were mounted to the island between 1917 and 1920 and on different occasions he was accompanied by other reef scientists, including Daly in 1919 and Potts in 1920. On all occasions there was as much interest in the geology as in the coral biology. Mayor continued his studies of what he termed the ecology of the reef, chiefly from his perspective of the range of species and their spatial distribution across the reef flats. As on Murray Island he reported that 'the largest number of corals [in density and species] was found to be growing in comparatively quiet water, about 200 feet shoreward from the region where the surges die out' at around 850 feet (260 m) (Hoffmeister 1925:4).

What troubled Mayor at that point was the identification of discrete species and the classification of coral groups generally. He noted, significantly for both coral taxonomy and evolutionary theory, that 'The species problem, difficult as it is in all biologic groups, is still more perplexing in such variable forms of life as the corals, where environment stamps its influence so effectively'. He then continued to lament that 'With our insufficient knowledge concerning the evolution of species, we can only use the term "species" in a more or less artificial manner for purposes of classification' (5). What Mayor did at the time was twofold: he produced an exhaustive table of then-identified coral species according to spatial distribution and location, and then had them transported to the United States National Museum. Second only to the British Museum in its collection, they were studied there by the leading American coral taxonomist, Wayland Vaughan.

Mayor made some amusing comments on the attitudes to the expedition of the local Torres Strait islanders who could not understand all the effort being expended without catching any fish. 'Several times the leading men of the village', he recorded, 'waded out to us on the reef-flats to suggest that the corals we were collecting were wholly worthless as food and to offer us fish and fruit'. The scientists, with mock humility, accepted themselves as imbeciles in the eyes of the islanders and it was the Canadian-born geological zealot Reginald Daly (they named him 'ma-ma', the 'rock') 'the most amusing imbecile among us' who really puzzled them by his incomprehensible actions in collecting totally inedible rocks (Mayor 1924:vii).

Of equal, if not more significant importance was the attempt in 1919 to settle the Darwin hypothesis by yet another drilling attempt which was the particular interest of Daly, recently appointed professor of geology at Harvard University. The attempt this time was made on two sites inside the harbour of Pago Pago itself, on opposite sides of the reef flats, and altogether four borings were made: one on the eastern shore of the harbour which reached a depth of 166 feet (50 m), and three on the western side, near the naval station, the depths reached being 68, 120 and 121 feet (20, 36.5 and 37 m respectively). The Mayor team had made excellent preparations taking a large amount of apparatus for laboratory studies (chemicals, glassware, a potentiometer and galvanometer, other electrical equipment of various kinds), dredges, an underwater camera and a 'diving hood', and a Davis-Calyx drill suitable for boring through soft limestone.

The drilling results were interesting but inconclusive. Mayor and Daly were able to establish that the original volcanic island of Tutuila had been eroded by the sea around its perimeter which subsequently submerged, 'while at the same time the island tilted so that the platform sank to a slightly greater depth on the southeast than along the north shore' thereby creating a fringing reef on the northern side, fused to the shore, and a barrier reef to the south, with an intervening lagoon. Mayor thereby demolished Semper's denial thirty years earlier of tilting as an explanation of the variant forms of reefs on Palau. As a result of the boring and the geomorphological investigations, it was confirmed that 'Tutuila, therefore, is consistent with the Darwin–Dana coral-reef hypothesis to the extent that a submergence of 400 feet has occurred since the corals began to form the old barrier reef'. Unfortunately, their inquiries did not confirm the Darwin–Dana hypothesis since the coral depth did not indicate that it had been 'built up several thousand feet from the slopes of a sinking island', nor did they consider sea level rise. Instead, as he recorded in his official report to the Carnegie Institution in 1920, Tutuila was 'found to be rooted on a broad, wave-cut platform which, slightly submerged, afforded favorable conditions for coral-reef growth' (Mayor, Posthumous Papers, ed. Hoffmeister 1925:3).

By the time the Carnegie findings were published the tragic World War I had ended and among the allies around the world there was a resolve that international peace and harmony had to be actively secured. Considerable idealistic initiative was displayed by United States President Woodrow Wilson with his vision of a League of Nations but the Paris Peace Conference of 1919 was less than successful: in fact it institutionalised old animosities that were to lead to yet another horrendous world war in 1939. The search for peace and international cooperation, however, was kept alive in various quarters, and in the following decade of the 1920s the Pacific region became an arena of such aspiration, which led to significant work in science generally, and reef science in particular, under the aegis of the Pan-Pacific Union.

CHAPTER 13

EXPLOITATION CHALLENGED: RISE OF ECOLOGY

Although issues related to the Great Barrier Reef in the late nineteenth and early twentieth centuries – controversy over Darwinian theories of evolution and reef formation, and the final decline in the pearling industry due to unrestricted resource use – attracted considerable specialised attention, apart from occasional reports in the newspapers, they rarely reached the general public. In the same period, however, a new perception of the Reef was being presented by Edmund Banfield (1852–1923), a journalist for the Townsville *Daily Bulletin*, whose writings were to have a profound and lasting effect on public awareness and attitudes, and indeed, on many scientists. They marked, in fact, the beginning of a new understanding of the Reef. No longer primarily a navigation hazard, a strange separate creation to be catalogued by biologists, or a frontier to be subdued and a resource to be exploited to extinction, it was now interpreted as one of nature's most diverse and beautiful creations, to be respected and preserved. The Reef was to come into international prominence as, quite literally, a unique natural phenomenon.

In 1897 Banfield, due to ill health, moved with his wife Bertha north from Townsville to Dunk Island, a small but attractive continental island, some 5 kilometres out from the mainland, rich with unspoiled rainforest, beaches and coral reefs. There, for the next twenty-six years, he wrote numerous articles for Australian newspapers, several tourist pamphlets, and four widely distributed books which brought a new interpretation of the Reef to an ever-widening readership. Banfield himself was to attain folk hero status, virtually unrivalled in the nation at the time. In place of the regular sequences of disasters, massacres and dangers of Reef travel and settlement, exemplified in such morbidly transfixing accounts as the fate of Eliza Fraser, Barbara Thompson and Mary Watson, the horrors of islander kidnapping, or the sinking of passenger ships such as the *Quetta*, Banfield introduced the public to a radically different understanding of the Reef through the genre of literary alchemy, transmuting the Reef from a domain of danger into what many imagined to be an idyllic retreat. At the same time, in his genuine commitment to the scientific study and conservation of the Reef's islands and wildlife, he distinguished his work totally from the romance novel tradition of adventurers on totally fictitious isles.

Close contact with untamed nature in remote places had become a favourite theme in popular reading since the eighteenth century, enlivened by the runaway success of Daniel Defoe's *Robinson Crusoe* of 1719 and further stimulated with the publication in 1812 of *Der schweizerische Robinson* by Johann Wyss, who constructed in Switzerland – from travel accounts of the period – a tale of a shipwrecked but resourceful Swiss family on a tropic isle. Translated into English in 1814 as *Swiss Family Robinson*, and numerous other languages, it rapidly became one of the most widely read books of the century, offering an intensity of exhilarating escapist adventure to the great masses of Europe and America for whom exotic travel had to be vicarious. Although a totally artificial and confused account of a tropic island called 'New Switzerland', where the family encountered monkeys, lions, elephants, a kangaroo 'discovered by the great Captain Cook' and even a 'duck-billed platypus', it is indicative of the way in which such regions were represented to the reading public of the period.

Banfield, by contrast, specifically rejected the style of Johann Wyss: 'We do not all belong to the ancient and honourable family of the Swiss Robinsons, who performed a series of unassuming miracles on their island. There was no practical dispensation of providential favours on our behalf' (Banfield 1908a:13). Moreover, the Swiss family were ruthless hunters of wildlife, aiming not only at satisfying their own appetites, but also experiencing the perverted joy of hunting and killing for its own sake. Instead, Banfield turned to a radically different tradition established by two serious naturalists: Gilbert White and Henry David Thoreau. In his third and final book, published posthumously in 1925, Banfield explicitly acknowledged White as a major inspiration, both in literary style and attitude to nature, describing him as 'a man of genius who delights hosts of readers all over the world' and himself as 'a loving disciple' (Banfield 1925:34).

NATURE AND THE INNER SEA

The English cleric Gilbert White (1720–93) was vicar of Selborne in rural Hampshire, about 80 kilometres (50 miles) to the south-west of London and one of the earliest and most influential proponents of what later came to be called an ecological view: the desire to see beyond the obvious apparent cruelty and unpredictability of nature, and its ruthless exploitation by humans, into its unified complexity and beauty. In 1789 his collection of

110 letters, written between 1781 and 1787, was published as *The Natural History of Selborne*. An outstanding work of descriptive natural history, it became widely read throughout the century and profoundly influenced not only Banfield but, among many others, Darwin and Thoreau, and served as an exemplar throughout the nineteenth century, and into the twentieth, for naturalists to follow.

Distinguished by its elegantly constructed prose, *The Natural History of Selborne* established its reputation from the quality and practical wisdom of White's descriptions as he recorded the surrounding countryside in his parish: geology, vegetation and animal life, with a particular emphasis on birds and their habits. Necessarily, for a cleric of that era, it was set within the context of divine design and the assumption of that Platonising era that the universe is an interconnected unity or *plenum formarum*, in which the rich diversity of life is represented. White, a contemporary of Linnaeus, was strongly influenced by his seminal 1749 Latin essay *Oeconomia naturae* ('The Economy of Nature') which introduced the concept of nature as a circular chain of mutual dependency, a theme repeated by Linnaeus in 1760 in *Politia naturae*, where he saw each link existing for the sake of all others. Following that idea, White wrote, 'It is in *zoology* as it is in *botany*: all nature is so full' (White 1789:58). The infinite creative wisdom of God, he believed, had filled every space within the universe: the task of the naturalist was to understand and reveal the manifold wonder of that creation.

The *Natural History of Selborne*, however, is more than a collection of carefully crafted essays describing the interactions of nature. At that early juncture White was critical of mere collecting and describing since 'the standing objection to botany has always been, that it is a pursuit that amuses the fancy and exercises the memory, without improving the mind or advancing any real knowledge' and 'where the science is carried no farther than a mere systematic classification the charge is but too true'. The task before the natural scientist, he continued, was that of careful experiment: the botanist 'should study plants philosophically, [and] should investigate the laws of vegetation'. Not that White deprecated the useful contributions of botany. On the contrary, he stated categorically that 'vegetation . . . is of the utmost consequence to mankind, and productive of many of the greatest comforts and elegancies of life . . . [with] a vast influence on the commerce of nations' (192). Readily accepting the anthropocentric approach of Christianity given in Genesis 2:26 which asserts that mankind has received dominion over all creatures, he cited confirmation of that assertion in James 3:7 that all species – animals, birds reptiles and marine creatures – have been tamed for the use of humans (63). The way to the improvement of agriculture, however, was still that of the 'philosophical' – that is, scientific – approach: 'not that system is by any means to be thrown aside; without system the field of Nature would be a pathless wilderness: but system should be subservient to, not the main object of, pursuit' (192).

In the United States one of the most dedicated readers of Gilbert White was Henry David Thoreau (1817–62) whose works were to provide Banfield with a direct role model.

Banfield, in fact, consciously modelled his life on the ideas in Thoreau's biography *Walden*. That totally different kind of work raised in its many readers an intense awareness of a widespread, subliminal hunger for a lifestyle of unfettered freedom from the conventions and restraints of an increasingly ugly era, as the industrial revolution with its burgeoning population growth had begun to crowd the middle classes into the bland and stifling conformity of the ever-spreading suburbs, and the rural masses into the polluted, disease-ridden slums of the cities and factory towns.

Thoreau lived in Concord, Massachusetts, 32 kilometres (20 miles) north-west of Boston and became drawn into the transcendental movement initiated there by Ralph Waldo Emerson. Part of the nineteenth century reaction to the horrors of industrialisation and the quest for ultimate meaning to life, American Trancendentalists also drew upon the philosophical tradition of Platonic idealism. Following much the same ideals as those cherished by Linnaeus and Gilbert White, they believed that the universe was fundamentally good and that the search for absolute truth could be pursued, not by a slavish following of past dogmas, but by the use of innate human intelligence. Accepting the Platonic concept of divine – 'transcendent' – plenitude, they sought enlightenment in the contemplation of nature, at the same time drawing upon the ideas of such mystics as Emmanuel Swedenborg as well as aspects of Hindu, Buddhist and Taoist teachings that were becoming available in translations.

Very early in life Thoreau, having graduated from Harvard University, decided to become a nature poet. In 1845 he withdrew to self-imposed solitude for two years on the shores of Walden Pond outside Concord where he built a cabin and attempted a regime of self-sufficiency. His aim was to use as much of the natural environment as he could, and by minimising his material needs, in the Buddhist tradition, he sought to eliminate desire, and hence unhappiness. His primary purpose was to embark on a voyage of self-discovery. Comparing his quest to the recent United States Exploring Expedition of thirteen naval vessels that traversed the Pacific under Lieutenant Charles Wilkes in 1844, in which James Dana sailed as naturalist, he asked, rhetorically, 'What was the meaning [of that expedition] with all its parade and expense' compared with the more urgent need to explore the great unexamined interior of one's own self? 'It is easier', he asserted, 'to sail many thousand miles through cold and storm and cannibals, in a government ship, with five hundred men and boys to assist one, than it is to explore the private sea, the Atlantic and Pacific of one's being alone' (Thoreau 1854:213). The private sea of one's inner being is an evocative phrase that determined the course of Thoreau's life, and that of Banfield after him. The results of his exploration of his own private sea were set down in a fascinating introspective account of 1854 under the title *Walden, or, Life in the Woods*.

One's inner being, for Thoreau, is to be found in close communion with nature, virtually in immersion: 'I come and go with a strange liberty in Nature, a part of herself' (90). Most people, he believed, never seek their inner being – or else are too timid to do so – and by default follow the anaesthetising routine of daily life, serving the state as machines without ever making moral distinctions: 'The mass of men lead lives of quiet desperation ... A stereotyped but unconscious despair is concealed even under what are called the games and amusements of mankind' (10). The first step to inner freedom is to divest the multifarious trappings of daily humdrum existence; to begin 'every morning [with a] ... life of equal simplicity, and I may say innocence, with Nature herself' (64). Declaring his total divorce from conforming society, Thoreau asserted, in a famous aphorism: 'If a man does not keep pace with his companions, perhaps it is because he hears a different drummer. Let him step to the music which he hears, however measured or far away' (216). 'I will breathe after my own fashion' he asserted, 'If a plant cannot live according to its nature, it dies; and so a man' (234).

Thoreau, sadly, was unable to sustain his solitary retreat. The pressures of a suspicious and hostile community on a conservation minded lover of unreformed nature, along with the bitter winter snows in those cold, bleak latitudes were too great, and in 1847 he left Walden. His health, unfortunately, had been undermined by the privations of outdoor life in the harsh winters and he died prematurely of tuberculosis in 1862, barely forty-five years of age. His reputation and legacy, however, became even greater as the reading public, especially in his native United States, came to appreciate his attitude to nature, and his reputation there is now greater than ever. Probably almost entirely unknown to his American readers that legacy was translated to the other side of the world in the southern hemisphere – virtually an antipodean location – where it was given a complete, lifelong expression in the actions of Banfield: 'Walden, or Life in the Woods' became transformed as 'Dunk, or Life on a Tropic Isle'.

'BEACHCOMBER' BANFIELD: A DIFFERENT DRUM

Edmund James Banfield was born in Liverpool, England, and grew up in Ararat, Victoria, where his parents had migrated in 1856. The young Banfield moved into journalism and in 1883 arrived in Townsville. He soon became identified with the North Queensland separatist movement, formally known as the Townsville Separation League, and in particular with one of the leading protagonists, the prominent merchant Robert Philp (1851–1922) who had come to Townsville in 1874 as resident partner to a fellow Scot, James Burns, to initiate the new trading firm of Burns Philp and Company. A vigorous entrepreneur who expanded the company into coastal shipping, bêche-de-mer harvesting, sugar and mineral transportation, and provisioning for inland stations, along with Kanaka recruitment and importation as virtual slave labour, Philp believed that further expansion, and dominance, would be best effected if a new colony were created with a capital in Townsville. He also saw separation as an opportunity to be free of interference from an officious government in Brisbane which had impounded two of his labour recruiting vessels in 1884 for breaches of the Act of 1880 and had prosecuted several of the crew for various offences against islanders, including murder. Banfield, however, wrote supportively of the separation policies of Philp, with whom he had a lifelong friendship, although the movement never succeeded.

Having lost the sight of one eye in a childhood accident, Banfield's first close acquaintance with the Reef came when, needing to travel to England for a prosthetic replacement, he was offered a free return passage on the Burns Philp mail steamer *Chyebassa* in 1884 in return for a series of articles as their 'travelling correspondent' extolling the scenic qualities of the Reef and the voyage to England via the Dutch East Indies and the Suez canal. Philp's intention was unabashedly commercial: he sought to promote the Torres Strait route to Britain taken by his British India Steam Navigation Company in competition with other colonial shipping lines which took the southern ocean way. Like so many before him, and a multitude after, as the ship steamed north through the Inner Route, Banfield was captured by the Reef, overwhelmed by an ecstatic intoxication from which he never recovered. Originally a willing journalist, he later became fully converted as his own fascination grew with a deeper understanding of coral reefs.

On his return in 1885 Banfield's articles on the Reef were gathered into an early version of a tourist promotion pamphlet, while at the same time he actively supported in the press the Convention that met in Townsville in May 1885 to plan a campaign for the separation of the north into a new colony. In Brisbane the Griffith ministry attempted to subvert the separatist movement by creating an additional northern seat around Townsville which Philp, in his first attempt to enter parliament, won resoundingly in 1885, taking his seat in the Assembly the ensuing January 1886. Philp remained active in politics for the following thirty-two years and was premier from 1899 to 1917. The separation quest, however, was sidelined in the 1890s by the greater movement to federate the existing colonies, and although it lingers to this day, it ceased to be an effective political force.

A tall but lightly built man with constitutional health problems, Banfield was a quietly introverted personality who found adventure within the reaches of his own imagination and very early discovered a kindred spirit in reading Thoreau, whose account of two years of enjoyable solitude at Walden Pond gave him considerable empathic comfort. Having returned to his position at the Townsville newspaper, where he became sub-editor to the alcoholic Dodd Clarke, Banfield wrote a stream of articles on the Reef under the pen name of Rob Krusoe, a quite transparent identification as he began to explore his own private sea: the quest for greater freedom of the spirit.

In 1886 Bertha Golding, daughter of family friends who had farewelled the Banfields thirty years earlier, whom Edmund had met on his return visit to Liverpool, arrived as his fiancée and they were married in Townsville in August of that year. The pressure of journalistic work, and difficulties with the dipsomaniac Clarke who left all the editorial work to an increasingly depressed Banfield, undermined his already weak constitution; his weight fell to barely 50 kilograms (110 pounds) and he was diagnosed with some form of progressive wasting, close to pulmonary consumption (Noonan 1983:94).

In search of rest and recuperation he took several short camping breaks along the Reef coast north of Townsville. It was Dunk Island, however, some 200 kilometres (120 miles) north, that enthralled him. 'Having for several years contemplated a life of seclusion in the bush', he wrote in retrospect, 'and having sampled several attractive and more or less suitable scenes, we were not long in concluding that here was the ideal spot'. Discovering that its Aboriginal inhabitants had left, and that it had not been alienated from the Crown, with Bertha's encouragement he decided to set up a temporary holiday camp there and in 1897, with the help of a hired workman, a prefabricated hut and household equipment were moved across on a lighter. Then, to his immense joy, a lease was granted in 1900 which allowed continuing tenure: the holiday camp became their permanent home until Banfield's death twenty-six years later (Banfield 1911:11f.).

Dunk had been named in 1770 by James Cook when he logged it as a 'tolerably high island'. Lying offshore from Mission Beach it is an irregularly shaped continental island 5 kilometres long, with a maximum width of around 3 kilometres; a central ridge with four peaks runs for most of its length, the highest of which reaches 271 metres and was used as a radar defence station in the Pacific War of 1941–45. Now a national park, with an airfield and tourist resort and mainly undeveloped to this day, it is heavily covered in rainforest vegetation with some forbidding valleys. On the north-west side facing the mainland it has a sheltered cove with an adjoining sandy spit and coral reefs that Banfield named Brammo Bay, derived from a local Aboriginal word for butterfly. In September 1897, having chosen a site a little uphill from the bay, the one room cedar hut was built, to be augmented over the years with additions and improvements; the complete bungalow was finished on Christmas Day 1903, as home for the exhausted invalid and his wife. In the following months they were grateful for the help of a strapping mainland Aborigine called Tom, one of the island's original inhabitants, who canoed across to offer his services and told Banfield that its traditional name was Coonanglebah. Tom was to become a welcome worker and guide, as their Robinson Crusoe adventure began in a setting that seemed ideal.

From the outset, the Banfields' life was one of extreme frugality: food was grown in a kitchen garden, and with Tom's expertise, fish came from the sea, while minimal groceries and supplies arrived aboard the weekly mail steamers which by that time were part of regular traffic through Reef waters. Later, cows, goats, poultry and horses were brought over. Far from being an isolated Juan Fernandez in the South Pacific some 700 kilometres off the coast of Chile where 'Robinson Crusoe' (Alexander Selkirk) was marooned, Banfield was in regular contact with the mainland nearby that could be reached from Dunk by canoe or small craft quickly and safely in calm weather. Banfield later acquired a small powered launch – the *Nee Mourna* (twice replaced) – and built a flat bottomed punt to use as he developed a trading system with the mainland, exchanging garden produce for groceries. To provide an income to supplement the project of self-sufficiency, and to express his inner need to articulate his experience of a life of simplicity, close to nature Banfield continued with his journalism, sending a regular stream of articles to newspapers, chiefly in Queensland, singing the praises of the islands of the tropical Reef. Banfield actually was no recluse: on the contrary he was quite adept at promoting himself to an ever widening reading public, and clearly enjoyed the limelight that his articles generated. His health was gradually restored, his spirits became buoyant and the pulmonary diagnosis was found to be wrong. He simply needed, as Thoreau had written, 'to live according to his nature'.

Passing ships stopped over in Brammo Bay: Charles Hedley soon discovered Banfield and visited on his collecting expeditions; naval survey vessels frequently dropped anchor and the Banfields offered what simple hospitality they could. On 10 October 1903 a most unusual visitor came for a short stay: Sir Walter Strickland, a Cambridge graduate, globe-trotting gentleman naturalist and heir to a wealthy estate. Strickland was one of those eccentrics that only England seems to produce. He travelled lightly and abstemiously in third class, reflecting his close identification with Buddhism that was gaining ground among European intellectuals as a more satisfying explanation of the human condition than the then dogmatic aridity of conservative Christianity.

New ideas, generated by archaeology, anthropology, psychology, philosophy, philology, science and technology were undermining the old unquestioned, authoritarian order. In particular, the late nineteenth century was a period of ferment in religious ideas as Western enthusiasm for the literature and religious teachings of the East exploded after the discovery of Sanskrit literature and the foundation of the first university chairs of comparative philology and religion in the 1860s.

Although Banfield, as an active journalist, was aware of the growing volume of philosophical nature literature circulating in the intellectual counter culture, it is likely to have been Strickland who introduced Banfield to the specific ideas of Buddhism, especially to notions of the interrelatedness of all life and the sanctity of nature. A fundamental tenet of Buddhism is that human suffering and unhappiness are caused by the desire for material possessions, set out in the doctrine of the Three Lakshanas (literally, 'signs'): *anitya*, that nothing is permanent and all existence is subject to continued change; *anatman*, that all life is interconnected spiritually and no single being has a unique, individual identity; and that suffering, *dukkha*, comes from greed for material possessions, and happiness from the minimisation of such desire. Strickland advocated that position in his conversations which later became reflected in Banfield's numerous writings; two years before his death Banfield wrote in a letter to a friend that 'If one had to make a choice between the two religions I for one would gladly accept Buddhism. At least it is clean. Ours teems with reference to blood or as one of its Ministers said, "reeks of the abattoirs"' (cited, Croucher 1989:16).

Strickland was quite taken by the naturalism in Banfield's articles and encouraged him to bring them together for book publication, underwriting its financing with the London firm of T. Fisher Unwin. Banfield completed the anthology with new inclusions by late 1906, the short Introduction indicating his disarming aim 'to set down in plain language the sobrieties of everyday occurrences – the unpretentious homilies of an unpretentious man'. The ensuing 336 pages are anything but that: ranging through eight chapters, in 138 essays, he described his motives, the island, birds, coral communities, vegetation, reptiles, marine animals, including in Part II nearly 100 pages of notes on many aspects of Aboriginal life and culture, with a final affirmation of his own life of simplicity. Chapter IV records his impressive ecological observations on Dunk's 'Garden of Coral' at the time. In 1908 it was published in London under an intriguing title as *The Confessions of a Beachcomber*, with a dedication 'To the Honourable Robert Philp, MLA . . . by one who owes him much of his love for Tropical Queensland'. Significantly, the title page carried the famous quote from Thoreau – his evocative image of a different drummer – that was to remain Banfield's life theme. Below it was a second quote, from Henry Longfellow: 'Trust in yourself and what the world calls your illusions'. An immediate success, it has remained a perennial publication to this day.

Banfield's choice of title was itself a cheerful questioning of social assumptions. Originally a 'beachcomber' described a large wave rolling in from the ocean, as recorded around 1840 in the Merriam-Webster dictionary, receiving its transferred meaning later in Bartlett's *American Dictionary* as a 'settler on a Pacific island, often by disreputable means'. In 1865 Admiral William Smyth, of the British navy, extended the meaning of 'beachcomber' in his famous nautical dictionary *The Sailor's Word*, as one who 'loiters around a bay or harbour', and 'beachcombing' as 'loafing about a port to filch small things' (Smyth 1867:88). The derogatory connotations of the word were generated in the early nineteenth century when whalers and resource raiders roamed the Pacific rapaciously and large numbers of their crews came to live on the Melanesian islands, either inadvertently or as a result of shipwrecks – figuratively rolling on to the coast like beachcombers – or sometimes as impressed convicts or slaves who seized an opportunity to desert. Yankee whalers often had black slaves as cooks – the most menial position on the ship – who on occasions found desertion an attractive alternative, given that their colour would enable them to blend into the local population more readily. In many cases the intruders were tolerated since they provided welcome expertise in such Western ways as distilling alcoholic liquors, along with blacksmithing and boatbuilding skills using materials salvaged from wrecked vessels which enabled the construction of more seaworthy craft. In some instances they acted as mediators and facilitators for the island communities in their trading activities with the resource raiders (Ralston 1977).

At the same time, contempt for beachcombers was heightened in the popular press by missionaries who disapproved vigorously of their alcoholic habits and polygynous behaviour. In cases where they abused the women they were often killed, Banfield himself noting that 'A whack on his hardened head from the club of a jealous native is the time-honoured fate of the typical Beachcomber' (Banfield 1908a:55). The appellation 'beachcomber', consequently, would have struck a responsive chord in the reading public who knew it to be a pejorative term referring to the most disreputable stratum of society.

Banfield immediately captured a reading public with his audacious title, a clever use of inverted bravado by a self-professed 'unpretentious man'.

All Banfield's writings, his four major books and the hundreds of other uncollected articles, still make absorbing reading. His first commissioned work, *The Torres Strait Route from Queensland to England*, was begun aboard the *Chyebassa* and completed in England to promote the British India Steam Navigation Company. Published in the Townsville *Daily Bulletin* in 1884 it is a gem of travel literature, with a compelling story line that draws the reader along. It was his Reef articles, however, which captured attention and changed perceptions. Throughout all his writing the influence of White and Thoreau is evident. From Gilbert White he copied the techniques of lengthy lists of bird and plant names with their binomials, the citing of rainfall data and frequent quotes from major authors. From Thoreau he adopted and polished the technique of integrating himself into the descriptions, using the love of nature to heighten his aesthetic and intellectual responses by communicating to the world at large. Many of his hundreds of articles, especially those dealing with Aborigines, are riveting short stories in their own right, sensitively told with respect for the characters, and often of a humorous anti-climactic kind which maintain tension to the denouement. Very deliberately he sought to portray Aborigines as people with equal rights and natural dignity, although with the full range of human foibles: 'not as ethnological specimens, but as men and women – types of a crude race in ordinary habit as they live, though not without a tint of imagination to embolden better truths' (Banfield 1918:7).

In other articles he moved into the literary equivalent of a musical tone poem where the prose simply flows along with no particular structure, describing the beauties of the Reef and his responses to it. Some of his best writing comes in the opening chapter of the *Confessions* where he invites the reader to share in the ecstatic release when

> The blood tingled with the keen appreciation of the crispness, the cleanliness of the air. We had won disregard of all the bother and contradictions, the vanities and absurdities of the toilful, wayward human world, and had acquired a glorious sense of irresponsibleness and independence. This – this was the life we were beginning to live . . . [within a paradise] of the dark compactness of the jungle, the steadfast but disorderly array of the forest, the blotches of verdant grass, the fringe of yellow-flowered hibiscus and the sapful native cabbage, [which] give way in turn to the greys and yellows of the sand in alternate bands. The slowly-heaving sea trailing the narrowest flounce of lace on the beach . . . (Banfield 1908a:17–18).

Adopting Thoreau's rejection of social graces, Banfield proclaimed his freedom to the masses in the cities who were forced to dress in conformity with the sombre, heavy clothes of the era: 'the man who has to observe the least of the ordinances of style knows no liberty . . . How can a man with hoop-like collar, starched to board-like texture, cutting his jowl and sawing each side of his neck be free?' (171). For his part, as depicted in all the surviving photographs, Banfield went barefoot and habitually wore little more than a cotton undershirt, denim trousers and a large straw hat, apparently even when the island was visited on separate occasions by State Governor Sir Matthew Nathan and Governor-General Lord Forster and Lady Forster.

Again, following Thoreau's example, Banfield used the *Confessions*, and the sequels, *My Tropic Isle* (1911), *Tropic Days* (1918) and *Last Leaves from Dunk Island* (1925), throughout as an essay in self-justification, and to proselytise his reading public, many of whom were in England. In the Introduction to the first book he described his actions as conforming to the Buddhist 'wisdom of the sage who wrote – "If you wish to increase a man's happiness seek not to increase his possessions, but to decrease his desires"' which for him was best pursued in North Queensland where one 'can draw nearer to nature'. Even so, Banfield was no marginal fringe-dweller unable to cope with the increasing technological development of industrially dominated society. On the contrary, he was educated above the average for his times, extensively well read, a highly intelligent and strong personality. The life of voluntary simplicity was a conscious choice pursued with a definite moral purpose, to promote a new ethic of concern for nature.

RISE OF ECOLOGY: THE SUBVERSIVE SCIENCE

The Western society Banfield rejected had gone down the path of increasing materialism in the nineteenth century while science, profoundly influenced by the arid philosophies of positivism and pragmatism, had adopted an approach to nature of impersonal objectivity, an effective blind eye to unrestricted exploitation. The scientific quest of the period was for control – in effect, the conquest – of nature, not empathic understanding and cooperation. Throughout the nineteenth century, however, a different scientific culture was gestating, stimulated to a significant extent by Alexander von Humboldt who laid the foundations for ecological research and exerted a powerful influence on Darwin with his pioneer studies in plant geography and his innovative approach of perceiving all of nature in terms of interactive communities (Bowen 1981: 213f.).

In 1866 Ernst Haeckel (1834–1919), professor of zoology at Jena, gave Humboldt's concept – in a more limited form, stripped of its wider connotations – a new name, oecology. The origins of that term can be traced to the 1749 *Oeconomy of Nature* by Linnaeus in which he sought to reveal the relationships operating within the divine design. Oeconomy, from the Greek *oikos*, a household, and *nomos*, custom or management – later simplified as economy – suggested management of a household and, by extension, of the earth itself seen as the home of mankind. Haeckel, a strong proponent of Darwin's evolutionary theory of natural selection expressed as survival of the fittest, proposed a new branch of biology to study that process. With the Greek suffix *logos*, reason or law, he created a neologism, 'ecology' (*oekologie*) for a new rational science dealing specifically with the relationship of living organisms to their environment.

After Haeckel ecology was assigned a limited function within the biological sciences as a means of understanding plants or animals – but not humans – in their habitat, as for instance, in studies to provide maximum yield calculations. At the same time, with rising opposition to environmental degradation, the wider significance of ecological perspectives on a global scale in the Humboldt tradition was also revived, gaining strength and public support in the following century. In the commercial and industrial sector, however, ecology in that form was viewed as subversive, since it questioned the unrestrained exploitation of nature (270f.).

RESPECTING NATURE: CONSERVATION AND SANCTUARY

During the late nineteenth century an increasing groundswell of concern over obvious damage to the environment, and the consequent need to conserve nature, gave an impetus to Banfield. In Europe and America, as population increased rapidly and human impact accelerated forest clearing, the visible despoliation of nature reached alarming proportions. In response, a vigorous, even radical, version of ecology emerged to advocate conservation of nature in wilderness reserves, not only for their commercial value, but to maintain their integrity. It was chiefly in the United States, due to the writings of Thoreau and his contemporary, George Perkins Marsh (1801–82) that early conservation movements, using the new insights offered by ecology, first achieved international prominence.

In the most significant environmental book of the period, *Man and Nature: or Physical Geography as Modified by Human Action*, published in 1864, Marsh, a former Congressman and at the time American ambassador to Italy, appalled at the wastelands being created in his home state of Vermont by the ravages of the timber industry, asked the seminal question: whether 'man is of nature or above her?' Following his stimulus, along with that of Thoreau and conservationists such as the great wilderness advocate John Muir, in 1872 the United States Congress declared the world's first national park at Yellowstone as a conservation measure. Australia had the distinction of declaring the second: the Royal National Park on Sydney's southern border, and other countries followed. Their efforts, however, were focused on land protection, particularly forests. It would be a further century before the concept of marine parks was established.

It was in such a climate of thought that Banfield attempted to translate some of those ideas into action on Dunk Island. Not that he was conversant with the complex issues of current ecological thought: in fact, he denied any sophisticated knowledge, stating explicitly that 'I am not a professor with a mind like a warehouse, rich with the spoils of time, but a mere peddler, conscious of the janglings of an ill-sorted, ill-packed knapsack of unconsidered trifles' (cited in Noonan 1983:198). Indeed, he described himself as 'one who disclaims expert knowledge, who regards birds from the standpoint of aestheticism and sentiment blended with utility, who fears that Australia as a whole has not yet learned the worth of many species peculiar to the land and who cons the steadily-growing "extinct list" with dismay' (Banfield 1925:184).

He was far too self-effacing: he corresponded regularly with Charles Hedley at the Australian Museum in Sydney; with Frederick Bailey, the government botanist in Brisbane; and Douglas Ogilby, a former ichthyologist at the Queensland Museum, for information on, and scientific names of, the specimens he sent in for identification. As the section in the *Confessions* headed 'Early History' reveals, he had a good knowledge of the voyages of Cook, King and Moresby, and the scientific work of another seeker for inner freedom, John MacGillivray. Banfield was thoroughly familiar with MacGillivray's natural history account in the *Voyage of the Rattlesnake*, especially the part dealing with Dunk, and on the nearby mainland Tam O'Shanter Point where the ill-fated Kennedy Expedition unloaded for the exploration of Cape York. Moreover, he kept himself aware of developments in the current scientific literature, even citing an excerpt from the report of the 1913 Cambridge University expedition to Murray Island in Torres Strait by Frank Potts. Intimations of his understanding of ecological concepts of cyclic processes that maintain stability in natural systems are evident, in fact, in Banfield's final, posthumously published work, *Last Leaves from Dunk Island*, where he wrote that 'Nature's restorative operations are performed unceasingly, with never-failing design and often with the exhibition of wonderful power' (Banfield 1925:138). His approach, however, was largely intuitive and never fully thought through, remaining impressionistic to the end.

At the beginning of the *Confessions* he asserted that settling on Dunk was a deliberate choice: 'the idea of retiring to an island was not spontaneous. It was evolved from a sentimental regard for the welfare of bird and plant life' and to counter 'the offences which man commits against the laws of Nature [through] heartlessness and folly' in which he 'destroys birds for sport or in mere wantonness' (Banfield 1908a:92). As early as 1877, Queensland had passed the *Native Birds Protection Act* (41 Vic. No.7) in which the Preamble stated: 'Native Birds are disappearing rapidly from some of the Districts of this Colony and it is expedient [therefore] to protect them and their progeny'. Only the southern districts – Moreton, Darling Downs, Wide Bay, Burnett and Port Curtis – were specified. An amendment of 1884 (48 Vic. No.12) extended the provisions so that 'the term "lands" shall be construed to include any land covered by water or any waters within the territorial jurisdiction of Queensland' (Section 5). Section 6 of the amended Act provided for the appointment of rangers. Even so, as the attached schedule of species indicated, protection was not extended, *inter alia*, to raptors, sea birds or cassowaries. Moreover, distances across the colony were vast and enforcement difficult. Banfield decided, consequently, to create a sanctuary on his own initiative and planned 'that one of the first ordinances to be proclaimed [on Dunk] would be that forbidding interference with birds' (93).

He soon found himself involved in major conflicts of interest. For a start, simply building the hut, and later the house, involved clearing the land. Fencing a garden, fowl run and orchard increased the impact since he then had to confront the problem of goshawks swooping on the chickens, birds eating the millet and maize plots, flying foxes feeding on the fruit trees, pythons raiding the hens' nesting boxes for eggs, and sharks poaching from the fish traps that Tom had set up in the bay. Recognising from the outset that 'the Garden of Eden was not to be left entirely in its primitive state . . . it was firmly resolved that our interference should be considerate and slight' (41). Firm resolutions, however, are not always translated into practice: Banfield shot the unprotected goshawks and flying foxes and axed the pythons. Even more contradictory was his assertion that 'our sea-girt hermitage' would be 'a sanctuary for all manner of birds, save those of murderous and cannibalistic instincts'. Frigate birds, which he observed chasing terns to force them to surrender their prey caught in the sea, were described as 'tyrants of the upper air' while raptors, such as the falcon which 'swooped down on a wood swallow' and ripped it open, left him 'shocked at the audacity of the cannibal'. His response was 'A bullet [which] dropped the murderous bird with its dead victim fast in the talons' (61). Throughout all of his works runs the same selective view of nature with its sustained anthropocentric value judgments. He was, quite understandably, aware of the contradictions in which he found himself but, despite his allegiance to White and Thoreau, Banfield never explicitly appreciated that he too was a predator. On one occasion, in fact the only one it seems, he recorded that 'for this indulgence of my feelings I am aware, [that I have] laid myself open to blame' (Banfield 1911:177).

Despite his self-confessed failings, he effected considerable changes in the conservation of wildlife, chiefly birds, on Reef islands. One species in particular he championed: the nutmeg pigeon. Known today as the Torres Strait Pigeon, and more correctly as the Torresian (or Pied) Imperial-Pigeon (*Ducula bicolor*), it is a gentle, harmless bird which congregates in large numbers to nest and roost each night on Reef islands, and flies to the mainland each day in great flocks to feed in the rainforests. Banfield became incensed at the continuing senseless slaughter of these specifically protected birds which, due to some peculiar perversity in humans, were seen by so-called 'sportsmen' as targets for destruction. A stream of vituperative criticism ran through all of his writings for nearly twenty years directed at that 'wretched sport' in which shooters sought to kill as many birds as possible for nothing more than the thrill of destruction. 'It is not strange', he wrote, 'that men shoot 250 in an hour or so. The strange thing is that "men" boast of such butchery' (176).

In a personal campaign to prevent such practices on Dunk, widely publicised in his articles, quite early in his residence Banfield sought to have the island declared a bird sanctuary. He received a positive response from the Queensland Department of Agriculture and Stock. Dunk was gazetted as a protected reserve on 10 May 1905 and the notice was published in newspapers in the nearby coastal towns of Geraldton (later renamed Innisfail), Townsville, Ingham and Cairns as well as Charters Towers inland. Six weeks later on

24 June Banfield was appointed a wildlife ranger and provided with a number of large, printed calico posters, specifically provided for in Section 2 of the 1884 amended Act, for fixing to trees and other sites, warning of the prohibition (Noonan 1983:135f.). Dunk was nominally protected, but Banfield found himself continually having to patrol his domains since official prohibitions were rarely respected.

At a popular level, Banfield became a major advocate of conservation practices, advocating wildlife sanctuaries and the rights of nature. This was brought out explicitly in the first chapters of the *Confessions*, and progressively refined in his later writings especially after 1916 when he came into correspondence with Alec Chisholm, an active naturalist in a number of societies, and a journalist employed in the years 1915 to 1922 by the *Daily Mail* in Brisbane. Chisholm had become enthused by Banfield's writing from reading reviews of his books in the Sydney *Bulletin* and when in 1921 he met Banfield in person, struck up a friendship which lasted to the end of Banfield's life. After his death, Chisholm edited his final writings under the title *Last Leaves from Dunk Island* and meanwhile took up the cause of conservation in his *Daily Mail* column 'Nature Notes' from material provided by Banfield. He recorded in his autobiography that the stories coming from the beachcomber's pen were so compelling that one reviewer had written that 'Almost it makes one wish to go "a-Dunking"' (Chisholm 1969:336).

Banfield's conservation concerns were not focused solely on birds but upon a wide range of biota. In particular, he took up the cause of the dugong, a gentle mammal that grazes quietly on seagrass beds in shallow littoral regions – often known colloquially as a sea-cow – which was becoming seriously endangered by hunters. In Banfield's day it was still known as *Halicore australis*, thereby alluding to the mermaid legend. His article on 'The Mermaid of Today', the dugong, begins with the shock sentence 'The rapacity of the blacks is a rapidly diminishing factor in their extermination' (Banfield 1908a:162). Depredations by uncontrolled Aboriginal hunting, however, became exacerbated when the Japanese, in much greater numbers, hunted them commercially, and with more deadly technology, prompting Banfield to seek Chisholm's help in his columns to raise public awareness, stating in a letter to him (23 October 1922) that along with the pigeons there is another creature that urgently needed protection.

> There is one other favourite beast of mine that you might like to take under your shield – the dugong. Like the nutmegs the harmless creatures are fast being exterminated by the Japs who have a practical monopoly of the Great Barrier Reef. No sooner does a lugger anchor in the Bay, or anywhere in the neighbourhood than 3 or 4 dinghies will be chucked over-board for raid. As you know, dugong are shy and slow breeders and since they are perpetually harried in shallow waters they like the pigeons will soon be numbered among the extincts (cited Noonan 1983:209).

The reference to extinct pigeons was not to the Torres Strait bird, but to the American passenger pigeon (*Ectopistes migratorius*) that once flew in uncountable billions across the skies of the United States. Like the nutmeg, it was seen as a target for gun-happy hunters who literally shot it out of the skies into extinction in the same years that Banfield was living on Dunk. The last one died in an Ohio zoo on 1 September 1914, and a monument in Wisconsin gives its epitaph: 'This species became extinct through the avarice and thoughtlessness of man'. Banfield was only too painfully aware that the dugong could well follow the passenger pigeon.

He was also bitterly critical of the depredations of the so-called 'collectors' who shot tropic birds for their colourful skins for mounting, plundered their nests for egg collections, and stripped lagoon pools of corals for mounting in displays. While it was acceptable in the interests of science for some collecting to occur, and he instanced such reputable collectors as the eminent John Gould, he was hostile to the growing commercial trade of Reef collecting for personal gain. Writing in the *North Queensland Register* in 1912 he described 'the typical collector of the present day' as nothing 'but a mercenary individual trading as a man of science . . . a sordid trader to whom had been denied the bowels of compassion' (Banfield 1912; collected in Noonan 1989:144).

Banfield never stopped promoting the goals of conservation and the rights of nature throughout his life until on 1 June 1923, during a storm, he was stricken with internal pains which sent him to bed in severe pain. The next day he was dead from peritonitis. Bertha recorded the last moments in his diary: 'June 1. Ted very ill. Came down from the barn about 4 o'clock suffering agony'. Her next entry was tragically simple: 'June 2. Ted died about 12.45 pm today' (Noonan 1983:216).

On June 5 Bertha was able to hail a passing steamer, the *Innisfail,* whose crew built a simple coffin and buried him not far from his home, where his grave remains to this day. Over the grave a cairn of cemented local fieldstones was constructed with a plaque bearing the inscription: 'EDMUND JAMES BANFIELD (The Beachcomber) Born Liverpool England 4th. Sept. 1852 Died Dunk Island 2nd June 1923'. Beneath was inscribed his epitaph taken from Thoreau's famous words that had directed his life: 'If a man . . . hears a different drummer [let him] step to the music which he hears'.

Not only in his public campaigns for conservation but also in the world of science Banfield's work had a profound effect. Just sixteen weeks after his death, on 21 September 1923 a party of 19 delegates to the Second Pan-Pacific Science Congress held in Sydney three weeks earlier, including Governor Matthew Nathan, Charles Hedley and two professors of the University of Queensland who were both to be distinguished later for their contributions to Reef science – Henry Richards and Ernest Goddard – were taken on a post-conference tour of the Reef. One day of the sightseeing trip was given over to visit his grave as a mark of respect to one who 'did so much to enlighten the world to the glories of Queensland's coral strand'. The *Courier Mail* reported the occasion in respectful, yet vivid language, reminding readers that

> The beachcomber's confessions, in which he laid bare the secrets of his tropic islands, have reached every nature lover of our continent. When the Barrier Committee [of scientists] came to Dunk Island, Brammo Bay was glittering in the afternoon sun, and the glossy green of the forest spread itself in the cool shade. The feathery casuarina dropped in avenues along the beach, but the fair island was mourning for its master, the clever brain that pictured it so that all the world could see, will itself perceive no more. The pen has fallen from the hand that made Dunk the most famous of a thousand isles of the Queensland coast.

A deeply moving sentence recorded the final act of respect, and even reverence: 'Upon his grave was laid a handful of the wild flowers he loved so well, and the spirit of the "Beachcomber" walked with his friends under the palms' (*Courier Mail* Saturday 22 September 1923, p. 7).

Banfield always portrayed himself as an enthusiastic amateur for conservation practices. None the less, he was outstandingly successful in promoting the idea of wildlife sanctuaries: he was a pioneer in conservation of the Reef, and has the distinction of being the first person, in a paper submitted to the Royal Geographical Society of Australasia, Queensland on 'Dunk Island – Its General Characteristics', to project the dream of the Reef becoming

> a great insular national park . . . a park not to be improved by formal walks or set in order to straight lines or lopped and trimmed according to the principles of horticultural art, but just a wilderness – its primitive features preserved; its excesses unrestrained; its waywardness unapologised for. In such a wilderness the generations to come might wander, noting every detail – except in regard to original population – as it was in Cook's day and for centuries before (Banfield 1908b:63).

That was his legacy which came to be realised half a century later when the *Great Barrier Reef Marine Park Act* received royal assent on 20 June 1975.

CHAPTER 14

REEF RESEARCH AND CONTROVERSY: 1920–1930

THE PAN-PACIFIC UNION 1920

Following World War I a major new phase of Reef research occurred throughout the 1920s, stimulated by the efforts of Alexander Ford (1868–1945), a prominent newspaper publisher in Honolulu, who, in the same idealistic spirit that motivated President Wilson, dreamed of a fellowship of the Pacific nations, united in a common bond of 'friendly and commercial contact and relationship'. To that end he worked tirelessly to create a formal organisation to further his vision, which also sought to promote Hawaii as a centre of Pacific cultural and research activity. Ford's efforts were rewarded when in 1919 the government of the Territory of Hawaii, as it then was, incorporated the Pan-Pacific Union as a trusteeship of twenty-one nation members appointed by Pacific governments with a comprehensive charter 'to unite the races and countries in and about the Pacific in closer bonds of fellowship'. The central activity envisaged was promoting knowledge of their resources and opportunities by means of periodic conferences on a wide range of matters of common concern.

In those same years a separate movement had been initiated by William Morris Davis from Harvard University, one of the more accomplished of the foreigners invited to the British Association meeting in Adelaide and Melbourne in 1914. A world authority on coral reefs, Davis had taken the opportunity during his voyage to the meeting to survey numerous Pacific reefs between February and July of that year and went on to visit the Great Barrier Reef on a short post-conference trip (2–5 September) out of Cairns. On his return to the United States he proposed that more extensive coral reef work should be organised by all interested Pacific nations, on a cooperative basis. That proposal was pursued in 1919 when the American Association for the Advancement of Science set up a

Committee for the Exploration of the North Pacific Ocean and began negotiations with the Pan-Pacific Union concerning a possible conference.

Due to Ford's energetic lobbying the Hawaiian legislature in April 1919 appropriated funds for the Pan-Pacific Union to organise a 'congress' in either 1920 or 1921. In July of 1919 the dream of Ford, and the proposal of Davis, came together when the planning of the congress was undertaken by the Committee on Pacific Exploration of the American National Research Council. The first of the projected series would be held in Honolulu under the auspices of the Pan-Pacific Union, the main purpose being 'to outline the scientific problems of the Pacific Ocean region and suggest methods for their solution' (Foreword, Pan-Pacific Union *Resolutions* 1920).

Two major items were proposed for the agenda: 1) 'to take stock of our present knowledge' in anthropology, biology, geography, geology and related sciences bearing upon the Pacific, and, 2) to initiate 'desirable research programs to pursue the above topics', with an aim also 'to lay the foundation for a higher utilization of the economic resources of the Pacific' thereby acknowledging the 'commercial' aspect that had been required by the legislature (*PPSC Proceedings* 1920:v). In April 1920 invitations were sent to learned societies around the Pacific, including Australia.

The conference met in Hawaii from 2 to 20 August 1920, and out of a total of 65 participants, five Australians attended: Ernest Andrews, New South Wales government geologist; Charles Hedley of the Australian Museum; Leo Cotton, professor of geology at Sydney University; Henry Richards, professor of geology at the University of Queensland; and Carl Sussmilch, geologist and director of the Newcastle School of Technology. The program had five areas for discussion: race relations and anthropology in the Pacific; botany; fisheries; biological research stations; and desirable biological research for the future. The only Australian paper came from the irrepressible Charles Hedley, of which an abstract appeared in the *Proceedings* on a topic dear to his heart. Opening his talk in Section 4 on Biological Research Stations – in which other delegates described their already operating stations – Hedley apologised that 'There is not now a single marine zoological station in Australia'. He then described the collapse of the short-lived Miklouho–Maclay marine station in Sydney, and the problems of the pearlshell and bêche-de-mer fisheries which 'have especial need of investigation' which would be served by such a station (PPSC *Proceedings* 1920:243).

The Congress concluded with the passage of a number of resolutions of which two items were to have a major bearing on research into the Great Barrier Reef and to be pursued by Hedley, and a newcomer to the field, Henry Richards. In 'Part III. Biological Science' there was a stress on the need for a marine biological survey and the problem of depletion of resources. The conference indicated that 'the necessity for conservation of natural resources has become imperative, since, in the case of the Pacific Ocean, certain economic marine species have been exterminated and others are in peril of extinction or grave depletion. Measures for such conservation must be based on an exact knowledge of the life histories of marine organisms'. Land fauna were not ignored: the resolution also noted that in many places, due to human impact, many species were 'fast disappearing or likely to become extinct in the near future' (PPSC *Resolutions* 1920:31).

Part IV of the *Resolutions* dealt with geology and affirmed that, 'since it is in the interest of science and of value in the development of the natural resources of the different countries concerned' there should be a program of creating contour maps of all Pacific landforms, geological formations and mineral resources, the form of the ocean floor along with its submarine structures and sedimentary processes. An educational resolution was also passed, stressing the need to promote and train young men in science by means of fellowships and exchanges among institutions. Yet another significant outcome was the decision to hold the next conference in 1923 in Australia that was to become a direct stimulus to intensified Reef research.

A RESEARCH BODY FOUNDED: THE GREAT BARRIER REEF COMMITTEE 1922

Serious and sustained research on the Reef, directly stimulated by his attendance at the Congress, appeared to be gaining impetus with the pioneer efforts of Henry Richards and the foundation of the Great Barrier Reef Committee in 1922. Henry Caselli Richards (1884–1947) represented a new phase in the development of Australian science: a scientist from within the nation. Born in Victoria, he gained a science degree at the University of Melbourne in 1907. In 1911 he was appointed the first lecturer in geology in the newly founded University of Queensland and in 1919 its first professor of geology. An energetic and competent administrator as well as researcher, and member of numerous committees both within and outside the university, he laid the foundations for a sound department of geology, with a particular interest in the Reef.

Richards had been greatly enthused by the Pan-Pacific Scientific Conference and was deeply impressed by Davis whom he first met there; at the same time the papers presented at the Congress made him aware that Australia was lagging badly in coral reef research and that some form of coordinated effort was needed to catch up with other nations, especially the Americans. Following the suggestions of Davis, he began to explore ways of furthering research on the Reef and a year after he was appointed to the chair of geology an opportunity presented itself. Sir Matthew Nathan had arrived as the new State Governor on 3 December 1920, and, as was customary, accepted the presidency of the Queensland branch of the Royal Geographical Society of Australasia, Queensland (RGSAQ). Knowing that a high level patronage would be of great benefit, on 7 December 1921 Richards wrote to Nathan as president, outlining a program of Great Barrier Reef research to discover its 'proper structure and growth, its economic potential' and urging that 'we should do here what Mayor and Vaughan are doing in the Gulf of Mexico' (UQA S226/3).

In his capacity as president, Nathan circulated a memorandum to all members based on Richards' proposals. A week later, on 21 April 1922, with Nathan's encouragement, Richards read a paper to the Society entitled 'Problems of the Great Barrier Reef' which was published that year in the *Queensland Geographical Journal*. In that short but influential paper, Richards gave a comprehensive survey of the conflicting theories of coral reef research, followed by a very polemical section designed to stimulate interest in what he considered to be the 'problems'. Speaking as a geologist, the problems he identified, however, were mainly geological since 'the general condition of the Great Barrier Reef is not known. Is it in a static condition or one of elevation or of subsidence?' he asked, and then continued with a list of the still controversial theories proposed by Darwin and his opponents yet to be resolved (Richards 1922:51). His major suggestion was a complete geological survey and charting of the entire Reef.

Richards did not dwell exclusively on geological aspects: he also argued that the economic resources of the Reef should be considered carefully, especially since 'the exploitation of the economic wealth of the Great Barrier Reef has gone on and we stand idly by' (54). In fact, the final few pages stressed the economic aspects, relying to a considerable extent upon material gathered by Hedley, who had sent it to him in a letter. Conservation of natural resources had always been a prime concern for Hedley: in the 1910s, while seeking financial support for his various reef projects from the Advisory Council on Science and Industry, Hedley had continued to warn of the dangers from uncontrolled exploitation and in one particularly strong letter to Secretary Lightfoot on 18 August 1919, had reiterated his concern over the depredations of the Torres Strait pearl fisheries. To strengthen his argument he gave an analogy, writing that

> A rough comparison may be drawn between the shell and timber industries. At the commencement only the finest trees are cut. But as prices rise and facilities for transport improve, the timber-getter, if unchecked, would at last take every tree, and planting nothing, would complete the destruction of the forest. In both industries continuity can only be assured by control (CSIRO archives, Series 1: 68/186).

In his letter to Richards less than three years later, he continued the same conservation theme, which Richards read, verbatim, to the audience. Hedley began with the complaint that 'The Government regulations which control these fisheries allow them to be exploited by vagrant licensees down to a point at which exhaustion refuses the last profit'. Then with an implicit reference to the Japanese, Hedley went on to stress that 'A patriotic policy might aim to replace the present wandering and foreign population which subsists on our marine tropical products, by resident European fishermen', in turn supported by scientific cultivation through zoological research and legislative protection (Richards 1922:52). Richards then ended his speech by enlarging on the warnings of Hedley and, pointing out that in America scientific investigations were 'generously subsidised', asked rhetorically:

'What are we doing? Why do we not play our part? The exploitation of the economic wealth of the Great Barrier Reef by foreigners has gone on and we stand idly by. Is that right?' His final sentences gave a ringing call: the 'Royal Geographical Society is capable of making some definite move to point out the proper path! We have in this country men of training, energy and ability to carry out these desirable investigations. Let us see that facilities are provided for the work to be done' (54).

At the conclusion of that meeting a motion was passed to establish 'a committee to consider and report on how the investigation of the Great Barrier Reef on the lines indicated by Professor Richards can best be initiated and carried out'. Then followed a proposal to contact a wide range of scientific societies and relevant government departments throughout Australia, inviting them to 'appoint representatives to form a committee' of implementation, of which 'the nucleus might be located in Brisbane which is the nearest capital city to the Great Barrier Reef' and which would have 'more binding interests' (CSIRO archives Series 1:201/1).

With the papers from Richards and Andrews taking centre stage, the idea of furthering the Reef quest was greatly stimulated and Nathan immediately swung into action, writing on 14 June 1922 to George Knibbs, who had moved from his position as Commonwealth Statistician to become the first director of the Commonwealth Institute of Science and Industry (CISI), created by Act of Parliament in 1920. Nathan enclosed a copy of Richards' speech and outlined three areas of investigation arising from it which he recommended to Knibbs as worthy of CISI funding: a thorough charting of Reef geomorphology and a survey of island flora and fauna; a survey of economic resources, chiefly trochus, bêche-de-mer, pearlshell, sponges and turtleshell; and experiments on the growth of corals under varying conditions. He also included a list of 36 institutions which might conceivably be interested, including the CISI. Knibbs telephoned Percy Deane, Prime Minister Hughes' secretary, a week later, for advice and after further consultations wrote to Deane on 7 October 1922 to formally endorse the proposals of the Great Barrier Reef Committee and to state that he was prepared to represent the CISI and had so informed Richards. In concluding his letter to Deane he recommended that the 'matter be sympathetically regarded' by the Prime Minister, with the very curious final sentence 'I assume, of course, that the Barrier Reef is under the aegis of the Commonwealth Government' (CSIRO Archives, Series 1, item 201/1). On the contrary, fifty years later authority over the Reef was to become a matter of intense, often bitter, confrontation and negotiation.

Richards displayed enormous energy and contacted a number of authorities for cooperation, including the Navy, Queensland Marine Department, the Australian National Research Council, the Commonwealth Navigation Department, State museums, universities and scientific and naturalist organisations.

Of these Twenty-one institutions nominated representatives and the inaugural meeting of the preliminary committee was held in Brisbane on 12 September 1922. As a natural sequence of events, Nathan was elected chairman of the Great Barrier Reef Committee (GBRC) of the Royal Geographical Society of Australasia, Queensland (RGSAQ), and Richards the honorary secretary, since they were resident in Brisbane and had been the active promoters of the idea which Richards used to great advantage in furthering his geological interests. Both Andrews and Hedley from New South Wales became members. Two major needs had to be secured in the first days: finance, and specific research programs. Funding came initially from the Queensland State government which promised to match equally donations from other sources and with that encouragement the Committee was able to proceed.

In those same months, Nathan wrote to the Royal Geographical Society of London (RGSL) and the British Museum (Natural History) on the possibility of British cooperation. Museum director S. F. Harmer replied positively on 2 February 1923 stating that he had delegated Dr W. T. Calman, Deputy Keeper of Zoology, to liaise with the Committee. Three weeks later a reply came from Arthur Hinks, secretary of the RGSL, expressing strong interest and setting out a very detailed and impressive list of possible research areas: theories of reef formation, deep borings, intensive study of reef fauna and associated biota apart from corals, further pearlshell cultivation studies, conservation of green and hawksbill turtles, nesting habits of birds, and the recording of breeding and migration patterns of whales. Even the possibility of aerial photographic surveys was mentioned. Early in 1923 the sub-committees submitted recommendations for research which covered a wide range. In oceanography the need was expressed for studies of the chemistry of sea water, soundings, temperature, currents, tides and sea floor deposits. The sub-committee recognised, however, that those investigations were 'of such a magnitude that the Committee cannot hope to undertake [them all] but perhaps the Navy Department of the Federal Government might be induced to enter upon the work'. The coastal physiography group also brought forward an impressive list of desired studies, focused on the character and evolution of rivers and coastal valleys, the relation of the coastline and sand reefs to currents and prevailing winds, and preparation of a submarine contour map of the reef floor. The zoology sub-committee provided a very interesting list of desired projects separated into two areas: economic and pure. The former covered the traditional areas of fisheries, as well as the study of the life cycles of vertebrate and invertebrate fauna. The 'pure' area, as they designated it, was to be concerned with systematic, morphological and embryological studies, and a particularly advanced notion for the day, of the Reef as 'a living entity'. Botany was more modestly represented with recommendations for a study of the flora of continental and coral islands, along with algae and halimeda.

By far the longest and most comprehensive report came from Richards who sought the detailed geological investigation of all the unresolved issues of the past hundred years, including the general geological structure of reefs and their thickness; formation of atolls; true nature of erratic blocks; the vertical movement of lagoon and reef areas; the relation of reef openings to coastal rivers; and the general question of the origin of coral reefs. These were followed by a list of specific projects to deal with such issues, which included the urgent need to make borings 'at selected points on the reef proper and in the lagoon area ... [and that] the cores obtained should be carefully examined physically, microscopically and chemically' (Report of the Special Sub-committee to Consider the Programme of Investigations 1923).*

In April 1923 the combined suggestions of the sub-committees were consolidated into a three page paper by Richards, which set out the broad outlines of the 'chief functions' of the Committee which were twofold. Firstly, with a geological interest paramount, 'to inform the world why Queensland possesses along its coast line the greatest Barrier Reef in the world; what is the nature of the platform on which it is built, what is the nature and construction of the unique coral mass, how and when it developed'. Secondly, following equal pressure from numerous members for a biological emphasis, it noted that

> associated with the reef are many products of great economic value such as Pearl Shell, Trochus, Bêche-de-mer and Sponges, etc., and it is also the hope of the Committee to help towards a fuller knowledge of the development and growth of these products in order that the Commonwealth may utilise them in the most efficient and wealth-producing manner.

The most telling phrase in the paper is the short paragraph: 'The problems are great, the work will be long and funds will be needed' (Richards 1923: Au.Arch.). The Committee planned to use the volunteer input of the numerous scientists on the membership roll on their individual research and vacation projects, and so act as a coordinating clearing house and publications centre. At the same time it recognised that 'sustained and continuous efforts by men trained in physiography, geology and zoology are necessary in order that coral growths, the coastal islands, and the coast itself may be carefully investigated and measured and these structures understood'.

The fledgling society still had slender resources, depending heavily on unpaid labour. There was also the hope that essential hydrographic and oceanographic data would be supplied from naval surveys. At that stage the Australian navy was still in its formative years: in 1921 the Royal Australian Naval Surveying Service was founded, with HMAS *Geranium* as the first vessel so employed; to be followed in 1926 by HMAS *Moresby*, the first ship specifically fitted out for surveying, formerly HMS *Silvio* (Lack 1959:138). Until then, hydrographic work in waters of the Australia Station had been carried out by the British Royal Navy, which was still surveying, and to whom the Committee also turned for assistance. The agenda paper then concluded with a one page summary of the areas planned for investigation: physical and geological studies; suitable places for boring operations to examine the substrate; fauna, ecological and halimeda studies; oceanography; a bathymetric contour plan of the entire Reef region; life-history studies of pearl, trochus, and tortoise; other shelled animals; bêche-de-mer and sponges.

The proposals were then circulated for consideration by the numerous members of the

Committee throughout Australia and New Zealand. At that time the Commonwealth Lighthouse Service offered the GBRC two berths aboard its service ship the *Karuah* which was travelling from Cairns to Torres Strait. Richards promptly nominated himself and Hedley and they departed for a five week geological survey of the northern region of the Reef with the specific aim of studying 'the manner in which the foundation or platform, on which is built the Great Barrier Reef, achieved its present form and position'. Spending most of their time between Cairns and Torres Strait, they investigated eleven coral islands, five continental islands and two coastal headlands, from the Low Isles to Booby Island. While on Thursday Island they made the acquaintance of Commander Patrick Maxwell, captain of HM Survey Ship *Fantome*, from whom they sought information on the possibility of access to naval survey and oceanographic data. Later that year, on 3 October 1923, Maxwell sent a reply to Richards with the results of his inquiries, outlining the problems involved in supplying the information requested, chiefly that many of the projects concerning tidal and sea temperature data would be very costly due to the long periods of observation required, while survey findings were the property of the Admiralty who would have to be approached directly (Doc. A/72/4: Au.Arch.).

On their return Richards and Hedley prepared a report that was read to the GBRC on 13 July 1923, and published in the *Transactions* of the Royal Geographical Society of Australasia, Queensland, in which they reported that 'submergence on a grand scale has gone on' (Richards & Hedley 1923a:1). That inference, though, was only a calculated guess, and their final concluding statement made it clear that nothing definite was yet known about the formation of the Reef. They went on to hazard the theory 'that possibly the 2000 fathom trough of the Carpenter Deep has controlled the whole history of the Great Barrier Reef, and that pulsations undulating thence have thrust landwards, crushing back the coast and alternately raising and lowering its margin' (26). Richards was preparing the way for the next stage of geological investigation.

In addition to that account of the survey, Richards and Hedley prepared a second report from their observations for the GBRC, also read on 13 July 1923, and later published in greater detail in the *Queensland Geographical Journal* (No.3, 1923) which covered much the same material and brought out Hedley's biological and conservation concerns about the depletion of marine resources, especially in Torres Strait. Having consulted as widely as they could with men engaged in the pearlshell, bêche-de-mer and trochus industries, they reported that collecting grounds which had received respite for some years during the war had recovered quite strongly in some areas, as much as fourfold, indicating 'a natural restocking of grounds more or less worked out'. At the same time they counselled caution since the weight of opinion among those directly engaged was that more serious studies of natural processes should be undertaken. Yet again Hedley stressed the need for a marine biological station, to which Richards seemed to give some support. They reported that 'the question of establishing a Marine Biological Station naturally concerned us a great deal'. Various sites were examined and the Director of the Institute of Tropical Medicine in

Townsville offered assistance if one were founded there, but they concluded that 'since the material for investigation is much more accessible at Thursday Island [that] would be the best location' (Richards & Hedley 1923b:109).

For suggestions to the Committee on possible economic products which the GBRC could investigate, apart from the already heavily exploited pearlshells and bêche-de-mer, Richards wrote on 20 July 1923 to one of the members, Henry Tryon, who was also Queensland Government Entomologist. Tryon replied on 13 August with a detailed list of nine additional potential resources: phosphates (apart from guano); sponges; trochus shell; table oysters and ostreiculture; crayfish; certain species of reef corals for bleaching, painting and sale as curios; other kinds of decorative soft corals; sharks, chiefly for fins, oil and skin; and lime for cement manufacture (UQA S226 Box 3). The stage was being set for an all encompassing assault on what Tryon called the 'assets' of the Reef. Hedley had already stressed the need for what later generations would call 'sustainable yield' as evidenced in his letter to Gerald Lightfoot back in August 1919, and it seems that the warnings of the *Resolutions* of the First Pan-Pacific Conference on 'fast disappearing species' were not being taken seriously either.

By now the perennial and vexed issue of a marine biological station was receiving attention from the zoology sub-committee, and Tryon had been delegated the task of formulating a proposal. A week after his comments on economic products, and to follow up the biological aspect of Reef research, Tryon wrote to Richards on 18 August 1923 with a two page memorandum outlining the results of his investigations for such a station. His reply argued in favour of Townsville as the optimum location, listing ten points: central location on the coast; proximity to major passages in the Reef; 'easy accessibility to centres of learning (Universities)' (not identified); a major port of call; access to numerous valuable research islands; skilled artisans in the city who could service technical equipment; sufficient population, including tourists, who would make a proposed aquarium profitable; a proud civic minded community that would welcome and support such an institution; availability of a constant supply of fresh water; and an optimum location for economic exploitation of Reef resources (Tryon to Richards, 18 August 1928: Au.Arch.).

Richards did not warm to the idea. Moreover, his preference, which never seemed very strong, was that if a station were to eventuate, it should be on Thursday Island. By that time, however, he was uncomfortably aware that the total GBRC budget was only £5000 for the following five years, and forewarned that establishment costs of a station, based on overseas experience, would be around £10 000 with annual maintenance about half that, he made no effort to help realise the biologists' dream. Moreover, were he to so act, his

---

* From this chapter forward the narrative depends upon a large number of original documents in the possession of the author collected from various sources, in several cases from persons holding earlier documents, or from interviews with persons involved in recent decades. These are acknowledged as 'Au.Arch.'.

cherished goal of geological research, and sinking another bore on the Funafuti model, would possibly not eventuate. Richards was determined to keep geology as the prime concern of Reef science, and it appears that always in his mind was the hope of finding petroleum basins.

SECOND PAN-PACIFIC CONGRESS, MELBOURNE 1923

The opportunity for Richards to call for a full-scale study of the Reef, and push the case for a geological survey, came at the Second Pan-Pacific Congress (as the conferences were designated thereafter) which met in Melbourne for ten days between 13 and 22 August 1923, then moving on to Sydney for another ten days from 23 August to 3 September. There was a very large attendance from overseas, mainly from the United States and Japan, and, of course, a large number from Australia. Opening the Congress, Professor David Masson expressed the hope that it would help 'the public demand for science in Australia [to] grow' both in its fundamental and applied areas for the economic benefit of the whole community, and that Australian governments and sponsors would provide support comparable with America 'where the organization and endowment of scientific work are now on a scale that arouses universal admiration, not unmixed with envy' (PPC *Proceedings* 1923:I.18,25).

After the plenary sessions, the Congress separated into the eleven designated sections, from Agriculture to Zoology. Most of the members of the GBRC attended Section 7, Geology, which covered a wide range, with sub-sections on stratigraphy, palaeogeography, oil and water resources, and aerial surveys. A special sub-section entitled *Coral Reef Symposium* attracted eighteen papers altogether. Two came from Richards, his major one entitled simply 'The Great Barrier Reef of Australia' in which he reiterated his plea for more geological research into the still unresolved issue of the formation of the Reef, and a minor one on the need to improve the navigational survey of the Reef for commercial ships. In support Dr Yamasaki of Japan moved Resolution No. 2 of Section 7 which stated 'that special attention be called to the scientific and economic interest [in] the construction of detailed charts of the Great Barrier Reef of Australia' (PPC *Proceedings* 1923:I.37). Richards' main paper, however, was a plea for geological investigation that was largely a repetition of the material presented in his speech to the Royal Geographical Society of 1922, and the results of his joint survey with Hedley shortly before the Congress. Richards showed a good knowledge of the previous surveys, both throughout the nineteenth century and the first decades of the twentieth, and mentioned specifically the formation of the GBRC as an Australian committee 'representative of the Universities and of the scientific institutions and societies' to investigate the manifold problems so far identified (II.1104). Ever alert to the possibility of proselytising his audience, even in the short paper on a navigational survey of the Reef, he included the argument that 'a complete survey of the Barrier Reef would be of the most far-reaching significance in helping us to solve the scientific problems as to its origins, thickness, &c.' (I.707).

Determined to seize every possible promotional opportunity, Richards planned a post-conference cruise of the Reef for a select group of persons including Governor Nathan and five overseas scientists: four from the United States and the influential Cambridge geodesist, Colonel Sir Gerald Lenox-Conyngham. Included in the final group of 19 were three of Richard's close associates, Charles Hedley, Heber Longman (Director of the Queensland Museum), and the recently appointed (1922) professor of biology at the University of Queensland, Ernest James Goddard. Richards, through the influence of Nathan, was able to obtain free rail passes for the party from Brisbane to Mackay, and the services of the government steamer *Relief* for ten days, from 12 to 22 September. The party travelled to Mackay and then cruised north, first stopping at the Palm Island group, then as far as Cairns; returning south within the inner and middle regions they visited seven reefs and spent the last day on Lindeman Island in the Whitsundays, before returning to Mackay (UQA S226/2).

Richards was also aware of the value of press coverage and on the ship's return both Brisbane newspapers, *Courier Mail* and *Daily Mail*, were given lengthy 800 word press releases which they featured in full on the main page of each publication on Saturday, 22 September. The release was a glowing account of the cruise, excelling in florid hyperbole which each paper made individual by the use of their own banner headlines. In heavy block type the *Courier Mail* proclaimed: 'SCIENTISTS CHARMED'. Then in smaller capitals came the sub-banners: 'Beautiful Barrier Reef, Unforgettable Pictures, Tropic Forests under the Sea'. The *Daily Mail* went one better: 'AN ENCHANTED STAIRCASE. Glimpse of the Sea God's Hall. Brilliant Hued Forests of Coral. Work of the Barrier Reef Expedition'. Both accounts featured the pilgrimage to the grave of the recently deceased Reef conservationist Edmund Banfield on Dunk Island, bringing out the significance of the action where 'On Dunk Island they laid wild flowers on the grave of the "Beachcomber" (the late Mr Banfield)', in appreciation of the fact that he 'did so much to enlighten the world to the glories of Queensland's coral strand' (*Courier Mail* 22 September 1923).

The Reef was thereby brought vividly to public attention, and especially resonant to the pioneer State at the time would have been 'the prospect of garnering a rich harvest from the reef that will add to the material prospect of Queensland and the well-being of her people'. To maximise publicity Nathan gave a rousing speech to the Brisbane Rotary Club three days later which both newspapers also reported in lengthy detail. The *Daily Mail* headed its report under the headline '388 SHIPWRECKS – Great Barrier Reef – Governor's Arresting Facts'. Having alarmed his audience of influential business men with a lengthy recital of the years between 1891 and 1918 when '201 steamers and 187 sailing vessels were wrecked on the Great Barrier Reef . . . [with] 162 people being drowned', he moved on skilfully, pressing the priority of the claims of the GBRC, for more thorough surveys, and of course, research, not only geological but also biological. The obvious benefits would be improved economic productivity from 'an increased food harvest from the sea' along with other products such as pearls, tortoiseshell and sponges. Yet again the desire of the GBRC for a marine station came in for mention, as he pointed out that 'the location of [the] station and sub-stations has not yet been decided on, but the Murray and Darnley Islands in the north, the Palm Islands in the centre, and the Capricorn group in the south have been mentioned as centres of investigation' (*Daily Mail* 25 September 1923). The same day the *Courier Mail* reported Nathan's speech under the headlines 'QUEENS- LAND'S HERITAGE – Riches of the Barrier Reef – The need for keen research – Appeal by the Governor'. Richards had indeed acted wisely in seeking the patronage of the Royal Geographical Society of Queensland for his projected research program, and no better Reef evangelist could have been found than Matthew Nathan.

BORING MICHAELMAS REEF: GBRC RESEARCH AND CONTROVERSY, 1925–1927

At the end of 1923, in one of the most scurrilous episodes in the history of the Australian Museum, Hedley, despite his outstanding scientific record, had his employment terminated. In a dispute, nominally over retirement pensions in which Hedley had supported the disadvantaged staff, he aroused the ire of some of the more conservative members of the board, chiefly the government Auditor-General Frederick Coghlan. In a replay of the equally infamous episode of ejecting Gerard Krefft fifty years earlier, the board of the Museum thereupon resolved at a meeting on 2 November 1923 to force Hedley out. Its technique was to 'reclassify' the Museum structure and to effectively sideline Hedley who would thereby become redundant, with no specific duties. He could, consequently, be 'retired'. Coghlan had neither scientific qualifications nor knowledge; Hedley was the greatest invertebrate scientist the Museum had ever employed.

There is no doubt that Hedley was a thorn in the side of the trustees, in large part because he was a brilliant and energetic scientist pushing forward the frontiers of knowledge, and a dedicated conservationist. The Museum board, in contrast, was composed mostly of government appointees of decidedly authoritarian character whose main function was to exercise tight control over expenditure and the staff, which inevitably brought it into conflict with free spirits like Hedley, and the minority of qualified scientists on the board, at the time William Haswell, Edgeworth David and Ernest Andrews. The sacking of Hedley led to bitter debate within the boardroom. Taking the conservative side, and implicitly endorsing an intention to retain the display of 'horrible mounted impossibilities' so trenchantly criticised by Krefft half a century earlier, the trustees, led by the myopic Coghlan, stood firm. Hedley had to go. Hedley submitted his resignation immediately, while both Professors Haswell and Edgeworth David of Sydney University resigned from the board in protest and disgust (Strahan 1979:63–65).

That was not the end of the Museum's troubles: the generic problems arising from management by ignorant trustees and their determination to cling to the outmoded ideas of the previous century led to continuing conflict between government bureaucrats on the one hand, and scientific staff on the other. The Hedley episode was far from over: his sacking and the resignation of David, and Haswell (after thirty-three years of membership), continued to fester for the ensuing two years and developed into a *cause célèbre* which was reported with considerable verve in the Sydney press. The Sydney *Daily Telegraph* gave a detailed account of the turmoil for several days, noting that the government was considering a public inquiry into the whole issue of management, and was bringing the central issue to public awareness that the conflict was one of 'whether the Museum should be primarily for "show purposes"' or for 'education through scientific research' (8 August 1926).

In 1924 Hedley, actively recruited by Richards, left for Brisbane where he became Scientific Director of the GBRC on 1 April. Throughout 1924 he projected himself wholeheartedly into the work of the GBRC in close collaboration with Richards who was determined to steer the Committee's efforts in the direction of drilling; to repeat, as it were, and more successfully, the Funafuti projects and to either support or reject the Darwin hypothesis. On 4 November 1924 Richards and Hedley first raised the specific issue of planning a drilling project in a series of bores on the outer edge as well as a series down the middle of the Reef parallel to the coast. On 6 February 1925 the sub-committee considered a preliminary report which set out many of the considerations involved. Three locations were considered: Raine Island on the northern outer barrier, Green Island off the central coast of Cairns, and Masthead Islet in the southern Capricorn group. The last was preferred due to the proximity of Brisbane for supplies and possible technical support. In response to inquiries numerous potential difficulties were encountered: the various mining companies approached for estimates pointed out that conventional steam powered drills could not be used on cays without a copious water supply and large quantities of coal for the steam boilers; even when dismantled to their component parts for transport, drilling rigs were very heavy and could not be lightered easily on to the proposed sites, since engines weighed around 2.5 tons and the derricks, over 20 feet in length, were even heavier. Most prohibitively, diamonds for the drill bits were very costly and the GBRC would have to meet the cost of replacement.

At a meeting four months later, on 22 April 1925, Richards and Hedley, along with two other members of the geology sub-committee, the honorary treasurer W. M. L'Estrange (a geologist) and T. L. Jones, considered a discouraging report of 19 February 1925 from C. F. Jackson, Queensland State Mining Engineer, on even more complex technical aspects that had to be surmounted. Receiving endorsement to explore other possibilities, Richards then circulated a letter with possible bore sites to the Royal Australian Navy, all members of the Committee, and to major overseas reef scientists, including Wayland Vaughan and William Morris Davis in the United States and Stanley Gardiner in England. Vaughan recommended the Capricorn group as potentially the most promising while Gardiner, having been a member of the earlier Funafuti expedition of 1896, expressed some very serious reservations.

Gardiner was already well aware of Richard's drilling proposals from a report to the Royal Geographical Society of London by Lenox-Conyngham at a meeting held on 23 February 1925. Lenox-Conyngham had given an account of the post Pan-Pacific Congress cruise to the Reef, and after a very good description of the phenomena encountered, turned to the scientific aspects currently under question, especially from his own geological and geodetic viewpoint. He discussed at length the new concepts of isostasy and what he termed 'tectonic rocking' that were coming into the literature. The latter term (Gk *tekton:* 'carpenter', and by extension, 'construction') could well help explain the phenomena of elevation and subsidence and the 'problems' of Great Barrier Reef, therefore, would be an excellent site for further investigation. At the same meeting his Cambridge colleague Frank Potts described his own observations in the Torres Strait on the Mayor expedition of 1913 and urged that the proposed Reef boring project would benefit from biological studies as well (Lenox-Conyngham & Potts 1925: 314–34).

Attending the same meeting was Gardiner who spoke at length, chiefly from a negative point of view. A committed opponent of Darwin's reef formation theory, he made an extended criticism of the current state of geological theory, believing that neither Murray nor Daly had been taken sufficiently seriously, and stating that he doubted that efforts at boring would be of any real scientific value. Gardiner had little enthusiasm for the Reef as a natural phenomenon and expressed his opinion at the meeting that its age was 'within twenty to fifty thousand years' and it therefore did not need any further thought or investigation. Even further, he ventured the opinion that 'it is the greatest pity in the world that there is a Great Barrier Reef. Its existence is really a tragedy so far as the people of Queensland are concerned. It is a great nuisance to navigation. It is also a curse because it destroys 70,000 to 80,000 square miles of most admirable trawling ground' (333–34). Biological research, and its applied findings in resource exploitation was his preferred option. However, he warned Richards, if the drilling project were to proceed, then 'the boring will cost money [while] valuable physical and biological work will be equally inexpensive' (Richards: circular to GBRC 23 October 1925: Au.Arch.)

Richards then contacted the Institute of Mining Engineers for advice, as well as the Queensland and New South Wales Departments of Mines, and the Commonwealth Lighthouse Service in regard to shipping transport. Since all of the commercial estimates were beyond the means of the Committee other possibilities had to be sought. In the end, the Committee was offered free of charge by the Victorian Department of Mines in Melbourne a light, less costly, oil-powered (kerosene) machine, known as a Victoria Calyx drill, which had a hardened steel bit and was considered adequate for soft, crumbly coralline material, in distinction to the heavier equipment and diamond bits needed for penetrating the hard granites and basalts encountered on the mainland. The Department of Mines also stipulated that the Committee needed to pay the covering £500 insurance premium and the wages of its staff who would do the manual labour. Moreover, given the problems of a possibly unstable sub-structure, as Funafuti showed, bore casings would be necessary. In all, the projected cost was an enormous £2196 in addition to the insurance: more than half the available funds of the Committee. The dream of a series of bores down the outer edge and a parallel series down the mid-line had to be moderated to just one hole.

In a retrospective report to the Committee of 23 February 1927 Richards wrote that 'the purpose of the bore was to obtain quantitative information as to the thickness and nature of the coralline deposit, and further, to determine the character of the foundation on which the coralline mass was built' (Richards 1927:4). Having continued to consult widely with the navy and recognised experts overseas, it was decided to revise the proposal to drill on Green Island near Cairns, and move to Oyster Cay near Michaelmas Reef which could still be serviced from Cairns. Already well experienced at Funafuti – even if the timing of his report upset the British team – Hedley was to lead the field operations.

With all the apparent difficulties surmounted, the project moved into operation in April 1926 after the drilling equipment had arrived from Melbourne, and Hedley with foreman Thomas Hughes supervised its installation. Drilling commenced on 6 May, the technique being to pass the drill rod, to which was attached the core barrel with the bit on the end, down the casing, and by a rotary movement the drilling rod was turned 12 to 15 revolutions per minute. As material was loosened it rose inside the core barrel until filled, whereupon the rod was retrieved, the material removed, labelled and bagged, and entered in the drill log for date, depth in feet, sequence number of the sample, depth of the pull, diameter of the casing, and remarks on its general character. When the bit became blunt a new one was attached, the rod then lowered and drilling recommenced. It was necessarily a slow and tedious process, the logs showing that only short depths were drilled each day, and the core barrel removed – termed the 'pull' – once or twice a day. As the bore progressed, additional lengthening rods were fitted. An alternative technique also employed was percussive drilling, akin to piledriving for wharf construction, whereby the rod was raised and lowered around forty blows per minute, the loosened material being collected in the core barrel and held in place in a chip cup. If the material was very loose water had to be pumped down and the contents of the core barrel retrieved by a double-acting force pump, whereupon it was bagged and entered into the drilling log. Obviously, it was in a pulverised form and valuable fossils, vital for dating the stratigraphic column, were damaged, often destroyed.

On 13 June Richards visited the site and reported to the GBRC that most of the difficulties seemed to have been overcome although the results at the time were not so encouraging. In his report he recorded that the drilling had found 'more sand than anything else . . . and nothing in the nature of continuous solid rock has yet been obtained' (Richards to GBRC, 13 June 1926: Au.Arch.). In the first five weeks of operations, to 8 June, only 39 feet were drilled; by 12 June the drill and its casings had been driven down to 52 feet. Then came an encouraging improvement: Hedley cabled Richards from Cairns on 29 June with the news that 'Depth Saturday one hundred and fifteen feet stop Bored block symphyllis coral in which considerable solution and recrystalisation occur stop Sending specimen by Reid stop Consider this sign approach zone of consolidation stop Prospects hopeful' (Hedley to Richards: Au.Arch.).

Drilling continued until early August when a depth of 580 feet (177 m), the limit of the casings, was reached without finding bottom. The widespread press reports had created considerable public interest that led to alarm in Cairns when on 24 July 'a low, deep boom, reverberated through the town' and created panic that drilling was endangering the region. Hedley quickly moved to assuage the anxiety of the townsfolk that it was not related to the drilling but rather was simply due to 'slight movements in the earth's crust' (*Cairns Post* 27 July 1926). Within two weeks prospects had faded. A final 20 feet (6 m) were probed further without any change detected. The boring had to stop since no more funds were available for the additional casings required. The sequences recorded in the log kept by Hedley revealed a series of alternating layers of coral sand, below which were bands of mud around the 212 foot level after which the layers were variously 'foraminiferal sand ooze', 'consolidated foraminiferal sand' and 'grey green calcareous glauconitic ooze'. Alas, no firm coral strata, much less any sign of bedrock (Richards, 23 February 1927: Au.Arch.). Some years later, in January 1928, Richards gave a more detailed account of boring operations, and in 1942 an extended technical description, with illustrative drawings, of the drilling apparatus, and the scientific analysis of the cores were published in the *Reports* of the GBRC (Richards 1928:vii–xvi; Richards & Hill 1942:3–5,76–104).

The next step was to salvage the operation by proclaiming its positive aspects, which Richards did on 30 August 1926 at a meeting of the Royal Society of Queensland where he displayed an 18 inch length of core, 5 inches in diameter (460 × 130 mm) found at a depth of 12 feet (3.6 m). Reported in the Brisbane *Telegraph* newspaper, he put the best possible gloss on the disappointing result by commenting, rather fatuously, that 'boring to a greater depth would have proved of intense value as it would have been possible to ascertain the nature of the rock foundation'. Richards was forced to admit that much more work would have to be done 'before any weighty evidence could be adduced as to the correctness or otherwise of various theories which had been put forward concerning coral reef structure'. At the time, he pointed out, the various samples extracted were being distributed to experts for analysis: Frederick Chapman for the foraminifera and Dorothy Hill for the coral fragments. The molluscan debris was never analysed. All of the material indicated that it was of geologically recent origin, but did suggest that subsidence of up to 100 fathoms (180 m) had occurred (*Telegraph* 31 August 1926; Hill 1985a:4). The final cost of around £2950 had consumed nearly all of the GBRC funds.

The less than satisfactory outcome of the geological expedition – virtually a fiasco – and the exhaustion of funds made many in the Committee restive, especially the biologists who believed they had been ignored. Close examination of the surviving records reveals the single-minded intensity of Richards to make geology the primary focus of GBRC effort. It had been generally agreed that oceanography was clearly out of the question and that whatever data were available would have to be garnered from naval sources. Further, since it was a patent fact that marine stations were prohibitively expensive, Richards was able to persuade the Committee that biological contributions should take the form of individual research studies, which had to be, necessarily, of a desultory character. Moreover, to reassure all members, it was decided that research papers as they came in would be published in the *Transactions* of the Royal Geographical Society of Australasia, Queensland (RGSAQ), the official organ of the parent body.

When the first collection of scientific findings of the members was assembled, however, they appeared instead in a separate series designated *Reports of the Great Barrier Reef Committee,* published on 13 July 1925. Apart from a short five page paper on the 'Sea Birds of the Great Barrier Reef' by W. B. Alexander, all of the remaining 15 papers, running to 151 pages, were on geology and coastal physiography. As a consequence of the publication of that volume of reports, a major rift occurred between the parent body, the RGSAQ, and the Committee. The official GBRC historical record by Dorothy Hill, secretary from 1946 to 1955, is less than clear on whether the parent body authorised the publication of the papers, or the Committee took it upon itself to proceed without authorisation, but with the assumption that the RGSAQ would pay the publications costs without demur (Hill 1985a:4–5). Certainly the upshot was that, although the volume had been already distributed to members, the RGSAQ denied they had approved publication and refused to pay the bill of £345 for printing and distribution costs (UQA S226/1/1). There was a great deal of acrimony over the whole affair, and the doubt must remain that the parent body objected to paying for what seemed to be a journal of geological reports: nothing, from its point of view, had a geographical character. Eventually the Commonwealth Treasury came to the rescue with a special grant of £1000 in August 1926 for publication of GBRC reports. Concern over geology's dominance must have been reinforced from a RGSAQ perspective when, in the second volume of reports, published independently by the Committee in 1928, all of the nine papers printed, totalling 111 pages, were also on geology and coastal physiography.

Despite the Commonwealth grant the GBRC continued to press the RGSAQ to pay for the publication, arguing that, as its letterheads plainly showed, underneath the heading of 'Great Barrier Reef Committee' appeared the subscript, in parentheses: (Inaugurated by the Royal Geographical Society of Australasia, Queensland 1922). A bitter conflict was sustained for the ensuing years until a total impasse was reached in October 1928 when the Society informed the Committee that it 'declines emphatically to reopen the matter' of payment for the printer's account (UQA.S226/1/1) and in response, in an act of absolute exasperation – and perhaps not a little pique – both the State Governor Sir John Goodwin (who had replaced Nathan whose term had expired in September 1925) and Richards resigned from the RGSAQ. The press was delighted with the opportunity for sensational reporting: 'Governor Resigns: Differences in Local Branch of Famous Institution – History of the Rumpus' headlined the *Truth* on Sunday 7 October 1928, which was followed by similar accounts in the *Telegraph* on 9 October and in the *Daily Mail* on 10 October. Thereupon the GBRC totally divorced itself from the RGSAQ although, rather anomalously, it continued to call iself a 'Committee' for more than half a century until 1988 when on 1 January it adopted its present title of the Australian Coral Reef Society.

CHAPTER 15

THE LOW ISLES EXPEDITION, 1928–1929: PLANNING AND PREPARATION

ISSUE OF REEF BIOLOGICAL STUDIES, 1923–1927

Virtually no biological work appeared throughout 1923 and 1924. The years 1925 and 1926, in effect, were a low point in the development of Reef studies when a sequence of difficulties came together. On 25 September 1925 Nathan completed his term as Governor and retired to England as 'Patron' of the Committee where he worked on its behalf until his death in 1939. The Committee had now lost its most eloquent and influential advocate in Queensland. The ongoing fracas with the Royal Geographical Society of Australasia, Queensland (RGSAQ) was a major irritant that continued to destabilise the Committee, along with growing discontent that no significant biological work was in progress, especially in zoology which, apart from the work of Saville-Kent and Hedley, had still not developed any vigour. Then, on 14 September 1926 came the distressing news that Charles Hedley had died soon after he had returned to Sydney to farewell his wife Harriett and pack for the imminent Third Pan-Pacific Scientific Congress to be held in Tokyo. A lifelong asthmatic who had successfully found relief from the rigours of his native Yorkshire in tropic regions, he had contracted a chest complaint that exacerbated a heart condition to which he succumbed quickly.

His public image as a great Reef scientist was so well projected that in mid 1925, a year before his death, he was eulogised in a press report as one who 'knows the beauty of the Barrier Reef as a lover the dimples on his lady's cheek'. Then, in a passage referring to his prodigious intellectual energy and achievement, it continued, with a hint of foreboding, that 'his life will be deck-high with knowledge of the world and its beauty by the time he is ripe for the Ferryman's boat' (unidentified newspaper clipping, AM Archives, AMS 272/2/14). The allusion to the ancient Greek myth that after death noble souls are ferried by the god Charon across the River Cocytus to live for eternity in Elysium – assuming it has celestial reefs and molluscs – seems most appropriate. Cremated with Anglican rites, his closest friend Ernest Andrews, in company with several other Australian delegates to the Tokyo Congress, carried his ashes to Queensland where they were cast upon his beloved Reef waters. So slipped from the scene one of the early giants of coral reef science: past president of the major Australian societies – Linnean, Royal, Royal Zoological, Naturalists' – and Clarke Medallist of the Royal Society of New South Wales, then the nation's highest scientific accolade.

Suddenly, the GBRC was adrift. On Nathan's retirement Richards had succeeded to the presidency, which he held for the ensuing twenty-one years, with a medical practitioner, Dr E. O. Marks, as secretary for the same period. And this raises a major problem in historical reconstruction of those years since virtually all the correspondence passed across Richard's desk, and the bulk of the surviving material consists of those papers he chose to retain, which are now preserved largely in the GBRC archives of the University of Queensland. Very little other material from Richards is held in the general records of the university and apparently nothing that sheds further light on the GBRC.

It is clear, however, from Richards' surviving papers that he was deeply troubled, on the one hand, by the failure of the Oyster Cay boring expedition in July 1926 which had exhausted most of the Committee's funds, and, on the other, from pressure to balance geological research by strengthening biological studies. On 23 February 1927 he distributed a mimeographed 'Account of the Scientific Investigations of the Great Barrier Reef Committee, April 1922 – February 1927' which was little more than a sustained effort at self-justification. The almost total concentration on geology, and the fact that 'little has yet been achieved in the way of biological results' he wrote, was 'due to the fact that the Committee has been unable to obtain the help to any extent of any marine biologist except Mr Hedley, who during the 1924 and 1925 seasons was unable to organise any systematic zoological work owing to his many other duties . . . supervising the boring operation and looking after the interests of the Committee at Cairns and Michaelmas Reef' (Au.Arch.).

At that time, however, the faintest beginnings of biological work were beginning to gestate under the stimulation of Ernest James Goddard (1883–1948) who had arrived at the University of Queensland in 1923. Like Richards, one of the growing generation of Australian born scientists, in his case from Newcastle, Goddard had graduated with Honours in zoology and palaeontology from the University of Sydney in 1906, where he was immediately employed as a demonstrator. After a time in South Africa as professor of biology in Victoria College (later the University of Stellenbosch), he returned to Australia in 1923 to the chair of biology at the University of Queensland in 1923 where the bulk of his duties were necessarily in agricultural biology, given the heavy dependence of Queensland on primary industry. That aspect of his multifarious teaching, research and committee duties always occupied most of his time. Even so, as a close associate of Richards, he was soon inducted into the joys of Reef research and made his first foray in the May vacation of 1925 when he led a two week student and staff field biology expedition to Palm Island. From then on Goddard became increasingly enamoured of Reef research and in his final years was one of its most vigorous advocates.

The imbalance between the geological and biological study of the Reef had developed into a serious issue by 1926, becoming even more focused now that Goddard had evinced a strong interest. As early as 1923 he had made an unsuccessful submission to establish a university marine station either on Thursday Island or else at Townsville, although the University Senate later approved it in principle at meetings in 1925 and 1926 (UQA Ac.78/2, Hill Papers). Even so, difficulties remained and the following year Goddard promoted an alternative plan to purchase Dunk Island for £2500 for a marine station, that was reported enthusiastically on 7 June 1926 by the Sydney *Daily Telegraph*. His formal submission to the Senate of the University of Queensland was not endorsed, but he was encouraged to proceed if he could raise the necessary funds himself (Thomis 1985:169).

Goddard's proposals were opposed in GBRC meetings by Richards since he believed that they would consume most, if not all, of the available funds, which he had already decided should go to his own geological projects. In fact, Richards acknowledged as much in a letter to biologist Frank Potts of Cambridge, writing that 'I have been opposed to using our money in setting up a marine biological station at Thursday Island or Townsville for I feel nothing much would be done to help us with the problems of the reef which we set out to solve'. At the same time, Richards knew that greater effort by the GBRC had to be directed towards the biological side of Reef research, especially given the strong interest of the government in economic exploitation of the region. The Oyster Cay expedition had necessarily taken up most of Richards' time, as well as that of Hedley. When it was finished – only in the sense of failing to reach finality on the Darwinian theory, and having exhausted all the Committee's available funds – Richards had to face demands for more biological research. In a somewhat mean spirit, he even blamed the deceased Hedley for lack of progress in that area: 'I urged repeatedly on Hedley' he wrote, in the same letter to Potts, 'that he should draw up a plan [in biological research] . . . to get things in train for carrying on this work on Michaelmas Reef . . . while the boring was going on. For various reasons, I am sorry it was not forthcoming' (Richards to Potts, 16 September 1926: Au.Arch.). The 'various reasons', as he very well knew, and acknowledged in another context, were, as already noted, because Richards had burdened Hedley with most of the organising and fieldwork at the Oyster Cay boring site.

FIRST MAJOR BIOLOGICAL STUDY PLANNED: POTTS EXPEDITION OF 1927

Richards' strategy then was to approach Australian universities with a request for graduate students in biology to spend periods of time, up to a year, on individual research projects, preferably on some kind of outside scholarship, possibly assisted with a small stipend. The major problem was the lack of marine scientists. A decade earlier, when Hamlyn-Harris retired from the Queensland Museum in 1917, Lightfoot had written to Hedley asking whether he could nominate a Queensland replacement on the Tropical Marine Economics Committee of the Australian Council for Science and Industry (ACSI) since 'so far as the Committee is aware, [there is] no other marine biologist in Queensland' (5 October 1917). The best Hedley could suggest was Professor William Dakin at the University of Western Australia who could conduct research on the north-west shelf (AM Archives, AMS:272/1/10). Nothing had changed and marine biology stagnated: neither the universities of Melbourne nor of Sydney was able to nominate a young graduate since 'very few, if any, young marine zoologists [were] available to carry out the desired work on the reef' (UQA S226/1/1).

The other possible source for biological research was England, especially Cambridge, where Professor John Stanley Gardiner (1872–1946) was the dominant figure throughout England as head of the Zoological Laboratory at Cambridge and Director of Scientific Investigations of the British Ministry of Fisheries. Gardiner already had extensive experience in coral reef research in the Maldives, Laccadives and Seychelles, had edited the *Fauna and Geography of the Maldive and Laccadive Archipelagoes* (1902–1906) and was preparing his major study *Coral Reefs and Atolls* which appeared in 1931. He had, moreover, been on the Funafuti expedition of 1896 where he worked with Charles Hedley. Even more significantly, as events happened, Gardiner was a strong opponent of Darwin, a position he held throughout his academic life, and in all of his publications opposed Darwin at every turn.

So Richards contacted Matthew Nathan, who was still actively continuing to address learned societies in Britain on scientific problems of the Great Barrier Reef. In July 1926 Nathan attended the Third Empire Universities Congress, as representative of the University of Queensland, where he discussed Richards' request with Gardiner. In reply, Gardiner suggested his Cambridge colleague Frank Armitage Potts, Fellow of Trinity Hall, who had gained considerable reef experience in Torres Strait and Samoa on the Mayor expeditions, and in Florida. Nathan immediately contacted Potts who on 27 July 1926 sent Nathan a four-page letter of acceptance in principle with an outline of the kind of research he would like to undertake (CSIRO Archives, Series 1:201/8).

Potts acted rapidly. His proposal was an ecological study of the kind pioneered by Mayor, centred on 'an intensive investigation of a small area of nib reef to find out the relations between the various elements of the fauna and the different factors in the environment like temperature, salinity and amount of sediment in the water'. In addition, a study of the plankton in Reef waters would be an essential component, since, 'very little is known of tropical plankton and its distribution', believed to be the main food source of the polyps. For that aspect he suggested the appointment of Frederick Russell of the Plymouth Marine Biological Station. A third area he proposed was the study of crustacea, for which he recommended Professor Cannon at the University of Leeds. In addition, he mentioned that he had also 'tentatively approached two English biologists and mentioned the possibility of a cooperative investigation'. Even further, if funds were to permit, and the Committee willing, he would like to invite Professor Longley of Croucher College, Baltimore, 'whose knowledge of the biology of coral reef fishes is unrivalled' who could also, by means of his 'under-water photographic' methods, produce an invaluable series of records (Potts to Nathan, 27 July 1926: Au.Arch.).

The prospects seemed promising and once Nathan had relayed the tentative acceptance to Gardiner, planning for a serious biological expedition began. Utilising the old-boy network, Nathan wrote to Australian Prime Minister Stanley Bruce who had been a fellow student with Gardiner at Trinity Hall, informing him of the planned enterprise and soliciting Commonwealth funding. The request came at an opportune moment. In 1926 the Commonwealth Government, by passage of the *Institute of Science and Industry Act* (No. 20 of 1926), had reconstituted the CISI as the Council for Scientific and Industrial Research (CSIR) with an endowment of £500 000. George Julius was appointed chairman and David Rivett the chief executive officer while Gerald Lightfoot remained secretary. Most importantly for Reef research, both Richards and Goddard were appointed to the CSIR council. Richards immediately sought the assistance of CSIR for the new expedition, receiving encouragement from Julius who delegated Rivett to liaise with the GBRC and the British organising committee.

At the same time Gardiner wrote on 30 July to Richards recommending Potts as highly competent in marine zoology and 'a very charming person with plenty of tact and easy to get on with' (UQA S226/2/1/). Richards discussed the letter with the GBRC management sub-committee which approved a subvention of £1000, and then sent a cablegram to Potts, 'Invitation extended our maximum obligation one thousand', following it with a written reply to Potts on 16 September reminding him that they had met briefly at the First Pan-Pacific Congress in Honolulu. He then asked whether Potts would like him to contact Russell and Cannon, and if so, it would be done 'in as effective a way as we can devise' (Richards to Potts, 16 September 1926: Au.Arch.).

On 2 April 1927 *The Times* of London, under the column 'Imperial and Foreign News', published a lengthy account with the heading 'GREAT BARRIER REEF. MARINE BIOLOGISTS' EXPEDITION'. Including a line map of the eastern half of the continent showing the Reef, the opening paragraphs pointed out that most marine biology throughout the Empire to date had been of small, localised studies. Valuable though they had been, there was still nowhere that tropical studies could be conducted in a 'properly equipped laboratory open all the year round for the study of marine life'. That would change later that year, under the Cambridge expedition, 'to be led by Mr F. A. Potts, Fellow of Trinity Hall and Lecturer in Zoology in the University' accompanied by several experienced marine biologists, which would make 'a preliminary survey of the vast problems to be attacked'. Their scientific findings, it was hoped, would be 'exact and valuable and help to provide secure foundations for so much more continuous investigations into the marine riches of the Commonwealth'. At last a serious, sustained biological study of the Reef was about to begin.

THIRD PAN-PACIFIC SCIENCE CONGRESS, TOKYO 1926

In the same months that preparations were being made for the Potts Expedition the Third Pan-Pacific Science Congress was meeting in Tokyo in October and November 1926, with representation from eight nations; alphabetically: Australia, France, Great Britain, Japan, the Netherlands (for the Dutch East Indies), New Zealand, the Philippines, and the United States. Among the Australian representatives were Andrews and Griffith Taylor as well as members of the Australian Research Council, itself drawn from the senior membership of the Australasian Association for the Advancement of Science. In the final plenary session on 11 November three major resolutions were passed that were to have a direct and very influential bearing on Great Barrier Reef research and the expedition proposed for 1927. The most significant resolution was the seventh which read, in full:

> Whereas, Coral reefs are symbiotic entities whose origin and growth relations have received too little attention, and whereas Coral reefs differ widely and the methods of investigation are complicated and costly;
>
> Be it resolved, That this Congress institute a Committee consisting of biologists, oceanographers, and geologists to consider and draw up a plan for comprehensive investigation of the coral reefs of the Pacific Ocean (Pan-Pacific Science Congress 1928, *Proceedings*, s.v. 'Resolutions').

Thomas Wayland Vaughan, of the Scripps Institute of Oceanography in California, was elected chairman of the planning committee with the task of implementing the resolution. Two other resolutions, the fourth and fifth, set out his brief: to create three subcommittees from each of the participating nations on 1) Physical and Chemical Oceanography; 2) Fundamental Marine Biology; and 3) Fisheries Technology; and to ensure that standardised research methods were adopted with a maximum interchange of data. What is most interesting is the comment on symbiosis which was to become a significant aspect of research into polyp growth in the coming years.

Vaughan began organising as soon as he returned to the Institute. The American subcommittees included some of the world's leading coral reef scientists, chiefly Vaughan himself, along with William Morris Davis, Reginald Daly and Edward Hoffmeister. The British were led by Stanley Gardiner and Lenox-Conyngham, while of all eight nations Australia had the largest membership of 38 persons, of whom the major players were Matthew Nathan, and all of the GBRC executive, as well as Ernest Goddard and Wood-Jones of Adelaide. Henry Richards became chairman of the Australian marine biology subcommittee and Ernest Andrews chaired geology. No fisheries subcommittee eventuated, mainly because fisheries were already operating actively under the aegis of the Queensland government.

Early in 1927, while the Congress resolutions were being considered, the GBRC continued to organise the Potts Expedition. The site chosen was the Low Isles, a small vegetated coral cay – so named in the plural because it has an adjoining semi-submerged crescent shaped lagoonal cay then covered with almost impenetrable mangroves (*Rhizophora* and *Avicenna* spp.). Logged by Cook, and first investigated by MacGillivray and Huxley on the voyage of the *Rattlesnake* in 1848, it was reasonably close to a convenient service centre at Cairns some 65 kilometres to the south-west. Richards had negotiated with the Queensland Government for free rail transport for the scientists from Brisbane to Cairns, then by the Commonwealth lighthouse ships *Cape York* and *Cape Leeuwin* to the Low Isles, since a prefabricated lighthouse along with three keepers' cottages had been erected in 1878 on the main cay which required fortnightly provisioning. So far, in addition to those free services, cash funding consisted of the £1000 offered by the GBRC and £450 from the Royal Society of London.

By that time, however, the full impact of the resolutions of the Tokyo Congress was beginning to sink in, which had an immediate effect on Gardiner who had begun to have serious reservations about the competence of Potts to carry out what he now believed to be a poorly planned project. Almost at a stroke, that proposed expedition was doomed, especially since all researchers were required to present their findings to the Fourth Pan-Pacific Science Congress planned for Java in 1929. If Britain and Australia were to be involved in what was to be the next major coral reef expedition in the Pacific region – which all Pacific nations would be watching closely – any inadequacies or incompetence would be disastrous for British and Australian reputations, especially considering the painfully unsuccessful outcomes of both Funafuti and Oyster Cay. Gardiner moved – very diplomatically, but none the less, ruthlessly – to abort the enterprise. Moreover, Potts himself had made an early indication of a less than wholehearted commitment. At that time he had become engaged to a widow with two sons: his impending marriage and the care of two future stepsons weighed on his mind. He wrote to the GBRC from Cambridge on 24 November 1926 that he doubted his ability to commit more than six months to the fieldwork, and that he was worried about having to enrol the two lads in a boarding school while he was away. The best he could offer was the prospect that his research assistant could most likely spend a full year (UQA S226: Box 78/1).

Gardiner now had two tasks ahead of him: the first was to terminate the planned Potts expedition; the second, to plan for a more comprehensive venture to encompass the grander vision of Pan-Pacific resolution seven. The first was probably effected surreptitiously through the informal personal contact and telephone network. Suddenly, Professor Cannon of Leeds found that he was unable to commit himself after all, while the Marine Biological Association discovered that Frederick Russell could not be spared from his essential duties at the Plymouth Marine Biological Station. That indicated covertly to the Percy Sladen Fund (one of the potential donors), but none the less very clearly, disapproval of the project (Morton 1992:387). In order to enlarge both the scope and therefore the personnel of the expedition, Gardiner had to seek significantly increased funding and also, most importantly, find a new and competent leader.

In a letter of 24 September 1927 Gardiner, having been contacted by Vaughan, wrote to Rivett at the CISI advising that 'The Pan-Pacific Congress invites us to serve as a British Committee for the purposes enumerated on the enclosed sheet' which set out a detailed program of research to be undertaken. Then, in a masterpiece of diplomatic periphrasis, Gardiner indicated that 'there may have to be considerable reconstruction of the expedition both as regards the scientific objects to be sought and also with regard to questions of the technique of the attack'. Funding secured so far in Britain, Gardiner advised, consisted of the Royal Society of London grant of £450 and the £200 from the British Association (BA) which both bodies had agreed to have transferred to the new project (CSIRO Archives, Series 9: G12/[1]). By 1 November 1927, in addition to the original GBRC £1000, and the Royal Society and BA donations, an additional £600 had been raised from individuals in England along with £100 from the Zoological Society of London. Ironically – at least for Potts – the Percy Sladen Fund was not one of the contributors.

Even so, much more was needed and the most promising source of major support seemed to be from the Empire Marketing Board. Formed in Britain in 1926 with the express purpose of keeping trade within the Empire, it had a research budget of £1 000 000. Yet again Richards raised the prospect of economic benefits from the Reef as a major outcome of scientific research, and Rivett requested the CSIR liaison officer in Australia House, London, to seek assistance from the Board. On 3 March 1928 a coded cablegram was received in the Prime Minister's Department (then in Melbourne) from the Secretary of State for Dominion Affairs in London with the remarkably good news that the 'Empire Marketing Board [will] contribute £2,500 provided a like sum is made available by Commonwealth and Queensland Governments'. While recognising the primary objective of scientific research, the letter indicated an expectation of a significant British potential benefit: the 'Board's contribution' it stated, 'is designed firstly to assist the fundamental research of the British Marine Biological Laboratory in the interests of British fisheries investigations'. The cable also indicated that the funding was also being provided for the 'economic exploitation of industries of the reef' and would therefore be a 'wise investment for a country seeking scientific and economic development of its fisheries resources' and that 'the economic exploitation of the reef industries primarily of interest to Australia would justify [a] contribution of £2,500 from public funds in the Commonwealth' (CSIRO Archives, Series 9: G12/[1]). Richards pointed out to CSIR that the Queensland Government had already been more than supportive by providing travel facilities for Reef scientists, along with considerable funding amounting to £3000 on an equal contribution basis (Hill 1985a:6). The upshot was, through the agency of the CSIR, the Commonwealth agreed to match the Empire Marketing Board grant and, in addition, to waive import duty on the scientific equipment brought in, provided it was re-exported to Britain at the end of the project. A second expedition for a major study of the Reef was now underway and planned for a full year from July 1928 to July 1929. Over the following eight months the project continued to raise hopes for an important contribution to knowledge of coral reef biology, while the quest for funds had been quite successful: altogether £8580 had been raised which matched exactly the projected budget.

EXPEDITION REORGANISED: YONGE AS LEADER

Early in 1927 a possible alternative leader had come to Gardiner's attention: a promising graduate student at the Plymouth Marine Biological Laboratory named Charles Maurice Yonge (1899–1986). A science graduate from Edinburgh in 1922, Yonge (pronounced Young), had continued straight on – unusual in those days – to the degree of Doctor of Philosophy (PhD) on the physiology of digestion in marine invertebrates. That degree completed, for which he had been awarded a Carnegie graduate scholarship, he continued research on aspects of the physiology of the edible oyster (*Ostrea edulis*) which he completed successfully, and was awarded the higher degree of Doctor of Science (DSc) from the University of Edinburgh on 1 July 1927 (Morton 1992:384–86).

While still working on that second doctorate at Plymouth he had been approached by Frank Potts on 24 February 1927 on the possibility of joining the latter's expedition, but declined. Gardiner, learning through the scientific network of Yonge as a promising leader, then wrote to him and hinted at the possibility that he could be offered what at the time was a magnificent research grant: a Balfour Studentship of Cambridge University. Created in 1882 to commemorate the zoological achievements of Francis Balfour, who had died that year, it provided a handsome stipend for a beginning zoologist to prosecute research, within limits advised by the fund, of his own choosing for three years. The most prestigious award of its kind in Britain, it had earlier been held by Frank Potts himself. Gardiner's intention was to nominate Yonge for the award that could be pursued on the Reef project. After considerable negotiation, described in fascinating detail by Yonge's biographer, Brian Morton, it was only after Yonge had accepted the position – as Morton concludes from Yonge's own correspondence (Harper & Powell 1991), supported by innuendoes in the GBRC archives – that Gardiner then moved and 'completely killed' the Potts expedition. Gardiner was an arrogant administrator and in that episode 'one senses a none-too-subtle conspiracy against Potts as the leader of the expedition from all concerned . . . [that] just had to be sorted out in truly British diplomatic style' (Morton 1992:387–88).

The way was cleared for a change of direction when the GBRC received a short letter from Potts on 5 May 1927 regretting his inability to begin the project as planned – then only two months away – but indicating that he would be willing to continue to have an active role in Cambridge organising the new expedition. Following advice from Gardiner on 9 July 1927, the GBRC approved Yonge as leader of the revised expedition the following month on 4 August 1927. The Balfour Fund managers had attached a condition to Yonge's studentship: it was 'to carry out research on "the Feeding and Digestion of Invertebrates", especially, if possible, with reference to corals and other reef organisms' (Yonge 1930a:2).

By that time the BA had also set up a committee on coral reef exploration. To avoid duplication, on 24 July 1927 the English Barrier Reef committee and the BA committee merged into one planning body to organise the scientific program, enlarge the membership, and intensify efforts for increased funding. Over the few remaining months, at the beginning of 1928 planning for the expedition was intensified by Yonge in close collaboration with Frederick Russell of Plymouth, and a team of ten members – six men and four women – came together from Edinburgh, Cambridge, Plymouth, London and Millport (in Scotland), who were to leave England in the vanguard, to be joined later by others in the course of the year, enabling some of the first group then to return to Britain. A year previously Yonge had married Martha Jane Lennox (known as Mattie), a recent medical graduate from Edinburgh, who travelled with the team, ostensibly as medical officer to the party. Frederick Russell, now miraculously made available from Plymouth and appointed deputy leader, was accompanied by his wife Gweneth who, not being a scientist, was delegated general management and household duties, and T. A. Stephenson the zoologist, was accompanied by his wife Anne who assisted him as 'honorary zoologist'. The other three men – chemist and hydrographer A. P. Orr, zoologist G. W. Otter, and botanist Geoffrey Tandy – were unaccompanied. Sheina Marshall, a zoologist whose areas of research focused on phytoplankton and chemistry of the sea also sailed with the initial group; later, in March 1929 Sidnie Manton, an invertebrate zoologist who did meticulous work on the ecology of reefs, arrived to work alongside Stephenson in the shore party.

The British team was supplemented by several Australians for varying periods, mainly Frank Moorhouse, a science graduate from the Central Technical College in Brisbane, seconded from the Queensland government, who remained for the full year and then stayed on after. Five staff from the Australian Museum also stayed for shorter periods: conchologist Tom Iredale and ichthyologist Gilbert Whitley, as well as two young scientific cadets, William Boardman and Arthur Livingstone, and zoologists' clerk Frank McNeill. For general duties, arrangements were made with the Anglican mission at Yarrabah on the southern headland of Cairns Harbour, for an Aboriginal couple, Andy Dabah and his wife

Gracie, to join them as handyman and cook respectively, later to be rotated by Claude and Minnie Connolly. To crew the research vessel *Luana*, Harry Mossman and Paul Sexton, who were in their early twenties and also from Yarrabah, were engaged for the entire year. On various other occasions, as the need arose, other Aboriginal men were employed for various labouring jobs, chiefly on the boats. As Yonge wrote in his report to the GBRC 'all the aboriginal servants did excellent service to the expedition, and by their cheapness and efficiency contributed very materially to its success' (Yonge 1931:8). In a mimeographed progress report of 17 August 1928 Yonge recorded that the Aborigines 'are giving complete satisfaction, are easy to deal with, and their wages are only 22/6 (£1.2.6) per month with free tobacco' (Yonge:LIE.I.8/28:1).[*] In the same period the basic wage for white male workers was around £20 per month, approximately eighteen times greater – females receiving less.

Unlike most previous Reef research, the Low Isles Expedition was characterised by meticulous and comprehensive planning from the start, probably with the idea that the expedition's quarters could remain as the first purpose built marine biological station on the Reef. Prefabricated buildings were designed, delivered and erected on the site in advance by a volunteer Queensland naturalist, J. E. Young, assisted by labourers; these consisted of six timber huts with corrugated iron roofing and galvanised rainwater collecting tanks. The largest hut, 49 feet in length (15 m), was divided into individual bedrooms for the married members and one for the single women (both staff and visitors); the other men were accommodated in a second hut 40 feet (12 m) in length divided into two rooms, each with five bunks. The Aborigines had a smaller hut. The laboratory, which measured 35 by 18.5 feet (10.6 by 5.6 m) with a dining area at one end, was in fact the building used on Oyster Cay two years previously. It had been dismantled, transported and reassembled, and extended with a new kitchen building. A toilet and ablution block completed the constructions. Only a minimum amount of furniture was provided and the participants had to exercise considerable ingenuity in constructing rough but serviceable additions out of packing crates and empty kerosene cans. Lighting came from kerosene and petrol vapour lamps, and the Electrolux company provided a kerosene powered refrigerator at a discounted rate which, Yonge recorded, made life more bearable throughout the summer months in that demanding climate.

Great care was taken to include an adequate supply of scientific equipment. A considerable amount was shipped from Britain, including microscopes and bench appliances, trawling and temperature recording apparatus and a centrifuge for plankton analysis, as were laboratory semi-consumables such as glassware and chemicals, which were replaced as needed from Australian sources and taken by lighthouse ship to the site. For recording purposes the group was able to borrow from the manufacturers in London a professional-quality Watson half-plate camera and stands to photograph the various field sites, the coral colonies and specimens collected, which they processed in a dark room in one of the huts. In addition, an extensive library of scientific books was also brought and housed on shelves in the laboratory.

Essential to the project were suitable vessels. For the main tasks to be undertaken by the 'boat party' a 39-foot (12 m) ketch, the *Luana*, with a 26 horsepower engine, was hired on generous terms from its enthusiastic owner, Mr A. C. Wishart of Brisbane, who remained throughout the year as its skipper, assisted by the Yarrabah crew, Harry and Paul. A second craft, a small 12-foot (3.6 m) dinghy with a 2.5 hp outboard motor was bought for close-in work around the cay. As occasion demanded, other small craft were rented or borrowed.

In addition to their own resources and the assistance offered by Australian scientists and workers at the station, other agencies were more than willing to assist with what had developed into a great national enterprise. Supplementing the considerable cooperation extended by Captain Roskruge of the Commonwealth Navigation Department in authorising the use of the two lighthouse vessels, the Royal Australian Navy lent a 'Lucas sounding machine, sounding lines and leads, binnacle compass, station pointer, sounding sextants, Douglas protractor, patent log, Admiralty charts, and much other material'; the Royal Australian Air Force provided aerial photographs of the Low Isles for accurate mapping of the region and transect evaluation, while 'the Commonwealth Meteorological Bureau supplied standard instruments of the following types: barograph, thermograph, hygrograph, sunshine recorder, anemometer, and maximum and minimum and wet and dry bulb thermometers' (Yonge 1931:10). Such were the preparations that awaited the very excited adventurers who set out in May 1928.

EXPEDITION BEGINS: LONDON TO CAIRNS

The original ten members of the expedition sailed from Tilbury, London, on the Orient liner RMS *Ormonde* on 26 May 1928, creating a great impression in England. On the front page of the *Daily News and Westminster Gazette* (Monday 28 May 1928) a huge banner headline proclaimed 'EXPEDITION OF YOUTH TO CORAL SEA', with a sub-head 'Year on Weird Reef'. Perceptions in England of the Great Barrier Reef had changed little since the sensational reporting of a century before. The emphasis on youth continued in a second section headed, again in block capitals, 'Leader of 27: Wife of 24'. Indeed, the average age of the entire party was less than 30. In contrast to Potts, none had any prior experience of coral reefs nor even of travel south of the equator. Apart from Russell who had some fisheries experience in Egypt, what little experience they had so far came from marine studies in cool temperate waters. Both Orr and Marshall, in fact, came from the Millport Marine Laboratory in the Firth of Clyde, close to 56°N latitude. Since the group's knowledge of tropical corals came mainly from the literature, the headline could equally have read 'Year on Reef of Weird Novices'.

Travelling via the Suez route the novice party recorded highlights of the voyage with zest and the archives hold a number of photographs of various members enjoying the stopovers at Gibraltar, Naples, Suez, Aden and Trincomalee on their way to the Bight. On the arrival of the *Ormonde* at Sydney in early July they were met and entertained by members of the Zoological Society. A day tour of Sydney Harbour was provided, with a visit to the site proposed for the State's first marine station at Watson's Bay, one that was to have a short but unhappy career.

On 5 July the Zoological Society entertained them to an evening function where much merriment was made of the inexperience of the visitors. Reported the following day in the Sydney mass distribution tabloid *Sun* (Friday 6 July 1928), under the heading 'Barriers – Risks of the Reef – Marine Studies – Welcoming the Expedition' the opening words began: 'Tiger-sharks with great triangular teeth slipping silently along'. That threatening statement came in an era of numerous shark attacks on Sydney beaches. Until 1910 surf bathing was illegal – chiefly because neck-to-knee costumes were considered indecent exposure – but when the joys of the surf were discovered in the 1920s by the urban masses Sydney beaches were crowded throughout the summer seasons. And sharks struck: two fatalities occurred in late summer of 1922; attacks came again in 1924 and 1925, and four in March 1928. Altogether five deaths and three serious mutilations occurred. Public fascination and phobia were heightened by excessively sensational reporting. The ferocious tiger shark (*Galeocerdo cuvieri*) was often blamed although some cases must have been by other species of the same dreaded family of the Carcharinidae, known collectively as whalers. That journalistic report of the meeting was designed to captivate readers, already aware of the youth and inexperience of the visitors.

The zoologists, it stated, regaled the group with a further list of Reef dangers from poisonous-spined stone fish, man-eating gropers, giant clams that were said to imprison an unwary diver's foot and hold him under until he drowned, and rock-eels that 'bite to the bone'. Continuing its report of the mischievous antics of the zoologists, chiefly from the speech by Gilbert Whitley, the Australian Museum's shark expert and inveterate practical joker (described in its own *Newsheet* as a 'puckish prankster'), the *Sun* noted that many other reef features were described: 'members mentioned mosquitoes, sand-flies, malaria, sunburn that made the ankles swell, ticks, octopuses, and the danger of drinking unboiled water'. To ameliorate the visitors' concern to some extent, entomologist Dr Waterhouse advised that the many beautiful butterflies on the Reef coast were quite safe since 'no one has ever been bitten by a butterfly'.

The *Sun* had chosen to highlight the deliberate ragging of the 'new chums' in classic Australian larrikin style. The visitors took it in good spirit and Stephenson responded for

---

* Original papers relating to the Low Isles Expedition are in the author's archives and the coding employed is given in Abbreviations and Acronyms.

the group by commenting that 'not a few persons seemed to think, when they heard that you were going to Queensland, that your chances of coming back were remote'. Yonge gave a measured and sober reply to the bantering welcome and 'paid a high tribute to the welcome and cooperation given to the expedition'. What formed a significant contribution to the evening was his elaboration of the direction of scientific work in Britain where 'scientists are turning more and more from systematic to experimental zoology'. Whereas 'systematising work' was once the major focus, he informed them, it was now being superseded by direct field experimentation, which, of course, was a major aim of this expedition. When the evening ended, it was reported in the *Sun*, that 'apart from the jotting down [of] one or two suggested remedies for sunburn and mosquitoes, the members of the expedition seemed quite unaffected by the lurid pictures painted'. They were, at the time, unaware that coral cays, without surface water do not have mosquitoes.

The *Ormonde* continued north and at 9 a.m. on Monday 9 July 1928 docked in Brisbane to a major civic reception with Richards a prominent participant. The *Daily Mail* was far more restrained than other newspapers so far: its article was headed, in bold type, 'Treasure Hunt – Barrier Wealth – Economic Research – British Expedition Arrives' and gave an excellent descriptive account of the project. The opening paragraph, beneath a photograph of the smiling faces of Richards and Yonge, stressed the economic theme: 'Our objective', Yonge commented,

> is as much economic as scientific – in the long run, everything is economic. The aim of the expedition is to learn as much as we can of the potential resources of the Great Barrier Reef with a view to discovering how they best can be developed. The reef is regarded as a region of great potential wealth, and special attention will be given to commercial products.

Yonge made it clear, however, that science had to precede commerce since 'it would never be possible to develop the resources of the Great Barrier Reef until the fundamental conditions underlying the marine life of the reef and the surrounding ocean were thoroughly understood'. Then followed tributes to the efforts of Nathan, Richards and the GBRC, as well as to the BA and all those who had contributed to the planning and successful commencement of the expedition.

That evening the group was entertained to a civic reception in Brisbane and the dinner menu, typed on a neat, buff, folded card with a photograph of a reef scene on the cover, continued the larrikin theme begun in Sydney with a list of the courses to be served (with all the indications of Whitley's mischievous hand). Headed 'Coral Corroboree, Brisbane July 10th, 1928' the nine courses were: Extract of Chelonia midas, Cymbium commersoni, Halicore cutlets with glauconite globules, Sterna fuscata en casserole, Pandanus pie with ooze, Coral detritus also with ooze, Mesoglea rubra, Clypeasters and Ambergris, Sepia. Coral scientists everywhere would have recognised that, perhaps apart from the opening turtle soup, and the dugong cutlets, the menu was largely contrived gibberish, with a few revolting inclusions. By that time the bemused visitors must have become slightly accustomed to Australian rollicking humour, although it seems – at least from the surviving records – their British formality was never completely loosened.

Completing the journey from Brisbane by rail, they arrived at Cairns on 16 July, and in the company of the waiting Frank Moorhouse, in the launch *Daintree* travelled the 70 kilometres past the sparsely settled coastline to the Low Isles base where they began preparations for an initial reconnaissance the following day. The first major study of the Reef began almost immediately and lasted for twelve months and twelve days.

Throughout the entire length of the expedition there was a continuing sequence of short-term visitors, 15 in all: Richards and Marks of the GBRC, with its secretary, always known as Miss Todd, Director Heber Longman of the Queensland Museum, a number of GBRC members from other universities, as well as ichthyologist Theo Roughley – an economic zoologist from the Technological Museum in Sydney, who was himself to publish eight years later a popular account as *Wonders of the Great Barrier Reef* (1936). Another important visitor was Anna Frederica (Freda) Bage, biologist and principal of the Women's College in the University of Queensland, and first woman admitted to membership of the GBRC, who provided considerable hospitality for various members of the expedition when they were in Brisbane (Love & Grasset 1984:286–87). The first of all such visitors, however, was Charles Barrett, a journalist from the Melbourne *Herald* who wrote a number of glowing accounts of the expedition which were syndicated to newspapers in Sydney, Melbourne and Brisbane, and such major provincial towns as Cairns, thereby contributing both to widespread public interest and to a perception of the team as glamorous scientific pioneers investigating and revealing the mysteries of the Reef.

CHAPTER 16

BIOLOGICAL RESEARCH OF THE LOW ISLES EXPEDITION

RESEARCH ORGANISATION AND PROCEDURE

While the focus of the Low Isles Expedition for the year was biological, although there was, in addition, an independent three month geomorphological survey of the Reef north of the Low Isles, from August to October 1928, financed by the Royal Geographical Society of London and conducted by a small team of three, led by James Steers. The aim of that survey, formulated in England, was to find further evidence on the origin of Reef foundations, and in particular, the relationship of the coastline to submerged reefs, cays, and continental islands, with the aim of assisting the biologists.

As soon as the main Low Isles Expedition team had settled in, the first task was to make a survey of the layout of the Low Isles themselves. The complex consists of two irregularly-shaped islands joined together, some 1800 metres in length and 1200 metres wide, with the long axis lying in a roughly east–west direction. The smaller island, the vegetated coral cay on which the lighthouse had been built and the expedition huts erected, is on the western (mainland) side. The cay itself is of regular oval shape, covering some 2 hectares at low tide, with an adjoining sandflat to the south of around 16.5 hectares. To the east is the mangrove and submerged lagoonal area of much greater dimensions. In the constriction joining the two islands on the northern side was the anchorage and expedition site. Considerable assistance was rendered on 24 September 1928 when a Supermarine Seagull III of Flight 101 (Fleet Cooperation) from the Royal Australian Air Force amphibian base at St Bees Island, off Mackay, landed at the anchorage and Flight Lieutenant Packer, the leading RAAF aerial photographer, informed them that he had come – courtesy of the Ministry of Defence – to make an aerial survey for the expedition from which a photographic mosaic map of the complex would be constructed. When completed it proved to be of invaluable assistance, especially to the geographers. At that time, in fact, the aerial survey of the Reef began with three of the aircraft of Flight 101, which supplemented the surface hydrography of the *Fantome* and *Geranium* (as discussed in chapter 18).

Despite his comments to the Sydney meeting of the Zoological Society, Yonge appreciated that experimentation alone was not possible: collecting and systematising remained essential elements of the biological research process. In his later account, Yonge himself wrote that 'the three stages of description, observation and experimentation were all exemplified in the work of the expedition' (Yonge 1930a:97). Three separate parties, as they were termed, were formed: a boat party under the direction of Russell, assisted by Sheina Marshall, a zoologist from Millport Marine Laboratory, and A. P. Orr, which was concerned with plankton, sedimentation, hydrography, and water chemistry studies; a shore party led by T. A. Stephenson, assisted by his wife Anne, Geoffrey Tandy and zoologist Sidnie Manton (when she arrived from Girton College, Cambridge, on 26 March 1929 for several months), worked on the reef flats and immediate waters for collecting, and studies of benthic biota. Associated with the shore party was the leader's party, led by Yonge and centred on experiments in the laboratory and the aquaria which had been constructed nearby. Frank Moorhouse, who had been sent by the Queensland government to work specifically on economic products, concerned himself with studies of trochus, bêche-de-mer and sponges, although he also assisted others as the need arose. It was soon realised, in fact, that there were limits to continuous close cooperation and before long most of the members began to pursue their own research projects independently, or in loose collaboration. At the same time there had to be continuing exchange of assistance and joint effort, especially important on the boats when engaged in dredging, tow netting and water sampling.

The major fields of study were the Reef flats, the intertidal zone, and the surrounding waters of the Low Isles themselves which were traversed virtually daily with the outboard powered dinghy. The boat party, concerned with deep water studies, went further afield in the *Luana*, from the Low Isles region to the Howick Group, a range of about 2 degrees of latitude (16°30'S to 14°30'S), approximately 140 nautical miles. In addition eleven other boat cruises were conducted throughout the early months of 1929, eight in hired craft, two in the *Luana*, and one in which Orr and Otter travelled aboard the *Cape Leeuwin* through the outer barrier to Willis Island meteorological station on its regular supply run in order to take water samples of the Coral Sea for chemical comparison with Reef inner waters.

Public information on the early days of the expedition – in fact the only source – came from newspaper articles written for mainland newspapers by Charles Barrett, a Corresponding Member of the Zoological Society, who arrived in the first week for a short period. In a sequence of articles published throughout Australia describing activities over the first weeks, in an attractive and informative style devoid of journalistic exaggeration, Barrett combined both the human interest aspect of life on a coral cay which engaged the non-technical reader along with an absorbing account that gave a clear understanding of the science involved. To a reading public indoctrinated only with sensational Reef reporting, and with minimal understanding of scientific research and no awareness of the complexity of Reef life, his articles were a refreshing new development towards popular scientific understanding. His article for 28 July, the tenth day of the expedition, for example, is an extended, well organised piece that gave a thorough résumé of the entire program of research by all of the members, set against the historical and theoretical background of coral reef science of the time. Under the heading 'Coral Maze' and the sub-heads of 'Boisterous Day on Reef – Tide Hides Beauty – One of the World's Wonders' he described his experiences of a trip in the first week aboard the *Luana* to Batt Reef near the inner approach to the Trinity Opening to the Coral Sea.

Barrett had in mind the ordinary reader who would be drawn into the sense of adventure aboard a ketch that 'takes the seas like a swimming gull, in the process receiving a "proper dusting" crossing eight miles of open water', and which on arrival came to 'the light green lovely colour of shoal water' of the lagoon, edged by 'the ominous brown and dark yellow "isles" beneath the surface like the sunken shadows of clouds which mark the coral'. At the same time in the narrative he was quite skilled in presenting a deeper layer of scientific information, in this case, on the geological controversies of Reef origin. Well briefed by Richards and Yonge, who were quoted directly throughout the article, the various geological theories of Darwin and his arch-rival Agassiz were outlined, with Yonge being reported as saying that 'the weight of evidence . . . is distinctly in favour of Darwin's classic theory so far as barrier reefs are concerned' (*Daily Mail* 13 August 1928). A week later, a slightly reworded version of the same article appeared in the *Sydney Morning Herald*, this time headed 'On a Coral Isle – Barrier Reef Theories' with the telling sub-heading 'Vitality of Darwinism'.

His seventh article, written three days later but not published by the *Daily Mail* for several weeks, since the various newspapers spread his accounts over a much longer publication schedule, he dealt with the dominant biological aspect of the expedition. Occupying most of a full page, composed around a large format photograph (3¾ by 4¾ inches: 97 by 123 mm) of the Stephensons, wearing pith helmets, on a Reef flat hammering in steel spikes to mark a transect grid for an intensive study zone, the central concepts of the research to be conducted were well described. Again, obviously well briefed by Yonge, who explained that the purpose of the aquaria was to hold the corals that were to be manipulated in various experiments, Barrett gave an excellent, if simplified account of the ecological processes of commensalism and cooperation. He explained the concepts of living beneficially together from the example of a 'lively red-tailed midget' prawn that lived as 'a tenant in the body cavity of a giant anemone', and as a second example, that of another anemone which 'sheltered a pigmy fish, scarlet with white bands, in its visceral cavity'.

In the next paragraph he introduced the reader to another aspect of ecological life, that of competition. Although Darwin was not mentioned, his theory of natural selection comes out very clearly in the sentence that 'the struggle for life among the coral is as ceaseless as it is in the tropic forest'. A heightened fear is induced in the reader in the account of highly venomous sea snakes – a perennial, if generally misplaced, concern of swimmers and divers to this day – in which he wrote that 'about 53 species . . . have been described; more will surely be found on the Barrier Reef and elsewhere'. The uninformed reader could have been easily misled: no more have been described and today there are 51 species (including four species of sea kraits) with Indo-Pacific-wide distribution; only seven species of true sea snakes, the Hydrophiidae, are endemic to the Reef, although there are perhaps six or so other aquatic species (family Colubridae) in estuarine and mangrove waters (Heatwole 1987:3–8). For a journalist, however, such hyperbole is necessary to sustain reader interest. Barrett certainly did that and his articles kept the reader informed about the progress of the expedition and gave a growing understanding of the natural wonders and complexity of the Reef. His writing, in fact, stimulated public response to the extent that Sunday picnic outings were organised from Port Douglas and other coastal areas by enterprising entrepreneurs for the curious public to see the research activities for themselves, which forced the scientists to have protective fences erected around the aquaria and experimental areas.

Barrett's entertaining articles, however, describe the more lively accounts of the work of the expedition. After its return to Britain a considerable number of popular articles appeared, while more formal articles were directed to the scientific community in the official *Reports*. Historical reconstruction of the expedition's daily progress, however, comes from quite a different, more obscure source. This consists of five periodical reports and a final 'Economic Report' written by Yonge and sent to Brisbane where they were typed, mimeographed, and distributed to members of the GBRC and relevant persons and committees in Britain which had provided support. The first, an eight page document, was issued after one month on the island, dated 17 August 1928, which gave a good general description of the organisation of the site, equipment, construction, Aboriginal workers, visitors and broad details of research planning and organisation. Yonge wrote complainingly to the GBRC that he was quite distressed upon learning that a copy had been acquired (from a never identified source) by the Sydney *Telegraph* newspaper that printed a journalistic version on 19 September. Consequently, the ensuing four three-monthly reports were marked 'Confidential' and did not become generally known. In fact, they do not seem to have been used for subsequent historical research; the present account, in fact, is based on the recently discovered personal copies of Heber Longman, at the time director of the Queensland Museum.[*]

Those reports, totalling 51 pages in all, constitute an excellent record of the daily activities of the expedition, covering organisation and routine matters, the arrival and departure of various scientists, assistants and visitors, and the fundamental work of the boat, shore and experimental parties. Although they contained nothing of a confidential nature – unless some members of the public would have been appalled at reading of the low wages and demanding working conditions of the Aborigines involved – they do reveal the complexities and difficulties faced by that pioneering team, and give an accurate and familiar account of the way in which persons of varying backgrounds and interests managed to live and work together for a year on a tiny island of no more than 5 acres at low tide. In addition to Yonge's three monthly accounts, there is also a single seven page report by Steers dated 28 November 1928, covering the three month survey by the separate geographical expedition.

The three groups began operations immediately. The work of the boat party, led by Russell from Plymouth until his departure on 12 December, and then taken over by Orr, was organised around a number of stations in the surrounding waters, 24 being the maximum determined. Continuous plankton hauls and dredgings, along with taxonomic identification of both the phyto- and zooplankton were made throughout the year, with a concentration on their seasonal and locational variation. Equally important were hydrographic surveys, chiefly water sampling for chemical analysis, dissolved oxygen levels and pH (acidity/alkalinity) readings, together with continuing daily meteorological recordings of temperature, sunshine hours and barometric pressure using equipment lent by the Commonwealth Meteorological Bureau. Problems were experienced with the unpredictable nature of the tides, which also bedevilled much of the geographical survey of Steers, since, due to the complexity of the reefs and islands, they do not follow a regular diurnal rhythm. The shore party, whose areas were the island itself, the Reef flats and lagoon, and the immediate surrounding waters, first marked out the Reef flat where the corals become exposed at low tide with metal transect pegs for regular observations. Corals were then collected for identification and placement in the aquaria where growth and physiological processes were to be studied and manipulated. Early in the project one hundred concrete blocks were constructed and set out across the Reef flat to which were fixed living corals from three main genera – *Favia*, *Lobophyllia* and *Symphyllia*. These were measured and photographed periodically for growth and matched pairs were also placed in different locations to determine whether there was any effect on growth from different environments.

To assist the economic experiments of Moorhouse, in mid September the Aboriginal workers constructed a pen on the Reef flat some 550 metres south-west of the cay for his experiments in cultivating both the *Pinctada* species of pearl oysters. Mangrove stakes were driven into the flat to form a circle about 4.5 metres in diameter, between which long thin mangrove saplings were woven in a horizontal wicker-work pattern up to a height of around 1 metre. To prevent loss of the oysters, 450 having been collected and placed within the pen, loose coral rubble and dead clam shells were piled around the outside as a protective bulwark. For experiments on trochus cultivation two separate pens, containing 600 animals, were constructed from farmyard wire-netting on the Reef flat; these, however, did not prevent invasion and destruction by stingrays until the sides were strengthened and an additional wire-netting enclosure built over the top to prevent their entry at high tide. The junior assistant Otter was assigned tasks of helping generally with the shore party, and at the same time pursued his own research on rock-boring molluscs – lamellibranchs – chiefly the bivalve *Tridacna crocea*, and that dreaded scourge of all timber vessels and wharves, the wood-boring teredo 'shipworm', a mollusc of the family Teredinidae.

Of considerable historical interest is the early use by the shore party of a primitive diving helmet for descent down the face of the cay – an innovation since effective scuba equipment was not available until the 1960s – for the observation, collection and experimental placement of corals for settlement studies. Up to that time coral observation generally was limited to looking through a cylindrical metal tube with a glass lens at the end, inserted into the water from the side of a boat. A pearling-type Siebe-Gorman helmet and diving dress – or even the abbreviated half-suit – would have been too cumbersome, needing a mechanical compressor with a large tender and crew, and requiring considerable prior training. Yonge reported that use of a much simpler kind of diving helmet formed 'a valuable part of the collecting apparatus' and was suggested by one 'originally employed by Dr Mayor at the Carnegie Laboratory at Tortugas'.

Constructed before departure at Plymouth Marine Laboratory from photographs of Mayor's helmet, and recorded in the surviving Low Isles photographs, it resembled, in Yonge's words,

> a dustbin with a handle on top and windows of plate glass in front. At one side there is a brass inlet pipe, to which is secured at the end a strong garden hose which carries the air from the double-acting motor-car tyre pump which forces it down to the diver. A lifeline of thin manilla rope is fastened to the handle at the top (Yonge 1930b:99).

To the front and back flanges of the helmet, lead weights were attached to hold it in place as the air came from an assistant pumping away in the dingy, while the surplus and spent air escaped from the bottom of the helmet. It was, in effect, a personalised diving bell sitting on the diver's shoulders. The diver descended wearing a bathing suit and canvas tennis shoes while the sole occupant of the dinghy pumped manually. In case of emergency the helmet was discarded and the diver surfaced by free swimming, while the helmet was retrieved with the attached line.

The depth to which it was effective was recorded as 7 metres for around 45 minutes, although in practice it was used most often at depths of between 3.6 and 6 metres (Yonge:LIE.II.2/29:4). Sheina Marshall commented that it could be used for 'hours if needed' (Marshall 1930:88) but obviously, with only one pumper, a time limitation would have been imposed. The helmet's greatest idiosyncrasy came from the periodic fogging of

---

* The six Reports to the GBRC by Yonge in the author's archives are catalogued as follows: Yonge:LIE, followed by the sequence in Roman numerals from I to VI, followed by month and year, and finally page number.

the two angled glass viewing panels (preventing stereoscopic vision) which had to be cleared by the diver bending forward periodically to rinse the glass, thereby allowing water to rise and swish up from the open base.

It is important to note, however, that Mayor was not the first to use this diving bell type helmet. In 1817 Augustus Siebe adapted the open diving bell for individual use. Subsequently, in the 1840s a further development of that apparatus was devised by the famous French marine scientist Henri Milne-Edwards, which he used in the Straits of Messina off Sicily in the world's first scientific underwater collecting expedition in 1844. As described by one of his companions, Milne-Edwards used 'a metallic helmet, provided with a glass visor, [which] encircled the head of the diver, and was fastened round the neck by means of a leather frame supported by a padded collar . . . [connected] by flexible tube with the air-pump, which was worked by two of our men'. Despite some mishaps and awkward moments, Milne-Edwards 'returned from the bottom of the sea, his box . . . richly laden with Molluscs and Zoophytes' (de Quatrefages 1857:2.17–19). That description closely fits the apparatus used by Mayor in the Tortugas, and subsequently by Russell and others at Low Isles. The most unusual diver must have been Tom Iredale from the Australian Museum. Born in Workington in the far north-west of England, he never learned to swim, yet so devoted was he to conchology that he made numerous descents in search of shells. Iredale, in fact, over his long lifetime – 92 years – had a prodigious output, publishing over 360 papers on molluscs, describing and naming 2660 new genera and species.

Midway through the project there was a change of personnel. Frederick Russell and his wife Gweneth, along with Geoffrey Tandy, botanist from the British Museum, left on 12 December 1928. No more botanical work was done. To replace them, Elizabeth Fraser, senior lecturer in zoology at the University of London, and the oldest member of the expedition, close to the end of her academic career, arrived on 5 March, and Sidnie Manton three weeks later on 26 March 1929. An interesting aspect of the attitudes of the period is exemplified in the case of Manton, and the unusually large number of women in the expedition, unlike Australia where women had few opportunities for a university education and none in marine biology. A 26 year old from Cambridge, Manton had graduated top of the class but was denied the university prize since women in that period, although allowed to enrol and sit for examinations due to growing feminist pressure, remained officially unrecognised as graduates of the university (Love & Grassett 1984:286).

EXPERIMENTS WITH CORALS BY YONGE AND NICHOLLS

As the work of the shore and boat parties progressed routinely through the variations of the tropical seasons, they gathered a vast amount of data and specimens which were carefully studied, catalogued and stored for return to Britain and later analysis and reporting. Yonge, by contrast, busied himself with experimental work in the laboratory, the aquaria, and on the Reef flats to the south and south-east of the cay. Throughout he was assisted by Aubrey Nicholls, a final year honours zoology student of the University of Western Australia – and son of Professor Nicholls, head of the department of zoology there – who came to the Low Isles for field experience and to collect data for his examinations. Those examinations, interestingly, were conducted on the Low Isles in February – obviously by some kind of special arrangement – and were passed by Nicholls with first class honours.

To the casual observer the various activities throughout the year by the scientists may seem quite disparate and uncoordinated; on the contrary, however, they were all devised by Yonge and Russell before leaving Britain as part of an integrated biological research program to complement existing knowledge of Reef geology. At this point it may be necessary to reiterate that for more than a century most coral reef research in Australia and internationally had concentrated on geological issues which had led Richards to seek finality on what, in his numerous speeches and articles he called the 'Problems of the Great Barrier Reef'. Yonge recognised that while geology had made impressive progress and dominated most coral reef literature, it was fundamental to appreciate that reefs are much more than geological structures: they are complex ecological assemblages in which corals have the determining influence. Across the seas of the world there are innumerable non-coral rocky reefs, continental islands, archipelagoes and other geological forms, all in continuing process of change due to underlying earth movements.

Puzzles abounded, reaching back to the early occasional observations of Pyrard de Laval, Chamisso and Forster, for instance from their commentaries that on reefs corals seemed most robust in surf and other turbulent zones, and always on windward fronts. Coral taxonomy had become more developed as numerous genera and species were determined, and the work of Mayor on Mer, to take but one significant example, had sought to measure the spatial distribution of genera and species to discover an organising pattern. It was a scientific conundrum that seemed intractable. Geology only provided knowledge of the underlying foundations: understanding reefs themselves was a biological problem. In planning the expedition, therefore, Yonge and Russell knew that a much wider range of the possible parameters contributing to the construction and biological complexity of the Reef, as then understood, had to be investigated.

Starting from the biological fact that the coral polyp is a coelenterate, and in common with all members of the sub-phylum Cnidaria (in the phylum Coelenterata), it contains nematocysts – specialised stinging and attaching thread cells used to capture prey, first identified by Peyssonnel – Yonge inferred that it was, therefore, a carnivore. The research program, consequently, was built around the assumption that polyps captured food from the only available source, the vagrant zooplankton, within specific levels of sea temperature and light. Further, since adult reef corals are sedentary and have built immense limestone structures over geological aeons, the circulation and chemistry of sea water, and in particular the availability of calcium carbonate in the sea water to build the protective external shell, the corallite, were necessary features for investigation. Moreover, the observed vigour of coral growth in turbulent surf zones suggested that those were regions with the greatest supply of available zooplankton, reinforcing the carnivore hypothesis.

Working from the belief, then, that coral polyps are entirely dependent on external food such as zooplankton for nutrition and growth (heterotrophs), Yonge directed his research into investigation of the physiology of corals concerning feeding, digestive enzymes, calcium carbonate metabolism and their responses to variations in light, temperature, water chemistry (salinity, dissolved oxygen, carbonic acid), plankton density and variety, current flow and tidal fluctuations.

Yonge, in particular, had developed a profound interest in the relationship between polyps and what Mayer (1918:28) had called 'infesting zooxanthellae' found inside the tissues of corals and many other molluscs. Those algae presented a significant puzzle to early researchers. Were they an infestation, and if so, what effect did they have on the host? By 1928 there was a growing alternative conviction that such dinoflagellate algae were not an infestation but rather, had some kind of commensal or even symbiotic relationship to the coral. If so, what was it? Since algae depend on light for chlorophyll photosynthesis, do they limit the growth of corals to the photic zone? Almost a century before, Charles Darwin had made the acute observation from his soundings in the Keeling (Cocos) atolls, and from evidence supplied to him by others, that 'in ordinary cases, reef-building polypifers do not flourish at greater depths than between 20 and 30 fathoms' (36–55 m) (Darwin 1842:86). In Darwin's observation, of course, the algae had not been identified, but from the time of Mayor on Mer in 1913 possible linkages between algae and hermatypic corals began to be suspected.

Wood-Jones, in his research in the Cocos Group in 1910, several years earlier, had stimulated investigation when he denied strenuously that those algae made any contribution to either the growth or nutrition of corals (Wood-Jones 1910:241–42), and claimed that when corals in his experiments were deprived of light, the algae died and the corals, although bleached from loss of their coloration created by the algae, remained unaffected, a conclusion confirmed by Vaughan in similar experiments in the Tortugas. Once his arduous administrative duties had eased after the initial months Yonge was able to devote increasing time to the coral experiments, focusing on the effects of light and darkness, oxygen discharge, feeding and starvation, $CaCO_3$ metabolism and digestive enzymes.

Throughout the year experiments were conducted on corals in the aquaria. One, set up to test the carnivore hypothesis, consisted of feeding Favia and Galaxea corals with zooplankton up to 3 mm in length, and it seemed to confirm their carnivore nature with the discovery that the plankton were 'completely digested and the empty skeletons ejected, within a period of 12 hours or less' (Yonge:LIE.III.2/29:8). To determine more accurately the role of zooxanthellae a large number of experiments were conducted to measure their oxygen output, which would be beneficial to the polyp, by containing them for periods in sealed glass jars and measuring the oxygen levels at the end of specified periods. Other experiments were conducted on calcification rates and the role of enzymes in polyp physiology.

The most interesting experiments to investigate the relationship of those algae, which were to have a major influence on reef biology for decades to come, as early experiments on coral bleaching, were designed to prove that zooxanthellae are not essential to coral reef health. Experiments on the effects of both starvation and darkness on corals of the *Favia*, *Fungia*, *Galaxea* and *Psammacora* genera led Yonge to conclude that if

> fed they remain in good condition whether in light or darkness, paling in colour slightly in the dark, but when starved their tissues quickly begin to shrink, [and] undamaged algae are expelled in large numbers and the remaining tissue turns pale as a result . . . The only interpretation that can be placed on these results is that the algae are not and cannot be used as food by the corals, a conclusion which agrees entirely with the results of the feeding and enzyme experiments (Yonge: LIE.III.2/29:9).

To pursue that line of investigation, Yonge had 'a large light-tight box . . . cemented on the reef flat and a number of corals placed in it' which confirmed earlier findings that after several months they showed 'a high degree of paling but [were] still healthy, the death of the algae apparently not affecting the corals' (Yonge:LIE.IV.5/29:6). In his final report of 25 September 1929, Yonge stated categorically that 'corals kept for 5 months in the dark box on the reef flat survived to a large extent, those dead having been mainly killed by sediment' (Yonge:LIE.IV.9/29:7). There the research ended, inconclusively, until it was resumed after the Pacific War, by among others, Leonard Muscatine at the University of California in Los Angeles and Thomas Goreau at the University of the West Indies in Jamaica. Stimulated by Yonge's work, Goreau investigated the role of algae in polyp nutrition and the fact that they live only in the photic zone to a maximum depth of 100 metres. In an extensive series of experiments, he determined that the waste materials of the zooxanthellae, which depend on the intensity of the light reaching them, are recycled by the polyp into calcium carbonate, thereby facilitating growth (Goreau 1959, 1961, 1979).

REEF GEOLOGY: GEOGRAPHICAL SOCIETY SURVEY ATTEMPT

Although the Low Isles Expedition was fundamentally biological, the geological issues, despite more than a century of inquiry, remained unresolved. The companion program of the Royal Geographical Society worked closely with, but independently of, the main group. Led by James Steers of Cambridge and assisted by Michael Spender from Oxford, they began at Townsville in August, and then in mid October they were joined at Cooktown by E. C. Marchant, recent Cambridge graduate, for six weeks. Despite the geomorphological brief to investigate the relationship of the Reef islands and cays to the adjacent coastline, in retrospect fifty years later, at a conference on the 'Northern Great Barrier Reef', Steers confessed – in contrast to the planned program of the Low Isles group – that 'when Spender and I began work on the reefs we had no definite idea of what there was to do, or how we were to do it!' (Steers 1978:161). Events throughout the expedition confirm those comments and his additional observation that 'we had to find our problems as we sailed along the coast' (162).

Right from the start nothing went well: in fact, the Steers expedition was close to a fiasco. To conduct the planned survey of the region between the Whitsunday Islands around latitude 20°S and the Flinders Group at Cape Melville near 14°S, a considerable distance, they leased a small semi-open launch, the *Tivoli*, which proved to be too small: a lot of the equipment that had been brought to Townsville had to be left behind, and what could be placed aboard took up so much space they were 'forced to sleep on and around the gear, fuel and food' as Steers wrote in a short memorandum to the GBRC of 28 November 1928 (Au.Arch.). The intended major activity of mapping the coastline, continental islands and cays by means of a new technology of photo-theodolite survey had to be largely abandoned. They found the equipment was too heavy, being 'an extravagant load for each man, moving over the roughest country, in a tropical climate' and even when steep cliffs or continental island peaks were scaled, the dense tropical vegetation precluded lines of sight. In default they were forced to spend most of the two months together making only a visual survey which formed the content of Steers' contribution to the official *Scientific Reports* of the expedition (Transactions, RGSAQ, Vol.3, No.1, 1930).

Steers' conclusions, based almost entirely on observations, both from the boat and ashore, contributed little, if anything to established knowledge beyond what had already come from the surveys of Andrews, Hedley and Richards, whom he discussed and cited. Steers left for Cambridge in the first week of November, where it seems that the relative failure of his geographical survey was not well received. Unfortunately, there is a paucity of evidence, but an interesting inference may possibly be drawn from a letter by Steers to Richards of 30 January 1929 which is excessively lavish in praise, stating that 'Everybody here is being made to appreciate all that you mean to members of the Expedition'. There is no doubt that Gardiner and the biologists were less than impressed, although it seems that no outright hostility came the way of Steers and the geography department: in an oblique reference to that possibility he ended with the comment to Richards that 'I should add, between ourselves, that so far as I can see there is no likelihood of friction with the Zoological Department. You will understand my meaning' (UQA S226. Box 1, item 1).

While Steers was back in Cambridge attempting to put the best possible construction on his misadventures, the other two remained to make topographical and physiographical maps of the Low Isles and of Hope Reef to the south on the outer Reef near Michaelmas Reef, until 9 January when Marchant also left. To compound their distress, the Bausch and Lomb recording tide gauge that had been shipped from England and had passed through customs in Townsville, remained on the wharf due to a shipping strike that prevented delivery until after Steers had returned to England. Even when it had arrived, and was finally installed on the western side of the cay in February 1929, Spender, who was also assisting the shore party at the time investigating 'the ecology of the reefs and on the breeding, development and growth of corals' (Yonge 1930a:237) under the direction of Stephenson, found it difficult to maintain in a stable location on the edge of the lagoon. Spender stayed on to the end of the main project in July, helping others on the Low Isles and also making a survey of the Three Islands off Cooktown which were considered to be particularly interesting. At its conclusion, the geographical expedition proved to have been almost worthless in contributing to Reef knowledge, and considerably expensive, running over budget, with the three members having to make up the shortfall. At least, however, in his final report Steers confirmed that so far no solution to the geological problems of the Reef had been found.

LOW ISLES EXPEDITION: RESULTS AND SIGNIFICANCE

By late June 1929 the end of the expedition's year was approaching and the process of winding down operations was commenced. As many routine observations as possible were continued, while, at the same time, various members of the expedition began to depart: the Stephensons and Elizabeth Fraser on 3 July, Sidnie Manton on the 16th, and Aubrey Nicholls on the 23rd. Final evacuation came on 28 July when the remaining six – the Yonges, Marshall, Orr, Colman and Spender – departed, leaving a volunteer, Mr H. C. Vidgen, to lock up the huts and supervise the Aboriginal workers in cleaning up the site. All the scientific equipment and specimens were crated for shipment back to England; the borrowed instruments were returned; surplus materials and furnishings were sold, bringing in £150, and to Yonge's immense satisfaction – and relief – the financial accounts were in credit by £200 from the original budget of £8580. Yonge hoped that the small surplus could be used to help finance the scientific reports, but since such a vast amount of data had been collected, he realised that a drive for funds to publish the numerous planned volumes would have to be started.

In March 1929, at a meeting of the GBRC in Brisbane attended by Yonge, 'the future of the camp on Low Island and the possibility of continuing the work of the Expedition after 1929 were the main subjects under discussion' as Yonge later recorded (Yonge:LIE.IV.5/29:1). At that meeting he formally offered the buildings and some unsold surplus equipment to the GBRC on behalf of the British Committee, which was 'most gratefully received'. Yonge was concerned to ensure, if at all possible, that the work of the group would be continued, especially on the economic side, stating that 'the creation of a permanent station on the Barrier would be the happiest possible outcome of this Expedition which would then leave a permanent impress on Australian marine biology – at present a seed struggling to germinate but within a few years to become, if present

Laboratory of the Low Isles Expedition

Library of the Low Isles Expedition

Aboriginal crew member on *Luana* and dredge

Freddie Russell on *Luana* with tow-net and depth recorder, 1928–29

Breeding pens, Black lip, Pearl Oysters

indications may be trusted, a flourishing plant'. Those final comments were to indicate the soundness and foresight of his recommendations.

Yonge had already communicated all his ideas and recommendations for the future to the GBRC and on his return journey to England via the Pacific he set them out in two small typewritten dispatches: a three-page *Recommendations for Future Work* and a ten-page *Economic Report* which were mailed from Suva. The former was a plea for the establishment of a permanent marine station at Low Isles which he believed should 'be maintained for several years if not indefinitely', while Torres Strait should also be considered since it 'will probably prove a most important centre for economic work'. Drawing on his brief survey of the region earlier in June 1929, he advised that Thursday Island, however, would not be suitable owing to the heavy deposits of mud, but that an appropriate location could possibly be found nearby for research throughout the region, particularly aimed at helping indigenous self-development in marine industries. Already, he noted, in a comment significant for its time, that Papuan Industries Limited was doing fine work in assisting the Torres Strait Islanders, and in a direct reference to that perennial source of discord, in respect to trochus harvesting 'there appears no reason why a further portion of the fishery, still dominated by the Japanese run boats, should not come to them'. A third promising region for research, he ventured, was the southernmost Capricorn Group, which he had visited on the way back to Brisbane, where 'it might be advisable to erect a small laboratory hut' on Heron Island (Yonge:LIE.VI.9/29:4–5).

Yonge's *Economic Report* surveyed the full range of commercial activities already being conducted in Reef waters with varying degrees of success, but he restricted his recommendations to the experience gained at the Low Isles. Despite the fact that the expedition had been largely funded for economic purposes, and Yonge had frequently made public statements to that effect, he offered no suggestions on the possibilities of pearl oysters, bêche-de-mer, fish, turtles, sponges and dugong since they were not found, if at all, in economically significant numbers in the Low Isles region, and consequently no research had been conducted to offer guidance. Even so, he added, the limitations of Low Isles waters should not preclude sustained efforts to maximise the economic potential of Reef waters generally. The lingering suspicion remains that there was never any real intention of investigating the economic potential of the Reef – apart from the work of Moorhouse – to honour the original commitment of funds, and that the frequent references to commercial possibilities were little more than genuflectory gestures to the main funding bodies.

In contrast, the trochus breeding experiments conducted by Frank Moorhouse were very promising and Yonge recommended that they should be strongly encouraged. While farming the edible oyster was not a viable proposition in waters outside Brisbane's Moreton Bay, greater efforts should be made to discover the detailed life history of the gold lip pearl oyster (*Pinctada maxima*) along with experiments in commercial farming and scientific control to conserve that ailing industry. Bêche-de-mer were ruled out as a commercial proposition since Yonge stated that their life history was a mystery and harvesting therefore should remain an opportunistic venture. Significant possibilities for further exploitation, he suggested, indicating his total insensitivity to the need for protection of Australian marine animals, were dugong for oil, green turtle for food, and the hawksbill turtle for its valuable shell, although so far the life histories of all these animals were still far from understood and no serious work had been conducted on techniques of cultivation. In a curious twist of logic Yonge pointed out that the green turtle was already 'an animal which can soon be seriously depleted in numbers as a result of indiscriminate fishing' but with proper controls, and scientific knowledge, the current commercial processing in the Capricorn Group could be improved and even extended to profitable factories in the Torres Strait. Likewise, in respect to dugong 'rigid methods of control will be necessary as the dugong can easily be exterminated in any area if unrestricted fishing is permitted' (Yonge:LIE.VI.9/29:7–8).

The Queensland government was obliged to act on Yonge's recommendations since it had already invested heavily in the expedition, and on 11 September 1929 it convened a conference to formulate a plan of action for scientific and economic work in marine research in conjunction with the Queensland Department of Marine and Produce. Five major objectives were proposed: researching the life history of trochus, the green turtle and the Murray Island sardines, along with experiments on the reproduction of the common sponge, and the commercial farming of the gold lip oyster with a view to extending its habitat as far south on the Reef coast as possible. In a memorandum from the Chief Secretary to the GBRC of 4 October 1929 other longer term scientific objectives were also listed, mentioning unspecified 'work of a purely scientific nature', along with a search for additional commercial products, and the better organisation and marketing of products for existing activities (Au.Arch.).

To implement that plan the government appointed Frank Moorhouse to the position of 'Marine Biologist' within the Department of Harbours and Marine, to be stationed at the Low Isles, using the existing expedition buildings as the research base. In addition, he was to have with him an assistant marine biologist 'who shall be a graduate of a university'. Professor Goddard, an active member of that planning committee, suggested a final year student in his department, a Mr Fison, as an ideal appointment. In cooperation with the Chief Protector of Aboriginals (as the official was then designated) additional unskilled labour could also be recruited. To confirm those arrangements, Under-Secretary Watson informed the GBRC that 'the Government, having given careful consideration to the scheme, decided to proceed with the proposals submitted' (Au.Arch.).

The Low Isles Expedition thereby effected another major achievement: it stimulated the foundation of the first purely Australian research station on the Reef and the appointment, in Frank Moorhouse, of the first Australian Reef scientist to its management. With that decision by government, what Yonge had described as a struggling seed had begun to germinate. Its development into a flourishing plant was still to be a tortuous process.

Back in Britain, the arduous task of preparing the eagerly awaited research reports began. In addition to Yonge's five three-monthly reports and his *Economic Report* of 1929, two series of publications eventuated. The main series were the official *Reports* published in six volumes by the Natural History division of the British Museum between 1930 and 1950. Then came a belated seventh volume on 'Crustacea, Decapoda and Stromatopoda' by Frank McNeill of the Australian Museum in 1968, which brought the total to 62 separate reports. Of these, 32 were written by the original participants: 25 by individuals and seven jointly authored. The remaining 30 reports, all with a strong systematic character, were written by others to whom the collected materials and field notes had been sent.

The second series consisted of occasional articles sent to various journals by individuals, along with more popular accounts to semi-scholarly magazines. Michael Spender, for example, published articles in the *Geographical Journal* on 'Island Reefs of the Queensland Coasts' and 'Tidal Observations on the Great Barrier Reef' in 1930 and 1932 respectively. Sheina Marshall sent a paper to the Royal Philosophical Society of Glasgow that was published in its *Proceedings* in 1930 under the title 'The Recent Expedition to the Great Barrier Reef of Australia'. Numerous similar articles appeared in the years immediately following, while further reflections on the expedition continued to be published even into the 1970s and 1980s.

In this latter group by far the greatest output came from Yonge himself. These included a large number of very similar promotional pieces – often promoting himself – sent to various places: universities, colleagues, scientific societies, newspapers and so on. Yonge even contracted to write a series of three 'exclusive' articles for the *Sydney Morning Herald* which were published on 9, 10 and 11 July 1929 at the conclusion of the project, in which he pointed out that the main reason the members of the expedition came from Britain was simply because in Australia 'there are few men trained in this kind of work at present' (*Sydney Morning Herald* 9 July 1929). In his considerable contribution to Volume 1 of the official *Reports* Yonge authored six papers individually on experimental physiology, and wrote four jointly with Aubrey Nicholls – three being very influential studies on the great mystery and controversy over the significance of zooxanthellae in coral polyps. In addition, Yonge also contributed two general articles to *Nature* in 1929, one on 'Progress of the Great Barrier Reef Expedition' and another version of 'Final Report on the Great Barrier Reef Expedition'. He also published a number of other accounts of the broad details of the expedition, the most accessible versions being those in the British Museum *Reports* (Yonge 1930a) and *Reports of the Great Barrier Reef Committee* (Yonge 1931). His most widely read and extended account of the expedition came in a popular, very well written semi-scientific monograph entitled *A Year on the Great Barrier Reef*, published simultaneously in London and New York in 1930 (Yonge 1930b), which enjoyed a wide readership and exerted considerable influence in academic as well as public circles.

The significance of the expedition cannot be overstated: it was the greatest marine science venture on a global scale since the *Challenger* oceanographic expedition more than fifty years earlier. Due to the enormous spread of popular press coverage, the Reef had become, in a real sense of the word, an international celebrity. Equally, within the scientific community, the massive volume of research data and the continuing issuance of the *Reports* commanded the attention and respect of all marine scientists. For subsequent investigators those findings were to become the datum point for Reef research to the present day: in the words of former GBRC secretary Patricia Mather, the conclusions reached by Yonge had 'a profound effect on coral reef science for the next 45 years' (Mather 1998:153).

An equally significant outcome of the expedition lay in the example provided by the range of research activities and the interrelationships sought. The explicit biological objectives, first outlined in a single page by Yonge as the planning document for the Balfour Committee and the GBRC, were 'to examine a sector of the Great Barrier Reef off Cairns, from the shore to the open ocean . . . studying the associations of plants and animals' along with investigation into the physiology of growth and reproduction of corals, seeking to understand the 'relative importance of plankton and commensal algae' (Au.Arch.).

The implementation of that approach was well expressed in a report *The Structure and Ecology of Low Isles and other Reefs* by T. A. Stephenson, leader of the shore party. The purpose of the expedition, he wrote, was to turn coral reef science away from its preoccupation with geography – meaning geology – and to turn 'the centre of interest back towards the biological side'. In a surprisingly forward looking view for the times, Stephenson stated that the specific aim of his shore party was an ecological study

> towards the elucidation and problems which have a direct bearing on ecology (i.e. towards a study of conditions and food-supply in the sea, and the feeding and metabolism of corals, of the growth and breeding of marine organisms and so forth) . . . [in order that] we should acquire a knowledge of what organisms form the bulk of these populations, and in what manner they arrange themselves with respect to one another and to their environment.

With a stress upon accurate measurement and instrumentation, he indicated that the scientists were seeking 'a true conception of the inter relation of the parts of a reef, both in the horizontal and vertical senses; and to describe accurately the distribution of organisms on it, especially as regards their zonation according to level and their relation to states of the tide' (Stephenson et al. 1931:19–20).

Biology up to that time was still being driven by the application of Darwinian principles of natural selection and the associated struggle for existence, even within genera and species. Ecology, however, was not considered a subversive science on the Low Isles: on the contrary, as Stephenson implied in his report, ecology provided a framework within which biological relationships could be studied and understood. Within that context, the scientific results of the expedition were to provide a major advance in understanding the complexity of Reef organisation and dynamics.

It is to Yonge's credit that he organised his colleagues so efficiently, and it is important not to discount his great achievements in the light of subsequent developments such as the discovery of the heterotrophic nature of scleractinian corals – that is, that they take in food from outside sources. Yonge's earlier assumption that polyps were carnivores has been revised with the confirmation that they are autotrophic, that is, predominantly self-sufficient in supplying nutrition from their own biological processes through symbiosis with zooxanthellae. A number of subsequent crucial experiments have demonstrated that there simply 'was not enough plankton to satisfy these [supposedly] voracious carnivores' (Mather 1998:154), thereby reinforcing the cooperative aspect. None the less, Yonge's pioneer work at Low Isles set in train a line of inquiry that was to take coral reef ecology forward into even greater understanding of the complexity of ecological relationships. At the same time, the conceptual limitations of that era are well exemplified in the broader theoretical framework: biology was still seen as an instrumental mode of understanding the Reef for economic ends. The funding on which Yonge and the team relied, some 60 per cent, was provided to secure knowledge for commercial exploitation, an aim with which – whatever his private thoughts may have been – he explicitly concurred. This is a dilemma increasingly confronting scientists – that is, those with a moral conscience – as they seek corporate funding for research.

In terms of the original resolution of the Third Pan-Pacific Science Congress, the Low Isles Expedition was an outstanding success: its achievements were far beyond anything envisaged three years earlier in Tokyo. In retrospect it justified Gardiner's intervention in revising the Potts plan into a far more exhaustive expedition. While the numerous reports took decades to work their way through the coral reef research network, one important result had a profound effect. Yonge's innovative work on the function of algae in coral polyp tissues, such as the removal of metabolic wastes, and the recognition that they were not an infestation but had some more beneficial relationship, began a new phase of coral research. Described by Veron as 'fraught with unknowns', research into symbiosis was followed through in such areas as calcification of the coral skeleton (corallite) and nutrient cycling, from Goreau in 1960 to researchers in the present day (Veron 1995:96). The famous coral physiologist Tom Goreau, who acknowledged the influence of Yonge in stimulating his own sequence of highly significant discoveries, has summarised the contribution of Yonge very succinctly with his statement that 'the modern study of the physiology of coral symbiosis began with a series of elegant experiments by C. M. Yonge on the Great Barrier Reef Expedition of 1929' (Goreau et al. 1979:113). Once the early experiments of Yonge led the way to confirmation of symbiotic relationships in reef-building corals, the earlier Darwinian framework of competition and 'the struggle for survival' became expanded with the addition of the complementary concept of cooperation in nature. A new field of investigation was opened up.

CHAPTER 17

FROM DEPRESSION TO WAR: TOURISM, CONSERVATION AND SCIENCE, 1929–1939

NATURALISTS, TURTLES AND EARLY TOURISM

In the mid 1920s, encouraged by the proselytising of Banfield, the Reef had increasingly gained a popular image as a fascinating, exotic realm. Following his lead, and with a growing interest in GBRC Reef research as extensively reported in the popular press, a number of individuals began organising small trips to a few of the easily accessible locations – under the banner of 'naturalist expeditions' – which grew into the beginnings of a tourist industry. Those early efforts are not well documented but possibly the first was conducted by E. F. Pollock, a councillor of the Royal Zoological Society of New South Wales, which sponsored the trip, to the Capricorn and Bunker island groups in the southern part of the Reef in November and December of 1925. Pollock gathered together a party of active members of naturalist societies – including Anthony Musgrave and Gilbert Whitley of the Australian Museum – to travel to various locations in a chartered vessel, both observing and holidaying for several weeks, a venture that he repeated on at least five more occasions.

One of the advertised diversions was turtle-riding in which a noose on the end of a long rope was placed around the neck of a turtle returning to the sea just after she had laid her clutch, with the rider standing on the carapace and attempting to stay on as long as possible while the reptile struggled to escape and swim away, dislodging the rider like a bucking horse. The cruelty of that activity was apparently not appreciated at the time and it was never explained how the noose was removed from the neck of the unfortunate animal. Another 'attraction' was a visit to the North West Island cannery which was also reported in the press: 'The turtle-canning company which operates on this island is at present in full swing. About

25 turtles, weighing between 2 and 3 hundredweight [100–150 kg] each are treated daily' (*Sydney Morning Herald* 16 December 1925). On their return Musgrave and Whitley published a report of the expedition in the *Australian Museum Magazine* focusing on the green turtle with a descriptive account of its nesting habits. The most significant part, however, came at the end where they raised two significant issues: the cruel methods of capture and the decrease in numbers. Even then they gave a warning that 'unless drastic measures are undertaken, the species in the long run will be come extinct' (Musgrave & Whitley 1926:336).

After his second expedition to Lord Howe Island in 1926, Pollock's third expedition of 1927 went from Bundaberg to the North West, Heron, Hoskyn and Fairfax islands. It received wide publicity in the *Sydney Morning Herald* between 3 December 1927 and 21 January 1928, in a series of fifteen lengthy articles by Elliott Napier, one of the journalists who accompanied the group. Explaining that there were 21 men and 9 women, he reported, without mentioning the women, that the men were all professionals, including two lawyers, five medical practitioners, a magistrate, and Melbourne ('Mel') Ward, well known in marine science circles as a collector for the Australian Museum.

Following the seductive lead of Banfield, Napier employed a breezy, highly informative journalistic style to recount his experiences. The articles covered a range of Reef aspects at a very generalised level, with a strong emphasis on two features, most highly visible in the days before underwater observation: bird life and turtles. Other topics dealt with the geology of island and cay formation and related aspects of botany and zoology. The three articles on green turtles went into considerable and accurate detail, including one on the capture and slaughter of nesting females which come ashore at night in the summer months to lay their clutches of around 100 eggs. The most disturbing aspect of capture was the practice whereby the turtles, having been turned upside down on the beaches, were left helplessly there until morning, their flippers beating vainly in the air, often left all day in the blazing sun until the time came for them to be decapitated on the spot and left to bleed. At the time of their visit, in marked contrast to that of December 1925, turtle numbers, as predicted by Moorhouse, Musgrave and Whitley, were in noticeable decline. Napier observed, perhaps in response to their concern, in his article headed 'Turtle Soup – Turtle Riding' that 'the loss of a hundred and fifty females every week, together with all their prospective progeny, small as it may be in comparison with the total turtle population of these waters, must tell in the long run' (Napier 1928:138).

So popular were Napier's *Herald* articles about the Pollock expedition that they were collected, revised slightly and, augmented by a chapter on the early history of European discovery and surveying, were published in 1928 under the title *On the Barrier Reef* with the intriguing sub-title *Notes from a No-ologist's Pocket-Book*. Napier displayed an impish sense of humour throughout his writing that must have captivated his readers, his self-description as a 'no-ologist' reflecting that skill. Since he was in the company of well-informed, if amateur, zoologists, ichthyologists, ornithologists and conchologists, and even a birds' egg collector whom he designated an 'o-ologist', he had no hesitation in describing his own ignorance as the specialty of 'no-ology', stating in the Preface that his book 'aspires to no scientific value'. Notwithstanding such self-abnegation, the first edition of November 1928 was an outstanding success, followed by a second printing in 1929 – doubtless made prominent by the current Low Isles Expedition. Third and fourth reprintings followed in 1932, a fifth in 1933, a sixth in 1934 and a final seventh in 1939. As the immediate successor to Banfield, Napier helped create what became a well-established literary genre: romanticised accounts of the Reef which, taking the reader far from the cares of a decade of international turbulence, presented a colourful marine fantasy world of peace and natural beauty.

The Pollock expeditions were almost immediately followed by those of E. M. ('Monty') Embury, a teacher at the Manilla Technical High School in the northern tablelands of New South Wales, to North West Island in 1926, the year that Pollock took his group to Lord Howe. Embury led six annual trips to North West Island: the advertised attractions included a visit to the turtle factories along with the so-called 'sport' of turtle-riding. On the seventh expedition the party camped in the former turtle-canning factory, which by that time had become defunct since so few animals remained to be captured, and it was there they met Noel Monkman on a new project of underwater photography, a skill that was to make him famous.

Embury's tours proved very popular and in 1931 he planned an expanded operation that he advertised as 'Embury Expeditions' in association with the Great Barrier Reef Angling Club which he had also helped to found. Having failed in an attempt to lease Musgrave Island, and then Masthead, mainly due to the opposition of the GBRC, he was successful in securing a lease on Hayman Island in the Whitsundays where he was able to have a small resort built (probably the first on the Reef), such as it was. A small caretaker's hut of corrugated iron sheeting was constructed along with a campsite for 150 persons, a rough golf links, a tennis court and a netted swimming enclosure on the beach. The following year, he expanded his plans which were to create on Hayman what he described as 'The Great Barrier Reef and Whitsunday Passage Biological Station'. To be financed by tourists and angling club members, its stated aim was to offer research facilities for overseas scientists. To give scientific credibility to the government, from whom he needed approval – and always under the suspicious scrutiny of the GBRC – he enlisted the support of zoologist Frank McNeill and palaeontologist Harold Fletcher of the Australian Museum, to accompany and inform the visitors. The facilities provided are quite interesting: in addition to tents, stretchers and hurricane lanterns, a 70 foot long (21 m) iron roofed mess hall was constructed, to be used in the evenings for dances and lantern slide talks. During the day, there were Reef flat walks at low tide and cruises around the waters. Clearly, the 'Biological Station' with no laboratory facilities to attract overseas scientists was a simple resort masquerading as a research facility.

At this time, immediately following the Low Isles Expedition and the final report to the GBRC in September 1929, a staggering blow fell on the world economy. In late October the New York stock market began to falter; on 29 October 1929, known as 'Black Tuesday', there was a massive crash which began the worldwide economic depression of the ensuing decade. The immediate collapse of the American economy was profound: its Gross National Product sank by a third, a quarter of the labour force was unemployed. All industrialised nations were immediately affected, especially Australia which was heavily dependent on overseas borrowing for infrastructure development and the export income derived from its major primary products of wool and wheat. Overseas markets shrank rapidly: wheat prices fell by 60 per cent, wool even more disastrously by 75 per cent. Factories closed, unemployment surged to encompass 40 per cent of the workforce by 1932, and for most of those still with jobs, average wages fell by 17 per cent. Queensland was less heavily hit than the other States since the prices of its major exports of beef and sugar held up better in overseas trading, but it too entered a period of financial stringency in which its government, like those of the other States, struggled to reduce public debt.

Science was profoundly influenced, research became limited, and throughout the 1930s the steady progress of Reef research achieved during the 1920s ground almost to a halt. The Fourth Pan-Pacific Congress of 1929 was held in Bandung, as scheduled, and nine Australian delegates attended, including Andrews and Goddard, but very little attention was given to coral reefs: none of the Australians contributed papers and the Low Isles Expedition was not even mentioned. The Congress was preoccupied with agriculture and economic recovery of the Pacific region. As Reef research stopped, in 1930 the University of Queensland reluctantly decided that the proposed marine biological station on Dunk Island would have to be deferred. It was a devastating blow to Professor Goddard who had worked so hard to establish it.

Moorhouse had been retained temporarily as government economic biologist on the Low Isles trochus cultivation project, where he also carried out a short term project on the life history and habits of the green turtle about which almost nothing was then known. For eleven weeks on Heron Island, between 31 October 1929 and 16 February 1930, he made a close study of the nesting habits of a sample of 1755 adult female turtles, noting egg numbers, survival and growth of hatchlings and issues of conservation. Three soup processing factories had been operating on both Heron and nearby North West Island since 1924, and as, in that short time he had recorded a noticeable reduction in turtles found on the beaches to supply the factories, he warned of the serious consequences of unregulated capture (Moorhouse 1933). He also discovered that each island had a specific population of turtles, and that as they were caught and killed, so returning numbers inevitably declined.

Citing the numerous regulations already in place in the West Indies, the East Indies, Fiji and the Seychelles where conservation measures had been taken earlier, in contrast to Queensland where there were no regulations whatever forbidding the capture and slaughter of green turtles, Moorhouse made a specific recommendation for conservation. He proposed, in drafting legislation for the consideration of the government, total prohibition of any attempt to 'purchase, kill, or attempt to export, between the dates of 30th September and 30th November of each year, any turtle of the kind known as the Green Turtle (*Chelonia mydas*). Penalty £10 for each animal found in possession' (20). Absolutely nothing came of either his report or his recommendations.

In 1931, criticising the lack of government support and funding, Moorhouse resigned to accept a position in the Pacific islands. He returned to part-time Reef research in 1933 under the aegis of the GBRC, but his position with the GBRC did not last and in 1935 he moved to Adelaide as an inspector of fisheries. For the next decade there was no government marine biologist in Queensland.

Despite the rigours of the Depression, Reef excursions continued to expand for the moderately affluent. In 1930 the Mackay Chamber of Commerce appointed experienced Swiss tourist operator F. Treudhardt as Tourist Manager of Mackay Tours Limited, and on 19 December the first group left for a 21 day 'summer camp' on Lindeman Island for gentlemen, with '50 beds in Bungalows for Ladies', in what the promotional brochure described as 'The World's Finest Playground'. Listed in the attractions were fishing grounds for rod and line, coral reef excursions, marine gardens, dugong hunting, island cruising in motor boats, and night recreations of listening to the wireless (radio), music, piano and dancing. The weekly tariff of £6.10.0 – around twice the average weekly wage of those employed in the period – included accommodation, meals, and fares to and from the island from the port of Mackay, along with daily excursions within the Whitsunday group; a small shop on the island stocked such extras as tobacco, confectionery and fishing tackle (Brochure, Au.Arch.).

At the same time the Queensland government seized the chance to promote tourism on the Reef for those wealthy Australians who were accustomed to travelling abroad. The Queensland Government Tourist Bureau was created as a publicity office, and by 1932 also as a travel booking office, advertising the fact, in the words of Director A. E. Cole, that the adverse overseas exchange rate for Australian currency 'which restricted overseas travel from Australia provided Queensland with a favourable opportunity to lure Southerners to her sunny winter climate with the slogan "Cairns before Colombo"' (Cole 1933:269).

Determined to promote the State as 'the playground of the nation' the Tourist Bureau began a continuing program of promotional literature in which the attractions of the Reef were highlighted throughout. Even Professor Goddard, who was now deeply committed to Reef research, contributed a five page article on 'The Economic Possibilities of the Great Barrier Reef' repeating the themes already developed by Richards, Yonge and others in seeking funds for the work of the GBRC. While the possibilities he suggested may have been in accord with the times, even in the early 1930s it must have been quite distressing for many to continue reading his account of the economic possibilities of the edible green turtle, considering the cruel methods of capture and slaughter. The failure of the factories on North West and Heron islands at the time and the research results of Moorhouse sent to the GBRC seem to have escaped Goddard when he wrote that 'I am strongly inclined to regard the possibilities of this industry as one of the most important assets of the Barrier Reef, and to consider that a properly organised effort to develop it and to test out the product on American and European markets would meet with success' (Goddard 1933: 221–22).

In 1935 the Bureau began a major campaign to entice tourists to Queensland, employing the 23 year old Townsville-born neophyte writer C. B. (Clement) Christesen – founder in his more accomplished years of the literary magazine *Meanjin* – as official publicist from 1935 to 1942. His first effort was a lengthy 271 page rhapsody of the beauties of the State. Organised into ten chapters, each dealing with one of the major regions, they had alluring titles such as 'Mackay and Sugar Valley: Sugar Cane and Silver Shores' and 'The Mellow South Coast: by Shore and Sunset'. The most over-written was the tenth, 'The Great Barrier Reef: Isles of Romance', which regaled readers with 22 pages of lush hyperbole. One entirely typical passage is enough to illustrate its style:

> The ozone-saturated air spiced and washed clean by the vast expanse of Pacific seas; the opulent, languourous warmth; the flamboyant splashes of colour, and delicate pastel tints; opaline waters softly splashing on silver beaches, or breaking with musical tinkle on coral-strewn strands; submarine bowers of unbelievable loveliness – lucent gardens of Neptune which rival those of the Hesperides, the daughters of Oceanus (Christesen *c*.1936:243).

Covering, in its more restrained passages, the history of exploration, the biology of reef formation and its economic potential, the Bureau with this launched into a major marketing exercise which it was hoped would lure the anticipated 'wonder-struck tourist' (243).

Reinforcing the tourism drive at the same time, quite adventitiously, came *Wonders of the Great Barrier Reef* in November 1936 by Theo Roughley which continued the same romantic theme established by Napier a few years earlier, although in more accurate descriptive detail. Professionally employed as Superintendent of Fisheries for New South Wales, Roughley wrote an account at greater length of his vacation travels through the Reef. In addition, for the first time in the popular literature, his book introduced a new development with the inclusion of 52 colour photographs, mostly full page, along with 30 black and white photographs and some line drawings. Not since Saville-Kent's *magnum opus* of 1893 had colour illustrations appeared in a Reef book, although unlike Saville-Kent's hand-drawn lithographic plates, Roughley's illustrations came from the colour transparencies of photographs he had taken during his travels. *Wonders* enjoyed the same sustained success as Napier's *On the Barrier Reef*: a reprinting was rushed through a month later in December 1936 and eight more printings appeared over the following decade until July 1947. Despite wartime shortages and difficulties, reprints appeared in August 1941, in July and again September of 1943, and July 1945. Clearly the public appetite for escapist Reef romance was growing, and indeed, has continued to grow without any sign of satiety.

The following year, 1937, the Queensland Government Tourist Bureau approached Henry Richards for background information to prepare a new promotional brochure and this prompted one of the most astonishing displays of professional ignorance in the history of Australian marine studies. At the time the Reef was often called, with an allusion to Venice, the 'Grand Canal', while the term 'Great Barrier Reef' had become so entrenched in scientific, cartographic and popular usage that it was thought an interesting beginning would be a comment on the origin and significance of the name. Unbelievably, no one in the tourist bureau had any idea of its provenance, nor, despite their wide inquiries, could they find any source of information. The obvious place to seek an answer, then, was from an expert on the Reef, Professor Richards, chairman of the GBRC. The Tourist Bureau wrote to Richards who, even more incredibly, had no idea. Quite simply, no one, including the executive of the GBRC, had any notion of how Queensland's most prominent natural feature had ever received its name.

Then followed a ludicrous episode: Richards wrote to Rear Admiral Edgell, Hydrographer of the Royal Navy at the Admiralty in London, stating that

> I have been rather knocked over by this question as it had never occurred to me, and all endeavours which hitherto I have been able to make to find out have been in vain. I think it likely that you more than anybody would be able to find the answer to such a question. If you are able to do so, we would be deeply grateful – it is when one comes to think of it a matter of more than ordinary interest (UQA 226/2/1).

Indeed. Unfortunately, we do not possess Edgell's reply which could make fascinating reading, but this episode makes abundantly clear the profound limitations of the kind of positivist and presentist scientific approach employed throughout all Richards' work on the Reef and his lack of any comprehensive knowledge of its historical background. Clearly, he had never read Flinders' *Voyage to Terra Australis* nor studied in its *Appendix* the charts of the east coast labelled in large capitals GREAT BARRIER REEF. Beyond George Santayana's oft-quoted aphorism that 'Those who cannot remember the past are condemned to repeat it', the case of Richards is even more pathetic: he seemed never to have known the past in his own discipline to be able to forget it.

GROWTH OF THE CONSERVATION MOVEMENT

Shorn of research funds and therefore of project initiatives by the exigencies of the Great Depression, the GBRC, still a band of dedicated enthusiasts, concentrated its efforts on another area of growing Reef concern: conservation. Throughout the nineteenth century the prevailing ethos in Australia was primarily one of unrestrained expansion and exploitation: clearing the forests; massacring native wildlife, either to provide pasture for imported animals or as 'sport'; and reshaping both vegetation and fauna through introduction of exotic species in acclimatisation programs. Increasing interest in conservation issues, however, was already being generated. As early as 1877, recognising that 'Native Birds are disappearing rapidly from some Districts of the Colony and it is expedient to protect them and their progeny', the Queensland government began one of its first conservation measures with the *Protection of Native Birds Act* (Preamble: 41 Vic. No.7). Stimulated in large part through the example set by John Muir and George Perkins Marsh in the United States, the foundation of Yellowstone National Park and the subsequent declaration of the Royal National Park in Sydney, there had been a steady, if small, growth of environmental and amenity awareness.

In Queensland small parcels of land began to be set aside as nature reserves and by 1900 seven such parks with a total area of around 25 000 acres (10 000 ha) had been declared. The Reef, however, was not seen in the same light. All early national park movements in Australia were terrestrial and the Reef remained virtually inaccessible to the public so that for most persons their experience was entirely vicarious, gained mainly from the writing of Banfield, along with descriptions or journalistic accounts, of disasters. The GBRC, indeed, came to function as a clearing house of information on Reef concerns and in turn a lobbyist to government departments on corrective measures, although it was always careful not to identify with the policies of any particular political party. Moreover, it had become very territorial in its attitude to the Reef, regarding itself as the one source of knowledge, and therefore its protector, constantly wary of intrusions into its domain that might introduce degrading activities. So the GBRC found itself in a difficult conflict situation: while it had been extremely active in promoting the Reef as a location of enormous scientific significance and great economic potential – witness the endless proselytising of Richards, Goddard, and Yonge himself during his expedition, all of whom advocated vigorously its economic exploitation – it then had to confront and help resolve the inevitable problems and protests that arose from such activities.

Always prominent were complaints about ruthless resource stripping throughout Reef waters by Japanese crews as they sought dugong, trochus and bêche-de-mer. Banfield in his day was a frequent complainant to both the GBRC and government, and apart from his letters criticising their relentless slaughter of dugongs, he also fulminated in one letter, 2 March 1923, about their excessive collecting of trochus and indiscriminate felling of native trees on Dunk and other islands for fuel to boil and smoke the bêche-de-mer (UQA Box 78 (1)). That refrain continued. On 7 June 1926 the Sydney *Daily Telegraph*, drawing from the experiences of the Australian Museum members who took part in the disastrous Oyster Cay drilling project, reported that 'the slaughter of dugongs by Asiatic peoples [read Japanese] constitutes a black chapter in the history of Australian fauna'. A week later they reported Tom Iredale's complaint that crews of Japanese luggers were ravaging colonies of nesting sea birds on the island by collecting their eggs for food (*Cairns Post* 15 June 1926). Ten days later Gilbert Whitley was quoted to the effect that 'Asiatics are gathering large quantities of trochus shell' and by cutting down their nesting trees for firewood 'are creating tremendous havoc among sea birds' (*Daily Telegraph* 25 June 1926).

By 1930 pressures on the Reef had increased greatly as it became a major shipping channel. Oil spills had become a matter of international concern and as early as 1926 a conference on oil pollution at sea initiated by the United States was convened in Washington. In building a file of relevant information, the conference contacted the Commonwealth Department of Navigation for information about Australian waters; in turn, the Queensland office, through Captain Roskruge, responsible for Reef waters, asked the GBRC for any knowledge it may have. The Committee advised that it was unaware of any problems, but that it would support any control measures introduced. In fact, prevention of oil pollution at sea was a late international development: the world's first Convention for the Pollution of the Sea by Oil (Oilpol Convention) came in 1954 although Australia did not sign until 1962 following Commonwealth legislation on jurisdiction over territorial seas. Most interestingly, due to markedly changed perceptions of the Reef as a natural asset, Australia's interest in the Oilpol Convention when adopted was mainly in securing powers in international law to prevent pollution of the Reef (Burmester 1984:444–45).

The main issues concerning the GBRC at first were not potential oil pollution but rather mining, animal harvesting and resort development. At the time, there was no overall supervisory body, nor was the Reef seen by government as a particularly sensitive region. When control measures were introduced they came as a result of responses to isolated episodes, and were managed by separate authorities: islands by the Lands Department, although animals and birds on them were the responsibility of Agriculture and Stock; fish and oysters by Fisheries; mining for guano and coral for fertiliser by the Mines Department – for deposits above the high water mark – and Harbours and Marine for those below low water. In addition, the Commonwealth controlled lighthouses and navigation. The GBRC, consequently, had to deal with six separate authorities, not all of whom were at times cooperative, or even sympathetic, to conservation concerns.

The chief mining issue was the extraction of coral from fringing reefs for processing ashore into lime fertiliser. The extensive coastal sugarcane plantations, reaching mainly from Mackay to Cairns, required considerable quantities of lime, while the reef, composed of limestone (calcium carbonate) was seen as a conveniently accessible source. In 1926 Harbours and Marine issued a licence for coral mining on Upolu Banks, 32 kilometres (20 miles) to the north-east, close by Oyster Cay where the drilling was taking place at the time. The GBRC protested vigorously, chiefly because the cay was a declared sanctuary and mining activities were very disruptive of nesting birds, while the removal of material threatened stability of the cay. Not until 1934 was the licence revoked. Even so, further coral mining leases – despite continuing agitation for their banning – were issued for Hutchinson reef in the North Barnard group off Innisfail, a major cane growing region.

Animal harvesting was a particularly difficult issue. At the time the waters of the Reef were international territory, and only those within the three mile limit came under Queensland's jurisdiction. The Japanese, therefore, were entirely within the law fishing in Reef waters provided they remained outside the limit. Fishing within territorial waters around the islands however was an offence, as well as landing, cutting down trees, building smoke houses for curing bêche-de-mer, and taking bird eggs. Ever since the late nineteenth century the Japanese, like the Chinese before them, had been bitterly opposed by Queenslanders, and only tolerated in the pearling industry due to their expertise and the international treaty between Britain and Japan. The great difficulty was enforcing control since the islands and cays were numerous and patrols few. And, as the report on turtle canning and the rapid decline in populations on Heron Island in 1930 by Frank Moorhouse illustrated, Australians were often similarly guilty of rapacious behaviour.

Equally disturbing to the GBRC were continuing requests for island leases to establish resorts. Although primitive by modern standards, mainly composed of camp sites, the Committee was receiving reports of vandalism by visiting groups, souveniring of coral, and even reefs being dynamited by fishing expeditions. Embury, for example, was strongly opposed when he applied for leases on North West and Masthead. Already camping groups had created considerable damage on Lady Musgrave, the second most southerly of the cays, and it was feared that further leases would only accelerate the damage across the entire Reef, at least on those islands and cays reasonably accessible to the mainland, as increasing numbers of motor launches and other vessels were appearing, with opportunist owners seeking clients.

Perhaps the most unpleasant aspect of the Committee's attitude to use of the islands came in the issue of Noel Monkman. At that time, as a freelance biologist, Reef enthusiast and highly skilled microscopist, Monkman was pioneering photomicroscopy of Reef marine life. Noel Monkman (1896–1969), a musician, and his Russian–Polish born concert pianist wife Getela, known always as Kitty, had decided to abandon the stage and move to Green Island in 1929, where they had established their own small private marine station to pioneer underwater cinematography. Monkman had a Cairns plumber make him a Low Isles style 'dustbin' helmet with a window visor, and from a dinghy, with air provided from a double action tyre pump supplied by 'Kitty power' – she recalled later that he always knew when she was tired, as his air ran out – he produced the first underwater films of the Reef. Five of them – 'Coral and its Creatures', 'Strange Sea Shells', 'Birds of the Barrier', 'Secrets of the Sea', 'Ocean Oddities' – received world-wide distribution, although in Britain the censors cut what they considered the rather indelicate footage of the turtle laying her eggs (Monkman 1975:30–32). Monkman was to go on to make for Commonwealth Films Limited in 1937 the adventure story of a wrecked, pearl-laden lugger entitled 'Typhoon Treasure', along with training films for the Royal Australian Air Force during the Pacific War.

Monkman's innovative filming recording his profound interest in and knowledge of Reef biology had come to the attention of Richards who sought his assistance in some of his own projects, which Monkman gave freely. As the friendship developed, Monkman asked in 1931 for a return favour, due to Richards' position in the GBRC, requesting the use of the buildings on the Low Isles for a brief period to do some underwater filming since the tourist crowds at Green Island were increasing to such an extent that he was finding it difficult to work in the peace so essential to his projects. Richards agreed, and simply told Monkman to write for the key when he was ready to use the buildings. Monkman was puzzled and upset, after making the request, when Richards replied that he could not grant their use as the Department of Harbours and Marine, the official custodians, did not allow access to private individuals. Monkman, understandably, became angered when, upon checking with the Department, he was advised that they knew nothing of the matter. In fact, the records of the GBRC reveal that Richards took Monkman's request to the Committee, and at the meeting of 14 July 1931, it rejected his request because it believed that he wanted the use of the islands for commercial and not scientific purposes (UQA S226, Box 1, Item 1). In a bitter riposte, Monkman wrote to Richards that 'the temple of science is a very noble structure, and the humble artisan who attempts to set one brick upon that structure, should be helped, not hindered, by its high priests' (Monkman 1975:65). The GBRC rejection implied, of course, that Monkman was a mere 'populariser', a very sad reflection considering the fact that today Monkman is being recognised as a significant pioneer in Reef cine- and photomicrography. In 1961 half the footage of RKO Radio's film of Rachel Carson's best selling marine book *The Sea Around Us* was made by Monkman. Today James Cook University is perpetuating his achievements in the 'Monkman Collection' at the campus, and in the Noel and Kitty Monkman Marine Research Station on Green Island.

Any attempt to understand the conservation issues of the Reef must be considered against the background of Queensland politics throughout the 1920s and 1930s. Always the most culturally backward state in Australia, Queensland had the lowest levels of general education (as measured in a series of long-term surveys by the Australian Council for Educational Research). With the notable exception of a few politicians, such as Samuel Walker Griffith, members of parliament were poorly educated and generally came from industrial or farming backgrounds. The overwhelming drive was to clear the land for agriculture, a major consequence being the destruction of forests and the extermination of wildlife. The Labor Party, which held office for most of the interwar years, with its ideology of agrarian socialism for the common farmer, had little interest in conservation. While the GBRC was attempting to influence the government to implement at least a minimum of management control over the Reef, the assault on the land was continuing at a rapid pace. In one particularly atrocious episode, in the single month of August 1927 over a million possums and around 600 000 koalas were killed in a great drive, their skins being used in the fur trade and their slaughter actively encouraged by Lands Minister Thomas

Foley as a major effort to make the fur trade one of the major industries of the State (Fitzgerald 1984:77).

Clearing the land, in fact, was to become one of the major issues of the period as the conservation movement began to grow. The depletion of forests became highly contentious when Foley, himself a timber merchant, introduced a Bill into parliament in 1929 designed principally to make leasehold of rural lands more secure in order to stimulate rural development. An aspect of the 'improvement' of land was set out in Part V, seeking to provide 'Encouragement of Ringbarking and Clearing of Useless Vegetation'. When the Bill received assent on 5 December (20 Geo. V, No.15) it immediately brought Foley into conflict with his Director of Forests, E. H. Swain, who was attempting to introduce enlightened management practices including reafforestation. A Royal Commission of 1929–32 was appointed to inquire into the industry and, not surprisingly, it made no recommendations for conservation. Immediately after, Swain was dismissed as obstructionist, the forests were put under the direct control of the Minister for Lands and the felling of the forests continued unabated, a disastrous policy with profound effects on the Reef that is still in operation at the beginning of the twenty-first century.

How did the Reef fare against that background? Initially, not too well. It did, however, have a more glamorous public image given its vice-regal patronage and the prominent public figures associated with it. The first glimmer of hope came in 1930 with the foundation of the National Parks Association of Queensland (NPA) and the North Queensland Naturalists Club in Cairns in 1932. The stimulus to create national parks was an initiative of New South Wales, not Queensland, begun in the premiership of John Robertson who led the campaign for retaining large tracts of land in as close to pristine condition as possible for the passive enjoyment and recreation of the ordinary working population. In Queensland the movement grew more slowly but the declaration of Lamington National Park in 1915 was a real milestone.

Many members of the GBRC joined the NPA: Henry Richards was in the chair at the foundation meeting. Due to the GBRC's considerable prestige, and the fact that it had generated an enhanced public perception of the Reef as a special place, and although there was still a redneck approach to land issues, the NPA and the GBRC were able to work reasonably amicably with the various government departments. One of the first items on the agenda was the thorny matter of island leases and the government was pressed for an acceptable policy of management, especially regarding extractive industries, Japanese resource stripping, and resort development. The first significant initiative was taken by Richards in 1933 when he convened a meeting of the GBRC and the NPA with the heads of the relevant government departments along with the shipping companies to begin a process of consultation in order to plan ahead cooperatively (GBRC minutes, 13 September 1933: UQA S226/1/1).

As the 1930s progressed the results began to add up. While no definite policy was articulated concerning resort developments some progress was made on Japanese intrusions when, as a result of submissions made to the government by Sir Donald Cameron, a member of both the federal parliament and the GBRC, the Commonwealth agreed in November 1934 to provide a launch patrol in the Torres Strait and Thursday Island region. Protection of wildlife also began to receive more attention: in May 1933 some islands were proclaimed sanctuaries under the provisions of the *Animals and Birds Act* (1921) while a series of Orders in Council (enforceable executive decrees that are not contrary to existing laws and statutes) increased protection for the Reef. In 1933 taking coral along foreshores was prohibited, in May 1935 all islands were declared wildlife sanctuaries and by 1939 one hundred islands had been declared national parks. Monitoring of the regulations was strengthened by seeking community involvement with the aid of honorary inspectors given legal powers under the *Fish and Oysters Act*, and honorary rangers under the *Fauna and Native Plant Protection Act*. In addition, the Tourist Bureau was requested to begin advising visitors about the regulations and encouraging their observance (Hill 1985a:12–13).

GEOLOGY REVIVED: THE HERON ISLAND BORE, 1937

By 1935 the most severe effects of the Depression were moderating and Richards was getting restive: the total cessation of Reef science had prevented any further geological investigation and he began to explore ways to revive the scientific activities of the GBRC. In 1935, while on study leave in England, he spent time at Cambridge enlisting yet again the support of Gardiner, Steers and Lenox-Conyngham for a British team to come to the Reef, and in effect, to take up where the Low Isles Expedition left off. He also had earnest discussions with Sir Matthew Nathan, officially retired but still politically active and influential.

Following talks with the Cambridge group, plans were formulated for several expeditions to begin as soon as funds could be found. On 5 June 1935 Richards wrote from England to E. O. Marks, secretary of the GBRC, that agreement had been reached for a project in the Capricorn islands to study 'the limiting conditions as far as the growth of coral reefs are concerned in this region', with close attention being given to 'the parts played by "Forams" and "Lithothamnion" in reef building . . . while the study of "symbiotic algae" associated with the corals merited further study' (CSIRO archives, Series 9: G12[4]). A second team would come to study the physical-chemical processes of reef formation, examining cementation, creation of 'beach rock', dolomitisation, and sediment deposition; and if possible, drilling some 'shallow bores'. Understanding such processes was vitally important in advancing geological knowledge: how did sand and coral debris come to be cemented together in the beach rock found along the edges of cays? what were the relationships between calcite ($CaCO_3$) and its subsequent combination with magnesium into the harder dolomite ($MgCO_3\ CaCO_3$)?

Richards was carefully manipulating the GBRC into further geological investigations to pursue his own particular interests. And always at the back of Richards' mind, it appears, as Dorothy Hill, his student and later colleague and successor to his chair of geology recalled to me, was his hope of finding oil. As she mentioned in that interview, for geologists, 'providing information for oil companies is our bread-and-butter' (Pers. comm., 14 June 1988). In the 1920s and 1930s, however, the oil companies were not particularly active in Australia and most exploration was conducted by State government geological surveys. D. J. Mahony of the Geological Survey of Victoria reported to the Pan-Pacific Congress that there was 'no evidence to support the view that petroleum is likely to be found in Victoria, and much more geological investigation is required' (Mahony 1923:1251–52). The most promising areas in the early 1920s seemed to be in Papua and New Guinea, and also in central Queensland, reaching from Warwick to Dalby and beyond (1275). The Reef was not considered a prospect, but on the coast south from Maryborough to Tewantin 'there are certainly strata which offer encouragement' (Jensen 1923:1276).

A third expedition from Britain was also undertaken by Steers, accompanied by F. E. Kemp, a graduate student, to continue his earlier geographical survey of the Reef that had been arranged concurrently with the Low Isles Expedition in 1929, and proved less than successful. For three months, until the end of August 1936, Steers made physiographic studies of the cays and islands of the Reef, reaching as far north as Cape Direction (12°52'S), due west of Bligh's entrance of 29 May 1789. The results of the survey, along with 40 charts of the traverses which gave a better understanding of the structure of cays and sedimentation processes, appeared in two articles, the first in 1937 in the *Geographical Journal* (Vol. 89, 1937, 1–28, 119–46), the second the next year in *Reports of the Great Barrier Reef Committee* (IV.1938, 51–104) (Hill 1985a:9).

Geographical work of the kind carried out by Steers, however, was marginal to Richards' interests: he wanted much closer investigation of the Reef. By 1936 the GBRC had managed to build a bank balance of nearly £2000 (£1948 at the audit of 29 November 1935). Richards persuaded the Committee that there were sufficient funds to finance further exploration and moved to secure the advantages offered by the Cambridge group to send out a small team for a shorter period, say three or four months. So, secretary Marks wrote to Sir Donald Cameron for assistance yet again in lobbying the government. Writing on 9 August 1935 Marks gave a brief and very positive resumé of the funding of the Low Isles Expedition, with extensive reference to the generosity of the Commonwealth, soliciting assistance in seeking yet again financial help for a 'considerably smaller party and less elaborate programme, to work in the Capricorn group, off Gladstone'. To moderate the request, he added that 'The expenses should be on a correspondingly smaller scale'. Seeking a pound-for-pound subsidy for funds raised by the British group up to £2500, along with waiver of customs import duties, he ended with an effusive expression of gratitude to Cameron for his 'continued personal interest and support' (CSIRO archives, Series 9, G12[4]). Cameron approached the federal government and four months later, on

10 December, the secretary to Prime Minister Joseph Lyons advised Marks the request had been favourably considered and that 'subject to the sum of £2500 being raised from other than governmental sources, the Commonwealth will provide a subsidy of £1250 towards the expenses of the proposed investigation'. In addition, customs duty on imported scientific apparatus would be waived, and if required, 'any application by your Committee for the loan of naval equipment for the purposes of the expedition will receive favourable consideration by the government' (CSIRO archives, Series 9, G12[4]).

In anticipation of a positive response, planning had proceeded as soon as Cameron had been approached and, fortuitously for Richards, news came that the group intending to study the physical and chemical processes of reef formation had not been able to find adequate funding. It suited Richards' purposes quite well and he did not inform the Commonwealth of the cancellation. Only some months after the offer of funding for the physical-chemical group had been made did Richards advise the office of the Prime minister. In fact, it took nearly eight months for the GBRC to consider a response when Richards wrote personally to Lyons on 21 August 1936, advising him that in respect to the planned physical-chemical investigations 'there must be a postponement of this part of the programme of activities'. However, in that letter Richards offered a more than satisfactory alternative: the other project of a deep boring could be brought forward and 'to concentrate on that work rather than postponing it till some future date', going on to add, with the customary deference to Britain, 'We trust . . . that you will not find it necessary to review your favourable reception of our application for financial help' since 'the British authorities in particular are anxious that we should carry out the boring' (CSIRO archives, Series 9 G12[4]). Richards had always anticipated Commonwealth agreement and previously, on 17 April 1936, the Committee had already sent out a letter of appeal urging donations with the persuasive argument that 'This scientific investigation of the unique coral mass – the greatest the world has ever known – is of the highest importance and much remains to be done' (Au.Arch.). Now, more than a century after Darwin published his theory on coral reef origins, Richards was again to expend all the GBRC funds on a futile attempt to prove that theory.

The original plan for investigating the substrate of the Reef had been to make three bores, one each in the north, central and southern zones, as close to the outer edge as possible. In 1926 Oyster Cay on Michaelmas Reef had been selected because it was in the central region, and the northern area was logistically difficult. The second, then, would be in the southern Capricorn region on Heron Island, so chosen, as Richards pointed out in his *Report*, because it was accessible to the mainland port of Gladstone, had a supply of fresh water, subsidised accommodation at the resort run by Captain Christian Poulsen of Barrier Reef Tours (who had bought the lease of the defunct turtle factory in 1932 and opened a tourist resort there in 1935), and the island could be serviced by the Commonwealth Lighthouse Service ships (Richards 1938b). The technology of the time prevented a bore being made any closer to the outer edge of the Reef.

The decision to bore on Heron having been approved by the Commonwealth, plans to repeat the Oyster Cay operation continued to mature, but now with the wisdom of hindsight. Larger diameter casing pipes were to be used, and as Richards wrote in his *Report*, the idea was to reach at least 1000, even 1200 feet (300–365 m). Through the generosity of the Victorian Department of Mines they were able to obtain yet again the loan of a Victoria Calyx drill and even the services of Thomas Hughes who had supervised the Oyster Cay drilling. Scientific supervision of the project was placed in the hands of J. B. Henderson, a mining engineer who had recently retired as Queensland Government Analyst. Captain Poulson arranged accommodation while a range of free services kept costs to a minimum: Queensland Rail carried the drilling equipment to the Port of Gladstone, the Commonwealth provided the lighthouse steamer *Cape Leeuwin* to take all 23 tons of equipment across to Heron Island, while the Gladstone Harbour Board contributed the services of a large punt on to which, in four separate relays, the gear was unloaded from the *Leeuwin* and ferried across the Reef flat to the drilling site.

To the great interest of some seventy onlookers, mostly Victorian private school girls enjoying a vacation at the resort, after a brief opening ceremony by Mrs Henderson at 10 am on Monday 17 May 1937, drilling commenced. It continued for three months until Friday, 13 August – if there was any significance found at the time in the date – when, having reached 732 feet (223 m) it was declared impossible to continue and was reluctantly stopped.

Despite careful preparations, the major problem was the extremely loose nature of the material through which the drill had to penetrate. Wider diameter casing – 8 inch (20 cm) – had been obtained, but as the drill descended the casing had to be progressively reduced; at 510 feet (155 m) it was replaced by 5 inch casings (12.7 cm); at 668 feet (203 m) to the minimum workable diameter of 4 inches (10 cm). Most of the way down only very loose, incoherent sand and rubble was encountered; altogether, as the log of the bore indicates, only eight bands of coherent siliceous limestone varying in thickness between a few inches (5–10 cm) and 30 inches (76 cm) were found (Log reproduced in Richards 1938b:Pl.XV).

It was a great disappointment and as Richards wrote, 'The log of the bore ... [is] essentially the record of a repetition of that established ten years ago by the boring at Michaelmas Cay some 770 miles further to the north-west along the "reef"' (Richards 1938b). No hard substrate: the foundation on which the cay rested had not been reached and no confirmation of the Darwinian hypothesis was achieved. Richards blamed the failure on the lack of a suitable, more specialised under-reamer bit which would have allowed the use of a wider diameter casing and therefore the attainment of a greater depth. At that point, he was forced to comment in his report that it would be impossible to finance a second hole on Heron: all drilling efforts by the GBRC thereupon ceased, never to be resumed. No great satisfaction was gained although there was some determination of age of the lowest fossils as coming from the Plio-Pleistocene boundary (2 mya). The final cost to the GBRC was £2190, a sum they could ill afford to lose for no meaningful result whatever. Reef research stopped entirely after the abortive drilling of Heron in 1937, and the GBRC lost its great ambassador in England when Sir Matthew Nathan died in 1939.

Throughout those years, the ominous events in Europe and China were casting a pall over the world as war threatened. Germany's rearmament and belligerent attitude towards France and Britain, and the Japanese invasion of China in 1937, were to reach a crescendo in September 1939 with the German invasion of Poland, and in December 1941 with the Japanese bombing of Pearl Harbor.

CHAPTER 18

THE PACIFIC WAR AND ITS AFTERMATH

Throughout the 1930s the Reef in the north was a scene of ongoing conflict with Japanese workers who had resumed aggressive resource stripping in the pearling industry, practices that had been arousing antagonism in Queensland ever since the 1880s. Tension in Australia was rising, evident in continuing denunciation of their impact in both the parliament and the press, exacerbated by a heightened fear which had been building up for decades. Japan's aggressive moves to gain military and economic hegemony over Asia, now combined with the growing threat of a rearming Nazi Germany, was creating alarm in Australia over what seemed an inevitable drift to war.

Hostility over Japanese depredation of Reef marine resources and towards Asian immigration had been of much longer standing. When the *Immigration Restriction Act* of 1901 implementing the 'White Australia Policy' was passed by the first federal government, although persons engaged in the pearlshell and bêche-de-mer industries were exempted, allowing access to Japanese divers, its provisions continued as a serious irritant. Even though Japanese citizens were not permitted to emigrate, and Japan itself opposed all foreign immigration – an effectively racist policy – the *Act* of 1901 was seen as an affront to national prestige and Japan's standing as an international power (Sissons 1972:193f.). Despite such objections, Australian hostility over access to Reef resources persisted. In the *Report* of the Royal Commission appointed to inquire into the working of the pearlshell and bêche-de-mer industries, Chairman Mackay recommended that 'Steps should be taken to reduce this vast preponderance of aliens, and to have this outpost [Torres Strait] guarded by a hardy population of loyal and patriotic Australians' (Mackay 1908:LX:58).

Concern was intensified after the World War I peace settlement when Japan, having received all of Germany's Pacific territories, and Australia, gaining a League of Nations mandate over German New Guinea, came to share a common maritime border in Micronesia. The Australian military high command assessed the strategic situation as one in which Australia's only 'potential and probable' enemy was Japan which, coveting the marine resources of Australian tropical waters, might conceivably use the White Australia policy as a *casus belli* to invade the relatively empty north (Hayman 1979:9). The general Australian sentiment of the time was well summarised in the parliament by Senator John Miller from Tasmania: 'The problem of the Pacific is the problem of Japan'.

Australia sought protection and in 1919 invited British Admiral Lord Jellicoe to visit and prepare a defence plan. Jellicoe recommended a massive naval base at Singapore for the British, Indian, Australian and New Zealand navies, operating as a grand unified fleet. The strategy, however, was seriously flawed when, in the Treaty for the Limitation of Naval Armament of 6 February 1922, Japan, to preclude any surprise strike on its own territory, secured an agreement for 'no fortifications' of naval bases by Britain and the United States any closer than Hawaii and Singapore. In 1923 at the Imperial Conference a decision was made by Britain to proceed with the Singapore base, while Australia would have only a small militia, trained to sustain a six month resistance to an invading enemy, such as Japan, until the great battle fleet arrived to effect a rescue.

JAPANESE INVASION FEARS: NAVAL SURVEY OF THE REEF

Australia, for its part, was sceptical of British resolve to defend Singapore, and was convinced – quite correctly – that the British government had no real idea of its sense of isolation in the South Pacific. Despite assurances by the British government in response to Australian pressure that it would begin to strengthen the Singapore base, Britain remained dilatory and Australia recognised the imperative to act independently. By 1921 the Commonwealth Parliament had prepared a plan for Australian defence, leading in 1924 to preparations to increase the stockpile of armaments and ammunition, and the upgrading of military equipment. The general tone of thought in the period was best expressed by Prime Minister Stanley Bruce in opening the debate on the Defence Appropriation Bill: 'We intend to keep this country white and not allow its peoples to be faced with the problems that at present are practically insoluble in many parts of the world' (Hansard, House of Representatives, 1924:1702).

A significant focus of that marathon debate of twenty sessions was the concern expressed by Josiah Francis, member for the Queensland seat of Moreton in the Brisbane metropolitan region, of the vulnerability of the Reef coast as the likely route of an invading force and that Australia was totally ill-prepared, having only 'charts of the reef [which] still bear the inscription "Soundings and reefs by Captain Flinders" . . . [which] must necessarily be hopelessly incomplete . . . [while] little is known of the direction and set of currents in the vicinity of the reef'. It was therefore essential, he argued, to begin an up-to-date survey, expressing the final sentiment 'that [this Bill] will be the beginning of an effective naval defence force for Australia' (2137). Senator Wilson repeated a well-voiced anxiety, especially in the Torres Strait area, that many of the crews of Japanese fishing luggers were actually naval officers collecting intelligence and that, consequently, 'Japan has probably better information as to the Great Barrier Reef than we ourselves possess' (3342). Senator Wilson was able to move successfully for a 'Barrier Reef survey vote' of £100 000 annually for the provision of a naval sloop carrying two seaplanes for aerial photography, then, of course, in its infancy. One of the provisions of the *Defence Equipment Act* (No. 18 of 1924), consequently, were funds 'for the survey of the Great Barrier Reef'.

Prior to the development of aerial photography, traditional sounding surveys of the Reef continued throughout the 1920s and 1930s, described in Commander Hardstaff's comprehensive survey manual (Hardstaff 1995). One and a half centuries after Cook's description of the Reef as a bewildering 'labyrinth', navigators were still continuing the task of charting safe passages, setting down, as it were, a metaphorical Ariadne's thread as a means of providing sailors with a secure pathway. Leadsmen continued to stand in the chains and measure depth in fathoms by swinging a lead line, reading them off from pieces of leather and variously coloured cloth inserted at 2 fathom intervals into the lay (twisted strands) of the line. The situation throughout the 1930s was little changed; virtually all of the Reef surveys were still limited to the main shipping channels and those areas frequented by fishing vessels: most of the maze of patch and ribbon reefs remained uncharted, and local information on safe navigation passages was often jealously guarded for commercial advantage. The very small number of naval survey vessels – three at the most in that period – were, in fact, kept constantly busy updating the major channels due to changes in the seabed from currents, reef processes and sedimentation, and the continuing discovery of previously uncharted obstacles.

For oceanic sounding when intercontinental telegraphy cables were being laid down in the 1870s and 1880s, mechanical techniques had been devised using strong cables on power-driven winches, such as the patented Baillie Sounding Machine carried aboard HMS *Challenger* on its epic global survey, and the Lucas Automatic Sounding Machine employed by the Telegraph Construction and Maintenance Company. For the Reef, with its shorter traverse distances and shallower depths, much less complicated devices were used, such as the Somerville Sounding Gear which consisted of a metal weight towed behind the ship at a very slow speed on a steel cable which followed the undulations of the seabed, being effective at depths up to 66 feet (20 m). As it rose up and down fluctuations

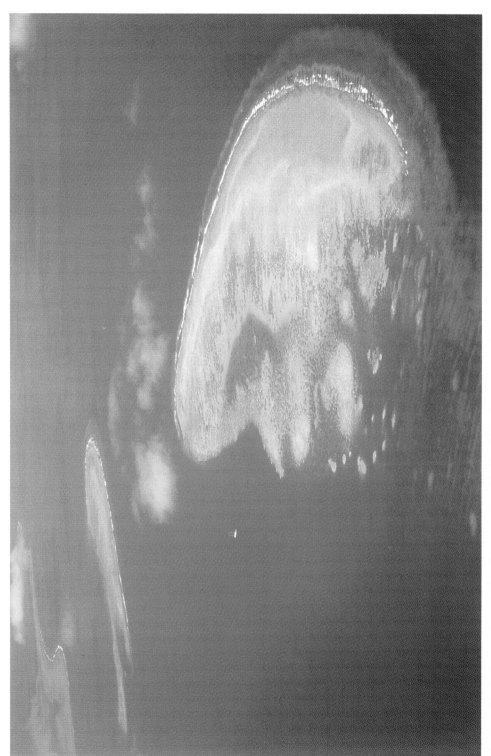

Patch reef with characteristic shallow lagoon and rubble flats behind a curved reef front. Photograph: Andrew Elliott.

Ribbon Reef in a northern section of the Reef: the drop-off into the Coral Sea on the right plunges to 2000 metres.
Photograph: Leon Zell.

Continental Islands. Dent and Hamilton Islands in the Whitsunday Group, one of the most popular tourist areas of the Great Barrier Reef.
Photograph: Wendy Craik.

Large coral patch formations of Pompey Reef. One of the sites of the first aerial survey of the Reef, conducted in 1929 by Flight 101 of the Royal Australian Air Force.
Photograph: Leon Zell.

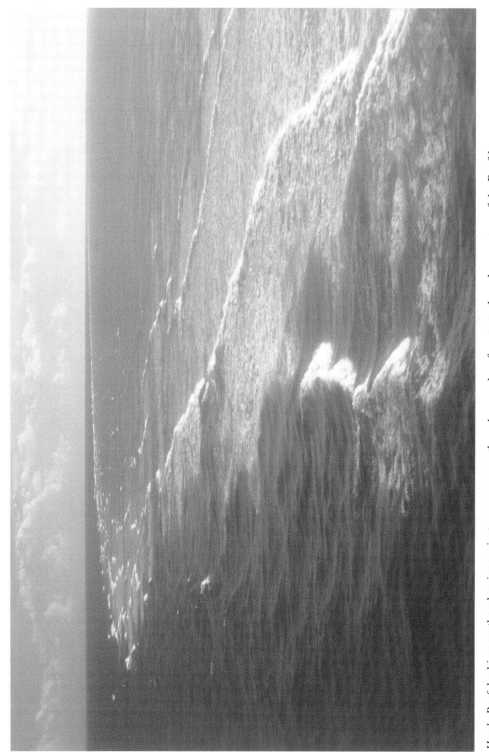

Hardy Reef, looking south: a barrier against ocean surges, the submerged reef protects the calmer waters of the Reef lagoon. Photograph: Leon Zell.

Shute Harbour: On the coast looking out across the continental islands of the Whitsundays, most of which are national parks administered by the Queensland Parks and Wildlife Service. Today, a major focus for Reef tourism. Photograph: Leon Zell.

Queensland Parks and Wildlife Ranger, Jeff Illiff, patrolling the Reef near the Low Isles, famous as the location of the 1928–29 English biological expedition, now a popular daytrip site. Photograph: Andrew Elliott.

Coral coring. Using a compressed air drill a core is extracted from a large clump of Porites for laboratory analyses of the deposition layers to identify past climatic and weather cycles. Photograph: Mark Johnson.

Fantasy Island. Moored near John Brewer Reef in 1988 as a day trip attraction, it was destroyed in a storm ten days later. Photograph: Andrew Elliott.

could be read off a graduated line hanging vertically below the rear deck. In 1921 the *Fantome* employed another device, known as a 'Submarine Sentry', to mark shipping channels through the Reef at two standard depth settings of 30 and 40 feet (9 and 12 m), although it was effective at depths down to 240 feet (73 m). It was, in its basic principles, simply a metal plate towed below the ship on a steel cable in a horizontal position at a set depth with a small rod protruding down which, upon contacting an obstacle, made the plate rise upwards and sound an alarm, following which more detailed examination could be made to determine and chart the depth of the obstacle (Hardstaff 1995:3). As ships became larger so more sophisticated technology came into use in the early 1930s when various kinds of acoustic echo-sounding methods were first employed.

AERIAL PHOTOGRAPHY: AN INNOVATION IN REEF SURVEY

By the early 1920s it was believed that conventional sounding methods could be supplemented by the newly emerging technology of aerial photography. Aircraft at the time were relatively ineffective for offensive action, but, since warships were now able to fire beyond the visible horizon, they were becoming employed as spotters for gunfire ranging and reconnaissance. It was decided, therefore, to form a seaplane unit for naval cooperation which, among other tasks, would be able to make a defence survey of the Reef. The first such aircraft used for that purpose were Fairey IIID seaplanes: open cockpit biplanes with permanent floats which in practice were cumbersome to launch since they had to be physically manipulated on to a trolley from their shore hangars and trundled down a ramp into the water, afterwards being retrieved by the reverse process.

In 1924 experiments were made in locating one on a specially built afterdeck on HMAS *Geranium*, carried aboard with its wings folded horizontally along the fuselage. For survey work it was lowered into the water by a special derrick, but in practice it had very little strategic value, being limited to sheltered waters, requiring the ship to remain anchored until its return. A trial aerial survey of the Reef was made of a few outer patch reefs east of the coast between Cairns and Innisfail by Flight Lieutenant Mustard and Flying Officer Swinbourne, where three previously unknown hazards were located and photographed from a standard height of 10 000 feet (3000 m) to produce a composite mosaic: they were named Mustard, Swinbourne and Raaf Shoals. It was a relatively simple procedure whereby 'the flights were controlled by dead reckoning and the reefs plotted by compass bearings and sketched on to field boards for transfer to the ship's collector tracing. The reefs were later positioned by astronomical observations made by landing parties from the ship' (Hardstaff 1995:7). The results, however, were mixed: as the official report stated, 'The use of an aeroplane in connection with this survey demonstrated that for reconnaissance purposes and for locating winding passages it was invaluable . . . on the other hand, the photographs conveyed little information about depths which were all sounded' (5). Although another limitation was that while numerous reefs in a given area were too far apart for the separate photographs to be joined together, they did enable the preparation of charts of individual reefs from a sequence of overlapping exposures (Jones 1989:33).

Given the serious limitations of the Fairey IIID seaplanes, and – wisely in retrospect – still distrusting British intentions to fortify Singapore adequately, the Australian government decided to create a Fleet Air Arm equipped with a new generation of improved naval cooperation aircraft. Following passage of the *Defence Equipment Act* in 1924, orders were placed in Britain for the Supermarine Seagull III amphibian, designed by Reginald Mitchell, later to gain enduring fame in aviation history for his legendary Spitfire. A high wing biplane with a single engine housed between the wings, the Seagull III was essentially a small flying boat, fitted with wheels for flying from airfields, and equally for travelling down the launching ramp under its own power, which were then semi-retracted sideways under the wing for take-off and landing on water. It carried a crew of three: a pilot in a separate open cockpit in the nose of the craft while immediately behind the engine was a dual open cockpit which held the observer, also acting as photographer, and behind him the radio operator who was also responsible for the swivelling machine gun. With a maximum speed of 85 miles per hour (137 km/h) and a service ceiling of 9150 feet (2788 m), it had an airborne endurance of four and a half hours. Six aircraft arrived with serial numbers A9-1 to A9-6 and were given a special designation as RAAF Flight 101 (Fleet Cooperation), an unusual joint venture between the Royal Australian Navy and the fledgling Royal Australian Air Force, founded only in 1921.

The first flight was over Papua in October 1927, on behalf of the Anglo-Iranian Oil Company in an aerial geological survey to search for possible oil-bearing formations. However, due to unrelenting pressure in the parliament by Senators Kingsmill and Bamford, and Member for Moreton Josiah Francis for the survey of the Reef to be underway as an urgent defence measure, it was decided to base the Seagulls at Bowen to begin with the Whitsunday Islands and the outer edge of the Reef north as far as Flinders' Passage east of Townsville. The ground crew arrived on 26 August 1928 and three Seagulls a few days later. A three berth hangar and support facilities were built, with a concrete ramp leading into the harbour near the place where Captain Sinclair had landed from the *Santa Barbara* in 1859 and proclaimed Port Denison. It was one of those aircraft, with Flight Lieutenant Packer aboard – as officer responsible for coordinating aerial photography – that arrived at the Low Isles in 1929 to make a photographic survey for the expedition. The other three Seagulls were sent to St Bees Island, 30 kilometres (18 m) north-east of Mackay where a camp was set up in a bay as a base for a survey of the Hillsborough Channel, from Slade Point in the south up to Cape Conway opposite Lindeman Island.

Experiments were also made at carrying a Seagull aboard a naval survey vessel, and in 1926 one was placed on the *Moresby* – one of the navy's three survey ships along with the

Geranium and *Herald* after the *Fantome* had been retired in 1924 – for a survey of the Flinders Passage region, the main outlet to the sea for the port of Townsville. Another survey involved three Seagulls working in cooperation with *Moresby* exploring the outer edge of the Reef from Elliot Island in the extreme south to Flinders' Passage in the mid region, also searching for unknown passages. Again, as for the *Fairey* seaplane, it was a cumbersome operation since the amphibian had to be lowered to the surface and afterwards retrieved by a hoist; even then it could only be used in calm conditions, not being able to handle surface winds of more than 20 knots. As a consequence, only a few surveys were made.

Then, due more to political pressure than to defence needs, in a highly confidential decision, the Australian government decided to commission the construction of a carrier for the Seagulls, which created a sensation when, at the opening of Federal Parliament on 10 June 1925, Governor-General Stonehaven announced the decision in his speech. Both the navy and the air force were caught by surprise, and the Chief of the Air Staff, who had the responsibility for training the pilots, learned of the decision in the morning newspaper. The carrier, HMAS *Albatross*, was laid down at Sydney's Cockatoo Dockyard in April 1926 and launched in February 1929. When the Seagulls were transferred to the *Albatross* no further attempts were made to locate them on the afterdeck of the smaller ships. By the time the *Albatross* was launched the aerial survey of the Reef had been completed, and the vessel with its aircraft moved to other waters until 1933 when the ship was refitted for other duties and the Seagulls relocated aboard the heavy cruisers *Canberra* and *Australia*.

WAR COMES TO THE REEF

From 1933 on, when Adolf Hitler became chancellor of Germany and began the process of rapid rearmament, all Europe began preparing for war. This came when, after annexing Austria in March 1938, he invaded western Czechoslovakia in March 1939 and Poland on 1 September the same year. Responding to a security guarantee Britain came to Poland's defence and declared war on Germany two days later and the European conflagration known as World War II began, lasting for nearly six years.

Throughout that period Japan too was displaying an increasingly belligerent attitude. In 1933 it left the League of Nations, and in 1934 denounced the Washington International Conference on Naval Limitation of 1921 as an unequal treaty since it was allowed only smaller tonnages than Britain and the United States. The following year, in the closing months of 1937 Japan invaded China and continued to expand south in a strategy to secure the oilfields of the region. In 1940 it occupied French Indo-China. With Britain fully preoccupied in Europe, Japan moved swiftly in a policy of rapid expansion into Southeast Asia and the adjacent Pacific region. On 7 December 1941 its great carrier-borne air armada destroyed the American battleship fleet in Pearl Harbor and two days later the

British warships *Prince of Wales* and *Repulse* were sunk in Singapore. Australian Prime Minister Curtin recognised the patent fact that Britain could not defend Australia. Depressing confirmation of that reality came seven weeks later when, on 15 February 1942, the supposedly impregnable bastion of Singapore was taken and the Australian 8th Division marched into barbarous captivity.

Suddenly, Australia was militarily naked. The three battle-ready forces of the 6th, 7th and 9th Divisions were in North Africa, and only a militia with no armour remained. The navy had only two capital ships in Australian waters, while the air force was practically non-existent with neither front-line fighter nor bomber aircraft. In a still highly controversial memorandum, known as a 'Defence Appreciation', Lieutenant-General Iven Mackay advised the government on 4 February 1942 that he had so few troops that in the event of a Japanese move against Australia he would be forced to concentrate all available resources in the south-east sector to defend the major cities of Sydney, Melbourne and Adelaide, and the industrial zones of Newcastle, Lithgow and Port Kembla surrounding them, writing that 'We must confine ourselves to the defence of that portion of Australia which must be held to enable the major part of our war effort to carry on' (document in Chapman 1975:252–53). There was a belief in Queensland that it was being sacrificed to the enemy and that the southern States were seeking to defend themselves south of what MacArthur later called the 'Brisbane Line'. Australia had to be defended by seasoned troops and Prime Minister John Curtin, to Churchill's anger, ordered the 6th and 7th Divisions of the Australian Imperial Forces back from Africa to defend their own country, leaving the 9th – the legendary 'Rats of Tobruk' – behind for the time being.

Far from signing a peace treaty with Japan after the devastation of Pearl Harbor, the United States mobilised for war and the Pacific became an arena of the greatest naval battles the world has ever seen. Following the collapse of the Philippines and the escape of General Douglas MacArthur to Australia in April 1942, he was appointed Supreme Commander of all military forces in the South Pacific and an American war command was established. The United States committed the 41st Infantry Division and huge quantities of equipment and personnel – army, navy and marine, all with their individual air forces – arrived in increasing numbers in North Queensland, where Townsville and the Great Barrier Reef became a centre for Australia's defence.

When the Japanese began bombing Darwin in February 1942, and then Townsville to a lesser extent, Australians knew that the war had arrived; Queenslanders were only too aware they were in the front line. Abandonment of North Queensland had been politically impossible and in an astonishing war effort Australia mobilised rapidly. By mid 1942 the first Australian military units began arriving along the Reef coast and the numbers grew into a huge influx of servicemen and women stationed in a very large number of bases as far north as Torres Strait, the Gulf, and Port Moresby, with Townsville becoming the command centre and the islands of Torres Strait in the front line, bearing the brunt of Japanese air raids.

Civilian populations were evacuated from many of the most northerly regions, especially Torres Strait where women and children were moved to the mainland. The indigenous islanders were also compulsorily evacuated on 28 January 1942, with less than two hours notice, and transported to an Aboriginal reserve at Cherbourg, 272 kilometres inland from Brisbane. Despite their customary treatment as second-class citizens on Thursday Island and in the Strait, it was at Cherbourg they discovered, and also became the victims of, the appalling practices of Aboriginal apartheid under the Queensland State government policy of 'protection' (Osborne 1997, *passim*). Many of the islander men volunteered for the armed forces, particularly the Torres Strait Light Infantry, often serving bravely in the face of the continuing Japanese bombing raids on nearby Horn Island where there was a major allied air base. Despite promises at the time, they were not accorded repatriation and other benefits, particularly the opportunity for skilled technical training, given to the white Australian soldiers. As large numbers of black American servicemen arrived, always being assigned the most menial work, Australian Aborigines were horrified at the treatment meted them under American segregation policies and saw their own condition as part of a universal racist practice based on belief in white superiority.

Cairns became the major naval base for both the Australian and United States fleets and immediate steps were taken to make the Reef safe from infiltration: the converted merchant ship HMAS *Bungaree* mined the entrances to all Reef passages except Trinity off Cairns, the major convoy exit to New Guinea and points north, between 1942 and 1943. In the same period the Royal Australian Air Force established 20 radar stations along the coast, five on Reef islands and the remainder on capes and headlands to warn defensive bases of impending air attacks. Garbutt Field in Townsville became the main airbase for the region.

Japanese strategy was to move south and seize all of New Guinea and the Solomons as their southern perimeter; whether they intended to invade Australia is still a matter of dispute. Just before he was hanged as a war criminal in 1946 General Yamashita wrote in his memoirs that, although he could easily have conquered the nation, 'I never visualised occupying it entirely. It was too large. With its coastline anyone can always land exactly where they want' (Chapman 1975:256). Probably the army had planned to invade the northern coastlines, where all of the Reef ports were the most vulnerable: the Japanese navy, however, declared it logistically impossible. In any case, the matter remains conjectural since in two great aircraft carrier battles, firstly in the Battle of the Coral Sea from 3 to 8 May 1942, when the Japanese invasion fleet destined for New Guinea was routed, and then a month later at Midway, 3 to 7 June. When the Japanese carrier fleet was devastated in their attempt to take the American base, their advance was checked. For the ensuing three years the Japanese forces were then relentlessly pushed back to their homeland and to the horrors of nuclear devastation at Hiroshima and Nagasaki on 6 and 9 August 1945.

By mid 1943 North Queensland and the Reef region had moved from a forward operational base to a replacement area, security having to be maintained by continued naval patrols and air force reconnaissance, much carried out by RAAF Squadrons 8 and 20 in Catalina flying boats based at Cairns. Japanese submarines, estimated by RAAF Command as probably three or four, patrolled the eastern coast of Australia and were able to inflict some damage. In the most destructive year of Japanese submarine activity, 1943, 11 ships were sunk between Fraser Island and the far south coast of New South Wales; another four were damaged while three that were attacked emerged unscathed. The worst case was the torpedoing of the hospital ship *Centaur* – brightly lit, white-painted and Red Cross marked, following the international convention – at 4 am on Friday 14 May 1943 off Cape Moreton. Within two minutes the ship burst into flames and sank, taking 268 men and women down with it. The Reef, however, remained a genuine barrier and no enemy vessels, surface or submarine, were able to penetrate its natural and naval defences.

By the end of the Pacific War in August 1945, the Reef was perceived in a totally new light: thousands of servicemen and women had experienced it directly, many aboard naval vessels and cargo ships, others as coast watchers, radar operators, aircrew on patrol and offensive missions, and in the hundreds of army and airforce units stationed along its entire length and in Torres Strait. They were to bring back stories, photographs and memories to their homes and so intensify interest in the Reef throughout the nation and in America.

CORAL REEF PROBLEM SOLVED: BIKINI AND ENEWETAK

The defeat of Japan in August 1945, however, did not end world conflict: at the Yalta Conference in February 1945, when victory was clearly in sight, Britain and the United States agreed, reluctantly, that the Soviet Union could have a post-war sphere of influence in Eastern Europe. Stalin went much further and ordered Soviet forces to occupy the ravaged nations along its European borders. Events were becoming ever more ominous for the democracies and at a speech in Missouri on 5 March 1946, Winston Churchill gave the sombre warning that 'from Stettin in the Baltic to Trieste in the Adriatic, an iron curtain has descended across the continent'. The term 'iron curtain' resonated around the free world, nowhere with more alarming effect than in the United States which, having emerged from the conflict as the world's most industrially powerful nation, stepped up its defence preparations as an era began of even greater tension than had existed in the 1930s. The military 'hot' war was replaced by the more sinister 'cold war'. Still the only nuclear power at the time, the United States, with its ingrained fear of socialism and the Soviet Union, initiated an accelerated program to improve the destructive capabilities of its nuclear weapons. In 1946 it began a series of 45 separate bomb tests in the Marshall Islands on the atolls of Bikini and Enewetak.

The Marshall Islands are a chain of 1225 atolls and reefs in Micronesia – east of the Philippines. Named after a British ship's captain who visited them in 1788, they had been a German protectorate between 1885 and 1914 when they were seized by Japan which fortified them in the 1930s in preparation for war, holding them until they were taken by United States forces in September 1944 in a devastating barrage of firepower that destroyed much of the small area of land surface. In 1945 they became a United Nations trusteeship in 1945, and given the razed landscape and the fact that the small population of only a few thousands had been evacuated, their trustee agreed that they seemed an ideal place to experiment with the detonation of nuclear devices and identified two of the atolls for destruction. Two atomic bombs of the Hiroshima variety were exploded on Bikini in 1946; in 1947 the testing site was moved to Enewetak ($11°N \times 162°E$), an atoll surrounding a nearly circular lagoon approximately 32 kilometres (20 miles) in diameter.

In preparation for the detonation of what eventuated as a series of 43 of the more horrendously destructive thermonuclear ('hydrogen') bombs between 1948 and 1958 by the United States Atomic Energy Commission, underground investigation of the substrate was made on Bikini. In 1947 three test holes were drilled there by the United States Geological Survey, the deepest reaching 775 metres, and a larger number on Enewetak between 1950 and 1952. Of these the deepest, coded as K-113, E-1 and F-1, descended to 390, 1287 and 1411 metres respectively. Both E-1 and F-1 located a volcanic foundation, each retrieving a 5-metre olivine (igneous) core. In the words of the official report, based upon the geological analysis of Harry Ladd, Joshua Tracey and their co-workers on the atoll (Ladd et al. 1953), 'The confirmation of a basaltic foundation beneath Enewetak Atoll substantiated Darwin's subsidence theory of atoll formation' (Ristvet 1987:39). In stark contrast, the previous efforts at Funafuti, Oyster Cay, Heron, and that of Agassiz in Fiji, paled into insignificance. It is a sad irony of history that Richards, who spent his professional career attempting to solve 'the problem' of the Great Barrier Reef, did not live to see the results. Worn out by the excessive demands during the war when academic staff were heavily overloaded, Henry Richards died of cardiac disease on 13 June 1947, aged only sixty-three. None the less, his lifelong conviction that Darwin had the correct hypothesis that coral reefs had built upon subsiding solid foundations received a posthumous confirmation.

Reef science was to become a major beneficiary, and at least some of the efforts at forging swords were employed in creating ploughshares when the United States decided to establish a research station to study the marine ecosystems, believed to have been massively disturbed during the war and by subsequent nuclear testing. In conjunction with the University of Hawaii which exercised supervision, the first stage was installed on 3 June 1954, designated the Enewetak Marine Biological Laboratory (EMBL). Given security demands, only United States male scientists with high level clearances were given access, but, surprisingly, and to the immense benefit of reef science, the authorities accepted the principle of academic freedom and the results were allowed to be published in the standard journals. Over the ensuing years a large number of studies were conducted by many visiting scientists, particularly into ecosystems and wide-scale lagoon metabolism – building on the Reef researches of Mayor's ecological survey in 1913, and Yonge's Low

Isles experiments on coral symbiosis in 1918–29. Of the EMBL studies, those by Howard and Eugene Odum on material flux and energy flows were to provide significant new avenues for investigation and came to exercise a major influence on coral reef research world-wide, including those on the Great Barrier Reef (Odum & Odum:1955).

With the addition of Laboratory No.2 in 1961 and No.3 in 1969, along with much greater infrastructure support, in 1969 EMBL was redesignated as the Mid-Pacific Marine Laboratory (MPML) and continued to operate until 1982 when the cleaning up of the atoll had been completed and it was restored to its indigenous inhabitants as part of an independent nation, the Republic of the Marshall Islands, in free association with the United States.

In marked contrast to the United States, the war years took a much greater toll on Australian research. Having weathered the economic depression only to enter six years of fatiguing war, and constrained by shortages of most commodities, by the end of 1945 the nation was exhausted. The Australian economy remained basically dependent on primary products, the existing industrial sector was minimal; most manufactured goods were imported, even the buttons made overseas from pearlshell and trochus collected in Reef waters. The national basic infrastructure was badly run down and wartime rationing of food, clothing and motor fuel continued.

QUEST FOR A MARINE RESEARCH STATION: THE BACKGROUND

Despite the limitations in that period of economic stringency, a few enthusiasts began the process of reviving Reef research. The major difficulty was still the lack of any field station. For the previous seventy years governments had been urged to support Reef research in order to construct a reliable data base concerning its ecology, dynamics and resources. As early as 1876, due to the indefatigable efforts of an internationally renowned visiting Russian scientist, Baron Nicholai Miklouho-Maclay (1846–88) Australia's first research station had been constructed at Lang Point at the entrance to Sydney Harbour, but in the hysteria surrounding a possible Russian invasion, it was resumed by the army in 1885 as a defence facility. In Queensland in the same period there was no interest in marine science, and the only scientific research on the Reef was being pursued intermittently by Hedley and Haswell in Sydney. The departure of Haswell from Queensland in 1880 created a vacuum that took more than thirty years to fill.

Even so, voices continued to be raised in Queensland in support of a marine station on the Reef itself, a quest begun in the previous century. While working as Commissioner of Fisheries Saville-Kent in his capacity as President of the Royal Society of Queensland had used his Presidential Address for 1890 to make the first explicit plea for a permanent marine biological research station on the Reef, followed in 1910 by Ronald Hamlyn-

Harris, Director of the Queensland Museum. An enthusiastic proponent of both Reef research and a permanent station, he toured the Reef in 1914 on a collecting expedition. Stopping over in Cairns after visiting Dunk Island, he stated at a public function that it was a shame the Reef was such 'a neglected quantity'; indeed, people in Brisbane 'had no conception of the great possibilities which existed here'. Taking up Saville-Kent's theme he said that it was vital for a marine biological station to be established to provide the opportunity to study the Reef from an economic aspect. 'If Germans or Americans had it within a thousand miles' he continued, 'they would carry out such science as would place them [as] the foremost nations of the world'. In those days of the bare three mile territorial limit he noted, in the context of the hostility to Japanese exploitation of the Reef – chiefly pearl and trochus shelling – that it was 'aggravating' to see 'many foreigners coming here day after day taking away these resources'. At that function in Cairns a reporter asked about the likelihood of the Queensland Museum establishing a station at Cairns, or at least on the Reef. 'Not at present' was his reply:

> There are only one or two of us really keeping the matter alive. I am hammering at it, and though I have no immediate hopes, I have hopes for the future and trust some day to see a station established. It would be of immense scientific value, and you can safely say that it would place Queensland among the primary scientific nations of the world (*Cairns Post* 26 May 1914, p.8).

The most persistent, and amazingly energetic, proponent of a field station, chiefly from an economic rather than a purely scientific, much less a conservation point of view, was Charles Hedley when, following his forced retirement from the Australian Museum, he became research director of the GBRC in 1924. All such proposals, as previously noted, were opposed by Richards who sought to concentrate reef research effort and funds under his personal direction in solving the geological problems of coral reef origins. Even when the first real research station on the Reef came about rather adventitiously in 1929 in the form of the small collection of prefabricated timber huts donated by the departing Low Isles Expedition of 1928–29 to the Queensland government, it was not, however, destined to last. Closed during the first years of the great economic depression, it was destroyed on 3 March 1934 when a tropical cyclone devastated the Low Isles, wrecked the buildings, severely eroded and reshaped the cay and smashed much of the coral cover. Although Moorhouse was able to salvage some of the materials from which he constructed a small laboratory measuring 27 feet by 9 (8.2 by 2.8 m) the former renown of the Reef's first marine station was gone forever. Exacerbating that disaster, the continuing economic depression of the 1930s and the inexorable drift to World War II had far more serious effects upon Reef research, and the quest for a permanent research station had to be shelved yet again.

REEF RESEARCH STATIONS: 1948–1973

In the immediate post-Pacific War years, reef research was revived by Ernest Goddard who, although burdened with administrative duties, continued as the driving force to create a marine station on the Reef – a goal for which he had been working for more than twenty years, despite Richards' strenuous opposition. Described by his colleagues as 'obsessed' with the need for a permanent research facility – and deeply disappointed with his failure to establish one on Thursday Island, or Palm, or Dunk – Goddard was determined to establish one on Heron Island, chosen because of its relative proximity to Brisbane and the port of Gladstone. Tragically, at the point of retirement, aged sixty-five, and just seven months after the death of Richards, Goddard also had a coronary occlusion and died on Heron Island on 16 January 1948. Shocked and saddened, his colleagues set up a memorial fund to establish the station he so ardently desired for the use of students and scientists, not only from Australia but from around the world.

Still minimally funded and staffed at the time, the University of Queensland would not accept any involvement. Alone, the GBRC launched the appeal and by mid 1951 some £6471 had been collected: £3750 from the Queensland government, £1250 from the Royal Society of London, £517 from the original Goddard Memorial Fund, £500 from the GBRC itself and the remainder from other donors. The lease on Heron Island had been held by the Poulson family since Chris Poulson had bought the defunct turtle-canning factory and developed a resort there several years later in 1936. After negotiations between the family and the Queensland Department of Lands, the small 8 hectare (20 acre) island was subdivided into a 2 hectare lease for the resort, 2 hectares for the research station and the remaining 4 hectares for a national park (Hill 1985a:15–16). Although, like Richards, Goddard did not live to see the consequences of his endeavours, the Heron station became a beneficiary.

With part of the funds collected a simple building of timber and galvanised iron, measuring 5.4 m by 3 m (18 by 10 feet), was erected, with nearby toilet and water tank, intended solely for overnight accommodation. It was ready for use by the end of 1952. The hut, of course, was designed merely as the initial stage: plans continued in 1953–55 for a laboratory wing and by 1956 a building for that purpose – unequipped – along with a maintenance officer's residence had been built. The search for further funding continued for more than a decade: the GBRC, as an independent organisation of Reef scientists, continued the struggle to develop the Heron Island research station. By that time the Fisheries Division of the Council for Scientific and Industrial Research (later CSIRO) had already established a limited facility on Thursday Island.

Throughout the 1920s and 1930s an additional problem constantly voiced was the lack of suitable science graduates to engage in Reef research. In large part that occurred because

Australian universities were small, inadequately housed, staffed and funded – and still colonial in their outlook – were dominated by British staff and focused more on European studies than Australian research. In the first post-war election campaign in 1949 Robert Menzies, leader of the federal opposition Liberal Party, which he had founded in 1944, in coalition with the Country Party, promised to abolish all wartime controls still in operation and to introduce immediately an expanding free enterprise, market driven economy with a heavy emphasis on industry and manufacturing, based on the new American slogan of 'Research and Development'.

Winning government on that promise in 1949, industry under Menzies promoted a rapid expansion of industry. Full employment was soon available for unskilled and semi-skilled workers, but a major obstacle to greater development remained, due to the lack of highly skilled artisans and professionals in all areas. Menzies recognised that the problem lay in the appallingly low standards of Australian education: most children left school at age fourteen – in rural areas often at thirteen – and the university and technical college sectors were seriously underdeveloped. Acknowledging that the States lacked the capacity to expand university and technical college funding – their obligation under the Constitution – Menzies realised that the Commonwealth had to provide supplementary funding for higher education.

In 1957 Sir Keith Murray, chairman of the British University Grants Committee, was asked to lead an investigation. His report was devastating, revealing that less than 5 per cent of Australian youth entered university, less than two-thirds passed the first year examinations, and barely one-third graduated in minimum time. Barely half of those who enrolled ever finished their degrees: 3 per cent of the population. 'The high failure rate', he wrote, 'is a national extravagance' (Murray 1957:35). And the gender imbalance was extreme: less than 1 per cent of girls entered university, half of those never graduated, and those who did were almost entirely in the area of Arts and Social Work. Almost no women went into science, medicine, dentistry, law, engineering, agriculture or veterinary science. Science, on which all the technical faculties depended for their foundation studies, was badly taught with large classes, no tutorials and a serious shortage of laboratories. As predicted, the parliament agreed that the Commonwealth could enter the field of funding education – readily accepted by the States – and in 1959 the Australian Universities Commission (AUC) was created to provide supplementary funding, to expand material infrastructure, create new universities, recruit better qualified staff and expand university education beyond the preserve of the wealthy by providing Commonwealth scholarships for students from less affluent backgrounds.

At the same time, given the first class marine biological station on Enewetak, which was an exemplar to Australians, it was clear that any serious tropical marine research had to be far more generously supported. In a remarkable quest for further funding the GBRC, supported by the University of Queensland, applied successfully to the Rockefeller Foundation for a grant. It was overwhelmingly successful: a then enormous sum of

£23 000 was offered. In the joint submission, the university indicated it would accept responsibility for management and disbursement of the fund.

As it transpired, however, the university wanted total control and a seemingly endless series of negotiations with the GBRC continued: several agreements were drafted, considered, and rejected. By this time, in the late 1960s, both State and federal governments wanted the university to be the sole authority since only incorporated institutions could legally be the recipient of public funds. Eventually a resolution was reached in 1970 whereby the University of Queensland made and received all submissions and grants, with control vested by Deed of Agreement in a twelve member board, six from each organisation (Hill 1985b:187–202). At last, after a century of unremitting effort, Australia's first permanent marine biological station came into being on Heron Island. Even so, it was not in continuous operation: research was piecemeal, conducted by individuals and small groups, usually in vacation periods, or else in short, specific purpose expeditions.

SYDNEY'S REEF STATIONS: ONE TREE AND LIZARD ISLANDS, 1965–1973

While those complex and protracted developments were taking place to establish Heron Island, a new and much smoother development was in train to create a totally different kind of research station on the Reef. In 1964 Frank Talbot, a 34 year old ichthyologist from South Africa, came as Curator of Fishes to the Australian Museum. By 1966 he was appointed Director and the runner-up, Donald McMichael, Deputy Director. Determined to expand the horizon of the museum beyond collections and systematics into a wider conceptual field governed by ecology, Talbot was well aware of the sophisticated research being conducted overseas, especially the work on energy flows by the Odum brothers in the 1950s at Enewetak, and the initial innovatory work in the mid 1960s by Don Kinsey on Australian reef productivity. Dissatisfied with the unintegrated researches coming through in the increasing number of short expeditionary studies, Talbot became intrigued to seek the possible linkages that must lie behind such seemingly stochastic results. Could they, he wondered, be drawn together as facets of system processes on reefs that could lead to more explanatory general theories? Mindful of the Low Isles research experience of a long term study in the one place, to date the only one ever done on the Reef, he believed that intensive long term study of an individual reef where researchers could collect continuous data would be invaluable. He began looking for a possible location.

Talbot's search for a suitable site took his team to Michaelmas Cay near Cairns, Holbourne Island near Bowen, and to the uninhabited One Tree Island, near Heron and relatively close to Gladstone. One Tree, named by Cook for its distinctive feature at the time, it has a special character with an almost closed lagoon, the advantages of a commercial launch service from mainland Gladstone to nearby Heron, facilitating supplies.

One Tree seemed a serious possibility, so, beginning in 1965 Talbot and colleagues mounted an expedition with aluminium outboard-powered dinghies and five camping tents on the beach for their initial venture. Believing that One Tree would be an ideal location for his long term ecological studies of fish, Talbot, as Director of the Australian Museum, arranged in 1969 for the Museum to secure a lease and in 1971 construction of permanent buildings began. Today these consist of a caretaker's residence, a scientists' lodge with kitchen, dining room and adjoining alcoves for up to twelve bunks, two laboratories (one wet, one dry), and two equipment and maintenance sheds. In 1980 an account of the history of the discovery and charting of One Tree, earlier expeditions, the founding of the station and research subsequently done there, was compiled from contributions by 16 scientists and edited by Hal Heatwole as *A Coral Island*.

Early in the 1970s, concerned that cooler winter temperatures in the southern part of the Reef might be affecting fish breeding times, Talbot thought that his research might be better conducted at a larger Reef station further north in the tropics. Having received intimations in 1971 of possible funding for a larger station, plans were devised to transfer One Tree to the University of Sydney, a choice easily made due to the ready enthusiasm of Peter Sale, an outstanding ichthyologist at Sydney who first visited the station in mid 1974 and, like so many others, was captivated by both its beauty and its research potential. Further negotiations between the Museum and the university led to the transfer of the lease to the University of Sydney in 1975. After two disastrous earlier efforts, that university at last also had a marine research station with a predictable future.

Around the same time, in 1969 the American space agency NASA, in conjunction with the United States Navy, conducted an underwater experiment in the United States Virgin Islands called Tektite, designed to study the effects of prolonged living in an isolated environment, prior to launching the Skylab Program. Four scientist aquanauts lived in a submerged complex for two months, going out daily in scuba to conduct experiments while all of their vital functions and activities were monitored and recorded by remote sensing. Two years later, in 1971, the experiment was repeated – designated Tektite II – with 50 scientist aquanauts, one of whom was Frank Talbot. While in the United States he made the acquaintance of the wealthy philanthropist Henry Loomis who became intrigued at the idea of a Reef research station and, after lengthy discussions, offered US$100 000 as an initial grant if a suitable location could be found in the northern part of the Reef.

While on the cruise of the *Marco Polo* in 1973, the venue for the Second International Symposium on Coral Reefs, Talbot became convinced of the enormous possibilities of Lizard Island. Although other sites were considered, Lizard was chosen and the Queensland government not only agreed to grant a lease, but, as a result of a direct personal request at an interview with Talbot, the then Premier, Bjelke-Petersen – at the time in implacable opposition to the Federal Labor government of Gough Whitlam and determined to keep all of the Reef a sole Queensland preserve – gave a government grant of $10 000 towards establishment costs. Events moved quickly and the first stage of the station was built the same year, 1973. So the Lizard Island Research Station was instituted under the authority of the Australian Museum as a first class field centre for a limited number of scientists at a time, with good accommodation and excellent laboratory facilities.

Incredibly, considering the vigorous activity in Europe and North America in founding marine research stations, Australia still lagged far behind: it was almost moribund. With one of the world's longest coastlines, and despite the efforts of a sequence of dedicated activists, it took more than a century for Australian governments to recognise the value to an entirely ocean bordered nation of marine scientific research and the correlative essential function of a marine station. By 1975, however, events had begun to quicken when the Great Barrier Reef became thrust into the centre of national environmental controversy in which the role of marine science became an essential element.

CHAPTER 19

A NEW PROBLEM: THE CONSERVATION CONTROVERSY, 1958–1972

CONSERVATIVE ASCENDANCY: UNRESTRAINED EXPLOITATION, 1953–1968

By 1958, the three Reef island research stations – Heron, One Tree and Lizard – were still, in world terms, small, under-resourced, and limited to vacation and short term projects by individuals or small groups, yielding at best, the random unrelated results that had bothered Talbot. The British Low Isles Expedition of 1928–29 thirty years earlier had made the only major in-depth study within the Reef since Mayor's in 1913, which Australians still had no prospect of emulating. Nor did any of the Reef stations enjoy the advantages of the Enewetak Marine Biological Laboratory (EMBL), nor the lavish funding provided. At the end of the 1960s that situation was to change markedly; Reef science was to enter, for the first time, a new phase of development as the Commonwealth and Queensland governments were forced to recognise, after pleas for assistance for almost a century, that it was a seriously neglected area that had to be remedied.

Yet, the stimulus to research and funding came, not as a positive response to scientists' requests, but negatively when the Reef became the catalyst for what a national newspaper called the 'most sustained public campaign in memory on a conservation issue' ever experienced in Australia after people had mobilised in their thousands under the slogan 'Save the Barrier Reef' (*The Australian* 24 December 1969). What exactly, was there to save, and how did the campaign originate? Two separate, but increasingly interacting, factors were major contributors: the decades of unregulated development following World War II and the rapid growth of the oil industry along with the sudden rise of worldwide environmental awareness in response to the increasing frequency of oil spills and mounting damage to the marine environment.

The dramatic change of government views regarding Reef science occurred, quite unexpectedly, when Australia became engaged in a quest for self-sufficiency in oil. In the 1949 federal election campaign, as part of his election pledge to abolish all wartime controls, opposition leader Menzies promised to end fuel rationing immediately, although there was a major difficulty since all petroleum products were imported and, given the economic costs of the war, such an expensive item was a serious drain on fiscal reserves. There was, at that time, wide public support for oil exploration which it was believed would usher in a new era of abundance for all in contrast to the wartime shortages, and the nagging sense of isolation and apparent backwardness, heightened by unfavourable comparisons with the United States made by American servicemen during the Pacific War. In the widely accepted, enthusiastic program to develop the nation, oil was promoted as an essential element in strengthening heavy industry and stimulating the flow of desirable consumer goods, of which the automobile was high on the list. Abundant, cheap oil was seen as vital to what was expected to be the good life.

In the immediate post-war years the Australian Motorists Petroleum Organisation Limited (AMPOL) began prospecting in Western Australia and in 1953 national expectations were heightened when it announced a strike at Rough Range, just inland from North-West Cape, near the Ningaloo coral reefs. In addition to Ampol, a large number of overseas oil companies clamoured for exploration permits and formed a powerful political lobby group, the Australian Petroleum Exploration Association (APEA). Results came soon, and in 1964 the Esso–BHP consortium found natural gas reserves in Bass Strait from their Barracouta drilling platform. Further finds of both gas and oil in commercial quantities were soon discovered at the nearby Marlin platform in 1966, at Halibut the following year, and then at Kingfisher. Soon a total of eleven drilling platforms were operating in the Strait off the Gippsland coast, to be connected with Melbourne and later to Canberra, Sydney and South Australia by undersea and underground pipelines. Finds were also made in Western Australia between Perth and Geraldton, offshore from the original Rough Range area, where another eleven sites came into production, and in the Timor Sea on the North West Shelf, offshore from Wyndham.

The State of Queensland was in the forefront of the post-war development thrust. In 1957 the Country Party, with the support of the minor Liberal Party, began an unbroken period of 43 years of government, from 1957 until it fell apart in 1990 after imploding from a decade of gross corruption. Queensland was the most decentralised State in Australia, in large part due to much of its settlement extending north to Cooktown along the coastline of the Reef, the main artery of all economic activity, as promoted vigorously in the post-war years by the Country Party. Led by Frank Nicklin, and strongly buttressed by a system of electorates heavily weighted in favour of Country Party seats, and also by the lack of a parliamentary upper house of review, it began a populist appeal to 'Develop the North'. Offering support for the rural sector with the allure of personal gain, expressed in the maxim of 'capitalising profits and socialising losses', petroleum exploration was a high priority.

Oil is generally believed to be of organic origin, being found in sedimentary deposits laid down in the Mesozoic Era of 'abundant life', the Triassic, Jurassic and Cretaceous periods from around 230 mya to 65 mya, and to a lesser extent in the following Tertiary period. The Great Artesian Basin, which encompasses most of Queensland west of the Great Divide from the Gulf of Carpentaria to the central northern part of New South Wales, is just such a formation. As early as 1927, following the report of government geologist H. I. Jensen in 1923 that the basin 'possesses rocks of the right lithological nature, folded in the right manner to yield oilfields of great magnitude' (Jensen 1923:1275) the Queensland government had encouraged drilling in the inland region near Roma. A sizeable flow of gas was discovered, but after a rush of speculative investment the flow faltered and stopped, and in the depression years the wells were closed down altogether.

Encouraged by the *Commonwealth Petroleum Search Subsidies Act* of 1957 which contributed 50 per cent of drilling costs – a subsidy expanded in 1959 to all geophysical surveys – exploration activity in Queensland accelerated rapidly. With new surveys and improved technology, several companies, chiefly Australian Associated Oilfields (AAO), Oil Drilling and Exploration (OD&E) through its subsidiary Exoil No Liability, and Artesian Basin Oil, began new drill sites in the Roma and nearby Surat Basins. In March 1960 AAO found gas (an indicator of oil, formed in the trapped strata as some of the hydrocarbons become dissipated) yielding around 35 000 cubic metres per day; by May the flow had increased to around 180 000 m³ per day. In the same period a modest flow of oil was reported around 2600 barrels per day (413 000 litres) which was, at least, an investment stimulus.

Queensland, however, seemed to be missing out: the flows at Roma failed to fulfil the great expectations held for them. The Nicklin government, secretly, began issuing more prospecting licences which now included the Great Barrier Reef. From 1962 until 1968, six companies – Shell, Ampol, Tenneco, Gulf Interstate, Australian Gulf Oil, Australian Oil and Gas – began actively exploring Torres Strait and the eastern seaboard of Cape York south to Princess Charlotte Bay, as well as all the Reef coastline from Cooktown to Fraser Island, by magnetic and seismic methods. Their persistence was based on a strongly held belief that the Reef province was a similar oil-bearing geological formation to Bass Strait and the North West Shelf of Western Australia. Test drills were made on the Reef in the Capricorn Channel at Wreck Island in 1959, Capricorn 1 in 1967, and nearby Aquarius 1 the following year; in May 1968 in the far north-east Torres Strait at Darnley Island a test drill was made at site Q1P. All results were dry (Wilkinson 1988:285).

REEF RESOURCES: DEVISING A LEGAL REGIME

By 1968 the quest for oil ceased to be simply an industrial activity and had assumed a highly political and controversial character in which the Reef was centrally involved, and was to remain so throughout the next decade. To understand the tortuous events of that period it is necessary to set out briefly the complex political and legal situation at the time regarding offshore oil exploration. Throughout the early mineral exploration years Australia had no clearly defined legal regime respecting the seas; indeed, clarifying maritime law had become an international problem since the Truman Proclamation of 1945 when the United States declared sovereignty over the entire continental shelf of their nation with the sole right to exploit its natural resources. In 1950 El Salvador, and later in the Santiago Declaration of 1952, Chile, Peru and Ecuador followed suit, all asserting sovereignty over a 200-mile coastal zone. Clearly such actions required a sound basis in international law. Following a request in 1950 by the General Assembly of the United Nations, its International Law Commission met in Geneva to devise an acceptable regime, producing in 1958 the Geneva Convention on the Law of the Sea. It comprised four separate conventions covering 1) Freedom of the High Seas; 2) Fishing and Conservation of Living Natural Resources; 3) the Territorial Sea and the Contiguous Zone; and 4) the Continental Shelf. Australia signed the Convention on the Continental Shelf in 1962.

A serious deficiency remained, however, in that the respective rights of the Commonwealth and the Australian States concerning the exploitation of the continental shelf were not defined. Nothing in the *Australian Constitution Act* of 1900 offered a definition. Yet the exploitation of petroleum resources in the Gippsland Basin in Bass Strait, and the North West Shelf – and any future offshore discoveries – required a firm legal basis, particularly since potentially there were considerable profits to be made. In 1965 Minister for Resources Spooner set out a loosely formulated Common Code for sharing royalties gained from the continental shelf between the Commonwealth and the adjoining state. The underlying philosophy was expressed by Prime Minister Harold Holt, who, having succeeded to the office when Menzies retired on 26 January 1966, set out the Liberal approach to Commonwealth–State relations in an election speech a month later on 20 February. The Liberal Party, he stated, seeks 'a true federation based upon a spirit of cooperation rather than a strict definition of powers'. At a series of meetings between the Commonwealth and all six States an Offshore Agreement 'relating to the exploration for, and exploitation of, the petroleum resources . . . of the continental shelf of Australia' was reached and signed personally by the Prime Minister and all six premiers on 16 October 1967. This attempted to express that cooperative spirit whereby all seven parliaments would enact identical legislation (known as 'mirror' legislation) under the title *Petroleum (Submerged Lands) Act* in which royalties from the exploration companies were to be divided between the Commonwealth and the state whose shelf was being exploited in the ratio of 4:6.

Despite considerable disquiet in the Commonwealth parliament at the speed with which it was pushed through, the Bill was passed on 7 November with the crucial words in the Preamble stating that 'the Governments of the Commonwealth and the States have decided, in the national interest, that *without raising questions concerning, and without derogating from their respective constitutional powers*, they should cooperate for the purpose of ensuring the legal effectiveness of authorities to explore or/and to exploit the petroleum resources of these submerged lands' (emphasis added). The high-minded philosophy of cooperation was to prove a source of endless debate and litigation for the ensuing decade, and reflected the lack of any clear understanding of sovereignty over resources on the continental shelf. The Senate had expressed serious reservations with the Bill, and in order to have it passed, it was agreed that immediately after passage the Senate would set up a Select Committee to make a full investigation of all issues related to oil exploration and exploitation, and in particular to seek a clearer constitutional basis for ultimate authority over the continental shelf (Hansard, Senate 1967:2337).

Before the Agreement had been reached, and confident that Queensland had total sovereignty over all of its territorial sea and the water of the shelf, the Queensland Cabinet acted. Under the direction of Ronald Camm, Minister for Mines and member for Whitsunday, and with the encouragement of Minister for Works and Housing Johannes Bjelke-Petersen, in early 1967 – again without any public announcement – it zoned all of Torres Strait and the entire Reef region (with a few small exceptions) along with the adjoining coast down to and including Brisbane into 17 areas, covering more than 80 000 square miles (207 000 km^2). Prospecting rights for six years were given secretly to six companies, dating from 1 September 1968. On 25 October 1967, following the Offshore Agreement reached with the Commonwealth the previous week, Camm introduced a 'mirror' Petroleum (Submerged Lands) Bill into the Queensland parliament in which he affirmed that oil exploration throughout the Reef would continue and if successful would be fully exploited and oil transported to land terminals by underwater pipelines. In answer to critical questioning from the Labor opposition concerning spillage into Reef waters, Camm asserted that pipelines in the Persian Gulf had never caused any problems and that Clause 97 of the Bill 'provides that a pipeline licensee shall properly safeguard his pipeline to prevent any possible escape of petroleum' (QPD 25 October 1967:1129).

Six months later, in May 1968, Camm engaged American geologist Harry Ladd (who had supervised some of the drilling on Enewetak in the early 1950s) as a consultant to report on the Reef's possibilities. In the incredibly short space of one month, by air and launch, Ladd stated that he had examined the Reef from Heron Island and the Swains in the south to Cape York and Torres Strait. In his short Preliminary Report he passed quickly over fishing and tourism and commented mainly on geology, his professional field. Both limestone and siliceous sands were recommended as worth considering, but his emphasis was on petroleum. So far, he commented, although 'neither oil nor gas has been discovered in reef drilling done to date . . . the outlook for such discoveries is promising' (Ladd

1968:35). The geological formations, he wrote, 'encourage the belief that oil and gas resources exist beneath the reef' and asserted that 'if such resources are discovered they will certainly be developed, no matter where they lie in relation to existing surface reefs', with care being exercised to ensure they would be 'properly controlled' (37). Giving great comfort to the government would have been his assurances regarding any public protest: 'in Australia many individuals and some organisations are unalterably opposed to any form of reef exploitation. These conservationists are well-intentioned and they feel strongly about attempts to exploit the reef's resources by mining or drilling'. In a diversionary tactic, he directed attention to other threats to the Reef with a comment that would become only too true in the following decades: 'Some of them, however, appear to be at least partially blind to the dangers involved in what might be called the peaceful exploitation of the reef' and listed expanding tourism, collecting corals and shells, and unrestricted fishing (42). His overall conclusions were favourable to the government, stating that 'if and when oil and gas resources are discovered the Queensland Department of Mines will be in a position to recommend regulations that will permit development without excessive damage to the existing reefs' (47). His final words expressed strong support for 'controlled exploitation without destroying the essential values of the reef' (48).

Premier Nicklin had already retired in August 1967, and after a short succession of two other premiers, a year later Bjelke-Petersen, born in New Zealand in 1911 of Danish immigrants, became premier. A keen oil speculator, with a less than successful record, he actively encouraged the oil search in Queensland, having himself acquired an exploration permit in 1958, which later was to embroil him in both a legal and political controversy. Ampol, in consortium with Japex (Japan Petroleum Exploration) was granted a lease for zone Q12P covering the central section of the coast from Bowen to Mackay and for some distance to the south. The premier himself had a significant interest in another company, Exoil No Liability, which had been granted a lease of Q8P in the Princess Charlotte Bay and he was determined that exploration of the Reef would go ahead. When the Opposition and environmental groups became aware of what they believed to be a sinister plan, an explosion of concern was triggered that would open a decade of violent controversy, led initially by a small band of dedicated volunteer conservationists and pursued in the federal parliament by three Queensland opposition members: Senator George Georges, William Fulton, Member for Leichhardt, and Rex Patterson, Member for Dawson.

ENVIRONMENTAL ANXIETY FOR THE REEF

Degradation of the natural environment had been steadily accelerating since the industrial revolution of the nineteenth century, but the effects at first were considered isolated episodes. One of the earliest disasters in the twentieth century was in the Meuse Valley in Belgium in 1932 where more than sixty persons died as a result of highly toxic gaseous fluorine emissions from fifteen factories in the region. A more widespread environmental disaster occurred in London in 1952 when over 6000 persons died of respiratory failure one winter from a similar katabatic inversion that trapped the sulphur-laden, particulate discharges from domestic coal fires inside a blanket of smoke and fog, creating the phenomenon known ever since as 'smog'. Equally portentous were the increasing number of crippling neurological diseases and deaths in Minimata in Japan which by 1956 had become an international scandal, later discovered to have come from emission of methyl mercury wastes into the water from battery factories along the foreshores, which were absorbed by the shellfish, a major component of the local diet.

Around the world similar incidents were increasingly reported and before long numerous such environmental disaster flashpoints began coming together in a pattern of connection that reached a focus in 1962 in Rachel Carson's *Silent Spring*, the greatest indictment yet of the insidious threat of industrial chemicals in accelerating worldwide environmental degradation. No work before or since has so triggered international awareness and concern and it will remain as the definitive indictment of humankind's heedless destruction of the natural habitat of all life. Fortunately for the world, it was written by a highly competent ecologist at Woods Hole Marine Laboratory in Massachusetts, already an acclaimed author for her three perceptive works on the marine environment – *Under the Sea Wind*, *The Edge of the Sea*, and *The Sea Around Us*.

Silent Spring was published only two years before Carson's death from cancer in 1964. Immediately entering the best-seller lists, and remaining in print to the present day, it rapidly gained world recognition in part, ironically, because it was pilloried with massive protests from the industrial and agri-chemical industries that were being exposed. Drawing together a vast armoury of data from the literature of industrial chemistry, Rachel Carson (1907–64) revealed that the wholesale application of the burgeoning array of organocholorine and other chemicals into agriculture, along with poorly regulated waste disposal systems, was polluting not only the air and soil, but the waters flowing inevitably into lakes and the seas. 'What we have to face', she wrote, 'is not an occasional dose of poison, which has accidentally got into some article of food, but a persistent and continuous poisoning of the whole human environment'. When she began writing, her biographer and publisher has recorded, 'the term "environment" had few of the connotations it has today. Conservation was not a political force. To the public the word "ecology" was unknown, as was the concept it stood for' (Brooks 1985: preface). From her initial stimulus those became household words, beginning the global environmental revolution that rapidly reached Australia and found expression in the formation of environmental groups across the nation.

Environmental concern over the Reef, as recorded throughout these chapters, had been gestating slowly, beginning with the pioneer work of Saville-Kent, Banfield and Hedley, and after the 1930s chiefly from the urging of the GBRC, but nothing that happened before matched the levels of anxiety in the mid 1960s. These originated from three seemingly separate sources that happened to come together in the same few years: a major outbreak of the Crown of Thorns starfish infestation, a proposal to mine the reefs for agricultural limestone, and the intention of the Queensland government to allow drilling for oil exploration over the length of the Reef. The eruption at the same time of the political scandal after the revelation that the premier and cabinet ministers held undisclosed financial interests in the companies planning oil exploration enterprises on the Reef also added fuel to the fire. In Queensland two voluntary groups were to play major activist roles in Reef conservation: the Wildlife Preservation Society founded in 1963, and the Littoral Society of Queensland founded in 1967.

The first serious environmental issue concerning the Reef was an outbreak of a starfish infestation by the Crown of Thorns (scientifically, *Acanthaster planci*), that began silently and unexpectedly around 1960 when a few were noticed, but not recognised, on the corals around Green Island, a popular Reef resort. At the time, it was the premier tourist destination out of Cairns, with a small resort and the compelling attraction at the time of a great novelty: an underwater viewing chamber. In 1953–54 two enterprising businessmen, Lloyd Grigg and Vince Vlassoff, had built a steel 70 ton viewing chamber in Cairns, then arranged for it to be towed to the island and sunk off the end of the jetty in an area rich with coral formations and their attendant tropical fish, where it remains to this day. With an enclosed staircase going down, tourists were able to go under water and observe the attractions of reef life through the porthole windows.

By that time tourism on the Reef had begun to increase, stimulated in part by the airlines which promoted an active campaign to fly holiday-makers to central locations – mainly Cairns and the Whitsundays – where they had begun to build small resorts. Hotels and holiday accommodation in coastal towns also increased steadily and day-trip operators met the need for tours to nearby reefs. Whereas in 1947 the Queensland Tourist Board reported an annual average of 5000 visitors; a decade later it had increased to nearly 60 000 and in 1963 more than doubled to 125 000, the majority arriving by air (Claringbould et al. 1984:10,12). It was from the underwater observatory on Green Island that the Crown of Thorns were first observed on the Reef when attendants were puzzled by those strange and unusual echinoderms on the corals outside, reaching 60 centimetres across with a radial pattern of up to seventeen arms, covered in long spines up to 5 cm in length which were later found to be highly toxic when they punctured human flesh.

Soon larger numbers were observed and as they increased exponentially some alarmist reports claimed that they were approaching 'plague' proportions. And, as sightings increased it also seemed they were spreading south as the starfish were found on yet more reefs, previously thought to be free of them. When a starfish left a coral formation, it was noticed that a white patch remained, assumed to have been a bleaching of the coral from sunlight deprivation on the animal's 'resting place' (see Raymond 1986). While there is no doubt that the range of sightings had increased with the swift rise in popularity of scuba ('self-contained underwater breathing apparatus') that in the same decade had brought many more underwater observers, along with the few scientists of the period, so maximising the likelihood of sightings, at the same time a pattern of genuine expansion in range and in numbers was believed to be emerging. *Acanthaster* outbreaks had been reported in the Ryukyus in 1953, then further south at Okinawa in 1955, at Rabaul in 1962 and then at Guam in 1965 where scientists at the university marine station began a research study of its behaviour.

By 1965 numbers had risen alarmingly on the Reef, and the white patches were found to be feeding scars where the starfish had literally sucked the polyps out of their limestone structures by inverting their stomach over them. The once-living coral did not regenerate readily: the bleached structures became covered in brown filamentous algae giving a dreary, lifeless appearance. Soon the premier tourist reefs off the major centres of Cairns and Townsville were losing their visual beauty, which was precisely what tourists had come to see. Alarm spread that the entire Reef could be doomed, later exacerbated by sensational reporting of the kind given in Theo Brown's *Crown of Thorns: the Death of the Great Barrier Reef?* (1972) and Peter James' *Requiem for the Reef* (1976).

Dealing with the outbreaks became a huge problem. Cutting them up underwater is the worst approach since, as in the fable of the sorcerer's apprentice, each separated arm regenerates as a new animal. Removal altogether for disposal on land was then tried and initially the management of Green Island employed a scuba diver to collect the starfish; in fifteen months he collected 27 000 specimens without making any noticeable reduction. The infestation then became a major operation for a while, using volunteer divers to remove the animals physically. That proved a tedious and time consuming process, and eventually it was found that injection of a biocide from a hypodermic apparatus (an idea borrowed from the drench guns used for cattle) was effective. At first copper sulphate was used, which, however, was toxic to surrounding marine life; later, formalin was found to be more effective since it is biodegradable in water and leaves no toxic residues.

The problem was brought to the attention of Queensland's Director of Fisheries, Harrison, and he contacted Robert Endean of the University of Queensland, a biologist with an interest in marine toxicity who had already become curious about the character of the venomous spines. Endean was commissioned to undertake a two year study with funding to employ his graduate student Robert Pearson. By 20 April 1968 Endean had sufficient data to compile a report that offered two possible explanations, the chief theory being breakdown of predator–prey relationships through heavy poaching by Taiwanese and Japanese fishing fleets of the giant triton, *Charonia tritonis*. A second, and perhaps additional possibility, stemming from the pioneer work of Rachel Carson, was that toxicity in Reef waters from organochlorine contaminants washed down with agricultural run-off from farms and canefields had built up in the planktonic food chain and begun to kill off the natural predators of starfish larvae, allowing the larvae to multiply to extravagant numbers. Pearson's report was forwarded to Harrison, who sent it in turn to the Premier's Department. Absolutely nothing was heard of it for twelve months.

The Queensland government was in no mood to deal with the starfish problem: it was facing an election in 1969, and the Opposition's revelations on the secret leasing of the Reef for oil prospecting had become the leading media issue of the day. As part of the preliminary skirmishing leading up to the election, Opposition leader John Houston exposed the fact that the Premier and a number of cabinet members had financial interests in the oil exploration companies (*Courier Mail* 3 March 1969). Bjelke-Petersen himself had a 49 per cent interest in Artesian Oil that he had formed in 1958 by applying to the Department of Mines for a non-transferable prospecting lease for the nominal sum of £2. In a lengthy political controversy in which even his colleagues pointed out the conflict of interest, the premier eventually tried to defuse the uproar by relinquishing the shares: he transferred them to his wife (Whitton 1989:14).

Bjelke-Petersen brushed the starfish infestation aside, denying any threat to the Reef or the tourist industry at a press conference: 'There is no real cause for concern . . . expert evidence is that there is no vast plague . . . the coral has rejuvenated . . . [and] we are not going to set up an elaborate enquiry into the problem' (*Courier Mail* 8 September 1969). Relying entirely on tourist operators who by that time had become fearful of the adverse publicity for their trade and presented the most optimistic version of the alleged 'plague', the premier had not indicated the source of his 'expert evidence', but the problem did not go away. In November of that year the Australian Academy of Science convened its own committee of highly reputable marine scientists: Elizabeth Pope of the Australian Museum; Frank Talbot, Director of the Australian Museum; Don McMichael, Director of the New South Wales National Parks and Wildlife Service; J. M. Thomson, professor of zoology at the University of Queensland; and Professor Cyril Burdon-Jones of Townsville University College under the chairmanship of geneticist Professor R. J. Walsh, which commenced what became a lengthy, comprehensive and independent study of the problem.

The Crown of Thorns issue by this time was not arousing the same intensity of concern as the rapidly escalating problem of oil pollution on the Reef, now a rising concern as a result of the growing number of ships on the Reef passages, and a dramatic increase around the world of shipping disasters as the world's fleets underwent major changes. In those immediate post-war years a considerable quantity of oil came from Middle East fields, all of which when bound for Australia and Japan travelled by tankers through the Suez Canal. Following its nationalisation in 1956 by President Nasser, and the disastrous Anglo-French invasion, the canal was closed for a year. Since its shallow depth only allowed tankers with less than 10 metres draft to pass through, shipowners turned to building larger vessels that were more economical per ton/mile, and also, by using the oil-loading terminals in the Persian Gulf they avoided the increased canal transit fees. By 1960 the standard smaller tankers of less than 25 000 tons deadweight (dwt) were being phased out by so-called supertankers in the range of 25 000 to 60 000 tons dwt, while even larger ships, the Very Large Crude Carriers (VLCCs) in the range of 65 000 to 205 000 tons dwt were being designed, facilitated by the newer technologies of high tensile steel fabrication and steam turbine propulsion. Already, pollution of the seas had become a serious matter as a consequence of smaller tanker accidents and spillages from drilling platform blowouts resulting from the rapidly increasing exploitation of the continental shelves of the world's seas for oil. In response, the first international regulation appeared in the 'Convention for the Prevention of Pollution of the Sea by Oil', known as Oilpol, in 1954 which came into effective operation in 1958.

Despite such attempts at control, public protest that with greater size came greater risks was brought to a crescendo on 18 March 1967 when the 121 000 tons dwt VLCC *Torrey Canyon* was wrecked off the coast of Cornwall and before the television cameras of the world 118 million tons of oil were spilled on to the prime English vacation beaches. The greatest environmental disaster in maritime history to that date, it created consternation of the highest order. Loss of wildlife was enormous with more than 20 000 seabirds killed by the oil slick which, in a bird watching Britain, with a vigorous Nature Conservancy movement, was seen as a profound disaster. The calamity could not have come at a worse time for the Queensland government, nor at a better one for the environmental movement. Fear of the additional massive environmental damage that could come from oil spills during drilling was also heightened among supporters of the Reef.

In the same period that the Reef was being clandestinely leased out to oil companies, the *Torrey Canyon* disaster was international news and the *Acanthaster* starfish was creating rising alarm in tourist locations, yet another cause for environmental concern came in mid 1967 when the Queensland Department of Mines received an application from Donald Forbes, secretary of the Cairns Canegrowers' Executive, to mine Ellison Reef, 35 kilometres north-east of Dunk Island, for coral to be taken ashore for processing into lime fertiliser. Sugarcane is heavily lime dependent, and politically the governing Country Party of the time was equally cane farmer dependent. It seemed a foregone conclusion that the application would be approved since most of the Reef there consisted of 'dead' coral, a claim consistently voiced by oil exploration geologists. Technically, of course, all coral reefs are mostly comprised of inert limestone which, like the dead timber inside the tree trunk, is essential for the living veneer of coral and algal communities along with the myriad forms of life that constitute a coral reef ecosystem.

Immediately the Wildlife Preservation Society and the newly founded Littoral Society were roused into vigorous opposition and lodged an appeal in the Innisfail Mining Warden's Court. They were aware that the seemingly simple application to mine Ellison Reef could have the direst of outcomes: it meant that a precedent would be set for mining the Reef. If limestone, why not oil, and consequently spills and devastation? The lead was taken by president of the local branch of the Wildlife Preservation Society, John Büsst, an artist who lived on the nearby mainland at Bingil Bay, with strong support from the poet Judith Wright and rainforest ecologist Len Webb. Joining forces with the newly formed Littoral Society, led by Eddie Hegerl, a campaign was planned, as graphically described by Judith Wright in *The Coral Battleground* (1977), her account of the crusade to save the Reef.

At the time the GBRC did not directly oppose the granting of the lease, but adopted a more measured approach. On 26 September 1967 the honorary secretary of the GBRC, Dr Patricia Mather of the Zoology Department of the University of Queensland, wrote to the mining warden that while 'this Committee is concerned with the promotion of scientific knowledge and conservation and exploration of the area' at the same time 'we have not lodged an objection to the present application as we have no information at present which would suggest that the Ellison Reef under consideration is significant biologically'. The Committee's lack of objection, however, she added, should not be taken to mean that the present application if approved would set a precedent for further mining in the Reef. The GBRC, in fact, was currently giving consideration to forming an overall plan for conservation measures to control exploitation of the whole Reef area, which had been a concern of the Committee since 1927 (Au.Arch.).

Despite the GBRC's refusal to act, many objections were heard, with evidence from a large number of persons with relevant expertise: the warden eventually, and reluctantly, in a climate of strident environmental protest, recommended against mining. A long delay ensued. After six months and further sustained public campaigning, during which the government was preparing for the election, Mines Minister Camm decided in early 1968 against the proposal. Even more curiously, the Reef was not in Queensland waters, nor was it covered by any relevant State legislation. The first round in the struggle had come down on the conservationist side, but more conflict was to ensue.

In August 1969 Camm hit out strongly at the anti-drilling lobby, commenting that he could not understand why there was 'such a fuss' since drilling had occurred previously without any public comment, failing to point out that the reason was due to the clandestine nature of that exploration at the time. The government, believing they had eliminated the troublesome issues of limestone mining, starfish infestation and parliamentarians' financial interests in oil exploration companies, moved on to continue with the plan to drill for oil, and the first location chosen was off Mackay in the area known as Repulse Bay. Ampol–Japex had arranged for a self-propelled drilling rig, the *Navigator*, to travel from Texas in the Gulf of Mexico, due to commence drilling in October 1969.

COMMONWEALTH CONCERN FOR REEF CONSERVATION

During this period the Australian Government was under considerable stress. In a tragic and still inexplicable event, on 17 December 1967 Prime Minister Harold Holt disappeared while swimming in a particularly boisterous surf in Port Phillip Bay, and his body was never recovered. At that time the oil drilling controversy in Queensland had become a national concern, and by early 1968 was beginning to preoccupy the coalition government in Canberra when a new Prime Minister, John Grey Gorton, assumed office on 9 January 1968. Gorton was to experience a tumultuous incumbency for the next three years that was to have the greatest significance for the future of the Reef, since during his term of office a major change occurred in both Commonwealth and Queensland attitudes to conservation and management. Specifically, Gorton initiated two significant issues: firstly, an intense period of debate and at times unbelievably hostile confrontation with Queensland concerning the sovereignty and control of the Great Barrier Reef; secondly, in order to put Reef debate on as fully informed a footing as possible, he proposed the foundation of Australia's first full scale scientific marine research institute.

Throughout 1968 the plight of the Reef was continually raised in the federal parliament, chiefly by Queensland Opposition members Fulton and Patterson. The perennial concern over Asian raiding of the Reef was also beginning to boil over, particularly in respect to Japanese and Taiwanese fishing. A major incident had occurred in November 1967 when the crew of a Taiwanese clam fishing vessel were discovered on Green Island, provoking a major outrage, as reported in heightened fashion in the *Cairns Post*, 3 November. By that time poaching of clams in territorial waters had become serious as all three species, *Tridacna gigas*, *T. derasa*, and *Hippopus hippus* were disappearing (and, indeed, later in 1983 they were listed by the International Union for the Conservation of Nature and Natural Resources as an endangered species). In Taiwan the waters had already been stripped bare, and the northern Reef was their last remaining source of supply. The federal parliament was prompted to introduce legislation to control the problem, since amendments made the previous year, 1967, to the *Commonwealth Fisheries Act* of 1952 protected green snails, bêche-de-mer, trochus and pearlshell, but not clams.

Following the authority conferred by the Geneva Convention on the Law of the Sea of 1958, a Bill for the Continental Shelf (Living Natural Resources) was introduced in early 1968. At the beginning of the debate Douglas Everingham, coalition member for Capricornia, had complained of the 'imbalance of certain fauna' [clams] due to foreign fishing, the theme being continued by both Fulton and Patterson who were exceptionally vigorous in debating the clauses. Fulton continued to hammer the theme of 'the government's apathy towards and neglect of one of our most valuable assets, the Great Barrier Reef'. He saw the encroachment by the Crown of Thorns starfish as a result of triton loss due to destructive fishing 'from Asian countries'. Not only did it require protection from Taiwanese poaching of clams, and Japanese trawling for prawns in the Gulf of Carpentaria, he continued, it also had to be protected from 'thoughtless exploitation for oil' (Hansard, H. Reps. 1968:152). With the Offshore Agreement based on the loose cooperative sharing of continental shelf resources in mind, Fulton pointedly asked Attorney-General Nigel Bowen 'Who has control over the Great Barrier Reef? The Australian Government or the Queensland Government? ... Is the Reef part of Australia's continental shelf? Is that recognised internationally? . . . if it is owned by Queensland, has Queensland asked for Commonwealth help in policing it?' (411). Bowen's reply was a clear indication of the very uncertainty concerning sovereignty that Fulton was attacking: the Reef, he replied, 'is not part of the mainland territory of Australia' (412).

On 30 May Patterson continued to criticise what the Opposition believed to be the government's supine attitude over Asian intrusions, and commented that navy and air force patrols were 'a farce' and any attempt to determine territorial boundaries (12 nm at the time) was impossible due to lack of reliable hydrographic data (1788). He then moved 'that legislation be enacted to assert Australian territorial control over the waters of the Gulf of Carpentaria and the waters of the Great Barrier Reef' (1790).

Bowen was in an impossible position. His reply indicated that the government was aware of the uniqueness of the Reef and the value of its living resources, but, under current international law it could not seal off the Gulf (where foreign trawlers, including the giant Russian factory ship *Van Gogh*, were stripping prawns relentlessly) since international law was 'in a state of flux', while the Offshore Agreement constrained the federal government from acting unilaterally. In fact, he added, it was currently having discussions with Queensland for joint legislation to exploit the mineral resources of the Reef (1792). That reply prompted Fulton into an impassioned objection: 'drilling for oil anywhere along the reef could absolutely ruin it . . . spillage in Barrier Reef waters would inevitably occur if drilling took place in the area. Oil could very well kill all the marine life and the reef would eventually die' (1798). On 21 November the Federal government finally passed the Bill as the *Continental Shelf (Living Natural Resources) Act*, No. 149 of 1968, based on the Geneva Convention of a decade before. Under international law the waters of the Reef province were the high seas open to all, but the new Act gave Australia control over the extraction of sedentary species, that is, those whose habitat was confined to the sea bed. Although the Commonwealth would not declare sole sovereignty, Australian authority, in the words of the Act, was now legislated 'for all the parts of the Australian continental shelf and all parts of the continental shelf of an external Territory and applies to all persons, including foreigners, and to all ships, including foreign ships' (Section I.8).

A year into Gorton's prime ministership, Reef issues continued to be serious concerns when the disastrous Santa Barbara oil spillage occurred on the popular beachside holiday coast of California 120 kilometres (75 miles) south of Los Angeles which at the time was registering one million vacationers a year. The State had prohibited drilling on Californian land, but after a territorial dispute between California and Washington, the United States Supreme Court ruled that the offshore channel between the mainland and the islands of San Miguel, Santa Rosa and Santa Barbara was federal territory. Subsequently, the federal government granted oil exploration leases for the channel. On 21 January 1969 a massive blowout occurred in which, for ten days, oil gushed out across the waters of the beautiful bay with a slick spreading 25 kilometres (15 miles) to the shores, coating the silver sand with heavy black oil and killing thousands of sea birds. The tragedy was made all the more dramatic since colour television had arrived in the United States in 1960, and the visual impact was heightened immensely across the networks of the nation. In the subsequent congressional inquiry the famous oil wildfire fighter Red Adair was called as a witness and gave the startling evidence that he had already attended fifty blowouts so far, of which eleven had been from offshore drilling rigs (US Congress Report 1969:273). The parallel with the Great Barrier Reef and a similar conflict of interest was quickly drawn: *Torrey Canyon* and then Santa Barbara gave powerful ammunition to conservationists.

Environmental degradation of rural Australia had already become a serious concern to scientists in the Council for Scientific and Industrial Research (CSIR) by 1962, at the time of Carson's writing. Particularly active was Francis Ratcliffe who had earlier drawn attention to the problem much earlier in his 1938 pioneer work *Flying Fox and Drifting Sand*. In 1964, following the example of the World Wildlife Fund founded in England in 1961 and supported by Prince Philip, Ratcliffe and some of his colleagues began planning for a similar body in Australia which was formally founded in 1965 as the Australian Conservation Foundation (ACF) (see Broadbent 1999). It was designed from the start as an influential organisation: the Governor-General was its patron and Sir Garfield Barwick, Chief Justice of the High Court, its president. Its Constitution declared its aims were to see that the lands and waters of the Commonwealth are used with wisdom, that efforts should be made to preserve its distinctive vegetation and fauna, 'and that competing demands upon them are resolved in the best long-term interests of the nation'. Evidently, the hope, never expressed explicitly, was that it would influence government decisions on conservation rather discreetly through the non-partisan presentation of environmental issues.

When John Büsst and Judith Wright joined the ACF, however, it was with the deliberate intention of using it as an activist body to save the Reef. At the annual general meeting in October 1968 Büsst spoke passionately against any mining activity and presented a motion that was seconded by Wright and carried by a large margin:

1 that the Commonwealth Government take immediate control of the whole Barrier Reef area to the 200 metre mark;
2 that an immediate moratorium on all mining for at least five years be declared in the whole Reef area;
3 that since the control and safety of the Great Barrier Reef is a matter of international concern which can no longer be left in parochial State hands and must be raised to both a Commonwealth and international level, a national committee, under the auspices of the Academy of Science, and with power to co-opt international scientific advice, be set up to determine the future of the Great Barrier Reef, probably the most important scenic and recreational area in the world, and certainly with the potential to become the most important marine biological laboratory on the face of the planet.

As a politician and Attorney-General Barwick had been in the centre of much critical dissent; now as Chief Justice he was attempting to present a new persona of dignified, judicial gravity. Overt political controversy had to be avoided; dispassionate reasoned discussion was to be the way. It was decided, accordingly, to organise a public forum for discussion of all relevant issues at Sydney University in six months time on 3 May 1969.

The symposium drew a very large audience of over 500 persons from scientific disciplines, law, fisheries, mining, petroleum exploration and production, and tourism. While all speakers affirmed the 'priceless value' of the Reef and its magnificence as an 'outstanding natural phenomenon' as would be expected, none the less partisan views prevailed. The longest, most impartial and challenging contribution came from the distinguished Australian jurist, Sir Percy Spender, former leader of the Australian delegation to the United Nations 1952–56, and president of the International Court of Justice 1962–67. His lengthy and impressive paper 'Great Barrier Reef: Legal Aspects' was a *tour de force* of legal research and presentation: it was unanimously acclaimed the highlight of the day. Beginning with colonial law and moving through the relevant cases and statutes to the present, Spender carefully and methodically demolished any claim Queensland might have to Reef waters with the conclusion that, in his view, 'no part of this belt of sea is an any legal sense, international or domestic, part of Queensland, or any other previous colony'. To ensure that there could be no further dispute, he added the comment that 'It should be in the interest both of the Commonwealth and all the States to have these issues authoritatively determined . . . before the High Court of Australia' (Spender 1969:37–39).

Significantly, a number of speakers raised the idea of some kind of advisory or regulatory authority to coordinate the conflicting interests on Reef resources. At the conclusion of the symposium discussion turned to the motion of John Büsst which had been passed at the AGM six months earlier. Clearly it was too politically confrontational and Barwick was aware that governments resented being directed to any particular course of action. The same ideas were retained but Barwick urged that they should be expressed in more moderate language whereby the Commonwealth and Queensland governments were to be informed that the foundation recommended a joint 'advisory body or commission on the Great Barrier Reef' be set up 'to examine and report on all proposals for development of the resources of the Reef, particularly those likely to have an adverse effect on the environment . . . and to formulate principles for assessing developmental proposals that would provide a basis for an overall plan on Reef development and conservation' (ACF 1969:3–4).

Queensland for its part continued to press on with the plan to drill the Reef and waited for the *Navigator* to arrive in Repulse Bay. Gorton, however, was becoming increasingly sympathetic to the conservationist cause. The late Prime Minister Holt had been a keen snorkeller and vacationer on the Reef, where he had a holiday home near Bingil Bay and relaxed there at times in the company of John Büsst, who persuaded him of the need for its protection. Similarly, while Gorton was vacationing on Dunk Island in July of 1969,

Büsst came across from his home on the nearby mainland to present the conservationist case. Gorton was already sympathetic to the cause and attempted, within the limits of his powers, to assist. Other issues, however, were taking the national stage, chiefly the tremendous wave of unrest sweeping the country over Australian participation in the Vietnam conflict that was beginning to seriously divide the coalition government. Moreover, the government was coming to the end of its term of office and Gorton – who was not entirely popular with some elements of his cabinet due to his dominant, non-consensus style, and his strongly centralist views that put the national interest ahead of States' rights – now had to move warily.

By September the obduracy of the Queensland government prompted the more activist conservationists to form a 'Save the Reef' committee under the presidency of Senator George Georges, one of the most vocal and politically effective parliamentary opponents of the coalition who raised in the Commonwealth Parliament the issue of Exoil's one million dollar subsidy from the Commonwealth and 'the questionable activities of oil companies in Queensland in which the present premier of Queensland is interested' (Hansard, Senate, 4 March 1969: 203–204). The ACF Conference convened to Save the Reef was attended by Mines Minister Camm who proposed the novel view that the only way to save the Reef was to rapidly exploit its oil potential by Australian companies before foreign companies could move in and take it all. At that meeting Camm put the Queensland view unequivocally when he asserted that 'We in Queensland claim the Barrier Reef and believe we own it' (Wright 1977:91). Indeed, it is easy to understand his impassioned view since, as this work has recounted, the Reef has shaped all of Queensland's history: in fact, from the beginning the Reef gave it an identity markedly different from all the other colonies, and, after federation, the States. The tragedy is not that Camm took justifiable pride in the Reef, but rather that he was prepared to risk endangering it irreversibly for short term financial gains without making a full investigation of all the issues involved.

A month later when the federal election campaign began the Reef had become a national issue and *The Australian* newspaper, at that period taking a radically critical stance towards social issues, was keeping the debate well before the public. By this time the federal coalition government of the Liberal and Country parties was being pressed hard by the renewed vigour of the Labor Party led by the dynamic leader of the Opposition, Gough Whitlam. In the election campaign of 1969, Whitlam's policy speech, delivered on 1 October, stole a march on the government by announcing that the Labor Party, if successful, would immediately suspend all mining and drilling on the Reef. Opening the coalition campaign on radio and television a week later Gorton committed his party to make a substantial effort at 'understanding the biological and physical resources in and beneath the sea around us' and to that end it 'would establish an institute of marine science at Townsville'. That pledge was reported briefly in *The Australian* the next day: 'Research in the area of the Great Barrier Reef will be one of the first priorities of the new Institute, he said. The capital cost is estimated to be of the order of $3 million' (9 October 1969).

Gorton was genuinely concerned to redress the century of neglect of marine science in Australia so stridently condemned by Saville-Kent seventy years before, and by Frank Talbot in the ACF symposium just five months earlier who stressed that 'the whole science of ecology . . . is in its infancy. We cannot yet understand simple natural systems. We cannot yet even comprehend the problems inherent in a system as complex as the Barrier Reef' (ACF 1969:68). The decision to locate the proposed marine science institute in Townsville instead of Brisbane, as favoured by the GBRC which had been campaigning for Reef science for fifty years and wanted to upgrade the Heron Island station, was a tactical political decision for Gorton. The federal electorate of Herbert (Townsville) was a marginal seat and Noel Bonnet, the sitting Liberal member, held it by the slimmest of numbers.

In the election of 25 October 1969 the coalition retained the seat of Herbert, but suffered the greatest swing against it – 6.8 per cent – since the early 1930s in the depths of the Great Depression. Then came the most startling development which took all of the parliament by surprise. Almost immediately on being returned to office with a slender majority, Gorton decided to follow the lead given by Spender at the ACF symposium and legislate for Commonwealth sovereignty over the Reef – a proposal that he knew would certainly be challenged and finally decided through proper constitutional means by the High Court. The cabinet was divided, with the 'States-rights' faction opposing the move. They were, however, still a minority and Gorton pushed ahead. When the proposal was later debated in the federal parliament in May 1970 it created a furore from both sides. Queenslander Peter Delamothe, Attorney-General and Acting Minister for Mines, accused his leader of 'an arrant, and I believe, considered breach of faith on the part of the Commonwealth' (Hansard. H. Reps. No.7: 1897–98). Meanwhile, opposition member Patterson moved to censure Gorton for failing to honour a previous commitment to consult with the States before legislating on the territorial sea and the continental shelf (No. 8: 2246).

By the end of 1969 alarm over the possible destruction of the Reef was still high, fuelled by what some considered the extreme views of Dr Endean that the *Acanthaster* infestation was a 'plague', threatening the Reef's survival. That threat, along with the oil drilling controversy and the lack of definitive government action, led *The Australian* to editorialise on 24 December that 'The Great Barrier Reef has been the object of a shameful exercise in buck-passing throughout this year' while 'the response to the public's demonstrated demand has been insulting'.

In the meantime the *Navigator* still had not arrived and the conservation lobby had been actively searching for a means to stop the proposed drilling. Events moved very quickly the following January once John Büsst and Senator Georges, recognising at last that it was impossible to reason with the obdurate Queensland government, sought trade union assistance. Having previously received resolutions of support from the Amalgamated Engineering Union on three issues: to ban mining on the Reef, to have it declared 'a National Marine reserve for the benefit and relaxation of the Australian public', and for the constitutional issue to be decided by the High Court, Georges acted quickly. On 6 January 1970 he sent several telegrams to Ampol informing them that the unions were ready to impose black bans on all aspects of the drilling: the wharves would be closed to the *Navigator*, no labour would be available, no services or supplies would be given. AMPOL caved in: a week later, 13 January it announced a postponement.

The Gorton government set the seal on its concern for the Reef by announcing the same month, on 21 January, that legislation would be introduced into the next session of the Commonwealth parliament to establish the promised institute of marine science in Townsville, at an estimated cost of $3 million, to conduct research 'into ways of preserving and protecting the Great Barrier Reef'. Making the press release, which occupied most of the front page of *The Australian* the next day, Nigel Bowen, now Minister for Education and Science, stated that cooperation would be sought from American marine scientists under an exchange program, from the Australian Academy of Science and from eminent Australian scientists to help plan its establishment and research program. This would examine not only the devastation being caused by the Crown of Thorns infestation but also possible damage from oil drilling, and other human activity such as tourism, reef blasting, mining and ship traffic. The government, he announced, 'is determined to use every resource available to win the knowledge necessary to preserve and protect this unique possession' (*The Australian* 22 January 1970).

In such a tense atmosphere of heightened concern, Gorton had already approached Bjelke-Petersen to resolve the impasse on drilling and all other aspects of potential Reef degradation by holding an impartial inquiry. After much hostile blustering, and faced with federal resolve, Bjelke-Petersen had little option but to agree and on 28 January the two met to determine its terms of reference and composition. Japex had already decided to suspend drilling, and on 14 February 1970 it terminated the contract with the *Navigator*. Bjelke-Petersen, however, was unrepentant and in his Ministerial Statement to parliament he affirmed that 'The Government was of the opinion that the drilling of this well should be allowed to proceed, and no evidence has been brought forward to cause it to change its opinion'.

Less than three weeks later came the sensational news of the first oil disaster in Great Barrier Reef waters off the tip of Cape York near Tuesday Island. The fully laden Ampol 58 000 ton dwt tanker *Oceanic Grandeur* bound for Brisbane struck an uncharted pinnacle of rock on 3 March which ripped a 57 metre gash in the hull and released oil on to surrounding waters in a slick 10 kilometres long. Part of the cargo was later pumped into a relief tanker but around 1100 tons of oil remained, defying all attempts to disperse it with the two available detergents, Gamlen and Corexit (Oceanic Grandeur *Report* 1970). Since it occurred away from populated areas the damage lacked visual impact, but, none the less, the event had a dramatic effect on the Reef debate. Both Commonwealth and Queensland governments commissioned inquiries into the incident which led to a series of investigations designed to avoid, or at best minimise, such disasters.

One major consequence was the decision by both governments that the proposed inquiry on the consequences of Reef damage from oil pollution, whether from drilling or tanker accidents, had to be investigated thoroughly; it was therefore upgraded to a Royal Commission on 18 March. Technically, two Royal Commissions were established: one for the Commonwealth, the other for Queensland, in order to satisfy the legal requirements of both parliaments, but with the same three commissioners and the same terms of reference, sitting in parallel: a judge as chairman, along with a petroleum engineer and a marine biologist. Gorton announced the Terms of Reference in the parliament on 5 May 1970, the date from which the Letters Patent creating the Commission from both governments took effect. The commissioners were to take most of the next four years over their very onerous task.

CHAPTER 20

CRISIS RESOLUTION: FORMATION OF AN ENVIRONMENTAL MANAGEMENT AUTHORITY

Well before the Royal Commission had ever been considered, pressure for Reef protection had continued to grow strongly following the ACF symposium in May 1969, when a few of the more determined advocates came together to press for political action on the final resolution. In the following years, two complementary, interacting processes were in operation: on the one hand continued conservationist efforts were being exerted to mobilise public support for Reef protection; on the other, was the drawn-out contest between the Commonwealth and Queensland governments to reach an acceptable resolution of difficulties to enable political and legislative change.

FROM RESOLUTION TO IMPLEMENTATION: FIRST DRAFTS OF AN ACT

A leading protagonist throughout was Patricia Mather, acting officially in her capacity as honorary secretary of the GBRC, but in large part driven by sheer personal determination to see that the Reef would be saved from mining and any other form of blight. In preparation for the symposium, on 21 February 1969 she sent Chief Justice Barwick a résumé of the deliberations of the GBRC over the previous two years in examining the legal issues relating to protection. Having described them in careful detail she then presented the crux of the GBRC case: 'The most pressing requirements at the present [are] . . . to extend the protection afforded all marine flora and fauna to territorial waters surrounding *all* national parks . . . to prohibit mining of minerals below high water . . . [and] to investigate setting up of further national parks'.*

Immediately following the symposium Patricia Mather and Francis Ratcliffe continued their efforts to formulate an acceptable mechanism to implement the ACF resolution. In a letter to Mather, Ratcliffe suggested they could consider either the preference expressed in the ACF motion for some kind of an 'advisory body' or, in his own view, possibly a 'Commission with statutory authority' representing the various interest groups, of say seven members, and provided with what he termed 'administrative machinery'. From the beginning Mather was totally opposed to any kind of 'advisory body': in her view nothing less than an independent statutory authority would be effective. She replied to Ratcliffe on 16 May, pointing out that two considerations had to be resolved: if the 'advisory body' were small in number, then some way would have to be found to accommodate the large number of separate departments already involved in Reef matters since neither 'the Commonwealth nor State governments would set up an advisory body so small that representation of relevant interests were ignored or neglected and where, by legislation, the government departments were required to act on recommendations of that body'. The alternative, in her personal view, from which she never wavered, was for 'a Commission deriving some power and authority from State and Commonwealth legislation . . . composed of individuals of considerable independence, status and influence in the community' (Au.Arch.).

With that reply Ratcliffe attempted to draft a submission for Barwick to send to Gorton. Barwick was proving hard to contact. He was, in fact, still hesitating at the idea of a statutory authority, since, as Ratcliffe informed Mather later, he believed that 'if we asked for executive powers the Commonwealth and Queensland governments would be scared and we would probably get nothing at all'. Ratcliffe was also wavering and commented that 'I have convinced myself that he is right, not so much because of the possible scare but because I can see no practicable way of defining executive or administrative powers for a body that works between two governments and several departments'. Mather did not budge from her own conviction that a statutory body was the only way to go. Writing to her friend rainforest ecologist Len Webb regarding Ratcliffe's position, her emphatic comments stated that she had 'been *unable* to persuade Francis that a Commission with executive power does *not* have to work between governments *or* departments – but does its *own* work!' By that time Ratcliffe's chronic heart condition was taking its toll and he was unable to maintain the pace: he died the following year.

Mather was now the main driving force in these negotiations and set herself the task, while continuing in her full time work as a zoologist in the University of Queensland, of compiling a draft 'Bill for an Act relating to the Great Barrier Reef', intended for the guidance of governments, a daunting task that she finally completed in May 1971. As Mather stated in the Introduction to the comprehensive 25 page document, the aim was 'an attempt to reconcile conflicting pressures regarding resource use in the Great Barrier Reef area and to provide a mechanism ensuring an appropriate administrative machinery for an area which is the legislative responsibility of both the Commonwealth and State Governments'.

Throughout its 62 sections the draft displays an impressive grasp of the political complexities of the period, the numerous legal aspects to be considered and the need to provide input for the large number of government departments involved. The significant element is contained in Part 3 that set out the concept of 'A body corporate to be known as the Great Barrier Reef Authority . . . consisting of a Commissioner and two Associate Commissioners'. Two commissioners were to be appointed by the Governor-General, 'one of whom shall be a biologist nominated by the Australian Academy of Science'; the other 'nominated by the Governor-in-Council in the State of Queensland'. The aim was to balance the interests of the Commonwealth and Queensland, since the bill would be enacted by both governments, specifically to ensure that 'the rights of the State of Queensland are therefore preserved while the responsibility of the nation in regard to the area is established'. Two major aims were stated: to guarantee 'the legal effectiveness of authorities to explore for and to exploit the resources of' the Reef, and to 'coordinate and reconcile the multiple uses of these submerged lands and island territories and waters in a manner compatible with the conservation of the living coral reef ecosystem contained thereon and with the maintenance of the geological structures formed by the fossil reefs'.

While Mather was working on the draft bill, the ACF resolution and the work of the committee had come to the notice of Opposition leader Whitlam who raised the matter in the federal parliament, putting the details of the resolution sent to Barwick on the Parliamentary Notice Paper for 10 September. Placed for the attention of the Prime Minister, it recommended that the Commonwealth and Queensland should confer on the establishment of a joint advisory commission on Barrier Reef development and conservation, particularly in respect to the dangers from oil drilling (Hansard, H. Reps. 1969:927).

Gorton, for his part, was disposed to act, but without legal certainty over Reef jurisdiction he still had to deal with Queensland, particularly since most of the government departments concerned with its management were, and would remain, State bodies. The major problem, of course, was the determination of the Queensland government to retain total control, compounded by the vacillation of previous Commonwealth governments to act decisively. On 16 April 1970 Gorton decided to move: the Territorial Sea and Continental Shelf Bill was introduced into parliament whereby the Commonwealth would assert sovereignty over the Reef. It was immediately opposed by five State premiers and also by an increasingly nervous faction within the Gorton cabinet.

Six months later Gorton's proposal to resolve the issue of ultimate authority over the continental shelf received support on 24 September 1970 from the Senate Select Committee on Off-Shore Petroleum Resources which had been examining the constitutional validity of the 1967 Offshore Agreement legislation embodied in the *Petroleum (Submerged Lands) Act*. On that date it presented an Interim Report (tabled in final form on 8 December 1971) on the first of its terms of reference, viz. to determine 'whether the constitutional conception underlying the legislation is consistent with the proper constitutional responsibilities of the Commonwealth and the States'. The Report expressed grave reservations in that 'the large national interest is not served by leaving unresolved and uncertain the extent of State and Commonwealth authority in the territorial sea-bed and the continental shelf' (Senate *Interim Report* 1970:126). Gorton's proposal, however, while constitutionally correct, created major political difficulties since the coalition knew that it would create untold friction with the States, all of which, with the exception of South Australia, had conservative governments, none more so than Queensland. The central difficulty stemmed from the ineluctable fact that seventy years earlier the six separate colonies had never fully federated; each retained its separate identity and remained eternally suspicious of too much concentration of power in Commonwealth hands. The *Australian Constitution Act* of 1900 had specified a limited number of powers for the Commonwealth, but in the nature of evolutionary changes in the nation, increasing centralised power had already come to it and the move by Gorton was seen as yet another such trend in that direction.

Within the coalition moves were quietly planned to replace Gorton since it was feared that his centralising tendencies would unduly antagonise the States and lead to instability and electoral loss. Indeed, the right-wing faction was deeply concerned about the increasing political kudos going to the Labor Opposition which was promising to end Australian involvement in Vietnam if returned at the 1972 election. Further, Whitlam had gone on record asserting that the Commonwealth undoubtedly had the power to assume sovereignty over the Reef, with the implication that Labor would do so if elected to government. In a party room spill on 10 March 1971 Gorton was deposed, although he remained a committed centralist and never sought to hide his views. In an interview three weeks later he asserted that

> I put the creation of an Australian identity and national feeling and the concept of national benefit, as being more important than pandering to the susceptibilities of six provincial potentates. I did not disguise that . . . it would be unthinkable for an Australian government not to have overriding power over the maintenance of the Australian economy as a whole (*Sunday Australian* 4 April 1971).

In the same period, moreover, the States were becoming increasingly hostile to any suggestion of either Commonwealth control or environmental group pressure. In particular, Barwick was singled out for criticism, and his judicial impartiality was impugned. How, it was being suggested, could the Chief Justice become party to activist pressure on

---

* Unacknowledged quotations in this chapter come from Commonwealth Department of Environment Files 'Formation of the Great Barrier Reef Marine Park Authority', and also from information provided by persons interviewed and other documents in possession of the author.

environmental matters? Barwick himself had given some cause for concern in that he had begun increasingly to make comments on environmental and conservation issues. At a Waste Disposal Conference in 1971 he decried the 'mad scramble for material growth' and later criticised 'the materialistic get-rich-quick system in which we see all hands enmeshed'. On a number of occasions he expressed very forcefully the need for legislative and planning controls on the environment, all of which to fellow environmentalists seemed perfectly reasonable. Not, however, to New South Wales premier Tom Lewis, who, under cover of parliamentary privilege on 21 September 1971, drew attention to the fact that some of the matters of conservation that Barwick was advocating 'may have to be referred ultimately to him or his colleagues on the High Court'. The impropriety of Barwick's actions, Lewis asserted, as reported on the front page of the *Sydney Morning Herald*, 22 September 1971, place 'the Chief Justice in a very invidious position' (Marr 1980:235–37). Barwick sensed the difficulties for the environmental movement, and his own personal position, and resigned from the presidency of the ACF, his place being taken by Prince Philip, president of the World Wildlife Fund.

Throughout that period of political intrigue the Crown of Thorns infestation was worsening, having spread the entire length of the Reef, even to the outlying Swains, and was continuing to receive sustained media coverage. In March 1971, eighteen months after it began its investigations, and at the time of Gorton's political demise, Professor Walsh presented the Australian Academy of Science report, known as the Walsh Report. A cautiously conservative document, in fourteen carefully worded *Conclusions* it began with the statement that *Acanthaster planci* 'does not constitute a threat to the Great Barrier Reef as a whole'. Acknowledgment was made of damage to a number of reefs, although, it added, 'recolonization and regeneration of coral have occurred on all reefs that have been examined'. The one aspect that was to become highly controversial was the Academy's fifth conclusion that the outbreak could well be part of a natural 'episodic event that may have occurred previously'. Although it did go on to acknowledge that it could not ascribe any particular cause to the current episode, its report questioned the theory of reduced predator pressure coming from the wholesale removal of the giant triton by excessive collecting, but made no other mention of human impact on the Reef. What the report made painfully obvious – and recognised as significant for the future – was the low level of marine scientific knowledge in Australia, and it stated very clearly that 'knowledge of reef ecology' was highly inadequate, especially in respect to understanding this explosion of the Crown of Thorns starfish. The report ended with a recommendation that the Commonwealth and Queensland governments should jointly fund research under an Advisory Committee (*Conclusions of the Official Committee*, AAS; Appendix II, James 1976:83–84). The recommendations of the Walsh Report were accepted by the

Commonwealth and Queensland governments, who then jointly funded an Advisory Committee on Research into Crown of Thorns Starfish with a subvention of $330 000 over three years, to 1975 (Lassig & Kelleher 1991:42).

On 8 December 1971 the Senate Select Committee on Off-Shore Petroleum Resources presented its massive three volume final report after more than three years of exhaustive inquiry, during which it took evidence from 84 witnesses, including members of the Queensland Wildlife Preservation Society, the Littoral Society, and Professor Burdon-Jones, newly appointed to the chair of marine biology at what was now James Cook University. In respect to its seven terms of reference concerning the full range of issues on oil production, every conclusion stressed the fact that all activities were hampered by the uncertainty surrounding 'the extent of State and Commonwealth authority in the territorial sea-bed and the Continental Shelf' (*Senate Report* 1972, p.7; also paras. 6.335–6, 6.339). Equally central to its findings was its frustration from the lack of reliable biological knowledge of the marine environment, in particular the effects of oil spills on coral and marine life. The section on the Reef consequently indicated the need for 'Information to be gathered in the disciplines of hydrology, bathymetry, marine biology, geology, ecology, chemistry and generally in the marine sciences' (13.185:503).

The proposal to proceed with the Territorial Sea and Continental Shelf Bill, none the less, could not be abandoned, and the new prime minister, William McMahon, decided that it could, perhaps, be redrafted in acceptable form. After all, the constitutional issue was there, and it remained unresolved. McMahon, however, was of totally different temperament to Gorton, and vacillated on pressing the issue, preferring to govern by cabinet consensus. The Territorial Sea and Continental Shelf Bill was dropped to the bottom of the waiting list of bills, and when McMahon called an election on 24 October the Bill lapsed.

Even so, the McMahon government had made some important moves in legislating for environmental protection, having responded to mounting global concern for the environment by setting up a Department of Environment, Aborigines and the Arts in 1971, and the following year a separate Department of the Environment. That initiative occurred during the planning stages in 1971 for the landmark United Nations first World Conference on the Human Environment, to be held in Stockholm, which would be concerned with accelerating evidence of global environmental flashpoints and crises. The briefing document prepared for the Conference placed great stress on the problem of the rapidly increasing pollution of fresh water supplies and degradation of the oceans. Citing both the *Torrey Canyon* and Santa Barbara disasters, it emphasised the fact that

> it is all too often the coasts which chiefly suffer from the pollution caused by the drilling and transport of oil and from the even larger burdens brought to the seas and rivers in industrialized areas. Underwater drilling for oil is steadily increasing on the continental shelves . . . [While] blowouts do occur in land drills, they do so with far less disastrous consequences than would be the case in coastal waters which

could spread the pollutants with every force of wind and tide and current (Ward & Dubos 1972:275).

In its final meeting of 16 June 1972, participants issued the *Stockholm Declaration* of 26 principles which stressed the need for all nations to maintain the health of natural ecosystems and, in particular, in Declaration 7, to 'take all possible steps to prevent pollution of the seas by substances liable to create hazards to human health, to harm living resources and marine life; to damage amenities or to interfere with other legitimate uses of the sea'.

THE WHITLAM GOVERNMENT: A RADICAL DEVELOPMENT

The really dramatic change to Reef issues came with the election on 24 October 1972 of the Whitlam Labor government that began a sequence of breathtaking reforms in nearly every facet of Australian life. Whitlam was even more centralist in ambition than Gorton, and the issue of constitutional authority over Australian waters was basic to his program of placing the national interest ahead of previous coalition concerns to promote the interests of the private sector, dominated by large transnational corporations and wealthy shareholders. The petroleum industry was an early target for reform and Reef issues became inextricably linked.

The Whitlam government, as it had foreshadowed while in Opposition, soon prepared legislation to deal with the question of constitutional authority over the continental shelf and therefore the Reef. There had been, of course, considerable debate on the question for several years; wider support was now evident in the community, and indeed, among some of the former government members. Accordingly, a Seas and Submerged Lands Bill was introduced into the Commonwealth parliament by Minister for Minerals and Energy Rex Connor on 10 May 1973. The conservative Opposition was annoyed that the bill had come before the House, and continued to maintain that the proper procedure was to follow the earlier practice of cooperation with the States according to the Off-shore Agreement, although it did concede that the question of ultimate authority, at some stage, had to be settled. After some delays in the Senate it was passed with amendments on 4 December 1973. Thereupon all six States acting conjointly, with New South Wales as the nominal plaintiff, began an action in the High Court to determine finally the issue of sovereignty by challenging the validity of Sections 6 and 11 of the Act which asserted that 'sovereignty in respect of the territorial sea . . . is vested in and exercisable by the Crown in right of the Commonwealth' (6) and that 'the sovereign rights of Australia . . . for the purposes of exploring its natural resources are [also] vested in and exercisable by the Crown in right of the Commonwealth' (11). The case was to remain before the High Court for the next two years.

Unable to accept defeat after twenty-four years in office, the Opposition was determined to obstruct Labor's reforming initiative by whatever means available and the most fateful occasion came on 12 December 1973 when the House of Representatives passed the contentious Petroleum and Minerals Authority Bill designed to increase Commonwealth equity in and control over the mining sector. Immediately, the Senate conservative majority refused to give assent. Thereupon, as provided for in Section 57 of the Constitution, the Bill was passed by a joint sitting of both chambers on 6 August. As in the case of the *Seas and Submerged Lands Act*, with Victoria this time as nominal plaintiff, together with New South Wales, Queensland and Western Australia, proceedings were begun in the High Court to have the legislation declared invalid, arguing that the stipulations of Section 57 had not been properly observed. The action this time was intended not to settle a constitutional issue but as a political stratagem to obstruct the new government's legislative agenda. The case was therefore held up until June 1975.

In the meantime, while those two major issues which had a direct bearing on the Reef were before the High Court, the Whitlam government moved into other areas of its reform program that included several highly significant, forward-looking environmental initiatives. To begin implementing that agenda, in May 1973 Australia joined the International Union for Conservation of Nature and Natural Resources (IUCN) which the United Nations had created in 1948, membership of which entailed observing its guidelines for protecting and preserving wild nature and natural resources. By that time the concept of the 'national estate' was being promoted: apparently first used in 1943 by the English architect Clough Williams-Ellis to refer to the built environment worthy of preservation, it was popularised in 1960 by President Kennedy in the United States to embrace both built and natural environments, and publicised in Australia in the early 1970s by Tom Uren, Minister for Urban and Regional Development, when he was a spokesman on the environment and urban and regional affairs for the Labor Party in Opposition. Whitlam brought it well forward in his reform program, stating that he wanted to prevent the prophecy of novelist Kylie Tennant that 'the unborn Australian will ask for his birthright and be handed a piece of concrete' (Whitlam 1985:549).

By 1973 previous Commonwealth and State conservative governments had allowed much of the architecturally significant built environment to be razed and replaced by bland, severely functional concrete buildings, while the natural landscape was being relentlessly excavated and mined for raw materials. The Reef was clearly on the list since the Queensland government was in the forefront of destructive practices. In addition to the proposal to drill for petroleum deposits, it had already licensed the sand mining of the superb natural environment where the Reef begins in the south near the multicoloured beach dunes of Cooloola and Fraser Island, the largest sand mass in the world, holding magnificent stands of rare rainforest and crystal-clear perched lakes. Strident public protests were able to save Cooloola, but Fraser Island was ravaged in places until rare earth sandmining on the island was halted in 1982 by federal intervention under Prime

Minister Malcolm Fraser in refusing export licences for the mineral products.

Meanwhile, in order to begin the conservation program of the Labor government, which Whitlam had made part of his policy speech on 13 November 1972 – 'to preserve and enhance the quality of the National Estate' – a Committee of Inquiry into the National Estate was convened on 17 May 1973 under the chairmanship of Justice Robert Hope of the New South Wales Supreme Court, with seven other members who included the prominent activists Judith Wright, Milo Dunphy and Len Webb. The Hope Committee was guided by the recently established UNESCO Convention for the Protection of the World Cultural and Natural Heritage of November 1972, extending it from its international context to the specific Australian situation. With the committee sitting as an informal inquiry, and not as a Royal Commission, 14 sessions were conducted around the nation, 650 submissions were accepted, and an exhaustive 415 page *Report* was delivered to the Australian government in April the following year.

The Hope *Report* was a devastating exposé of the extent of both destruction and degradation of the built and natural environment to date. Its final section, 'Findings and Recommendations', was prefaced with a pungent paragraph on the vandalising activities of previous decades:

> The Australian Government has inherited a national Estate which has been downgraded, disregarded and neglected. All previous priorities accepted at various levels of Government and authority have been directed by a concept that uncontrolled development, economic growth and 'progress', and the encouragement of private against public interest in land use, use of waters, and indeed in every part of the National Estate, was paramount (Hope 1974:334).

The Reef was included in the inquiry (§§ 3.123–26) with observations that although 'threats to the continuing existence and health of the reef are many' the hopeful signs were that 'the Australian Government's announcement of the pending introduction of legislation to make the Great Barrier Reef a marine national park is most welcome', and the 'statement by the Minister for Minerals and Energy specifically excepting the area for oil-drilling is just as encouraging'. The *Report* declined to comment directly on potential damage from oil exploration on the grounds that the Royal Commission was yet to announce its findings. What was clearly needed, the *Report* stressed, was protection of the Reef from damage by Crown of Thorns, water pollution and direct human interference 'by declaration and policing of that declaration' (66) which, in the event that the Australian and Queensland governments were unable to do so, then 'if necessary by an international authority'. Both the Australian and Queensland governments were urged to accept 'responsibility to the world for the preservation, management and presentation of what is a heritage of unquestionably world stature'. Its strongest recommendation came from the same GBRC submission that Patricia Mather had already presented to the Royal

Commission emphasising the need to establish a statutory authority: 'we do no more than point to the obvious need and urgency for some overall responsible authority charged with control and administration of this most important marine possession' (§ 3.126:67). Those conclusions brought no joy to the Queensland coalition government.

PLANNING AN AUTHORITY

Throughout 1974 the Whitlam government sought to deal with the vexed issue of the Great Barrier Reef, and to explore means of implementing the best response for its optimum environmental management, particularly since all the indications were that it was now being regarded by the Australian public as one of the most precious features of the Australian natural heritage. That responsibility fell to Moss Cass, Minister in charge of the newly formed Department of Environment and Conservation, who delegated the task of formulating a solution of the Reef issue to director of the Commonwealth Department of the Environment, Don McMichael, previously director of the ACF 1967–69, and director of the New South Wales National Parks and Wildlife Service 1969–73.

In 1974 McMichael began the process. Early in the year it was recognised that since a large number of government departments were necessarily involved, the best way forward was to create an 'Interdepartmental Committee on the Great Barrier Reef', the IDC. Using Dr Mather's draft bill of 1971 as a starting point, it held its first meeting under his chairmanship in April 1974, attempting 'to examine ways of implementing the Government's commitment to protect the Great Barrier Reef as a national park'. It was well accepted that provision had to be made for the inclusion of Queensland in any management regime since all of the islands were State territory: at the same time, it was necessary to ensure that Commonwealth legislation overrode that of Queensland if inconsistencies were to arise. Four kinds of possible administrative structures were considered and the final conclusion was that the most effective would be 'a small incorporated Authority with executive power ... consisting of a full-time Chairman and two or three part-time Assistant Commissioners' with the rider that if Queensland were to cooperate it could jointly nominate either the Chairman, or else one of the Assistant Commissioners. To avoid appointment of persons with departmental interests or affiliations, all three had to possess recognised appropriate professional qualifications.

At the outset it was realised that the Reef could not be included in legislation being prepared for the Commonwealth *National Parks and Wildlife Conservation Act* since it was such a huge and complex area. Solicitor-General Maurice Byers recommended special legislation for the Reef alone, administered by an independent statutory authority. He also advised that it could not be designated a 'national park' since it fell well outside the strict guidelines set by the IUCN. In its definition national parks had to be 'relatively large areas where one or several ecosystems are not materially altered by human exploitation and occupation' and the biota are of exceptional beauty or scientific interest, and where 'the highest competent authority of the country has taken steps to prevent or eliminate... exploitation or occupation... and to enforce effectively respect of ecological, geomorphological or aesthetic features which have led to its establishment'. In effect, they had to be high value regions close to wilderness status.

There was another way to ensure protection, however, of which McMichael was already well aware: to classify the area not a national park but a 'marine park'. Although the concept of a park as an enclosed tract of land, generally for recreational purposes, had been in use since the thirteenth century, generally by privileged landowners, the idea of defining a body of water for the same purpose was a very recent development. The first seems to have been the declaration in 1935 of the waters surrounding the Dry Tortugas, two tiny islands at the most westerly end of the Florida Keys. The next move to declare marine parks came in the 1960s when the United States and British Caribbean territories had begun to dedicate other small areas around their islands as marine parks. Despite those early beginnings, when the IUCN convened its first world conference on national parks at Seattle in June 1962, representing 63 countries, marine parks did not appear on the program. One indefatigable proponent, however, Carleton Ray from the New York Aquarium, presented a novel paper on 'Inshore Marine Conservation' which was a plea for preserving selected coastal zone waters for marine ecological research, an especially imperative need given the fact that around the world scientists 'have a truly abysmal ignorance of the seas at present' (Ray 1962:79). In an eloquent appeal he stressed the urgency of the problem, and suggested a solution:

> The most productive areas... on earth, in fact, are those of shallow water near shore. The estuaries, inlets, bights, and marshes with good circulation have been called the nurseries of the sea. These are also disappearing the fastest through development, filling, drainage, bulkheading and pollution... We might say rather cold-bloodedly that the areas involved are being flushed down the recreational drainpipe.

His solution was simple: securing government and public recognition 'that inshore waters should be managed in much the same way as the land and fresh waters, utilizing sound principles of sanctuary, management, recreation, and fish and game legislation as an alternative to the *laissez-faire* which still largely exists' (79).

That concept was later promoted in Australia by Don McMichael, at the time director of the ACF, when a serious accounting of the growing degradation of the natural landscape was published by Len Webb and other concerned environmentalists in 1969. In that seminal work, *The Last of Lands*, McMichael's chapter voiced the same concerns as Ray, bringing attention to widespread ignorance of marine ecosystems and the continuing practice of regarding the sea as an inexhaustible resource for exploitation and recreation on the one hand, and an equally inexhaustible and unpollutable dumping ground on the other. McMichael argued for two concepts: marine reserves and marine parks. The former 'include areas reserved specifically for fishery management purposes and ... scientific reference or for the conservation of species generally' while a marine park 'should primarily be an area of scenic beauty, principally underwater, but preferably also in its coastal landscape' (McMichael 1969:118). After surveying the Australian situation and the minimal efforts already made – Green Island, Heron Cay, Wistari Reef and a few Queensland continental islands – he stressed that 'In view of Australia's greatest natural asset, the Great Barrier Reef with its potential for tourism, fisheries and minerals, and the extent and variety of the rest of its coastline, the establishment of adequate marine parks and reserves in this country should receive high priority' (120). The suggestion raised the following year in the McMahon government of a 'marine national park made secure against mineral exploration or exploitation' indicated the growing strength of the idea.

With that new operative concept as a guide, at its July 1974 meeting the IDC recommended to the minister 'that a Great Barrier Reef Marine Park rather than a National Park should be established' with the assumption that the pending High Court judgment would validate the government's *Seas and Submerged Lands Act*. Since a separate issue was under negotiation in respect to the Torres Strait boundary between Australia and Papua New Guinea, the marine park area would need to be limited in the north at the tip of Cape York. Moreover, the petroleum exploration permits were due to expire in September 1974 and therefore delaying drilling would provide time for further negotiations on protection for the Reef.

Considerable effort was put into exploring all possible legal problems and safeguards that needed to be in place, but of greater magnitude at the time was the problem of the large range of existing activities on the Reef that had to be considered and brought into a coherent management regime: navigation, defence, meteorology, commercial and recreational fishing, Aboriginal reserves, planned petroleum exploration and 'Northern Development'. So arose the concept of 'multiple use' which became, and remains, a basic principle of the Reef marine park management. Those issues now brought to a seemingly satisfactory conclusion, Cabinet decision 2383 of 24 July 1974 authorised the drafting of legislation, based on the 1971 document 'The Great Barrier Reef Act of 197_' by Patricia Mather, now termed the 'Great Barrier Reef Marine Park Bill'.

REPORT OF THE ROYAL COMMISSION 1971–1974

Throughout the following three months the Interdepartmental Committee continued preparations for the bill which were reaching finality when the Royal Commission presented its full *Report* to the two governments on 1 November. The commissioners were all highly experienced in their respective fields and were accepted by the Commonwealth and

Queensland as impartial investigators. Chairman Sir Gordon Wallace had been a judge of the New South Wales Supreme Court and had recently retired as President of its Court of Appeal; geologist Vincent Moroney from Canada had more than forty years experience in petroleum engineering and management; biologist James Smith was director of Britain's leading marine research institution, the Plymouth Laboratory of the Marine Biological Association.

Their terms of reference, however, were specific and narrow, restricting them, in summary, to five particular matters:
1 an assessment of the probability risk from drilling;
2 the probable effects of oil or natural gas leakage on reefs, coastline and the ecology of reef life;
3 whether areas existed that could be safely drilled without causing ecological damage;
4 the necessary safety precautions to be in place before and during drilling; and
5 the probable economic benefits if oil or gas were found.

There was no brief to consider damage from tanker grounding, or from accidental or deliberate oil discharges from vessels in transit through the Reef.

Even so, within those terms of reference a vast range of topics were encompassed. Taking three and a half years – 5 May 1970 to 30 October 1974 – to conduct its inquiries exhaustively, the Commission sat for 267 days and examined 95 witnesses, 21 from the United States, Britain, New Zealand, Canada and Singapore. The commissioners and a large entourage travelled to selected sites throughout the Reef by ship and aircraft, inspected drilling rigs already operating elsewhere, along with onshore gas processing plants. Altogether 18 256 pages of evidence and 465 exhibits were taken and distilled into a massive two volume *Report* of 996 pages accompanied by an additional 66 volumes of 18 256 pages of evidence transcripts. No more exhaustive inquiry could have been conducted.

Three weeks after the Commission reported, Whitlam and Bjelke-Petersen released a joint press statement on 22 November giving a brief summary of the findings. The risk of blowouts was real and 'some measure of chronic spills would occur ranging from small to substantial' it stated. A major conclusion was that no reliable data existed on the effects of oil spills on coral ecosystems, and that until such evidence is obtained, 'drilling should not be permitted *on* any cay, island or reef or national park or marine park when declared' (Prime Minister's Press Statement 378, 22 November 1974). The *Report* did not encourage prospects of oil exploration for the present.

The full *Report* is a monumentally comprehensive document which covers much the same ground as the 1971 Report by the Senate Select Committee on *Off-shore Petroleum Resources*, from which a considerable amount of evidence was taken and designated Exhibit 450. Of profound significance was the exhaustive inquiry into environmental and conservation issues. It is interesting that the Commission began with an extensive survey of the physical and biological characteristics of the Reef as the framework within which the five terms of reference would be examined. Again, as in the case of the Senate Select

Committee, a major problem was encountered, recurring throughout the length of the Report: a lack of detailed knowledge of marine ecosystems. Not long after the hearings began, the commissioners wrote, 'it became apparent that little information was available from anywhere in the world on the effects of crude oil on coral organisms at any stage of their life histories and on their food chains and reproductive systems' (Royal Commission *Report* 1974, Vol. I. Pl.2.1:10). 'Nobody knows' said Professor Des Connell in his evidence, 'what is happening' since 'very little work has been done towards understanding the population dynamics of reef-building corals', adding, with an implied reference to the 1928 Low Isles Expedition, 'most of the best work was done before 1940' (17).

It was, then, within that framework of extreme paucity of scientific evidence that the commissioners reached their five conclusions which were, in summary:

1 there would be a 'small to very small risk of blowouts' and 'some measure of chronic spills ... ranging from small to substantial';
2 regarding toxicity from floating oil: 'subject to weathering for some $1-1\frac{1}{2}$ days ... it is virtually non-toxic to marine organisms';
3 no areas safe from drilling accidents could be identified at present and therefore all drilling throughout the GBRP should be postponed and not permitted until results are known of both the short and long-term research recommendations;
4 suitable safety provisions should be made for the contingency of oil spills; and
5 economic benefits could not be adequately predicted due to the uncertainty of finding recoverable quantities while any positive economic benefits would have to be assessed against the 'consumption of irreplaceable resources, interference with the environment and its enjoyment by mankind, the risk of damage to corals and other marine organisms and to birds, and hazards to the tourist industry' (31–36).

In stressing the continuing problem of the lack of scientific knowledge the Commission quoted verbatim from Paragraph 13.185 (p.503) of the Senate Report on *Off-shore Petroleum Resources* headed 'Lack of Knowledge of the Reef', to the effect that 'There was wide agreement amongst all witnesses that present knowledge of the Reef was extremely inadequate whether in terms ... of marine biology or any other of the marine sciences' (Royal Commission *Report*, Pl.6.225:152). It also reiterated the same concern that the difficulties would be overcome in time with 'full scientific research including short and long-term experiments [which] will soon be commenced by the Australian Institute of Marine Science at Townsville which was heavily endowed by the Australian government in 1972' (Vol. I. Pl.2.13:1920).

Equally important were recommendations for the declaration of a marine park to cover most of the Reef province. In her witness capacity as Honorary Secretary of the GBRC Patricia Mather gave a cogent analysis of the problems of the Reef. The central difficulty she believed was that although there was a plethora of legislation concerning the Reef, there was nothing directly *for* the Reef. Many laws and departments were involved, and she listed Primary Industries, Forestry, Mines, Native Affairs, Tourism, State Harbours and

Marine, Primary Industry, National Development, and Commonwealth Bureau of Meteorology (Vol. I. PI.6.228:153). In his closing submission Peter Connolly QC, barrister representing the GBRC, the ACF, the Wilderness Society and the Save the Barrier Reef Committee, urged the adoption of Mather's draft bill as the model for the statutory authority for which she had laboured so long and so strenuously: 'although it may not be the last word on the framework for a great Barrier Reef Authority' he pleaded, none the less, 'let us have one hand' on it, and gather 'the best brains of the Commonwealth and the State . . . to bear upon it' (Vol. 2, Tr.65:1719). That plea was accepted by the Commission which recommended in its final 'Conclusions' that 'A special statutory authority should be established responsible to the appropriate Parliament for ecological protection and the control of research and development within the GBRP' (Vol. 1, 5.3.39:982). The Royal Commission *Report* completely closed down any prospect of oil exploration on the Reef for the foreseeable future.

At least for the time being, the Great Barrier Reef had been saved, and the question 'what was there to be saved?' received vigorous affirmation in the outpouring of evidence from an impressive array of authorities, all of whom stressed its unique character as one of the world's greatest, and irreplaceable, natural phenomena. Although giving evidence as a leading economist, former Governor of the Reserve Bank of Australia, and at the time Chancellor of the Australian National University, Dr H. C. 'Nugget' Coombs advanced the view that it was a realm of superb aesthetic experience and that 'Australia is in a real sense a trustee for the world in the protection and preservation of the Reef' and it should, therefore, 'be reserved from oil development' and 'placed under the control of a National Park Authority' (Vol. 1, 3.3.14:581). While those experts were a select few, the interest of the Reef for the public was evident by 1974 when the Queensland Tourist Bureau recorded over 640 000 tourist visitor-days on the Reef (Claringbould et al. 1984:15).

Within two days of receiving the *Report*, and with the marine park legislation prepared, Prime Minister Whitlam began negotiations with Queensland, writing to Premier Bjelke-Petersen on 23 September 1974 advising that the Commonwealth government was planning to introduce such legislation. The Premier replied two months later, 25 November baulking at the proposal and rejecting Justice Hope's comments in the *Report on the National Estate* regarding the Reef with the curt comment that 'the Committee was not in a position to apply itself adequately to a question such as the future of the Great Barrier Reef area'. Bjelke-Petersen, who had no regard for heritage considerations, warned Whitlam against any 'impulsive action' and although they should get their 'due acknowledgement, all other matters should be accorded their just priority' by which he meant, presumably, oil exploration of the Reef. He then referred to the pending High Court litigation on the *Seas and Submerged Lands Act*, pointing out that a judgment had not yet been handed down, and he would need to wait since 'it is not my government's wish to be associated with anything which might prove to be unconstitutional'.

Whitlam sought advice from the Attorney-General on whether the legislation to establish a Great Barrier Reef Marine Park could proceed given the reply from the Queensland premier. On 23 December the Attorney-General's Department advised the Prime Minister that the Solicitor-General had already noted that while the Royal Commission had been sitting the Queensland government had 'not felt itself constrained by this consideration, since a proclamation was made on 30 November 1974 pursuant to section 29(2) of the Queensland Forestry Act 1959–1974, creating a Queensland marine park area in the three mile zones along the east coast of Queensland and around each of the islands off that coast'. In fact, most likely in an attempt to pre-empt possibly adverse findings by the Royal Commission, the Queensland government on 31 March 1973 had already amended, but had not yet proclaimed, Section 29 of the *Forestry Act* to create 'Marine Park Areas' and designated two such areas: the region from Cape Tribulation to Dunk Island (16°S to 17°50'S) and the Capricorn region (23°S to 24°20'S). Given that the High Court was yet to hand down a decision on the constitutional validity of the *Seas and Submerged Lands Act* the Queensland legislation was itself completely unconstitutional.

Having received advice from the Solicitor-General that 'introduction of Australian parliament legislation to establish a Great Barrier Reef Marine Park would not be a detriment so far as the *Seas and Submerged Lands Act* was concerned' Whitlam replied personally to Bjelke-Petersen on 29 January 1975, writing that 'the Australian Government wishes to introduce the legislation as soon as possible'. His letter was very conciliatory, pointing out that

> We still regard co-operation between our two Governments as being an important factor in the operation of the Marine Park Authority, and it is therefore intended to redraft the proposed legislation to make provision for nominations to be made at a future date. . . . I repeat my suggestion that such nominations should be made jointly by our two governments.

The Queensland government was in a very insecure position, especially given the legal opinion expressed at the ACF symposium in May 1969 by Australia's pre-eminent jurist Sir Percy Spender that Queensland had not the slightest claim to waters of the Reef. A month later, on 24 February 1975, the premier replied to the Prime Minister in seven short paragraphs still resisting any agreement and commenting that 'The only official information I have in respect of the proposed Authority is contained in your letters. How it will be comprised and where and how it will exercise its functions are matters on which I should appreciate details before considering the question of nominations, either joint or several' going on to point out that 'the fact remains that substantial areas of the Reef will be Queensland territory, whatever decision is made by the High Court'.

Queensland's possession of the Reef islands, of course, had been recognised by the Commonwealth all along and the Interdepartmental Committee had been advised by the Attorney-General's Department in February 1975 that the powers to be entrusted to the proposed Authority 'fall short of the powers of management that might be exercised in a terrestrial national park' and therefore the 'Bill should bring out this distinction'. A major issue was the status of what were termed territorial waters and continental islands that Queensland claimed as an historic right of ownership, and for which it had been legislating and managing for more than a century. They were not so recognised, however, in international law which held that the sovereign state was Australia, and not the separate internal States; consequently, the Queensland declaration of marine parks around its continental islands was itself insecure. For the ensuing four months the numerous considerations of overlapping jurisdiction and management were examined and resolved and, despite the fact that the High Court had still not delivered its judgment on the validity of the *Seas and Submerged Lands Act*, the Great Barrier Reef Marine Park Bill was introduced into the Commonwealth parliament on 22 May 1975.

The bill as presented was based on the original Mather draft and Minister for Environment Moss Cass opened his presentation with the observation that it built on the foundations laid in the *National Parks and Wildlife Conservation Act*, the *States Grants (Nature Conservation) Act* and the still pending Australian Heritage Commission Bill. His short introduction drew on both the positive aspects of the Reef as 'a priceless heirloom which we must safeguard for future generations' and the findings that came out of the Royal Commission. Somewhat surprisingly, Cass received a good hearing and support from both sides of the house. Country Party Member for Gwydir, Ralph Hunt, speaking in debate on 5 June began with the words 'Today is World Environment Day. What an appropriate day for the Great Barrier Reef Marine Park Bill to be debated in this House. It is a bill . . . to protect . . . a significant part of the Australian heritage', stating also that 'The Opposition supports this attitude wholeheartedly and congratulates the Minister on the motivation that has led him to introduce the bill . . . the Opposition parties regard the Great Barrier Reef as potentially a marine park of world significance'. Even further, he expressed the hope that 'the Federal Government should co-operate with and assist the Queensland Government in the declaration and preservation of the Great Barrier Reef as a marine national park' (Hansard, H. Reps 1975:3424). Numerous Opposition speakers continued to affirm the bill, and following several minor amendments in both houses, the bill was passed on 20 June 1975. Despite that, the Queensland government refused to cooperate until the High Court decision on the *Seas and Submerged Lands Act* had been handed down.

Then, just three days after the marine park bill was passed, on 24 June 1975, the High Court delivered its verdict on the challenge to the legality of the *Petroleum and Minerals Authority Act* which had been passed by the joint sitting of the parliament on 12 December 1973. In a highly controversial decision, with some other justices expressing hesitancy and the opinion that it was really a political issue, Chief Justice Barwick held that it was not a valid law of the Commonwealth (Victoria v Commonwealth, 1975, 134 CLR 81). By that time the pace of Whitlam's reforms, and their very character, which threatened powerful vested interests in the States, had become deeply vexatious. From then on the Opposition increased its efforts to unseat the Whitlam government. Whitlam had to be brought down, just as 'John Gorton was brought down when he tried to extend the Australian Parliament's power over the adjoining seas and submerged lands at the expense of the power of the States' (Enderby, *Australian Quarterly*, 48.3, December 1976:43).

Matters came to a head in November when the Coalition, having gained control of the Senate by a devious procedure when two casual vacancies occurred, refused to pass the crucial financial Supply Bill. In an unprecedented and highly controversial action, since known as the 'Constitutional Crisis', Governor-General Kerr, without consultation with the Queen, dismissed the Whitlam government on 11 November, appointed leader of the Opposition Malcolm Fraser as caretaker Prime Minister and requested him to conduct new elections. Fraser won the election and began a reactionary program to undo, or at best radically restructure, many of Whitlam's reforms.

By that time, however, some of the Whitlam government's significant environmental initiatives had already been passed into law. Two major Reports had been completed – the Hope *Report on the National Estate* and the Royal Commission's Report into petroleum exploration of the Reef and three Acts designed to ensure greater protection for the environment, had been promulgated: the Great Barrier Reef Marine Park Act, the National Parks and Wildlife Conservation Act and the States Grants (Nature Conservation) Act. Its final achievement was the proclamation of the Australian Heritage Commission Act on 17 December 1975. Moreover its judgment on the issue of sovereignty over offshore waters was also confirmed when in December the High Court finally delivered its verdict confirming the validity of the *Seas and Submerged Lands Act*, passed by the parliament two years before on 4 December 1973.

In reaching their decision the Full Court had to determine the basis of the case which in their view depended upon the concept of Australia as an internationally recognised legal entity. Before federation the waters of the colonies were British territory; after federation it was the Commonwealth which came into possession and so, in effect, held sovereignty and could claim such by virtue of its standing as a 'member of the community of nations'. The legal basis they found lay in Section 51 (xxix) of the Australian Constitution whereby the federal parliament had power to make laws with respect to external affairs, that is, as a sovereign nation (New South Wales v the Commonwealth (The Seas and Submerged Lands Case) 1975 135 CLR 337). The High Court confirmed that the Commonwealth held title to all waters from the low-water mark of all land.

That decision was in effect a serious blow to the coalition parties since it validated the primacy of the Commonwealth over the States and was one more move towards greater centralism. Liberal Party policy had been defined by founder Menzies as a pragmatic process to deal with issues as they emerged. Usually throughout the post-war years this meant opposing the reforming programs of the Labor Party, using the hyper-emotive appeal of fear of extreme socialism or even the 'communist threat' to the Australian way of life. The ideology advanced was 'freedom of the individual', and that extended to freedom of the States to conduct their own internal affairs with the minimum of Commonwealth control. So emerged the Fraser concept of 'New Federalism' which received an extensive publicity campaign, and generated a tremendous literature of apologetic rationalisation.

A FORMULA FOR REEF MANAGEMENT: THE EMERALD AGREEMENT, 1979

The Reef was in the centre of the political program of 'New Federalism', and for the following four years, from 1975 to 1979, implementation of the Marine Park Act moved very slowly while Fraser and Bjelke-Petersen attempted to devise a formula whereby both the Commonwealth and Queensland were able to cooperate on procedures. Despite the rhetoric of maximising State control, it was recognised that the primacy of Commonwealth sovereignty could not be surrendered since Australia was bound by a large number of international treaties and conventions relating to the law of the sea. Moreover, it was also appreciated that international affairs were becoming increasingly complex and that unforeseen circumstances would continue to arise which would be in the realm of Commonwealth responsibility, continued Taiwanese illegal clam fishing being an obvious example. Negotiations then proceeded between both governments with the desire to realise the spirit of the marine park legislation, with Queensland fully involved in management procedures.

By June 1979 a short three page working paper had been prepared by Alan Griffiths, senior adviser to Fraser, which the Prime Minister could take to a meeting with Bjelke-Petersen to 'secure agreement on the main elements' of a plan of implementation. On 14 June that year to discuss implementation and management issues both politicians met in seclusion in the small town of Emerald, some 260 kilometres (160 miles) inland from Rockhampton. Several issues were discussed and the main agreements reached were that the boundaries of the Park as defined in the Act would remain, while 'Queensland [would] be assigned the day-to-day management role by officers of the Queensland National Parks and Wildlife Service and while acting in that capacity would be subject to the Great Barrier Reef Marine Park Authority'. Legal arrangements would be subject to future negotiations at a Premiers Conference with the Commonwealth in which the *Seas and Submerged Lands Act* would be amended to return control over territorial waters within the three mile limit to the States. In addition, it was agreed that the southern Capricornia section would be the first to be declared as a marine park, and that a Ministerial Council of four members to coordinate policy would be established, two from each government representing tourism, marine parks, environment and science. Science received particular mention and it was agreed that it would proceed 'on a timetable and framework acceptable to both governments'. Both Fraser and Bjelke-Petersen personally signed each of the three pages, and so a binding compact, henceforth known as the 'Emerald Agreement', came into force. It was significant at the time, however, that mining and oil exploration were not explicitly prohibited, but merely that they would not proceed until 'research on the short and long-term effects of oil pollution on the Reef ecosystem was completed and assessed' (Commonwealth Department of Science and the Environment statement 1980:8).

Two weeks later, on 29 June 1979, at the Premiers Conference the Commonwealth and the States completed an arrangement known as the *'Offshore Constitutional Settlement'* for the resolution of the contentious and complex offshore constitutional issues which related to all activities on the continental shelf, at the time chiefly fisheries, and petroleum and gas exploration and recovery. In accordance with Law of the Sea Conventions, Australia declared a 200-mile fishing zone and indicated that it would amend the *Seas and Submerged Lands Act* (1973) since 'history, common sense and the sheer practicalities of life mark out the territorial sea, in particular, as a matter for local jurisdiction' although, at the same time, 'revision of existing petroleum mining arrangements is required to properly reflect the Commonwealth's paramount rights over the continental shelf' (Attorney-General's Department 1980:4). Title to the seabed beneath the territorial sea would be returned to the States and the legislative powers of the States in relation to coastal waters would be confirmed, including sole State jurisdiction over mining activities, while fisheries, which were mainly in the high seas, would be regulated by one set of agreed laws operating under Joint Authorities. Specifically, the *Great Barrier Reef Marine Park Act* of 1975 would continue in force, subject to the provisions already arrived at in the Emerald Agreement. In 1980, to reflect those arrangements, the Commonwealth enacted legislation in the *Coastal Waters (State Title) Act* which, as stated in its preamble, was 'An Act to vest in each of the States proprietary rights and title in respect of certain land beneath the coastal waters adjacent to the State and within the Sovereignty of the Commonwealth'. Assented to on 29 May 1980, a new phase began in the 'contentious and complex offshore constitutional issues' in conserving the Great Barrier Reef.

CHAPTER 21

A NEW ERA: RESEARCH BASED MANAGEMENT

REEF RESEARCH: JAMES COOK UNIVERSITY AND THE AUSTRALIAN INSTITUTE OF MARINE SCIENCE

Throughout the years of political confrontation between the Commonwealth and Queensland over the future of the Great Barrier Reef, the need for serious, institutionalised research had already been demanded for more than a decade by Queensland Senator Felix Dittmer, previously a marine science collector for museums. In his maiden speech of 27 August 1959 Dittmer had argued the case for more Commonwealth involvement in north Queensland, including 'the establishment of a large marine biological site on the Barrier Reef' (*Senate Parliamentary Debates*, 8 Eliz.II, V.S15, 354). Maintaining pressure throughout the following years, in the Senate Estimates Debate of September 1963 Dittmer questioned John Gorton, at the time Minister for Science, about allocation of funds for the much neglected area of marine research, asserting, in the flow of argument, that it was in need of considerable upgrading. 'I believe', he stressed,

> there is justification for the establishment in Australia of a marine biological research station. Off the Queensland coast is a formation which is unique. I refer to the Great Barrier Reef. By failing to explore the possibilities of the Great Barrier Reef we have not done justice to the scientific world . . . No tribute has been paid to this unique natural structure.

Sustaining the flow of forceful rhetoric, he went on to assert that even though 'there is a small threepenny-bit type of station at Heron Island, . . . the possibilities of this reef justify much more than that'. Further, he dismissed any suggestion that his demand was purely parochial politics: the reason for a large, well-funded research station in Queensland was simply because 'it just happens that is where the Great Barrier Reef is situated' (Senate Hansard, 25 September 1963: 833–34).

The cause championed by Senator Dittmer was taken up enthusiastically by Townsville University College, founded in 1960 as a northern outpost of academe, mainly as a teachers' college. Throughout the early 1960s the college grew slowly; by mid 1965, it had grown to 12 departments with 55 lecturing staff, expanding in 1968 with the creation of a small department of marine biology. In its funding submission to the Australian Universities Commission for the 1967–69 triennium it set out an ambitious program of development including a chair of marine biology with a focus on tropical marine research, and a biological sciences building. As a result of that successful application, Cyril Burdon-Jones was appointed to the chair on 1 January 1968 from his position as deputy director of Marine Science Laboratories at the University College of North Wales.

Burdon-Jones threw himself energetically into the task of creating a first rate marine research department at the Townsville University College and soon after arrival produced a duplicated document entitled 'A National Research Centre for Tropical Marine Science in Australia', dated 4 October 1968. It was breath-taking – for the times – in its ambitious scope. The introduction gave the usual glowing account, familiar since Saville-Kent's day, of the Reef: 'As a tourist attraction it could be unexcelled, as a potential source of food, pharmacological and toxicological products it remains undeveloped' (p. 2). Then followed the greatest wish-list for tropical marine scientific research yet devised in Australia, seeking a complement of 20 scientists, 30 technicians, and eight clerical/library staff; a site of 15 acres (6 ha) near coral reefs, foreshores, an estuary and a large town; adjacent to a good harbour, with its own jetty; accessible by road, sea and air; and 'close to a limitless supply of good, un-polluted, high salinity sea water of low silt, and organic content' (3). The proposed research range covered fisheries, pharmacology, biochemistry, physiology, hydrography, climatology, bacteriology, biostatistics, marine engineering, sedimentology, geochemistry, electron microscopy, plankton and benthic biology and ecology, and developmental biology. To house those manifold activities request was made for a large number of well equipped laboratories, field equipment, a capacious, well stocked library, living accommodation for visiting researchers and three vessels: a 6 to 7 metre 'fast tender', a 50 to 60 foot (16–18 m) trawler, and an ocean going research vessel of 100 to 120 feet (approx. 34 m) with a qualified crew including a captain with a 'deep-sea ticket'.

In vigorous pursuit of his objective, Burdon-Jones went to Canberra in 1968 and, sensing an opportunity from Gorton's election pledge of 9 October 1969 'to establish an institute of marine science at Townsville', went again to lobby for the institute to be located within his own university. There was, at the time, a definite conviction in Canberra, supported by Secretary of Science Hugh Ennor, that something had to be done: too many issues were attracting public attention – oil spills, *Acanthaster* infestation, lack of scientific knowledge – and the Burdon-Jones document clearly gave both impetus and a possible blueprint for action. The great obstacle, however, to such a national research centre being located in Townsville University College – to become James Cook University the next year, 1970 – was its status as a Queensland State government institution. Given the heightened conflict between Canberra and Brisbane while the oil drilling controversy raged, the Commonwealth was clearly not going to fund the proposal outright.

Having been returned to office at the general election of 1969, Gorton knew that the election pledge of 9 October had to be honoured. Following passage of the initial *Australian Institute of Marine Science Act, 1970*, planning commenced under the chairmanship of Dr Max Day, a member of the CSIRO Executive. When the Interim Council reported to David Fairbairn, at that time briefly Minister for Science, its recommendations followed closely the Burdon-Jones document of 11 October 1969, proposing that the national institute be sited on Commonwealth land at Pallarenda on Townsville's northern waterfront, staffed by 24 scientists, and equipped with two large vessels of 24 metres and 36 metres. The brief was most impressive, recommending that 'The research activities of the Institute should be directed initially to studies of the scientific problems of the Great Barrier Reef, the Coral Sea, and the coast and coastal waters of north Queensland' (Day 1971:6).

As planning progressed the Pallarenda site increasingly appeared less suitable. Originally a quarantine station, it had been occupied by the army during World War II, who remained in part occupation until their Lavarack Barracks on the south side of Townsville were ready. The suitability of the Pallarenda site, however, became questioned on a number of counts. Disadvantages soon became apparent: riverine discharge plumes from the inland nickel smelter were a major water quality contaminant, while the shallow waterfront precluded berthing for large research vessels. Another unanticipated problem was the intention to raise the status of Garbutt Air Force Base, adjoining Pallarenda, to that of an international airport. In the early 1970s airport radar control was coming into general use and interference with delicate scientific electronic equipment was discussed at the first council meeting of the new institute – known universally now by its acronym as AIMS – where it was decided that no risk of airport radar interference could be allowed. Although Townsville remained the general location, an alternative site had to be found. Some 32 kilometres south of Townsville a promising coastal site was found at Cape Ferguson on undeveloped Crown Land and early in 1973 it was recommended to the Council of AIMS as the preferred option. Council accepted that choice and requested the Commonwealth to procure a parcel of 255 hectares for the Institute from the Queensland Government. During its construction the Pallarenda site was occupied temporarily, using portable huts and laboratories.

A considerable infrastructure at Cape Ferguson had to be built: road works, water and electricity supply, as well as a large complex of buildings. On 25 June 1975 a contract for the buildings was signed and staff recruitment began with the appointment on 15 July 1975 of the foundation director, Dr Malvern Gilmartin, formerly Professor of Biological Oceanography at Hopkins Marine Centre of Stanford University. Within the next twelve months staff began to arrive, and on 16 September 1977 the complex was officially opened. The way was now ready to proceed with Australia's greatest venture into creating a tropical marine science research station.

Max Day advised Prime Minister Fraser that AIMS should not be a multi-purpose marine research station but should focus upon tropical marine science, directed specifically to the Great Barrier Reef. Fraser agreed, and commented that once AIMS was fully operational the Commonwealth would then be in a position to consider marine stations elsewhere in Australia. That decision, of course, was a profound setback to the ambition of Burdon-Jones to secure the institute as part of James Cook University; the only concession to the university was contained in the proviso that AIMS should work with it in close cooperation.

Unfortunately for Burdon-Jones, the failure of his ambitious plan to materialise meant that marine science remained a tiny department at James Cook, in 1974 with only two lecturers and two tutors in addition to himself. On a much reduced scale three research areas were devised: coral taxonomy, community structure and zonation, and evolution of reef ecosystems. For necessary fieldwork in all three areas, a small band of enthusiastic students and staff travelled in the research vessel *Kirby* to the rich coral reef systems of the Palm Islands group – Orpheus, Fantome and Palm Islands – some 80 kilometres by sea directly from Townsville, or else north by road to Ingham and then by boat some 20 kilometres due east. From the early expeditions, research groups began camping on the islands and by 1975 the long term advantages of Orpheus as a field station were recognised. John Veron, a student in coral taxonomy on a Commonwealth research grant, persuaded Vice-Chancellor Ken Back of its potential value, who in turn arranged for James Cook University to obtain a lease. Over the next few years prefabricated buildings were barged across from Ingham and up to twenty students alternated their research with half-day shifts erecting buildings while Inge Moessler – sole resident of the island – offered to help the group. So James Cook University failed to secure its grand institute, but at least gained an excellent field station.

TOWARDS A NATIONAL MARINE SCIENCE AND TECHNOLOGY POLICY

Just as two centuries earlier New Holland had been *terra incognita* for the *Endeavour* explorers, by 1970 the Reef was still virtually a scientific *mare incognita*. In the mid 1970s both the Great Barrier Reef Marine Park Authority (GBRMPA)[*] and the Australian Institute of Marine Science (AIMS) were generated out of a period of intense political turmoil and heightened public concern, when both institutions became the focus of constant media coverage, and a high level of public emotional support. As a result of their rapid creation, little thought was given to the way they would operate or what their relationship would be to other areas of pressing need in marine research or to their place within a national science program. Although Fraser and Bjelke-Petersen had reached a political accommodation in the Emerald Agreement in 1979 on the broad principles of managing the Reef through a Ministerial Council that would formulate policy for implementation by GBRMPA, there was still no research or management data base on which to proceed. That, of course, was precisely why AIMS had been envisioned as the centre of Reef research, which would eventually flow through into management, even if not explicitly stated in such terms at the time.

The overwhelming difficulty was that, in the same period, many equally deserving institutions – terrestrial, marine and maritime – were clamouring for governmental support in the national interest. The consequences of rapid industrial expansion of the nation in the 1960s had caught government unawares and, indeed, at the time when AIMS and GBRMPA were first mooted, there was no unified national science or technology policy at all, a glaring deficiency that had become increasingly embarrassing when viewed within an international context. There was no clear demarcation of areas of research, much less any coordinating mechanism. Before GBRMPA could proceed effectively, the government found it necessary to begin clarification of its research and management priorities, including those related to marine and maritime issues.

By the early 1970s that deficiency had become a serious concern to the Commonwealth which, following its wartime acquisition and subsequent retention of most taxation revenue, now had to balance the many additional demands for national funding and development. In March 1974 Minister for Science William Morrison circulated a discussion paper entitled 'Towards an Australian Science Council' which set out two broad objectives: to use science to raise living standards and promote trade, and 'to understand the natural environment and control despoliation of it so as to preserve its quality for future generations' (Morrison 1974:1). Morrison's Green Paper identified eight research objectives which included marine science and oceanography, and to implement them the new government created an Interim Science and Technology Council, later reconstituted as the Australian Science and Technology Council (ASTEC) on 19 April 1977.

At that time GBRMPA was little more than a paper organisation while AIMS was still fully absorbed in 'the formidable problems of its own establishment and in research on the tropical waters of north-eastern Australia' and had 'still not completed the establishment phase of its development' (Badger 1978:1A.120–21). The immediate needs were in the three fields of marine biology, marine geosciences, and oceanography, all of which required rapid development in terms of funding and equipment, with the Reef identified as a 'high priority region'. Marine science had become so complex, ASTEC reported in June 1979, that research and development now demanded a high degree of interdisciplinary coordination. To advise the Minister for Science, they recommended creation of a special management body designated the Australian Marine Scientific and Technology Advisory Committee, AMSTAC. Following its establishment, it subsequently became an independent statutory authority reporting directly to the minister. The initial step in managing the Reef, with adequate funding now assured, was to begin building the research data base.

In that year, when GBRMPA was beginning to wrestle with its first attempts at zoning, the ASTEC report made it clear that zoning, and Reef management generally, had to be based on scientific research, drawing on the expertise expected from AIMS, which itself 'should be given additional staff and support funding . . . to accelerate the research effort in this high-priority area' especially in view of the Reef's 'fragility in respect of human impact' (Badger 1979:1,9). Consequently, 75 per cent of the $400 000 allocated to national marine research for the coming fiscal year of 1979–80 went to AIMS and GBRMPA, amounting to $301 000 (Watson 1988:7,42).

Faced with that unexpected largesse, AIMS convened a workshop meeting during 12 to 14 October 1979 to determine priorities. The outcome was a decision to develop a research strategy directed towards understanding the 'central question': 'What affects and controls growth, maintenance and change in the Great Barrier Reef?' (AIMS Report: *Collaborations in Research on the Great Barrier Reef 1979*:1, private circulation). It was further resolved that AIMS would need to dovetail its activities with research coming from other bodies with a specific interest in the Reef, chiefly the CSIRO Division of Fisheries and Oceanography, and the Bureau of Mineral Resources (today Australian Geological Survey) which was making geophysical studies of Reef structure, along with various departments of James Cook University and several other universities with specific interests. From that time forward, Reef science became organised within the framework of an integrated national science policy. AIMS was designated an institution concerned with fundamental research while GBRMPA, as a management body, would draw relevant data from AIMS and from specially commissioned outside studies, but also conduct its own applied research, related mainly to in-house issues and management.

In the same period the government was forced into even more vigorous action in the marine science area at a national level with the impending declaration on 1 November 1979 of the Australian Fishing Zone (AFZ), as negotiated since 1972 under new provisions of the Law of the Sea, and the future planned declaration of an Exclusive Economic Zone (EEZ). With the extension of Australia's fisheries boundary to 200 nautical miles the nation would acquire responsibility for approximately 7 million square kilometres of its coastal waters, a considerable portion of which lay within the Reef region. Nothing that had gone before had prepared the government for such an event. To examine potential issues the Senate on 6 March 1979 instructed its Standing Committee on Science and the Environment to conduct an inquiry into Australian marine science generally.

Following widely publicised announcements, submissions were invited and public hearings planned for capital cities later in the year, beginning in Perth on 3 October. The

---

* The Authority is known universally by its acronym – pronounced 'Gh'broompah'.

initial findings were so disturbing that the Senate issued a *Progress Report* the following May 1980 which expressed a litany of concern, focusing on the central finding that 'Australia lacks a national plan for the marine sciences and technologies'. The report was devastating in its criticism that 'oceanography is pitifully inadequate' with a 'lack of co-ordination and an absence of expertise', that the nation is 'almost totally ignorant of the physical properties of the ocean off our coasts', that 'information on tides, currents, waves and navigational characteristics of the sea is inadequate' and that 'the influence of the ocean on weather and climate is also poorly understood'. Even further, 'marine engineering is also badly neglected' (Senate *Progress Report* 1980:9–10). One submission by Dr Brenton Groves, a marine consulting engineer, recommended strongly that the government 'define what Australia should be doing in marine science over the next ten years' and make a financial and moral commitment comparable to that given to tertiary education in the Martin Report of 1968 (Groves, Submission 26, Senate *Progress Report* 1979:398–9).

The Senate issued its final *Report* on 11 June 1981 with 44 recommendations for immediate action to improve marine science in the nation. Seven were accepted outright: those concerned with increased funding for oceanography, along with the study of fisheries resources and sustainable yields, and marine taxonomy, which, it was recognised, was very poorly developed with vast numbers of taxa scarcely known, much less described. Specifically accepted were issues of investigating and preventing oil spills and effluent contamination of the Reef, and the impact of tourism. In particular, regarding tourist management came the ambitious suggestion 'of the possibility of developing an extensive underwater viewing gallery . . . as a means of reducing this impact whilst at the same time promoting tourist interest' (Senate, Hansard 26 November 1981:2631). So, funding for a number of Reef research issues, to be administered through AMSTAC, was increased, and continued to grow, by 1985 reaching $1.35 million.

PLANNING THE GREAT BARRIER REEF MARINE PARK

The first three years, 1976–79, following passage of the *Great Barrier Reef Marine Park Act* marked a breakthrough for Reef science, as AIMS was becoming firmly established in its permanent Cape Ferguson site and James Cook University was developing its Orpheus Island research station – later to be widely used by AIMS scientists. Meanwhile, planning for the marine park was beset with numerous difficulties. The Queensland government was still deliberately obstructive, and, to compound matters, the *Great Barrier Reef Marine Park Act* (No. 85 of 1975) itself provided very little guidance for its implementation. The Act, to be overseen by broad policy directives from the four member Ministerial Council, provided for a three person 'Authority': a full time Chairman with 'experience and reputation in the field of biological conservation', assisted by two part-time members, one nominated by the Commonwealth and one by the Queensland government. To supply the necessary wider range of expertise the Act provided for a Consultative Committee of not less than twelve persons although in practice a larger number have participated, with nineteen members in the first year.

The purpose of the Act was defined in general terms as making provision for 'the establishment, control, care and development of a marine park in the Great Barrier Reef Region' (Section 5.1) by means of a zoning plan, to be effected 'as soon as practicable' with 'regard to the following objects': promoting conservation, allowing 'reasonable use' and exploitation of Reef resources with minimum effect, providing the public with opportunities for its 'appreciation and enjoyment', and preservation of some areas in their 'natural state undisturbed by man except for the purposes of scientific research' (32.7.a–e). Furthermore, the Region itself was dauntingly large. The boundaries, as determined in the Schedule attached to the Act, encompassed much of the Queensland coastline from the tip of Cape York at 10°41'S latitude south to 24°30'S (a little to the north of Bundaberg and Fraser Island), a distance of approximately 3000 kilometres. Seawards, the Park extended east from the low tidal mark on the mainland (excluding the 28 enclaves reserved to the Queensland Government for future harbour development) to the Coral Sea, following in general, but not exactly, the 200 metre bathymetric contour of the continental shelf. Within the Park were many of the patch reef formations, but the islands and cays above low water remained Queensland territory. The entire region was defined by the coastline and six geodesic baselines, enclosing an area of 339 750 square kilometres.

When Don McMichael, with a staff of eight, was appointed acting chairman of the Authority for a year on 1 July 1976, he was confronted not only with that generally worded Act and immense spread of territory but also the climate of abysmal ignorance of Reef biology, structure and oceanography as disclosed in previous inquiries. To manage effectively such a huge region the Authority decided to begin by dividing the Reef into a number of management sections, within which activity zones would be designated. The first, as decided in the Emerald Agreement, would be the Capricornia Section, an area of approximately 12 000 square kilometres containing 21 reefs and the vegetated Heron, Musgrave and Lady Elliot island cays, and chosen because they were already subjected to considerable tourist pressure. To create a scientific data base the Act provided for GBRMPA to conduct its own research, and to fund projects by other persons and institutions. As events developed, most research later came to be contracted out. McMichael, well experienced in the field, knew that it would necessarily be a long term process which had to be approached carefully, and that before any zoning plan could be introduced it would be essential to win public support by means of an extensive community consultation, education and information program.

During that first year the Authority was located in Canberra, although it was decided that Townsville would be its operational headquarters since, in addition to its central location on the Reef, it already had potential marine expertise in the fledgling institutions of AIMS and the James Cook University department of marine biology. In the meantime, to publicise the new Authority, four meetings of the new Marine Park Authority were held in Brisbane, and one on Heron Island within the planned Capricornia Section. Further meetings the following year were projected for the nearby coastal towns of Bundaberg, Gladstone and Rockhampton to inform the public and receive comments. In order to start zoning the Capricornia Section, a program of data collection was commenced, beginning with a survey of its geological formations and biota. A consultant marine biologist was also engaged to interview users of the region concerning their activities, while the Great Barrier Reef Committee, whose headquarters were on Heron Island, provided input.

At the time, the new Authority was in its earliest formative stage, and much of the initial planning was undertaken in consultation between GBRMPA and the newly formed Queensland National Parks and Wildlife Service who would be responsible for the day-to-day management not only of the lagoon but also the continental islands and cays which, as Queensland territory, were covered by Queensland legislation under the *National Parks and Wildlife Act 1975*. Moreover, although a statutory authority with a high degree of autonomy, GBRMPA operated within the portfolio of the Commonwealth Minister for the Environment.

The following year, beginning on 1 July 1977, Henry Higgs, Secretary of the Commonwealth Department of the Environment, continued as acting chairman and had to assume the onerous task of representing the Authority during the complex negotiations between Prime Minister Fraser and Premier Bjelke-Petersen concerning the thorny jurisdictional issues of Reef sovereignty. Once those compromises had been finally resolved in the Emerald Agreement of 14 June 1979 and the Offshore Constitutional Settlement two weeks later on 29 June, GBRMPA was able to move forward more confidently in the zoning and declaration of Capricornia. Higgs, however, was suffering ill health and in July was replaced as acting chairman by Graeme Kelleher. On 21 December 1979 Kelleher – an engineer and not a biologist as specified in the Act – was appointed chairman for a five year term with an office located close to government in Canberra. For the following fourteen years he held that position and under his direction most of the early developmental work of GBRMPA was effected. The administrative staff were located in Townsville under the direction of an executive officer and organised into five sections designated Planning, Research and Monitoring, Park Management and Information, Secretarial, and Technical and Administrative Services.

To create an initial database the Authority commissioned Edgar Frankel to prepare a bibliography of existing information and in 1978 he completed a mammoth research task when the *Bibliography of the Great Barrier Reef Province* was published, containing entries for 4444 books, 18 serials, 95 films, and a huge list of charts, bibliographies and periodicals. A major deficiency of the bibliography, which Frankel acknowledged in the preface, was that although over the years there had been 'a proliferation of literature, both popular and scientific, [it had] remained largely uncollected'. An even stronger point was made by Reef geologist Professor W. G. H. Maxwell in his foreword to the volume where he stated that in consulting it for references readers 'will discover the areas where our knowledge is deficient, fundamental areas such as hydrology, intertidal coastal ecology, taxonomy, subsurface geology, the impact of tourism'. He then made the telling point that 'most importantly, they will be aware of *the lack of synthesis of known data*, of researchers' reluctance to step beyond the boundaries of their own specialities and view the totality of the reef' (Frankel 1978:iii, emphasis added).

Those two features – lack of data synthesis, and researchers' reluctance to expand their understanding beyond their own specialties – were to hamper the Authority's planning in the years to follow. In all the 4444 monographs, none offered a comprehensive history of the Reef from the time of Cook to the present from which planners could gain a synoptic view of the Reef under human impact and exploitation over two centuries. Moreover, and throughout the zoning, planning and educational documents, there is no account of the complex history of efforts in the past by dedicated community groups and independent researchers to raise awareness and institute protection.

Preparations to declare the Capricornia Section included fish studies, benthic surveys, investigation into amateur and professional fishing activities, preparation of zoning strategies, and discussions with the tourist industry. An innovative public education program was introduced with the preparation of explanatory, illustrated brochures and maps that were widely and freely distributed. Public input was solicited and efforts were made to identify the needs and demands of various interest groups – even conservationists were considered as an 'interest group' – a wise provision given that the prime task of the Authority as defined in the Act is 'the conservation of the Great Barrier Reef'. Conflicts would inevitably arise, and the challenge was to determine the best possible compromises consistent with maintaining the ecological integrity of the Reef.

The zoning structure, once Capricornia had been completed, later described as a 'town planning approach', followed the Act quite closely, and provided for six zones and other designated areas, which reflected the broad requirements of Section 32 (7 a–e), to be implemented in stages and subject to a continuing review process. 'Reasonable use' and 'resource exploitation' were designated 'General Use A' and 'General Use B' which allowed nearly all activities; the latter excluded commercial trawling.

Most of the remaining waters, those surrounding reefs, cays and continental islands, were designated as Marine National Parks, also prefixed 'A' and 'B'. Marine National Parks excluded most extractive activities such as fishing, shell and coral collecting but allowed, under the permit system, camping and tourism: National Park 'B' was more restrictive in that it prohibited baitnetting, fishing and gathering. To put that into perspective, most of the park area – close to 95 per cent – remained open to ship transit as a major waterway, along with recreational activities of boating, fishing, snorkelling, scuba diving, commercial fishing and trawling and accounted for nearly all of the surface area. The most restricted zones were designated 'Scientific Research' and 'Preservation' where all activities are banned apart from research, which require permits. A special category, known as 'Seasonal Closure', was devised to cover unforeseen circumstances such as recovery from heavy use, or destructive weather events such as cyclones, and to allow regeneration of breeding colonies of all kinds of depleted species.

By August 1980 the first GBRMPA zoning management plan had been prepared and sent to both Commonwealth and Queensland parliaments – since Queensland had its own complementary plans for its islands, cays and territorial waters to enable uniform enforcement throughout the entire region – where it was approved to take effect from 1 July 1981. Since this was a daring, pioneer effort, the Authority very wisely adopted an encouraging, educative strategy, seeking to win over a public that previously had never experienced any serious restrictions or punitive measures. As Kelleher commented in his *Fifth Annual Report* (1980–81) to government,

> The zoning plan for the Capricornia Section is being introduced with a minimum of controls and regulations, and with the least possible disruption consistent with conserving the natural qualities of the area. It is hoped by adopting this technique, and by informing people about their responsibilities and obligations under the Plan, the Authority will receive their co-operation in conserving this part of the Reef for present and future generations.

Accepting the status quo of 'established and potential user patterns' however, while it may have seemed good public relations strategy at the time, created a precedent for user conflict that was to become increasingly exacerbated.

For the next seven years the Authority began progressively declaring Sections for zoning, and following the same public education and involvement procedures, always allowing long lead times between the proclamation of each section and the date of the regulations coming into force. The second section to be proclaimed, on 19 November 1981, was Cairns with an area of 35 000 square kilometres which had long been subjected to heavy usage by tourism, shipping, commercial trawling and fishing. Its zoning plan regulations came into effect two years later on 7 November 1983. Within the Cairns Section a small 3 square kilometre special Section known as Cormorant Pass was declared (later absorbed into the Cairns Section) in order to protect a colony of around sixteen huge, up to 90 kilogram tame cod known popularly as Potato Cod (*Epinephelus tukula*: in the same family Serranidae as groper and coral trout) which had become a major tourist diving attraction. In late 1983 it was also planned to declare a further five Sections, one in the Cape York region called Far Northern, two in the Townsville region known as Townsville and Central (for the outer waters), and two in the southern regions labelled Inshore Southern and the outer area of Southern which would surround the inner Capricornia zone already in operation.

In the light of experience and the increased flow of research information, zoning plans continued to be refined. A major change came in 1984 with a significant advance in Reef navigation when, following a survey between 1981 and 1984 by HMAS *Flinders*, a previously unknown passage into the Coral Sea that had eluded HMAS *Geranium* fifty years earlier was discovered north-east of Mackay, allowing a much shorter route to the open sea. Beginning in 1971 the great inland Bowen coal basin had been mined for the Japanese market and coal, brought by rail to Mackay, and loaded at a purpose-built giant wharf at nearby Hay Point. With the publication in October 1984 of naval chart AUS-821, named Hydrographers Passage, and the first transit by a bulk carrier a month later, the Authority decided to consolidate all of the park in the Townsville region into one Central Section, and in September 1987 to merge all three southern areas, the projected Inshore Southern and the (outer) Southern with the existing Capricornia into a single Mackay-Capricorn Section. The boundary of the Central Section was extended south to include all of the Whitsunday region which now, in the eastern part, lies along the shipping track of the Passage. When the last Section came into effective operation on 1 August 1988, the entire Marine Park containing 2904 reefs of varying formation, with a complete area of 20 055 square kilometres had been zoned. The two general use categories came to account for 94.96 per cent of the total surface area, the Marine National Parks constituted 4.85 per cent and the 'Scientific Research' and 'Preservation' areas a bare 0.19 per cent. The boundaries separating the four Sections were defined by medes (geophysical marker points on charts) and connecting lines, known as geodesics.

By that time the Authority had already moved into permanent quarters. After a decade of temporary accommodation, a three storey complex, including GBRMPA headquarters and the independent Museum of Tropical Queensland, was commenced on the waterfront of the Ross River in central Townsville. In September 1984 the Commonwealth and Queensland governments provided Bicentennial Funding for an aquarium to complete the complex of the Great Barrier Reef Wonderland, today renamed Reef HQ. The Aquarium was opened somewhat later on 24 June 1987 as both an educational and entertainment facility. Those purposes were accomplished through a large number of attractive displays, the main one being a large 20 metre acrylic walk through a viewing tunnel with a skilfully reconstructed self functioning reef ecosystem with its associated biota on one side, and on the other is a predator tank in which visitors can experience the exciting situation of sharks swimming alongside and above them. In addition, throughout the aquarium smaller individual displays are mounted.

MANAGEMENT THROUGH RESEARCH: THE ESSENTIAL FUNCTION

Effective management, the essential function of the Authority, demanded an accurate and continuing flow of research data, made all the more necessary by the continuing growth of Reef activities and the increasing impact of technology. To meet that need the Authority commissioned research projects in ever growing numbers. In the first research report to government covering the years from its beginnings in 1976 to mid 1982, projects funded increased every year, with a total cost of $825 400. From then on, management based research has grown exponentially. After ten years of funding, a total of 292 separate projects had been completed, which by 1988 accounted for more than one-third of all AMSTAC funds (Watson 1988:13). Ten years after the foundation of ASTEC, in 1988 came its enormous publication entitled *Australian Marine Research in Progress* which in 865 pages gave a synopsis on a nation-wide scale and details of the 1267 separate ongoing research projects, a very large proportion of which were being conducted by AIMS, GBRMPA, CSIRO, and James Cook University, specifically on Reef research.

A major advance in Reef management by science came in the early 1980s when satellite technology was first utilised. In five separate launchings between 1972 and 1984 the United States deployed the officially named Earth Resources Technology Satellites equipped with various kinds of cameras to gather information relevant to conditions on the Earth's surface. Already GBRMPA, following early experiments of Reef mapping by the Royal Australian Air Force in 1928 and during the Pacific War, had begun tentative moves in aerial photography in 1977 with a small project jointly conducted by AIMS, James Cook and its own staff to map the reefs between Townsville and Cairns. Then, from 1979 to 1983 a new project was devised in conjunction with CSIRO to determine the use and accuracy of 'Landsat' images and their cost-effectiveness compared with other forms of mapping. Since it was estimated that conventional mapping of the entire marine park would take between thirty and forty years, the Authority decided that Landsat imagery, although very expensive to purchase, would certainly be faster, and possibly cheaper. In 1980 GBRMPA joined in a program with CSIRO and the Australian Survey Office to develop a computer analysis system that could process the Landsat images through a dedicated program devised by Dr David Jupp of CSIRO named, now that GBRMPA had become enmeshed in the age of acronyms, as BRIAN: Barrier Reef Image ANalysis. By 1986 coverage of the Reef was complete, with the production of 133 maps on a scale of 1:100 000 and 22 on a scale of 1:250 000, satisfying National Map Accuracy standards. The cost of the image mapping was a mere $300 000, with ground surveys and equipment additional (Jupp 1986; McCracken & Kingwell 1988:41).

Such maps, however, did not provide for all possible information needs, and it remained essential for much research work to be performed on site location and relevant data coordinated. To that end GBRMPA entered the digitised information age in 1985 with the installation of computers and a database known as REEFLEX in which data obtained from BRIAN, other mapping, and the increasing amount of material coming from field investigations and remote sensing could be processed and made available for management. In the same year AIMS also entered the era of computerised data management with the installation of a huge central processing system, first using a VAX 11/785 that both replaced and supplemented the separate weaker and inefficient computers previously in use. Once installed, computing needs at AIMS grew exponentially, VAX clusters were installed and within the first four years, to 1989, disk files had grown tenfold, from 9000 to 90 000. Field data could now be entered into portable laptops in the field and later downloaded into the central processing units which, interconnected with CSIRONET, REEFLEX and other systems, provided a resource which ten years earlier could not have been imagined. Since then, computing power has continued to increase and a vast amount of field information is collected by remote sensing and transferred to central locations as more sophisticated computer complexes and data management systems continue to be developed and installed.

In the first ten years of effective operation, from 1979 to 1989, following the Emerald Agreement and a number of clarifying additional Agreements and Memoranda of Understanding, the Authority had made an impressive start. All the Reef had been zoned, the Townsville headquarters in the Great Barrier Reef Wonderland complex had been completed, four restructured divisions were operating effectively – planning and management, research and monitoring, education and information, environmental impact management – and a staff of 103 were employed, five in the Canberra office, 98 in Townsville. Research was continuing at an accelerating pace and good working relationships had been established with both AIMS and James Cook University, along with other Commonwealth and State government departments, as well with private enterprise, chiefly tour operators. A large stream of publications on all aspects of management, zoning, workshops and research findings had been published and circulated into the public arena. After the collapse of the Queensland Coalition government in 1989 the new Labor government instituted statewide electoral reform and began to minimise political control of the public service. Relations between GBRMPA and the Queensland National Parks and Wildlife Service, responsible for day-to-day management tasks, where there had been some areas of disagreement, improved noticeably, now under the new Queensland Department of Environment and Heritage.

A CONSERVATION CLIMAX: WORLD HERITAGE LISTING OF THE REEF

One of the environmental tragedies of Australia, as Phillip Toyne has evocatively stated in *The Reluctant Nation*, is that ever since the Australian government joined the International Union for the Conservation of Nature (IUCN) – today the World Conservation Union – in

May 1973 and signed the World Heritage Convention on 17 December 1975, State governments have resisted having their properties given World Heritage listing since it prevents further commercial exploitation. In the case of Queensland in the Bjelke-Petersen era, exemplified in the continued desecration of sites of outstanding natural beauty such as Fraser Island, Cooloola Sands, and Mt Etna limestone caves, World Heritage listing was absolute anathema and a hindrance to its perception of 'progress'. Although throughout the 1980s the Queensland government remained implacably opposed to the Commonwealth on many issues, and resisted tenaciously all other moves by the Commonwealth in conservation and environment protection in order to prevent any further restriction on State sovereignty, the passage of the 1975 *Great Barrier Reef Marine Park Act* and the subsequent Emerald Agreement led to a reasonably cooperative approach to marine park management.

Fortunately for future generations, the Queensland government was quite amenable to World Heritage listing of the entire region: the marine park, all islands and Queensland waters. In the previous decade, at a 1972 conference of the United Nations Educational, Scientific and Cultural Organisation (UNESCO) in Paris it was resolved that the world's outstanding cultural and natural features – the pyramids of Egypt, Chartres Cathedral, the Grand Canyon, Mt Everest, for example – should be preserved, and protected from degradation, through an International Convention for the Protection of the World Cultural and Natural Heritage. Australia signed the Convention in 1974 to prevent the destruction of much of the national estate as identified in the Hope Report. Nations had to follow a complex procedure to qualify. Areas nominated for World Heritage listing had to satisfy at least one of four natural criteria, in brief, being outstanding examples of: the earth's evolutionary history; significant ongoing geological processes and biological evolution; superlative natural phenomena; or containing the most important and significant natural habitats where threatened species of animals or plants of outstanding universal value from the point of view of science still survive. It was also stated in the submission to UNESCO that the Reef has 'over 30 historical shipwrecks' and 'on the islands there are ruins and operating lighthouses which are of cultural and historical importance' and areas of significance for indigenous cultural heritage (GBRMPA 1981:5–6). The nomination process took some time, and on 26 October 1981 was successful, the Great Barrier Reef having been included on the World Heritage list for all four natural categories and accepted as the most impressive marine area in the world. The long fight to save the Reef as an outstanding natural feature with unique qualities was finally recognised by the world at large.

Despite general Australian recognition of World Heritage values, desecration of the natural environment continued, exemplified in 1985 with conservation alarm at continued destructive logging in the magnificent rainforests reaching along the Great Divide of North Queensland. Although stunning in natural beauty and containing the richest diversity of terrestrial tropical plant and animal life in the continent – described as 'the richest jewel in the crown of Australia's natural environment' (Toyne 1994:64) – the Coalition government of Queensland was determined to keep logging the irreplaceable forest. Even worse, it began to construct a road through the most sensitive section adjoining the Reef at the Daintree River north to the Bloomfield, to enable real estate subdivision. John Veron of AIMS consequently warned Environment Minister Barry Cohen that siltation runoff from road construction was already entering Reef waters and damaging the fringing reefs and further land clearing could eventually destroy them completely.

In response to rising community calls for World Heritage protection, and to settle numerous similar controversies still raging over the destruction of some of Australia's few precious remnant forests, the Labor government, elected in 1983, passed the *World Heritage Properties Conservation Act 1983* to protect regions that met UNESCO criteria for World Heritage listing, specifically enacting that it is 'unlawful . . . to kill, cut down, or damage any tree on any property to which this section applies' or 'to construct or establish any road or vehicular track' (Section (10)(2)(h, j)). To save that region of unique, unsurpassable beauty from desecration, the Commonwealth also moved on 5 June 1987 to have it declared a World Heritage Area and so be protected by the *World Heritage Properties Conservation Act 1983*. The conflict continued to rage until 9 December 1988 when it was registered by UNESCO on its World Heritage List as the Wet Tropics of Queensland.

REEF MANAGEMENT AND PUBLIC RELATIONS

Within that context of continuing litigation between Queensland and the Commonwealth, the Great Barrier Reef Marine Park Authority was not experiencing a totally smooth run. No government agency, Commonwealth or State, was subjected to so much scrutiny in its foundation years. As zoning continued to be refined, conflict grew with user groups, notably in fishing and tourism, who became increasingly adept at attempting to influence management decisions in their favour. The Authority was constantly at a disadvantage in its laudable approach of maximum community consultation and its vigorous public education programs that inevitably encountered antagonism. Disregard of zoning regulations led to continuing difficulties since the Authority was consistently reluctant to enforce them strictly – in part because of the imprecise location of boundaries in the days before Global Positioning System devices became available – and preferred the voluntary acceptance approach, although developers of marinas, fixed moorings and tourist ventures of all kinds, when faced with Authority opposition, exercised their rights to litigation. Another serious area of concern came from deterioration of water quality from terrestrial runoff due to the high levels of nitrogen and phosphorus from coastal farming, particularly sugarcane farming. As executive officer Donald Kinsey in his Annual Report of 1989–90 observed, 'A substantial amount of time was required to meet the obligations under litigation to which the Authority was a party'. To keep ahead of litigation costs, GBRMPA came to require developers to pay the costs of environmental assessment which were becoming quite high as conditions to be met became more stringent.

A major litigation issue creating serious disquiet in the Townsville community at that time had arisen over GBRMPA approval given on 28 October 1988 for construction of a marina on Magnetic Island as part of a proposed condominium and resort complex at peaceful Nelly Bay. Supported by environment groups, local residents through their newspaper *Island Voice* began a strident and sustained protest, claiming that destruction of Bright Point – one of the island's most scenic headlands – to create rockfill for the marina wall had created visual blight, while dredging of the bay, as the ineffective silt restraining underwater curtains demonstrated, would irrevocably damage the fringing coral reefs. Even more, the protesters, including Senator John Coulter, alleged that Chairman Kelleher had been unduly favourable to the developers. GBRMPA was involved in the subsequent litigation in court and the protesters, represented by a young resident biologist Julie Walkden, lost their case, with legal claims and costs awarded against her, after a failed appeal to the Supreme Court in July 1990, of over $500 000.

On 18 December 1990 the company, Magnetic Keys Limited, went into receivership and when the project came to a halt it was revealed that no insurance to cover rehabilitation of the site had been required. Allegations against the Authority were quite injurious since it was accused of negligence in not exercising effective management over financial aspects of the development. Eventually Minister for the Environment, Ms Ros Kelly, ordered an inquiry on 13 February 1992, conducted by lawyer John Whitehouse, specifically to inquire into the 'procedures followed by the Great Barrier Reef Marine Park Authority in arriving at the decision to issue permit G88/462 for the Magnetic Quays marina development project' and whether the allegations of undue favouritism to the developers by Chairman Kelleher could be substantiated (Whitehouse 1992:1). Both Kelleher and the Authority were exonerated, but Whitehouse was critical of the practice of the Authority, which had occurred in that case, of indicating in advance that approval could well be given. He recommended that 'the Authority should eschew the issuing of "approvals in principle"' (67). Moreover, in respect to less than adequate records concerning granting approval, he reported that GBRMPA could improve its record keeping, referring to a curious absence of information in a critical period leading up to the granting of the permit. 'It would have been more helpful if the Authority's files could have shed more light on the events following the meeting on 26 October, 1988 up until the issue of the permit on 28 October, 1988' he wrote, in that there was 'no record of any discussion between the Authority staff following the conclusion of the meeting with the proponents on 26 October, 1988' (74).

Two other issues of widespread public concern, which proved somewhat embarrassing to the Authority, also occurred in that period, both related to tourist ventures on John Brewer Reef in a Marine Park A zone near Townsville. In 1985 the Authority, against community opposition, issued a permit for a 12 000 tonne six storey floating hotel to be constructed in Singapore and towed to a mooring in the lagoon of John Brewer Reef.

Strict conditions, requiring no damage to the reef and recycling or removal of all waste and effluent, were considered to have been met when the resort was finally moored in place in December 1987. It was, however, a financial failure from the beginning. Despite a huge publicity campaign by the operators, Four Seasons, with colourful brochures declaring it to be 'hailed as Australia's most exciting international playground', tourists failed to arrive. After a short period with occupancy rates hovering around 20 per cent, it was towed away to Saigon to become a waterfront hotel. Claims that no physical environmental damage was done to John Brewer Reef overlooked the blasting of an entry passage through the coral. Moreover it was condemned as visual pollution of the serene Reef skyline – comparable with an oil rig – and environmental critics were determined to keep both away from the Reef.

At the same time an enterprising Townsville tour operator, Doug Tarca, conceived a new tourist attraction: a large floating 'Fantasy Island' day resort to be moored nearby as a recreational facility for the floating hotel guests as well as day trippers. Having been granted approval by the Authority, it was constructed of crescent shaped, hollow concrete pontoons, that were towed from Townsville into position off Brewer Reef, linked together to form a closed structure with a large central pool, and moored to the seabed. The Doughnut, as it was labelled, was planned to have a wide range of facilities on its deck including a licensed bar, theatrette and child-minding centre, surrounded by potted 5 metre tall palm trees. In May 1988, ten days after the moorings had been completed, it sank in a storm. Not to be disheartened, Tarca declared it 'Fantasy Reef', believing it would become a magnificent diving and snorkelling facility. GBRMPA took a sterner view and ordered its removal, with Tarca to meet all costs. Three years later the last of the debris was finally brought ashore for disposal in a harbour landfill reclamation.

None of those projects – the Magnetic Island marina, the floating hotel, Fantasy Island – was shown to indicate negligence or dereliction of duty on the part of GBRMPA, but many critics saw them as at least unwise decisions that appeared to favour developers ahead of Reef protection, while the widespread media coverage heightened public awareness of the Reef's vulnerability. With the Reef always in the public view as a powerful symbol, even a sacred icon, the Authority constantly had to proceed with great care. Any imagined failure of management was likely to be magnified by the media; and one undesirable outcome was the continued sense of apprehension that Authority staff experienced as they performed their duties.

CROWN OF THORNS: CONFLICT AND CONTROVERSY

An outstanding example of difficulties experienced with the media appeared in relation to persistent outbreaks of *Acanthaster* starfish on the Reef after their original mass destruction of coral on Green Island in the 1960s had subsided. Ever since, the Crown of Thorns issue has remained one of the most intractable, publicly visible problems to confront GBRMPA, posing a major management and public relations challenge. It has also led to the greatest single expenditure of research. Of the initial three causes for public concern over the Reef, both oil exploration and limestone mining had been prevented from conducting operations within the Park. Only the *Acanthaster* outbreaks remained, and in the wider public mind their impact, as an indicator of threats to the Reef's beauty, biodiversity and its future, was a central issue.

Cause for alarm came again in 1979, then in 1982 when aggregations were recorded off Innisfail, and in 1983 with an outbreak mainly in the Central Section of the Reef between Lizard Island and Townsville. Then came media sensationalising of the outbreaks. On 16 October 1983 the mass tabloid Sydney *Sunday Sun* published a major 'news story' under the headline 'Reef Killer Plague' which was 'being covered up by a combination of silence and ignorance'. 'Return of the coral killers' it continued, in a survey of outbreaks since the 1960s. The next day the Melbourne *Age* in a full-page article asked: 'Are we heading for the Great Barren Reef?' and three weeks later the *Weekend Australian* account (5 and 6 November 1983) declared the 'Reef Park a farce in face of starfish threat'. Newspapers in Reef cities published the story in more defensive terms: the *Townsville Daily Bulletin* asserted that 'Experts dismiss increase in starfish' (19 October) and that there was a 'Hysteria warning on starfish' (8 December). For the following three months the press controversy raged, not only in the large metropolitan presses but also in Reef township papers.

The Authority, well aware of the problem, had already begun a methodical study of the phenomenon by collecting data from the entire Reef. From 1979 to 1983 most of its research budget, more than $500 000, had been spent on investigating the starfish problem, and by 1985 this was to reach $638 000. In December 1983 GBRMPA was in a position to release its first comprehensive report to the public entitled 'Crown of Thorns Starfish Survey Update'. The previous year it had begun a campaign to gather information from the public by means of survey questionnaires, and also commenced physical surveys. Towards the end of 1983 records assembled for 177 reefs from Cape York to Rockhampton indicated that no starfish were sighted on 56 reefs, on 11 they were common, and on 81 reefs they were uncommon. Aggregations, where the animals were in excessive numbers, accounted for 29 reefs.

The immediate step taken by GBRMPA to allay rising public concern – and press hysteria – was to convene a media conference in Cairns, epicentre of the outbreaks, with a tour of two affected reefs in the vicinity: Green Island and Michaelmas Cay, for the weekend of 4 to 5 February 1984. As a briefing preparation, the Authority prepared a detailed 'Media Information Kit' of eleven articles. It was, in effect, a major effort at damage control with a statement by Chairman Kelleher, a selection of news clippings including those quoted above, along with more sober articles on the biology of the starfish, its distribution, international views, and the research projects being undertaken by GBRMPA. The Authority attempted no cover-up, and the kit was a sincere effort to present the entire picture.

There was, of course, no doubt that the outbreaks were very serious and certainly were damaging sections of the Reef, particularly in the Cairns region where the City Council had been active in promoting the city's attractions with an international airport and its beautiful palm grove Trinity Bay parkland harbour front. These were featured in enticing tourist colour brochures and it had rapidly outstripped all other Reef locations as the city of arrival for most tourists. Since the economic viability of Cairns was seriously threatened, now that tourism had become its premier industry, the GBRMPA approach was to confront the problem as objectively as possible in the interests of finding a solution. To that end it created a special task force entitled the Crown of Thorns Starfish Advisory Committee (COTSAC). At the same time the Commonwealth government also moved to take prompt action: Minister for the Environment Cohen rejected any cover-up as alleged in the press and an inquiry was commenced by the Senate Standing Committee on Environment and Conservation to seek a solution.

Soon after came the deluge. Clearly there was a newsworthy story in the starfish outbreaks, and the subsequent media frenzy stimulated journalist Robert Raymond to begin his own investigations. Raymond had developed an interest in the Crown of Thorns phenomenon from its initial outbreak when he made a television documentary in 1969 entitled 'Life and Death on the Great Barrier Reef' that was significant in raising the starfish 'plague' to national prominence. Throughout 1984, following the GBRMPA statistical survey and media briefing, he interviewed scientists at AIMS and James Cook University, along with the chairman and several officers of the Authority including chief planning officer Richard Kenchington. Raymond also chartered the *Reef Explorer* and his journalist team visited several of the most damaged reefs: Lizard, Ribbon Reef No. 9, Green Island, Brewer, and Beaver Cay. The results of his inquiries came as a media bombshell. In the issue for 18 December 1984 Australia's major national news magazine *The Bulletin* featured his findings in its cover story under the title 'Barrier Reef Scandal: hushing up a disaster'. In a lengthy ten page illustrated article (pp.68–77) headed 'A conspiracy of silence over the destruction of the Reef' he reported a depressing scenario: at Lizard Island he found 'coral dead and rotting, crumbling to the sea floor', Ribbon Reef No. 9 was an 'atomic bommie' where 'big areas of coral [were] just devastated', 'the coral around Green Island has been literally obliterated' and the shattered reef flat revealed that 'an army has been advancing'. At Brewer Reef 'we found the encircling reef slope dead and hordes of starfish working their way across the reef top toward the area around the pontoon'.

Raymond's interview with Dr Barry Goldman, director of the Australian Museum's Lizard Island research station was reported to the effect that no direct work had yet been done on the problem. Dr John Bunt, then director of AIMS, in reply to a question whether 'any work had been done on the population dynamics of the starfish' advised that no work had yet commenced, in part due to the impossibility of tagging the creature and so monitoring its movements. Richard Kenchington, speaking for GBRMPA, expressed a concerned but scientifically detached view that much needed to be learned about the episodic outbreaks, although even so, 'we can be fairly comforted that there is not an overall catastrophe on our hands'. Raymond, however, was after bigger game, and wrote that although the 'plague' or 'infestation' was simply part of

> the struggle that has been going on silently, beneath the ocean . . . now, I believe, the struggle is about to erupt into the corridors of power in Canberra and perhaps onto the world stage. Before it is resolved it could affect the viability of the Queensland tourist industry, challenge the credibility of the Australian scientific establishment and bruise the reputation of one or two federal ministers (p.68).

In December 1985 the Commonwealth government announced that as a result of inquiries by the Senate Standing Committee on Environment and Conservation it had arranged for the greatest single marine research project on the Reef to be undertaken. With an initial grant to GBRMPA of $971 000 for the year 1985–6, funding was made available to coordinate management type projects and for AIMS to coordinate ecological investigations into the starfish outbreak. In all, 58 separate projects were commenced, involving the collaboration of some 70 senior scientists, some from overseas institutions, and for the ensuing four years over $3 000 000 was made available to GBRMPA to distribute for the research, of which AIMS received $1 997 960 (Moran & Johnson 1990:1–2). AIMS scientist Peter Moran had the demanding task of coordinating 33 separate projects and nearly 50 scientists, in four separate areas: population dynamics, prey and ecosystem context, predator–prey relationships, and technological methodology. Within the GBRMPA sphere Leon Zann had the task of coordinating other scientists in 25 management projects, structured around three central questions: attempting to determine if there was a geological history to outbreaks, testing theories on causes – natural or human – and the possibility of biological controls.

After the first year of intensive investigation no conclusive results had been reached. Out of the 225 reefs surveyed, 25 per cent were found to be affected, most in the Central Section of the Reef; a year later in June 1987 AIMS reported that 33 per cent of the reefs it investigated were seriously affected. Two months later, 28 August 1987, the *Sydney Morning Herald* reported that the Crown of Thorns 'still splits scientists', referring to the controversy arising from Dr Robert Endean's continuing assertion that the primary cause was the removal of the giant triton mollusc, its natural predator. By that time more than 300 scientific papers had been published, and scientists were still no closer to an acceptable solution. One interesting item of evidence claimed that coral coring indicated the starfish had been present on the Reef in earlier geological times, but was inconclusive on the extent of 'plagues'. The greatest dilemma facing management was the question of intervention. Only if it could be proved conclusively that outbreaks were anthropogenic, with collection of triton shells and pollution runoff from excessive mainland agriculture the two most popular hypotheses, could there be any valid reason for intervention. So far, commented Graeme Kelleher in 1987, 'attempts to eradicate the starfish on a wide scale have proved futile' and therefore the Authority had decided the best management approach was to channel funds into research in a quest for the so far elusive conclusive evidence (Kelleher 1987:14).

In 1989 a new three year study in 31 separate projects was commenced, funded at $2.75 million. Still no firm evidence had come in by 1992, although computer modelling of locations and the ecological context – spawning, recruitment, water circulation – had determined that in the decade of the 1980s outbreaks had occurred in nearly 21 per cent of the Reef. Where outbreaks had occurred there was an 11 per cent increase in dead coral cover, but there had been a slight decline in recent years, the majority of the infestations being in the Central Section between Townsville and Bowen. Even so, despite massive research and millions of dollars in funding for nearly fifteen years, still no acceptable research evidence had been found to settle the controversy. As a report to ASTEC by GBRMPA on the state of research in 1991 concluded: 'It seems very likely that there will be further outbreaks in the future. It is vital that research be continued to determine whether human activities cause or exacerbate such outbreaks' (Lassig & Kelleher 1991:58). If the starfish phenomenon investigations proved anything worthwhile, it is that the Great Barrier Reef ecosystem is astonishingly complex and so far defies human efforts to intervene in, and exercise control over, nature. Even worse: as nature becomes progressively disturbed by human activities, remediation may become even more difficult, if not impossible.

By 1990, despite the massive disappointment coming from the failure of the starfish investigations to achieve a final solution, a great deal had been learnt about reef ecosystems, and GBRMPA had achieved an increased maturity in its approach to the numerous problems of management. The coming final decade of the twentieth century was to be even more demanding as planetary environmental imbalance became increasingly manifest, and the goal of ecological sustainability was to test its management competence even further.

CHAPTER 22

THE REEF UNDER PRESSURE: RESEARCH AND MANAGEMENT

ENVIRONMENT AND ECONOMIC GROWTH IN THE GREENHOUSE DECADE

Over the final decades of the twentieth century the rapid development of industry and the pursuit of continued economic growth for corporate profits had accelerated changes to the entire world environment. The incredible expansion of organic chemistry had produced thousands of synthetic compounds for industrial manufacturing that have no counterparts in nature, and against which nature has no defences. In 1962 Rachel Carson's book *Silent Spring* had brought dramatically to world attention the issue of organochlorine contamination of American farmlands, and in turn, rivers and seas, from agricultural runoff. Carson was followed by a sequence of concerned critics – Lynn White Jr, Paul Ehrlich, René Dubos, Barbara Ward, among many others – who began constructing the pattern of connections of massive global environmental degradation that led to the United Nations Stockholm Conference on the World Environment of 1972 and the concept expressed in the title of its publication *Only One Earth* (Ward & Dubos 1972). That concern, however, was not taken seriously, and the quest for development under the mantra of economic 'growth' continued, not simply unabated, but with mounting impact.

A decade later increased loss of biodiversity and productive land, and the growing impoverishment of the less developed nations, had become so serious that in early 1984 the United Nations commenced an international campaign not just to arrest the trend, but to change direction entirely by establishing the World Commission on Environment and Development under the direction of Gro Harlem Brundtland, Norway's Prime Minister. For three years the Commission held hearings on all five continents and issued one of the most influential documents in United Nations' history under the title *Our Common Future*, known popularly as the Brundtland Report. The evidence her committee collected was indeed disheartening as it documented major global species loss, destruction of forests, and the expansion of blighted urban conglomerations as relentless population growth destroyed traditional farmlands, while transnational corporations increasingly pillaged natural resources worldwide for short term profits. Coral reefs were some of the most plundered regions which, it was revealed, 'with an estimated half-million species in their 400,000 square kilometres, are being depleted at rates that may leave little but degraded remnants by early next century' (Brundtland Report 1987:151).

The final meeting of the Commission in Tokyo on 27 February 1987 issued the 'Tokyo Declaration' which presented the central concept of sustainability as the only way forward for future global development. Despite almost overwhelming evidence of ever worsening world environmental conditions, it held out hope that 'it is possible to build a future that is prosperous, just and secure', but only by nations 'realizing this possibility depends on all countries adopting the objective of sustainable development as the overriding goal and test of national policy and international co-operation' (363). The report, however, became strongly criticised when it accepted the need for economic growth to continue for some time and moreover changed its original goal from ecological sustainability to 'sustainable development', which inevitably was taken by many powerful interests to imply that past practices could continue at an expanding rate.

Meanwhile, concern was mounting that environmental deterioration was now extending to the global atmosphere and oceans. Evidence of an unprecedented increase in atmospheric carbon dioxide had been available for some time. Until the year 1800, in the recent geological epoch, as revealed in air bubbles retrieved from ice cap cores in Greenland and the Antarctic, CO_2 concentrations had remained steady, around 280 parts per million for thousands of years. In the era of rapid post-World War II industrial development, concentrations rose to 315 ppm in 1957 and continued to climb, being recorded in the mid 1990s at 362 ppm (Vitousek et al. 1997:496). Other forms of pollution also began to increase. The proliferation of 'greenhouse gases' – chiefly carbon dioxide, nitrous oxide and methane – were evidently changing the earth's climate by raising global temperatures, increasing the already measurable rise in sea levels, and threatening worldwide damage, with coral reefs particularly at risk.

In response to these disturbing trends, the United Nations Environment Program (UNEP) convened a conference at Villach in Austria in October 1985 'to assess the role of increased carbon dioxide and other radiatively active constituents of the atmosphere (collectively known as greenhouse gases and aerosols) on climate changes and associated impacts' (Villach Statement 1985). The recommendations of the Villach Statement, along with confirmation in 1986 that the long-suspected impact of the new chlorofluorocarbons (CFCs) was now thinning the earth's protective ozone layer by allowing greater penetration of ultraviolet radiation, led the International Geosphere/Biosphere Program to investigate those issues the following year. With mounting evidence of an 'ozone hole' over Antarctica, an international agreement was reached in the Montreal Protocol in 1987 to phase out production of CFC gases, and in the Helsinki Agreement of 1989, following the first World Conference on the Changing Atmosphere held in Toronto in 1988, a firm target date was set for the year 2000.

OCEANS OF WEALTH? MARINE SCIENCE UNDER REVIEW

Throughout the 1980s Great Barrier Reef science and management were conducted against that rapidly escalating greenhouse background, at first with little close connection to outside world events. Research by AIMS and management by GBRMPA were chiefly determined by the Commonwealth government as their funding body, and to a lesser extent by the Queensland government, in consultation and through its own government departments. In both cases the overriding concern was to maintain multiple use of the Reef for economic benefit from fishing (including trawling) and tourism, with scant overt recognition or knowledge of its ecological limits, while Queensland also had a major interest in ports and associated shipping. In 1983 scientific research funding under the Commonwealth's new Labor government ceased to grow for the ensuing four years. AMSTAC was abolished in 1987 and research funding decisions passed to the new Department of Industry, Technology and Commerce (DITAC). Pursuing the government's goal of economic growth, the Commonwealth began to require scientific organisations to attract funds from industry and to become increasingly commercial in approach. In particular, it wanted AIMS to work in closer cooperation with GBRMPA in applied research related to management issues, and to seek commercial partners in joint ventures.

Consequently, in 1985 AIMS had a major change of direction when Dr Joe Baker, a chemist at James Cook University, who had earlier worked for the Roche pharmaceutical firm, was appointed director. AIMS was restructured into four main programs in the areas of coastal processes and resources, coral reef ecosystems, environmental and biotechnology studies, and coastal oceanography and marine systems analysis. In effect, most of its activities were moved into applied science and away from the fundamental research of its first decade when the basic need was to assemble data on the taxonomy and growth processes of the Reef's coral ecosystems.

Pressure from the Commonwealth to gain greater economic benefits from the considerable funding that had been going into the marine sector – $70 million in 1988 – prompted Minister for Science Barry Jones that year to establish a committee, under the direction of Professor Ken McKinnon, to review the organisation of Australia's marine industries, and marine science and technology capabilities. Three terms of reference were specified: to evaluate Australia's current strengths and weaknesses in marine science and technology; to provide advice on enhancing international competitiveness; and most significantly, 'to assess the industrial and commercial opportunities which are likely to arise in Australian and international marine industries in the Year 2000, with special reference to the opportunities dependent on marine science and technology'. In 1989 the committee reported under the ambiguous title of *Oceans of Wealth?* that marine science across the nation was unbalanced: too much effort went into marine biology, too little into the physical sciences and technology. Opportunities to increase wealth from the marine sector through fishing could be exploited, although it was recognised that the industry was beset with problems due to excessive, unsustainable catches and failure to find new resources.

The entire range of Australian marine and maritime activities was investigated, including offshore oil and gas, marine transport and shipping, shipbuilding, defence, marine biotechnology, and coastal engineering. It was concluded that procedures within those areas were unintegrated, variable in standards, and in need of all-round improvement. Some attention was given to environmental management as 'essential for protecting our marine assets from degradation or pollution, and for maximising returns from their resources', with the implication that there was room for significant upgrading (McKinnon 1989:9). The Reef came in for particular mention in regard to tourism where it was noted that 'visitor nights', the index of calculation, had increased by 50 per cent since 1982 and in order to counter increasing tourist pressure 'levies should be raised on marine tourism to help pay for environmental monitoring and research on the Great Barrier Reef and other attractions' (15). However, the primary need for Australia, the government was advised, was to accelerate and coordinate science and technology in order to extract maximum economic return from the marine environment.

The 'Summary of Findings' in *Oceans of Wealth?* concluded that 'Our oceans represent a new frontier for science and technology ... the scope of opportunities for wealth from the sea are only partly known ... but it will be by good management alone that Australia will have oceans of wealth in the future' (2). The McKinnon Report was directed specifically to government preoccupation with productivity and growth, and therefore focused on a strategy for maximising economic returns while concern for environment preservation received only the barest mention, and even then solely to sustain production. The report gives the impression that none of the committee had any deep understanding of, or at best no concern with, environmental and ecological issues at all. The feasibility of 'maximum economic return' was not investigated and no evidence was presented to suggest there had to be careful scrutiny of the extent of resources, and permissible levels of exploitation, to prevent species population collapse.

With that report as a starting point the Commonwealth began assuming almost total control over marine research, development and technology, beginning in 1990 with the creation of a Consultative Group on Marine Science and Technology and an announcement of $3.9 million funding over five years for a program to encourage research and development in Australian marine industries, science and technology. With the impending declaration of the Australian Exclusive Economic Zone that would create a marine region one and a half times greater in area than its continental landmass, it was considered essential to begin forward planning, setting the year 2000 as an initial goal. Australia, it was implied, could anticipate a continuing harvest of wealth from its oceans. Evidence from the Great Barrier Reef however, was sounding a strong note of caution in that regard.

CORAL REEF RESEARCH: INTO THE SUSTAINABILITY ERA

By 1990 there was clear evidence that human pressure was rapidly accelerating changes to Reef ecology, perhaps irreversibly. There was continuing dispute among scientists over the causes of Crown of Thorns outbreaks, for example, and the fact that despite all efforts and a massive injection of funds into research for more than a decade, eradication remained intractable. Reef ecology was still far from understood. As Frank Talbot stated at the ACF Conference in 1969, 'the whole science of ecology is in its infancy'. A decade later another warning came from Tom Goreau (1924–70), one of the world's leading coral reef scientists, famous for his research into the symbiotic relationships between zooxanthellae and coral polyps. In a posthumous article, co-authored by wife Nora and son Tom (Jr), both coral scientists, Goreau warned that reef ecology is governed by a 'complexity [that] can only be dimly grasped'. Going further, he commented that 'many marine biologists view with alarm the spread of tourist resorts along coral coasts in many parts of the world' and the consequent damage from overfishing, sewage discharge and construction of marinas, wharves and associated dredging and landfill dumping. 'The deleterious effects of human activities in an environment as complex as the reef ecosystem' he concluded, must urge 'caution against taking the stability and productivity of the reef community for granted' (Goreau et al. 1979:119–20). The unresolved Crown of Thorns problem, and the obvious degradation of many reefs gave a dramatic vindication of that warning, as the most publicly visible indication that the Reef ecosystem was becoming increasingly destabilised. The Reef, it was clear, can never be an isolated sacred icon comfortably insulated from the pressures of human use, much less the consequences of global processes, natural or human induced.

In a long-delayed response to the warnings of the Stockholm Commission of 1972 and the Brundtland Commission of 1987 the Commonwealth government was becoming aware that serious efforts had to be made to arrest environmental degradation and conserve resources for the future by embracing the concept of sustainability. Scepticism about the reality of climate change began to evaporate as data accumulated. In August 1990 when the United Nations Intergovernmental Panel on Climate Change (IPCC) held a preparatory conference in Sweden, the panel heard Peter Timeon of Kiribati, whose atoll homeland rises only 2 metres above sea level, make an impassioned plea for immediate reductions in carbon-based gases. That plea was repeated three months later at the Second World Climate Conference in Geneva when it was made clear that low-lying nations on tropical atolls such as the Maldives and Kiribati, and deltaic zones such as Bangladesh, could well be submerged in time as sea levels rose inexorably.

Yet at the same time, in Australia as in other developed nations with a high standard of living, successive governments remained committed to unlimited economic growth. In an effort to reap such imagined benefits which, as it was belatedly recognized, can only occur when productive regions are maintained in ecologically healthy conditions, a new emphasis started to appear in May 1990 when the Commonwealth government issued a Discussion Paper entitled 'Ecologically Sustainable Development'. Growth had to become environmentally responsible, a direction that was to receive strong international endorsement later at the Rio de Janeiro Intergovernmental Conference on Sustainable Development in 1992. Simultaneously the Commonwealth launched yet another approach to scientific research, primarily in the natural sciences and engineering, in what was announced as the Cooperative Research Centres Program, with a proposal to support up to 50 Centres nationwide with total funding over the ensuing five years of $100 million.

In 1992, in the third round of Cooperative Research Centres Program applications, a consortium comprised of James Cook University departments, AIMS, GBRMPA, the Queensland Department of Primary Industries and the Association of Marine Park Tourism Operators submitted an extensive project entitled 'Ecologically Sustainable Development of the Great Barrier Reef'. Four additional organisations joined forces in a supporting capacity: the Bureau of Tourism Research, the Queensland Fish Management Authority, the Queensland Commercial Fishermen's Organisation and the Queensland Sport and Recreational Fishing Council. In a lengthy 20 page submission the principal focus was described as an 'effort to work with industry and environmental managers in addressing strategic and tactical issues relating to the achievement of ecologically sustainable development of the Great Barrier Reef region while maximising the user opportunity spectrum'.

Five separate programs were proposed for the Cooperative Research Centre (CRC) project, arranged into 15 sub-programs, covering the areas of

1 Regional Environmental Status: assessing the overall ecological health of the Reef;
2 Operations: tourism and fishing impacts;
3 Engineering: site vulnerability to moorings and semi-permanent infrastructures;
4 Extension: 'the critical interface between research, education and user groups'; and
5 Education, which would train up to 23 graduate students a year in marine environment education.

The central aim of the proposal was to 'identify and respond to the needs of users of the GBR region and maximise the contribution of research to the cost-effective ecologically sustainable development and multiple use of the region'. This was accompanied by what the submission called a 'vision', defined as an attempt to 'maximise ecologically sustainable benefits to all Reef users, in a manner which conserves, in perpetuity, the area's unique biological and physical features' (Au.Arch.). The submission was successful and the CRC Reef Research Centre, as it was designated, received funding of $13.46 million for seven years, commencing at the beginning of 1993. Although ensuring permanent maximum yield remained the primary goal, here at least was a recognition that sustainability also depends on maintaining the marine environment and Reef ecosystem in as healthy a condition as possible.

A little over a century earlier, T. H. Huxley as President of the Royal Society of London had opened the Great International Fishery Exhibition in 1883 with a speech entitled 'The Inexhaustibility of Fisheries' in which he commented that 'the cod fishery, the herring fishery, and probably all the great sea fisheries are inexhaustible; that is to say, nothing we do seriously affects the numbers of fish'. Huxley, however, was vigorously refuted at the symposium by his brilliant, and at times controversial, protégé Ray Lankester who warned against over-exploitation, and urged the need to conserve spawning stock (Botsford et al. 1997:512). Even Lankester, a recognised authority at University College London on fossil fish, who became director of the British Museum of Natural History, could hardly have predicted the collapse of the herring and cod fisheries a century later.

Fishing activities, both commercial and recreational, are now identified as a central factor in precipitating what threatens to be a worldwide collapse of coral reef ecosystems. Mounting evidence is confirming the devastating effects on marine ecosystems as human predation makes serious reductions in fish life, particularly in respect to the larger, most sought-after species. As revealed in a United Nations Food and Agriculture Organisation (FAO) report, by 1995 22 per cent of recognised marine fisheries were over-exploited and 44 per cent were at their limit of exploitation (Vitousek et al. 1997:495). Trawling practices are particularly wasteful, and another FAO study revealed that worldwide some 27 million tonnes of bycatch are discarded annually, amounting to one-third of the trawled catch (495). We now have reached that crisis point where fish stocks worldwide are critically depleted and without the most scrupulous management and maintenance of a healthy environment, economic sustainability is impossible.

Evidence indicates that once the larger species of fish go, and habitats are further disturbed by other anthropogenic influences, invasive species, possibly including the Crown of Thorns, are able to move in and become consolidated. Consequential changes to the trophic structure (food web) of the waters allow pathogens, such as microbial outbreaks, to spread disease more widely, continuing the relentless destruction of the total environment (Jackson et al. 2001:635). The Great Barrier Reef, although one of the world's best managed reef systems, is no exception to those international trends, and a disturbing feature of the CRC submission is its emphasis on 'maximising the user opportunity spectrum'. A vast range of scientific data emphasises that sustainability will never be achieved while present extraction rates continue: 'sustainable development' will remain an empty phrase, a literary oxymoron with no relationship to the real world.

CORAL REEF SCIENCE AT THE TURN OF THE CENTURY

By the mid 1990s, scientific understanding of the Reef had advanced considerably, and the foresight of the Commonwealth government in founding AIMS as a major research institution was showing significant dividends. From a base of abysmal ignorance of coral reef ecosystems in the early 1970s, the combined efforts of the initial planners laid the foundations for a sound body of consolidated research. Studies of reef corals, mangroves and coastal processes have been accompanied by other equally important areas of research including the dynamics of food chains, and commercial ventures such as development of pharmaceuticals from marine natural products. Since it commenced operations in 1978, AIMS has developed into a first class facility with an international reputation for excellence and a number of world ranking scientists on its staff, achieving outstanding success in fields such as coral taxonomy and oceanographic studies.

Other institutions, including a number of universities, also contribute to the advancement of coral reef science, notably the University of Queensland with its Heron Island marine station (a monument to the persistence of Professor Goddard in the 1940s), the University of Sydney with its research station on One Tree Island in the Capricorn group and the Australian Museum with its Lizard Island Research Station in the Cairns section – One Tree and Lizard owing much to the initial efforts of Frank Talbot and his colleagues. James Cook University in Townsville, with its Orpheus Island station, often in use for AIMS research, is also actively involved in reef science.

With the burgeoning of Reef research it is possible in this study to mention only a few examples that have gained international recognition. A notable advance in coral science occurred in 1982–83 with the identification of mass coral spawning by Peter Harrison and his colleagues from James Cook University, based on studies on Orpheus and Magnetic islands. Although it had long been known that corals reproduce by budding and by release of planulae, they discovered, contrary to prevailing belief, that rather than brooding and releasing their young continuously throughout the year, many species of corals have an annual mass spawning event several days after the full moon in the months of October to December. In a frenzy of ejaculation, eggs and sperm are broadcast simultaneously into the water for external fertilisation after which the planulae, moved by currents and tides, may disperse widely before settlement and corallite building (Harrison et al. 1984).

The investigations of John (Charlie) Veron at AIMS in collaboration with other researchers led to a significant achievement in coral taxonomy with the five-volume *Scleractinia of Eastern Australia* (1976–84). For the first time all the region's reef-building corals had been brought into a comprehensive taxonomy and this provided a sound data base for further comparative investigation into corals of Western Australia and subsequently of the Indo-Pacific, including the epicentre of world coral biodiversity which extends from Indonesia to Papua New Guinea and Torres Strait. After publishing *Corals of Australia and the Indo-Pacific* in 1986, Veron went on to complete a remarkable monograph, *Corals in Space and Time* (1995), where he reported that his intensive observations of corals were challenging the established Darwinian concepts of species and the process of evolution. For a long time he had been puzzled by visible differences in the same species of coral found at different depths in the water column or separated by distance. He noted also their pattern of variance related to biogeographic distribution, as well as the frequent recurrence of hybrids among corals – observed earlier by Peter Harrison (pers. comm. Feb. 2002). Hybridisation, long recognised among plants as a source of new species, has been rarely studied among animals. Veron concluded that, since evolution can only be inferred from existing evidence, it was necessary to question the discrete nature of species. The central thesis advanced is the concept of vicariance in which species become separated by 'barriers' and then evolve separately, forming in effect, new 'species', which themselves over time and space may merge again, creating – not a tree of life, as Lamarck and Darwin had envisaged – but a reticulant pattern as proposed by the master botanist Robert Brown in that same period. Yet another great benefit from that 20 year research program came with the publication in 2000 by Veron of the world's most comprehensive coral taxonomic study, the three volume *Corals of the World*.

Equally impressive has been investigation into the oceanography of the Reef lagoon and surrounding waters. The achievements of Eric Wolanski and his colleagues have been of immense value in understanding the complexities of tides, currents, eddies, upwellings and riverine flood plumes that significantly influence life in and near the Reef lagoon. As nineteenth century navigators discovered, often to their cost, water circulation patterns in the Reef are unpredictably bewildering. With the publication in 1994 of their *Physical Oceanographic Processes of the Great Barrier Reef* the sustained researches of Wolanski's team received the accolade of a $1 million scientific award by the IBM International Foundation (Wolanski, in Bell & Veron 1998:136). Following that distinction, oceanographic research at AIMS has extended to intensive investigation of processes within the Reef ecosystem, recently presented in a second major publication: *Oceanographic Processes of Coral Reefs: Physical and Biological Links in the Great Barrier Reef* (Wolanski 2001).

One of the major applied tasks of AIMS today, in response to the disturbing problems of Reef degradation and climate change, is providing the world base for the Global Coral Reef Monitoring Network, undertaken in association with GBRMPA and the United States National Oceanic and Atmospheric Administration in collaboration with other overseas institutions. A range of tasks is in progress: studies of rising sea temperatures, coral bleaching and other consequences of climate change including effects of rising levels of carbon dioxide on the ability of hermatypic corals to build sufficiently strong skeletons (corallites) to enable continued growth of healthy reefs.

WARNING SIGNALS: A 'PARTICULARLY SENSITIVE AREA'

Meanwhile, against a global background of intensive application of technology and research in order to increase production from the sea, in 1989 a sub-committee of the International Maritime Organization (IMO) of the United Nations, known as GEEP (Group of Experts on Effects of Pollutants), had become concerned with the profound consequences of increasing human impact on marine ecosystems. GEEP accordingly sought to measure vulnerability in ocean areas with respect to pollution and marine system disturbance, and in 1990 established criteria for identifying and safeguarding sensitive areas worthy of preservation – those rare and threatened ecosystems that exhibit natural values, integrity and vulnerability. In the same year, reinforcing earlier international recognition of the Great Barrier Reef as a World Heritage Area of outstanding significance, the IMO declared it the world's first, and one of only two 'Particularly Sensitive Areas'. While that was an accolade of the highest order, confirming the widely held view of the Reef as one of the world's natural wonders, it also carried an implied warning. As a 'particularly sensitive area', parts of it require great care. And that increased the strains on management that were already becoming quite severe.

Failure to understand the causes of Crown of Thorns outbreaks and other problems, was bringing no joy to GBRMPA. In fact, as the 1990s progressed, it was in danger of becoming a rather beleaguered organisation with an increasingly impossible task as conflicting management demands were placed upon it. 'The Authority is under continual pressures from two opposing directions' Kelleher reported, citing the ever-present conflict between those exerted on the one hand from special interest groups 'to minimise regulations', and on the other from its mission 'to protect the ecology of the Great Barrier Reef'. 'Balancing the response to these conflicting pressures', he stressed, 'within the context of what is Australia's most comprehensive public participation program, is a tremendous challenge' (Annual Report 1992–3:3).

Difficulties for decision making had been built into the *Great Barrier Reef Marine Park Act* itself in Section 32 (7) where the Authority's responsibilities required it to 'protect the Great Barrier Reef while allowing the reasonable use of the Great Barrier Reef Region'. At the same time it was to regulate those 'activities that exploit [its] resources' in such a way as 'to minimize the effect of those activities on the Great Barrier Reef'. In addition, allowing 'enjoyment by the public' raised conflicts as the massively increasing volume of tourist traffic exacerbated pressure with the deployment of ever more and larger vessels, the construction of wharves, marinas, moorings and pontoons, and the impact of escalating numbers of snorkellers and shore based parties, with the site damage and waste that accompanies such operations. When the original Act was passed in 1975 there was no clear understanding of the difficulties that would inevitably arise in seeking to ensure the Reef's permanent conservation, much less any explicit recognition that the Reef is part of one ocean, which is itself part of one global ecosystem. Within that context, what could 'reasonable use' mean?

The Authority had already recognised an imperative to take stronger action in response to rapidly increasing pressure in the Cairns Section and decided to rezone it in 1989. In doing so, it met such determined opposition from various user groups that, as Kelleher noted, 'rezoning has proven to be a more difficult task than the original zoning. The Authority's freedom of action in modifying existing zones is inhibited by the expectations of the public that activities which they have carried out in particular areas over the past years will continue'. He went on to make the point that user groups, particularly in tourism and fishing, have become skilled in attempting to manipulate use of the Park to their own particular advantage, and 'now have developed a degree of sophistication which enables them to apply significant pressure on the zoning team, making resolution of contentious points more difficult' (GBRMPA Annual Report 1989–90:1).

One measure proposed to control the exponential growth of tourism and its inexorable pressures was to charge users of the Park in order to provide additional funds for increasing management costs. As early as 1988 GBRMPA began investigating the potential for user charges both to recover a proportion of management expenditure and require recognition of scarcity value by imposing a 'resource rent' which, it was hoped, would involve tour operators in valuing the resources accessed by them for their own commercial gain. A year later the McKinnon Report also recommended a levy on Reef users to defray some of the mounting costs of environmental management, and in response the Authority commissioned an investigation of these proposals by the Australian Bureau of Agricultural and Resource Economics (ABARE).

The statistics that emerged from its 1991 report, *Charging Users of the Great Barrier Reef Marine Park*, were overwhelming. Tourism had become the dominant commercial activity within the Reef, having increased sixfold over the decade of the 1980s. In the year 1987 tour operators had taken 1 450 000 persons to Reef sites. By 1990 the average had almost doubled to 8500 persons per day on 300 vessels and seaplanes of all sizes with a strong growth in the introduction of fast catamarans holding several hundred passengers per day trip. Resorts had increased in number to 26 on 16 islands and three cays, registering 2 million visitor nights. Fishing was also increasing rapidly, with 70 per cent due to recreational and tourist fishing from 36 000 registered small craft, averaging 690 000 visitor days per annum. For 1990, ABARE reported approximately 250 000 fishing trips and the extraction of 12 000 tonnes of fish (Geen & Lal 1991:16–17).

Charging users was proposed as an effective means of managing a scarce and limited resource. In the past, the Reef had been a 'commons', a public place, open territory, available to anyone who chose to use it. In effect it was a frontier to be freely exploited, and that is exactly what had occurred in the days of resource raiders who pillaged the whales, sandalwood, pearl and trochus shell, turtles, bêche-de-mer and dugongs. As population pressure grows, however, without prudent management the commons can become degraded and resources scarce, or even cease to exist, at least in any useable way. Enclosing the area and limiting entry to prevent degradation of public spaces is a common option, although in the Great Barrier Reef Marine Park where fencing is not possible, control must be exercised by alternative means such as coercive laws or taxes on entry. GBRMPA research discovered that many tourism operators were making a 100 per cent return on their investment and so charging users of the Reef was considered a fair solution since, in effect, they were a new generation of resource raiders. While a tax would curtail their previous liberty to do as they chose, it would maximise the options of others to share in what is becoming a scarcer resource every year.

When, in response, GBRMPA recommended in 1991 a small 1 per cent 'reef user tax' to be levied on commercial activities – charter, cruise and dive boats – it was vetoed by the then State premier Wayne Goss as politically unacceptable because of his election pledge in December 1989 of 'no new taxes'. The costs of Reef management, however, continued to escalate and since commercial operators were required to meet part of the costs through a scale of charges, the Commonwealth government was forced to reconsider charges for recreational users. In 1993, after consultation with the tourism industry, GBRMPA imposed its first user charge of $1 per passenger per day.

Even so, ideas such as the freedom of the commons remain strongly held beliefs, despite growing awareness that the Reef is at risk of the same tragic damage to the commons of water, air and lands that is occurring worldwide, due to uncontrolled exploitation. When, in the Commonwealth Budget of 1997 it was decided that a daily entry fee of $6 per person would be levied on all tour operators, such an outcry erupted that the Senate conducted a wide inquiry into charging the public for access to museums, art galleries and national parks. Its report, issued in July 1998 and entitled *Access to Heritage*, dealt with the essential question of charging users, within the Australian cultural tradition of free access to public lands, facilities, beaches and ocean waters and asked whether there are 'powerful or symbolic reasons why access *ought* to be free of charge, as a fundamental value statement' (Senate *Access to Heritage* 1998:130). It concluded that fees for access to parks would militate against low income earners, often the very ones in need of the 'outdoor temples of the wonders of nature' and that for 'many people this spiritual dimension warrants free entry to national parks as surely as it does to churches' (132). At the same time, in view of serious concern over rising management costs on the Reef, the report noted that fees for the general public should be levied only when there is clear evidence that they would not 'significantly discourage visitation', and that collection of entry fees would assist managers in 'their core mission of conserving our heritage for public enjoyment and for future generations'. For commercial operators, however, 'higher charges are acceptable to the extent of a fair contribution towards maintaining the resource from which they draw profit' (133). Consequently, the daily fee was retained, but reduced to $4.

A further issue of contention came from the increasing number of proposals to erect various tourist structures on the Reef. A valuable lesson had been learned from the economically disastrous Four Seasons floating hotel venture on Townsville's Brewer Reef, and so the Cairns rezoning of 1991 had a new 'no structures' sub-zone. Its declared intention was 'a guarantee to the public that the whole of the Great Barrier Reef will not become saturated with structures and a high proportion of the Marine Park will remain in a pristine condition' (GBRMPA Annual Report 1990–91:2). That statement raises an obvious question: what could 'pristine' signify in that context? Its meaning, from the Latin *pristinus*, is 'in original condition, unsullied or changed by any outside influences'. The Reef was anything but that, and the decades of relentless European exploitation had altered its character to the point that, when Sir Maurice Yonge returned to the Low Isles in 1978 after fifty years, he was visibly distressed at the condition of the degraded cay compared with the ecological riches of the site when he led the team that conducted the first full biological study on the Reef.

Degradation of the Low Isles is a significant case study of the relentless impact of development. Situated less than 20 kilometres offshore from the burgeoning Reef tourism centre of Port Douglas and the Mossman River estuary complex where expanding cane fields and farms have replaced the rainforest and mangroves that in 1928 had released mainly clear, filtered water to the sea, the Low Isles have been exposed to increasing levels of siltation and agricultural runoff. Higher levels of nitrogen and phosphorus, used widely in crop fertilisers on coastal farmlands, have been identified as a threat to coral ecosystems. Once leached from the fields by rainfall into Reef waters from the extensive coastal riverine systems, they provide nutrients for microscopic single-celled algae that multiply rapidly, altering water quality and ecological balance – a phenomenon known as eutrophication. One unexpected consequence, it has been hypothesised, might be that higher levels of algae provide an abundance of food for Crown of Thorns starfish larvae, leading to their rapid increase. As pressure from many other quarters mounted, GBRMPA was forced to abandon what had been described as 'passive' constraint and move to a more rigorous policy of framing and enforcing coercive regulations.

THE REEF AS A MARITIME SUPERHIGHWAY

If the Reef could be described as a maritime highway in the early decades of the colony of Queensland, by the beginning of the twenty-first century it had become a superhighway as shipping, trawling and recreational boating contributed ever greater pressure, with increasing illegal bilge and ballast discharge and shipping accidents continuing to occur. From the days of George Heath in the nineteenth century to the present, ensuring safety within the Reef has been a constant preoccupation of government and all maritime users. A central role in this throughout the twentieth century has been exercised by the

Commonwealth Department of Navigation and the Royal Australian Navy. Operating quietly in the background, the contribution of the Navy, through its division known today as the Australian Hydrographic Service, created in 1921 as the RAN Survey Service, has been of immense value in providing reliable charts and hazard information in its periodic *Australian Notices to Mariners*.

As traffic increased rapidly in the late twentieth century, with the advent of huge oil tankers and container ships, compulsory pilotage of certain classes of vessels had to be considered. In November 1987 the International Maritime Organisation (IMO) declared Resolution A.619 which required all vessels greater than 100 metres in length, and all vessels with dangerous cargoes including oil, liquefied natural gas and chemicals, to carry a licensed pilot while transiting the Reef through Torres Strait and the Great North West Passage. In 1989 the IMO extended that Resolution to include the Inner Route and the newly charted Hydrographers Passage. In addition, to provide a link with the IMO, the Australian government in 1990 created the Australian Maritime Safety Authority (AMSA) which was charged with responsibility for safety, navigation, search and rescue, marine pollution and other services to the maritime industry. In part that was a response to the increasing number of ships registered by Third World countries as a source of income and with no concern to enforce stringent safety standards. By then nearly 50 per cent of the world's fleet, most of which were ageing, were registered under such 'flags of convenience' and were contributing to most of the shipping accidents, as forcefully brought to notice in the House of Representatives' report *Ships of Shame* (1992).

In November 1990, following IMO recognition of the Great Barrier Reef as a 'particularly sensitive area', in order to minimise the risk of oil spills on the Reef, the AMSA tightened regulations by requiring compulsory pilotage for all vessels of 70 metres or more in length and for all vessels, regardless of length, which carry oil, chemicals or liquefied gas, in two areas: from Cairns through the Inner Route to the northern boundary of the Great Barrier Reef Marine Park (11°41'N) and through Hydrographers Passage from Hay Point, near Mackay north-east to the outer edge of the Reef at White Tip Reef. In addition, pilotage was also recommended for all kinds of vessels passing through the Torres Strait (GBRMP Regs. 1991; No.296, Reg.26A, Pt4A).

On 1 July 1993 the Commonwealth acted even more decisively, removing control of Reef shipping from the Marine Board of Queensland, assuming responsibility for all coastal pilotage, and amending the *Great Barrier Reef Marine Park Act* to require compulsory pilotage for all loaded tankers carrying oil, chemicals and liquefied petroleum gas, regardless of length, throughout the entire Reef passage. And to strengthen maritime safety regulations even further, following well established practice in air traffic control, in 1996 the Reef Reporting System from 14 VHF sites was introduced, requiring all vessels greater in length than 50 metres, and all vessels regardless of length carrying dangerous cargoes, to radio their position periodically to the monitoring office, REEFCENTRE, located near Mackay at Hay Point, operating on a 24-hour basis under control of the Queensland

Department of Transport. To further monitor compliance with reporting requirements, radar surveillance began operating in Torres Strait, off Cairns and Mackay, and all Reef entry passages.

At all times navigating Reef waters requires constant vigilance. In the period between 1979 and 1991 43 accidents were reported: 19 collisions with wharves, beacons and other vessels; 24 groundings in harbours and on reefs (Ottesen 1994:21–22). Due to negligence one of the most serious groundings occurred on 2 November 2000 when the southbound Malaysian container ship *Bunga Teratai Satu*, having left its pilot in Cairns, just one hour later ran aground on Sudbury Reef. There it remained fast for ten days, in the process crushing the coral cover across an area of 1500 square metres and a radius up to 25 metres, and further damaging it from the anti-fouling paint on the hull which contains the powerful biocide Tributyltin (TBT) to resist marine organism corrosion and so can kill any Reef biota it contacts. In January 2001 when a six-week clean-up by divers commenced, a total recovery period of several decades was estimated. As a result of pressure by GBRMPA, the Malaysian shipping company agreed to pay $1 million in clean-up costs, and on 5 February 2001 following a court action lodged by the Queensland Environmental Protection Agency, it was fined $400 000, the largest such penalty in Australian history. Ever mindful of such environmental threats, on 17 August 2001 additional compulsory pilotage by Commonwealth licensed pilots was introduced in the sensitive Whitsunday Passage between the mainland and Whitsunday Island for all vessels over 70 metres in length, and all vessels with dangerous cargoes (*Australian Notices to Mariners*, 17 August 2001, 458: AMSA 01/6140).

Equally important was reform of the pilot service. All Reef pilots were required to be licensed by the Commonwealth, and as part of those 1993 reforms the Prices Surveillance Authority (PSA) conducted an Inquiry into Pilotage Services on the Great Barrier Reef and effected a significant change in breaking the long-standing monopoly of the Queensland Coast and Torres Strait Pilot Service which had been seen as something of a closed shop. After the dust had finally settled, two separate agencies emerged, operating today as the Queensland Coastal Pilots Services and Australian Reef Pilots. In addition, Hydropilots Pty Ltd operates a helicopter service to carry pilots navigating the large bulk carriers through Hydrographers Passage to and from Hay Point to the Coral Sea passage at White Tip Reef. Prices became lower, more negotiable, and the PSA reported that they seemed to be working well (PSA Report No. 50, 24 September 1993).

WORLD HERITAGE PROTECTION: A TWENTY-FIVE YEAR STRATEGY

The 1990s were to become years of radical revision. Developments in world attitudes to global degradation of the environment, resource depletion and unsustainability of present practices had together begun to exert a more profound influence on Reef management. A major new emphasis began to appear in management documents, with more frequent reference to protecting the Reef as a 'World Heritage Area'. Whereas its heritage standing had been quietly accepted throughout the 1980s as a symbol of pride, in the 1990s the Authority brought that status forward to public notice in many of its publications.

In its 1981 submission to UNESCO for World Heritage nomination, GBRMPA had declared that 'The Great Barrier Reef is by far the largest single collection of coral reefs in the world supporting the most diverse biological ecosystem known to man'. The nomination criteria also stated that it 'provides some of the most spectacular scenery on earth and is of exceptional natural beauty'; provides 'nesting grounds of world significance for the endangered . . . green turtle and loggerhead turtle, as well as major feeding grounds for large populations of the endangered species *Dugong dugon*'; and it meets the criterion 'of integrity in that it includes the areas of the sea adjacent to the Reef'. A decade after the listing of the Reef as one of only fifteen World Heritage sites to have met all four natural criteria, GBRMPA faced the need to maintain that status.

By the 1990s massive increases in pressure, and in some quarters increasing user restiveness over the growing number of constraints that had to be imposed, led to concern about retention of heritage status. It was well appreciated within the Authority that the health of the Reef had to be maintained. A major continuing concern was water quality; other looming threats were potential loss of fish species due to excessive commercial and recreational fishing and the phenomenal increase in tourism, particularly in the northern Cairns Section and the Whitsunday Group in the Central Section which together receive 65 per cent of all tourists on only 15 reefs. As part of a constant educational program to inform the public more widely, to stress their role in helping to conserve its quality, and to persuade the numerous user groups not only of the Reef's vulnerability, but of its international significance, GBRMPA began to publicise increasingly its World Heritage Area status.

In a determined attempt, then, to become proactive in dealing with increasing management problems, and with genuine concern for the ecological stability and survival of the Reef, the Authority began developing a Twenty-Five Year Strategic Plan in August 1991. The central tasks were to implement the original purpose of the *Great Barrier Reef Marine Park Act* of 1975, and to comply with the World Heritage Convention, which enjoined Australia to ensure the protection, conservation and presentation of the Reef area and its transmission to future generations. Since 1975, however, Reef management had entered a context of vastly changed circumstances, created by rapid population growth along the coastline, increased agricultural output with greater use of fertilisers and biocides, greater shipping density and commercial trawling, and the exponential explosion of tourism and recreational fishing. In addition, over the years UNESCO had begun to tighten criteria for heritage listing and in 1992 published its revisions as *Operational Guidelines* which stressed the 'duty of care' aspect and required listing nations to exercise that responsibility by continuing 'to identify, protect, conserve, present and transmit to future generations all of their properties that are of outstanding universal value, regardless of their inscription on the World Heritage list' (see Lucas 1997:17f.).

Given the large number of separate interest groups that had become either financially dependent on the Reef or involved in tourism or recreational fishing, along with canegrowers, developers, conservationists, indigenous communities, and scientists for whom it is their field laboratory, it was considered essential that a strategic planning process to ensure preservation of World Heritage status had to begin with wide community consultation. And in that context GBRMPA began to refer to Reef users as 'stakeholders', with the goal of encouraging a sense of responsibility for cooperative, responsible management. While GBRMPA initiated the planning process and acted as project manager, the Authority decided that to embark upon what was foreseen as a highly contentious task, it would be best if the task of coordinating submissions and discussions from such a wide range of user groups were undertaken by an outside, independent body. It therefore commissioned an independent firm of consultants to act as 'facilitators'. For two years development of the Twenty-Five Year Strategic Plan continued, beginning with two separate workshops for government and non-government organisations, followed by formation of a planning team drawn from both groups to outline the first draft. Altogether some sixty separate organisations were involved, including the Aboriginal Co-ordinating Council (ACC) which presented a decidedly adversarial contribution.

The vision for the future, when ready for release, was presented in two separate sections: an immediate five year plan, and a long term plan. The function of the five year plan was to set relatively achievable goals within the longer 25 year context; every five years the situation would be reviewed in the light of experience, and if necessary the Twenty-Five Year Strategic Plan revised. Why a 25 year plan? Because it is 'for the next generation. They are the beneficiaries of today's long-term planning. Making decisions today that allow the next generation to benefit from and be responsible for the Great Barrier Reef World Heritage Area is a principle of *ecologically sustainable use*, which is a cornerstone of this plan' (GBRMPA 1994:6). From May to July 1992 the first draft plan was released for public comment; revisions and refinements followed and were brought together at the final planning team meeting on 1 June 1993.

CULTURAL HERITAGE: RECOGNITION OF INDIGENOUS RIGHTS

Meanwhile, a rising concern for recognition of indigenous heritage was claiming a rightful place in Reef management and participation was actively sought from Aboriginal communities. Although not advanced as a criterion for World Heritage listing, it had been recognised explicitly that the Great Barrier Reef 'contains many middens and other archaeological sites of Aboriginal or Torres Strait Islander origin' (GBRMPA 1981:5–6).

The ACC contribution to the GBRMPA Strategic Plan, however, began on a highly critical note, commenting that the first draft was 'not good enough' and that Aborigines wanted much more from the plan than simply being consulted. They required participation in decision making and freedom 'to pursue their own lifestyle and culture, and exercise responsibility for issues, areas of land and sea, and resources relevant to their heritage within the bounds of ecologically sustainable use and consistent with our obligations under the World Heritage Convention' (ACC *Draft Submission* 1992:1).

Further, they sought not just 'traditional use' but also economic use, that is, a share of income from Reef resources. The ACC appended to its Draft a former submission to the Resource Assessment Commission that argued 'prior ownership of the whole coastline and its resources should be recognised and compensation should be paid for alienated coastline and loss of resources'. In particular, it complained that Australian law does not recognise 'customary marine tenure', already widely accepted overseas by maritime nations in respect to their indigenous communities, which Aborigines also believe to be their right. The ACC submission sought, in the Far Northern Zone of the Reef, to have some areas adjacent to Aboriginal communities declared such that 'these seas are closed except for Aboriginal use' and that 'traditional' should not be defined in terms of hunting from canoes with harpoons, but extended to include powered boats and the 'incorporation of new materials', which were not identified. Its demand, to become the subject of much further litigation, was for 'traditional' to be redefined in terms of purpose rather than method for the use of community members (11–13).

Responding to indigenous demands, and changes in Australian law, a significant feature of management approach was incorporated into the Strategic Plan as a result of a 1992 High Court judgment concerning indigenous title to Reef islands. Ten years earlier, in 1982, Eddie Mabo and four other Torres Strait Islanders from the Murray Island group outside the formal northern boundary of the Park – the site of Mayor's historic coral studies in 1913 – commenced a legal challenge to claim traditional ownership of their island. Their suit was based upon evidence that their people had been in continuous occupation and cultivation of the land long before British occupation in the nineteenth century and that European settlement, therefore, had not extinguished their rights. In 1985, three years after Mabo lodged his claim, the Bjelke-Petersen conservative Queensland government attempted to defeat his suit by passing the *Queensland Coast Islands Declaratory Act* which asserted that the islands had always been part of Queensland. In 1992 in a six to one judgment the High Court declared that the Queensland Act was contrary to the overriding Commonwealth *Racial Discrimination Act* of 1975, that native title had not been extinguished and the doctrine of *terra nullius* – the dubious assertion made by British colonialists that prior to European settlement that the continent consisted of 'vacant land' – did not apply (Mabo v Queensland (1992) 66 ALR 408). Mabo and his descendants held valid title (see Brennan 1993:24f.).

From that point forward the Marine Park Authority incorporated the *Mabo* judgment into its planning, accepting much of the subsequent ACC submissions. The *Corporate*

Plan 1994–1999 stated that an explicit aim of management policy would be 'To provide recognition of Aboriginal and Torres Strait Islanders' traditional affiliations and rights in management of the Marine Park'. Even further, it expanded that recognition by asserting that indigenous persons were to be able 'to pursue their own lifestyles and cultures, and have responsibility for areas and resources relevant to their heritage, within the bounds of ecologically sustainable development'. To that end the Marine Park Authority (known as 'The Board') of three was enlarged, and Dr Evelyn Scott, of indigenous descent, was appointed a fourth member, specifically 'to represent the interests of Aboriginal communities'. A further positive consequence was that indigenous communities were to be drawn more responsibly into the management regime. That new situation was reflected in the rezoning of the Far Northern Section of the Park adjoining Cape York which included significant areas of Aboriginal participation. It was a fitting introduction to the United Nations declaration of 1993 as the International Year for the World's Indigenous Peoples. In the same year the Commonwealth passed the *Native Title Act* of 1993 expressly confirming indigenous rights to certain areas.

Emboldened by the judgment in *Mabo* and the *Native Title Act*, one Aboriginal community in the Northern Territory in 1996 sought to extend the range of native title rights to cover exclusive fishing and pearling rights to 3300 square kilometres of waters surrounding Croker Island in the Arafura Sea, near the Coburg Peninsula. With Mary Yarmirr, a traditional elder as nominal plaintiff, five clans launched an action in the Federal Court under the *Native Title Act* of 1993. Their claim, however, was rejected by Justice Olney who, while confirming both their traditional rights to access to these waters for historic fishing and hunting for personal and domestic needs, and their freedom to observe traditional, cultural, ritual and spiritual activities, ruled that in law their sea rights were restricted to non-commercial activities. Specifically rejected was any claim to 'possession, occupation, use and enjoyment of the sea and sea-bed within the claimed area to the exclusion of all others' (Action DG 6001 of 1996).

An appeal to the full bench of the Federal Court failed to reverse Justice Olney's ruling, declaring that native title has to be 'consistent with natural justice, equity and good conscience' and cannot be separate from, nor beyond the power of, the paramount common law (Action DG 6006 of 1998). A further appeal to the High Court, which delivered its judgment on 11 October 2001, upheld the decision of the Federal Court ([2001] HCA 56). While indigenous communities certainly have the right to unhindered traditional activities in the waters surrounding their lands, by a majority of six to one the full bench of the High Court determined that native title did not confer exclusive rights in Australian waters. Not to be deterred, on 27 November 2001, a claim was lodged by some Torres Strait Islanders for exclusive native title rights to their surrounding waters, outside the World Heritage Area, a claim that may take years to resolve.

GREAT BARRIER REEF MARINE PARK AUTHORITY MANAGEMENT REFORM

When the Twenty-Five Year Strategic Plan – the world's first intensive environmental planning program for a marine park on such an immense scale – was released to the public in 1994 it received extensive acclaim, both in Australia and abroad for its comprehensive and innovative approach to large scale environmental management. Entitled *Keeping it Great: The Great Barrier Reef – A 25 Year Strategic Plan for the Great Barrier Reef Heritage Area 1994–2019*, the 64 page booklet was distributed widely and freely. Its content was organised into three major sections: an Introduction setting out the motivation for the strategy in the context of preserving the World Heritage status of the Reef through close community consultation and enthusiastic public acceptance; the main central section on objectives and strategies which gave in greater detail the program to be implemented; and a final section of appendixes. The vision for the World Heritage Area was set out in six statements: a healthy environment, sustainable multiple use, maintenance and enhancement of values, integrated management, knowledge-based but cautious decision making in the absence of information, and an informed, involved, community. What possible objection could be made to those superbly normative statements? If difficulties were to arise, they would not come from public rejection of those values, but from their lack of internalisation of these in mindset and observance of them in practice, and that was where the real challenge to management came in the strategy.

In the accompanying *Corporate Plan* the four management objectives for staff of the Authority were set out as Maintaining the Ecology, Providing for Reasonable Use, Seeking the Commitment of Others, and Developing an Efficient and Effective Organisation. Therein lay the challenge. A degree of discontent had arisen as issues demanding resolution had accumulated, and the last objective demanded a careful approach to ensure that GBRMPA staff could move forward as a well integrated, committed team imbued with zeal to discharge their mission with a high level of morale. One outstanding characteristic of the organisation has been its educational commitment to informing and securing cooperation of the community: its publications are of the highest quality, and issued on an incredibly wide range of relevant topics with commendable efficiency and frequency.

The real threat to the effectiveness of the Twenty-Five Year Strategic Plan will come from the near impossible task of providing for reasonable multiple use integral with maintaining the ecosystem. Applications for tourist permits, at the time of the release of the Strategic Plan, had doubled and the Reef had become overcrowded and overused in many places, particularly on reefs within day trip distance from the main towns. Despite the best possible gloss presented in tourist brochures, many reefs remained degraded, not only from starfish devastation and brown algal bloom, but from sheer tourist impact and the destructive activities of which GBRMPA was only too well aware and was valiantly attempting to control: damage from anchoring, snorkellers' fins and reef trampling, illegal fishing, souvenir collecting, waste discharge. Transit ships continued the illegal process of emptying ballast oil and other wastes, and enormous destruction of marine life occurred both through overfishing and trawling, which scours the seabed. In the process, huge numbers of unwanted, but often endangered species are caught in the nets – including luckless turtles – that are hauled on board and then discarded over the side, under the euphemism of 'bycatch' after the desired species have been sorted.

Sea turtles have always been exceptionally vulnerable to capture. The threat of biological extinction – or at least economic extinction – led to the world's first turtle conservation legislation when in 1620 the Bermuda Assembly passed *An Act Agaynst the Killinge of Ouer Young Tortoyses* to prevent those profligate fishers who

> at all tymes as they can meete with them, snatch & catch up indifferentlye all kinds of Tortoyses both yonge and old little and greate and soe kill carrye away and devoure them to the much decay of the breed of so excellente a fishe the daylye skarring of them from our shores and the danger of an utter distroyinge and losse of them (cited Carr 1967:I).

In the Reef, two centuries later, that kind of thoughtless predation continued, despite the warnings of Anthony Musgrave and Gilbert Whitley of the Australian Museum in the 1920s, Frank Moorhouse in the 1930s, and Frank McNeill, also of the Australian Museum, in the 1950s. The turtle, mainly the green and hawksbill species, only escaped extinction when the development of plastic imitation tortoiseshell in the early twentieth century led to a collapse in the trade. By 1934 hawksbill shell had no economic market value.

In Japan, however, it was always sought, and following recovery from the Pacific War of 1941–45, its market interest became revived and intensified, and by 1969 Archie Carr, the world authority at the time, asserted that the pressure 'imposed by the growing demand for tortoiseshell . . . makes *Eretmochelys* [hawksbill] seem one of the most clearly endangered genera of reptile in the entire world' (Carr 1969:74–5). It was estimated that between 1970 and 1990 Japan alone had captured and processed more than 2 200 000 individual turtles. An indifference to conservation of endangered turtles, as to other threatened marine species such as whales and numerous kinds of fish, is a disturbing feature of Japanese willingness to buy hawksbill turtles for the 'tortoiseshell' trade, known as bekko, for use in the very highest grade of finished products for which there is no known plastic match (Broderick et al. 1994:123).

The cause of turtle conservation, within the Reef and worldwide, received an invigorating stimulus in 1976 when Col Limpus, a scientist with the Queensland Parks and Wildlife Service (QPWS) received a grant for a three year study of turtles. From then to the present day he has campaigned for their conservation, having become a critical opponent of the Japanese bekko trade that 'threatens survival of the species'. Although hawksbills are widely distributed throughout tropical waters, he pointed out that even today little is known of the their population biology (Limpus 1992:489).

Within Reef waters Dr Limpus has been studying turtles for nearly three decades. Nesting sites have been identified, carefully monitored, and where feasible, protected from predators – humans and feral animals – with the aid of QPWS rangers and volunteers, and publicised widely to enlist community support. Tagging turtles on the front flipper with coded identification has provided data on their extraordinary transoceanic migrations over a period of years and enabled him to create a conservation oriented data base that has earned international recognition. While turtles are still ravaged throughout South East Asian waters, the cruel and ignorant practices of exploitation that characterised resource extraction on the Great Barrier Reef in the 1920s and 1930s have been replaced by a belated recognition of the incredibly complex life cycles of these harmless and ecologically essential reptiles and the need for them to be conserved within the economy of nature. A recognised world authority on turtle biology and ecology, with strong support from the QPWS, Limpus has travelled to a large number of nations in the Indo-Pacific region advising on conservation measures.

One effort to minimise loss of the seriously endangered green and hawksbill turtles, urged by Limpus and recommended in a report issued in 1995 by CSIRO and the Queensland Department of Primary Industry from a comparative study of fish stocks in closed and open zones, was the introduction of a bycatch excluder device that had been developed in the United States. At first trawlers were encouraged to use nets with escape pockets in the mesh called 'excluder devices'. Both kinds, the turtle excluder device (TED) and the bycatch reduction device (BRD) are now mandatory on the Reef and represent yet a further effort to sustain the ecological biodiversity.

A major change in policy approach occurred in December 1994 when after fifteen years as chairman, Graeme Kelleher retired and was replaced by Dr Ian McPhail a month later, whose position was now designated chairperson. Whereas the formative decades of GBRMPA had been determined by engineer Kelleher, as chairman of the Authority, and Queensland public service administrator Schubert (also an engineer), its direction was now entrusted to a scientific professional whose expertise was in political geography and who had been previously director general of Environment and Planning in South Australia and before that director of the Commonwealth Environment Protection Agency.

Since periodic reporting on World Heritage status was a UNESCO requirement, McPhail decided soon after his appointment to commission an independent review of the Reef's condition in complying with the criteria for continued World Heritage Listing, and to organise a major workshop on Reef management that would help prepare for a comprehensive in-house review of the World Heritage condition of the Reef for 1997. Every aspect of the Reef was to be examined and reported on within two main categories of Environment and Management. In the year following the change of chairperson the Authority, in conjunction with the CRC Reef Research Centre, convened a National

Conference for November 1996 at James Cook University under the title 'The Great Barrier Reef: Science, Use and Management'. In his opening address McPhail set the theme by quoting Canadian environmentalist David Suzuki to the effect that 'humans do not manage natural systems, they can only attempt to manage the activities and impacts of humans in and on that system'.

The statistics that emerged served again to emphasise the pressures being imposed on the Reef, and to bring to public notice the warnings of Dr Harry Ladd in his report to Mines Minister Camm back in 1968 and repeated by Goreau a decade later: that even more potentially damaging than oil drilling would be the 'peaceful exploitation of the reef' by activities such as tourism, collecting corals and shells, and unrestricted fishing (Ladd 1968:42). Tourism revenues, it was reported, were approaching $1 billion per annum; commercial fishing $250 million; shipping had increased to 2000 large vessels transiting annually, of which 10 per cent were oil tankers; canegrowing on the coastal strip yielded Australia's second largest export crop. The necessary conclusion, McPhail stated, was that while management of the Reef was of a very high and rigorous standard, none the less it 'should be subjected to standards that are [also] appropriately stringent' (McPhail 1996:5).

By that time, however, the Authority was moving into an increasingly difficult management situation, chiefly as a result of the now well entrenched concept of 'multiple use', which, although expressed in the 1975 Act as 'reasonable use', had become stretched well beyond its original intention. Could it possibly have been imagined in 1975 that throughout the 1990s tourists would come in millions, generating billions of dollars in income? Or that ever larger and more efficient trawl nets would deplete fish stocks with more ruthless efficiency, encouraged by Commonwealth pressure for more effective technology to extract 'oceans of wealth'; or that the world demand for sugar would increase the number of cane farms and the use of greater quantities of fertiliser and biocides that inevitably find their way into Reef waters? And as the Authority had allowed the rapid growth of the tourist industry, could any kind of effective controls now be applied? No matter what rhetorical defence is offered, it remains incontrovertible that uncontrolled multiple use – at the current increasing pace of activity – is fundamentally incompatible with maintaining a healthy Reef ecosystem, and the policy of balancing conflicting user interests as a primary goal will inevitably lead to degrading of the Reef's biological communities which, after all, are central to the Reef's continued existence as a World Heritage Area. Under McPhail's leadership a major phase change in management philosophy occurred: whereas in the first twenty years the primary focus had been on ensuring a comfortable accommodation of demands for multiple use, the new emphasis was to place environment, conservation and heritage as primary responsibilities.

A WAVE OF CONCERN: MAINTAINING THE HERITAGE VALUE OF THE REEF?

In the same year, 1996, the review of the current heritage status of the Reef, commissioned by GBRMPA and funded by the Marine Park Authority, the Queensland Department of the Environment and the World Heritage Unit of the Commonwealth Government, was submitted by consultant Percy (Bing) Lucas, assisted by a team of three scientists from James Cook University, under the title 'The Outstanding Universal Value of the Great Barrier Reef Heritage Area'. The Authority decided to seek public comment and to consider recommendations regarding changes in management style expressed in the Report, some of which pointed out that future impacts on Reef values had to be taken into account in forward planning. While the Lucas Report concluded that listing as a World Heritage Area continued to be justified, it recommended that the Authority take more notice of the fundamental listing criteria in its planning and decision making processes, particularly in respect to the increasing pressure on resource use, and to be more cognisant of the aesthetic and natural beauty values of the Reef.

The Report was, in effect, a warning, particularly since UNESCO now publishes a 'World Heritage List in Danger', even though, as the Lucas Report indicated, it is not meant as a 'black list' but as a cautionary indicator. So far, none of Australia's fourteen World Heritage properties are on that list, although Kakadu's standing has been questioned due to uranium mining within an excluded enclave, and the United States bears the opprobrium of the listing of both Yellowstone and Everglades National Parks as endangered World Heritage Areas. Executive Officer McPhail made that issue clear when he reported that 'a clearly articulated policy on World Heritage attributes is an important prerequisite to more extensive coordination with Queensland and local governments on the planning and management of coastal development adjacent to the World Heritage Area' (GBRMPA Annual Report 1997–98:37).

After twenty years as the world's most ambitious experiment in environmental management of such an extensive area, amounting to almost 350 000 square kilometres and consituting more than 97 per cent of all Australia's protected areas and around 3.5 per cent of its Exclusive Economic Zone (Cresswell & Thomas 1997:iii), the time had come for a public accounting. In 1998 the Authority presented a report on the condition of the Reef which appeared in a lavish, full colour illustrated book *State of the Great Barrier Reef World Heritage Area 1998*. Planning for the report had been commenced at a World Heritage Area workshop in 1995, followed by the conference on Science, Use and Management of 1996 which supplied most of the material content, along with some that had appeared in the Lucas Report. The final production, originally anticipated for release in 1997, was organised into two sections dealing separately with Environmental Status and Management Status.

In assessing environmental status, twelve areas were surveyed: water quality, mangroves, island plants, seagrasses, macroalgae, corals, Crown of Thorns starfish, fishes, birds, reptiles, marine mammals, inter-reefal and lagoonal benthos. Each section of several pages described the state of the category, pressures being currently exerted and planned responses. In the management section eight key areas were identified (management proper, fisheries, tourism, threatened species, indigenous issues, water quality and coastal development, shipping and oil spills, and monitoring) each being reported in terms of an overview, followed by current status and a summary.

The focus of the report was to identify trends in Reef use for the purposes of planning and management, to which end data were being increasingly assembled, particularly from surveys of all twelve environmental categories, past and present. Overall it is an impressive work, meticulously prepared and presented, representing the first synthesis of information on the state of human pressures on the Reef environment and management responses. Even so, the disturbing conclusion drawn from those surveys was the inadequacy of the data and the verdict that, 'As a result, for most environmental categories, it is not possible to say with certainty if they are in a satisfactory or unsatisfactory condition' (Wachenfeld et al. 1998:1). In his foreword McPhail contrasted World Heritage status as having changed from its original 1981 designation as 'a prize or badge of honour' into 'an international obligation' which has made management an increasingly onerous task given the massive accumulation of stresses and difficulties. In identifying weaknesses the 1998 report had a positive function. At least, he wrote, it will 'provide a guide to where we should be going in the future in order to ensure the Great Barrier Reef World Heritage Area keeps its status as the premier World Heritage Area'.

CHAPTER 23

THE REEF AS HERITAGE: A CHALLENGE FOR THE FUTURE

At the beginning of the twenty-first century renewed emphasis on the Reef as a World Heritage Area has come to dominate planning and management, with a heightened recognition of the social and economic implications of World Heritage listing and the interrelationships between natural and cultural heritage. Underlying this is an increasing awareness, formed during the final decades of the twentieth century, that the Reef and its hinterland must be regarded as a dynamic ecosystem: dealing with environmental issues as localised problems on an individual basis is no longer an adequate basis for policy. Faced with continued population growth, mounting tourist traffic and rapid expansion of urbanisation and commercial infrastructure, forward looking planners recognise that the entire area – from the watershed of the Great Divide, through the coastal strip and across the waters of what is called the 'lagoon' to the edge of the continental shelf – must be understood and managed with a recognition of the high levels of connectivity within what is in effect a single ecosystem, if optimum environmental balance and sustainable resource yields are to be maintained. Activities in the rainforests, on coastal farmlands, in urban developments, in factories and industrial complexes on the foreshores of harbours and bays all exercise a direct, mutually reinforcing impact – literally, from 'Divide to Drop-off'.

Earlier problems such as the aborted Magnetic Island marina development and the floating hotel are now understood, not simply as minor isolated incidents, but as indicators of a wider pattern of irreparable damage if allowed to multiply throughout the 2300 kilometre length and 350 000 square kilometres of the marine park. All environmental matters of profound concern to management – maintaining water quality and biodiversity, ensuring sustainable fisheries, minimising tourist impact and shipping accidents – must be appreciated as interrelated factors influencing the future of an already unstable ecosystem. We are now witnessing the first stages of a concerted effort to understand the ecological complexity of the Reef on an unprecedented scale, and to develop and implement sound management policies. Moreover, with evidence mounting on climate change and the potential impact of global warming on coral reefs worldwide, the reality is now being faced that the whole Reef system is subject to global atmospheric and oceanic conditions: it cannot be regarded as a separate entity, a reality not always recognised in the past.

MANAGEMENT REFORM AND THE 'DUGONG WARS'

After more than a decade of Reef management under Labor governments since 1983, a new Coalition government in Canberra in late 1995 launched into a rapid program of expanding policies of economic rationalism, and in line with conservative ideology, promoting economic growth, maximising private ownership and corporate wealth, while constraining expansion of the public service and publicly owned government enterprises. The new administration soon moved to put its own Reef policies in place and an early project was an inquiry into the cost effectiveness of GBRMPA in discharging its responsibilities, initiated by Minister for the Environment Robert Hill and commissioned to solicitor John Whitehouse, who had conducted the earlier Magnetic Island marina inquiry.

The Authority by this time had become a very complex organisation dealing with permits and licences, an extensive range of legal services and responses to problems, along with planning, research, monitoring, and education. Although his findings were generally favourable Whitehouse criticised procedures for issuing permits which he believed were bureaucratically complex, leading to an unacceptably large backlog of applications. The main problem, however, was the huge volume of applications and the extensive range of activities that needed approval: tourism, moorings, waste discharge, research, conservation and traditional hunting, for which a total of 854 permits had been issued in 1995–96 for 1.5 million visitor days. Astonishingly, future projections were for 10 million visitor days per annum. Recommendations were made in the report for conducting more effective research, especially commissioned research, along with improvements at the senior executive service level, including a restructuring of senior management to handle more efficiently the increasing complexity of tasks (Whitehouse 1997).

In recognition of the need for important policy reforms that had not been implemented in previous years, and for more holistic forward planning, following several external reviews and an intensive internal assessment, the executive of the Authority reorganised GBRMPA into four 'Critical Issues Groups' on 1 July 1998. The position of chair of the Authority had already been combined with the former position of executive officer of GBRMPA and was now redesignated chief executive officer with the main office in Townsville – not Canberra as formerly. To share the administrative load two executive director positions were created, each responsible for two of the Critical Issues Groups: one taking Tourism and Recreation, along with Conservation, Biodiversity and World Heritage; the other dealing with Fisheries, and Water Quality and Coastal Development. The numerous small in-house research projects were wound down as financially inefficient and GBRMPA moved towards the role of 'research broker' using outside commissioned investigations as needs arose. In addition, there were eight subsidiary groups dealing with other aspects of marine park activities. Although World Heritage is located with Conservation and Biodiversity, it remains the overarching management concept: all GBRMPA policy is predicated on the effective maintenance of what is the largest World Heritage property listed by UNESCO, amounting in square kilometres to almost a third of all its listed areas. World Heritage has become, figuratively, the prime meridian from which all Reef management and planning decisions and actions must now proceed.

A crucial management issue in preserving World Heritage status today is a problem that directly affects marine biodiversity: preventing continued deterioration of water quality from commercial and recreational use, as well as land based discharges. By the 1990s polluted waters within and around Australia were becoming a major national issue, and the rapidly increasing contamination of the marine environment was given full recognition in 1992 when the Commonwealth Resource Assessment Commission appointed Mr Justice Stewart to inquire into the health and future prospects of the coastal zone in respect to current practices. Defining the coastal zone as those regions within local government areas with a boundary 'abutting the coast', a zone that forms an outlet for the 'natural drainage basins' of the continent, the investigation revealed that such areas constitute 17 per cent of Australia's land area, measuring approximately 1.318 million square kilometres, and containing 86 per cent of the total population. Moreover, the area defined as coastal zone experienced 50 per cent of all population growth since 1972. Australians, in fact, are migrating to the coast in greater numbers. Queensland's coastal zone has attracted more than 2.6 million people – 85 per cent of the State's population – while a quarter of its people live in the tropical regions alone. The Reef was singled out for particular mention as the 'largest area of coral reefs in the world', requiring significantly greater efforts to raise awareness within local government authorities, especially in regard to urban development and discharges into the sea (Resource Assessment Commission 1993:7). In order to protect biodiversity, on World Environment Day, 5 June 1992, Australia signed the International Convention on Biological Diversity, a commitment that required the establishment of a National Reserve System, including a National Representative System of Marine Protected Areas.

Subsequently, the Senate moved to examine problems of marine pollution in Australia and on 26 June 1995 instructed its Committee on Environment, Recreation,

Communications and the Arts to plan a comprehensive inquiry into the commercial and recreational uses of the coastal zone, with specific reference to land-based discharge of biological nutrients and their effect on water quality and marine biodiversity. In particular, the Committee was briefed to report upon 'the adequacy of existing Commonwealth, State and Territory legislation to give effect to Australia's obligations under the United Nations Convention on the Law of the Sea, and other international treaties, to address land-based and ship-sourced marine pollution and its effects', and to examine ways of improving management policy and maximising local community involvement (Senate Parliamentary Paper 1997:v).

Commencing the inquiry in mid 1996, the Committee published its Report in 1997, again identifying the Reef as a matchless biome that is now an 'Area at Special Risk', due chiefly to the immense volume of contaminating discharges entering its lagoon, mainly from unrestrained land clearing and farming practices that increase soil erosion and sediment transport into river systems and Reef waters. From the beginning of colonial land settlement in 1860 it was estimated that soil loss has increased fourfold, and that by the end of the twentieth century more than 50 per cent of Queensland had been stripped of its original vegetation for agricultural purposes (Haynes & Michalek-Wagner 2000:428). Drawing evidence from a 1992 GBRMPA document, the Senate Report cited data from a study which calculated that 15 million tonnes of sediment, 77 000 tonnes of nitrogen and 11 000 tonnes of phosphorus entered Queensland waters in 1990. It also brought into focus the disturbing discovery that it had taken only 40 years, since Queensland began its postwar campaign to 'develop the north', to reach volumes of that magnitude, and noted that those discharges came almost entirely from grazing (80 per cent) and canegrowing (15 per cent) (Senate Parliamentary Paper 1997:50). Queensland landholders are continuing to bulldoze woodlands and forests faster than all other States and territories combined. In 1999 Queensland accounted for 90 per cent of all land clearing in Australia with the stripping of 400 000 hectares, and satellite imagery now reveals huge, muddy plumes of silt reaching the mid-reefs of the Great Barrier Lagoon (Talbot 2001:336).

The Senate Report yet again drew attention to the fact that marine pollution was increasing rapidly and destructively, mainly from the organochlorine class of highly stable artificial chemicals, including the notorious DDT and Agent Orange, characterised by persistence with a very long half-life survival, rapid mobility and high biological activity. Nevertheless little was still being done to minimise both discharges and impacts. Even worse, there had been minimal effort in monitoring water pollution, and any studies that had been done were all limited to specific sites and could not be extrapolated into a comprehensive analysis of the scale of pollution. Since the commencement of data collection on polluting discharges in 1990 the situation has continued to worsen, as revealed in studies commenced in early 1997 on biocide discharges from agriculture.

One inevitable consequence of discharges and eutrophication is the impact on seagrass meadows. In Reef waters the effect on dugongs, a particularly vulnerable mammal, has been devastating. A shy, harmless herbivore that grazes on those seagrasses, and reproduces only a single calf at minimum intervals of three years at best, the dugong has shown an appalling decline. Data from continuing survey studies by Professor Helene Marsh, Australia's leading expert in that field, revealed that in the region of the Marine Park from Cooktown south to Hervey Bay, numbers had fallen by some 50 per cent from around 3500 in 1986–87 to approximately 1700 in 1994 (Marsh 1996:1). Her conclusions gave a dismal account of human activities in the Reef: polluting discharges have poisoned large areas of the precious meadows which she described evocatively as 'an underwater Serengeti . . . by far the most important dugong habitat in the world'. Tragically, traditional hunting, gill netting by commercial fishers, shark netting for swimmers' protection and defence training are all responsible for situations in which hapless dugongs are caught and drowned. Marsh's incrimination of gill netting was violently opposed by the Queensland Commercial Fishermen's Association and led to a particularly bitter dispute dubbed by some of the GBRMPA staff as the 'dugong wars'. Protection of dugongs became a major concern of the Ministerial Council of GBRMPA and after careful assessment of seagrass areas, particularly in Princess Charlotte Bay, the Starke River area north of Cooktown, the Hinchinbrook Channel and Shoalwater Bay, in December 1997 a series of 16 Dugong Protection Areas were declared, accompanied by a vigorous public education and continuing research program. Since dugong numbers in Shoalwater Bay are close to collapse, a voluntary moratorium was negotiated there with indigenous hunters.

THE OYSTER POINT CONTROVERSY

In the same period, concern over both Reef water quality and threats to the endangered dugong was heightened in a serious controversy that led to a lengthy Senate inquiry from May 1998 to September 1999. In 1996 Senator Hill had approved the resumption of a huge resort development at Oyster Point, near Cardwell, despite strong environmental protests and evidence of potential threats to water quality and dugong grazing grounds in the adjacent Hinchinbrook Channel, a highly protected wetland area. When Australia signed the United Nations Convention on Wetlands of International Importance in 1991, first formulated at Ramsar in Iran on 2 February 1971, among the 78 Australian sites listed three are in Reef waters at Shoalwater Bay, Bowling Green Bay and Corio Bay. These contain extensive meadows with more than 30 species of seagrasses, some of the largest in the world, which are not only essential grazing grounds for green turtles and dugongs, but also important nurseries for a wide range of marine biota, prawns and fish. Australia therefore assumed an international treaty obligation to protect and conserve those Ramsar Wetlands (as they are known) although by then a considerable area of Reef seagrasses had already been killed by land-based discharges, chiefly organochlorine biocides such as diuron, used in the canegrowing industry.

The Oyster Point resort development, which had never received a comprehensive preliminary environmental impact assessment, had been commenced in 1988 by the Tekin development company which cleared the mainland site, adjoining the Hinchinbrook Channel. In its preparations numerous breaches occurred, including removal of mangroves without approval, none of which resulted in prosecutions. Then the company failed, went into liquidation and the site was left in a seriously degraded condition. It was next bought by a new developer, Cardwell Properties, which lodged a new application in 1994 for a 'comprehensive, integrated resort' and resumed construction work (Senate *Report* 1999:10). By that stage conservationists became seriously alarmed and commenced agitation as the 'Friends of Hinchinbrook' to have the development stopped. The issue was further clouded by conflicting scientific advice on environmental impacts and the desire of the Queensland government to see the project continue as part of its drive for further development. In its comprehensive inquiry into the controversy, the Senate Committee reported that one of the most unsatisfactory aspects of the entire controversy was the inadequacy of environmental review procedures, demonstrated in 'the lack of a thorough up-front environmental assessment *as an input to a decision on whether to grant approval*, and the secrecy which surrounded the [Queensland] government's deliberations at the time [that] have been the major causes of subsequent objections by environmental groups' (15; Senate's emphasis).

When the project recommenced a public outcry ensued, despite assurances by the developer that all environmental safeguards were in place, and that approval had been given by the Queensland Co-ordinator General and Cardwell Shire Council in consultation with GBRMPA, resulting in the signing of a Deed of Agreement in September 1994. Protests continued to mount, warning of possible leaching from acid sulphate soils into the sensitive seagrass waters. When subterranean soil with a high iron sulphide content is disturbed and comes into contact with air, the sulphur combines readily with atmospheric oxygen creating sulphuric acid which, when leached by rainfall and tidal movements adjacent to Reef waters makes them highly acidic. In turn, the lowered pH level (indicating a higher acid content) in the waters enables bonding with dangerous heavy metals – chiefly mercury, arsenic, cadmium, copper and zinc – discharged from other land-based activities. These enter the food chain and, acting synergistically with organochlorine herbicides, lead to poisoning of seagrasses and loss of habitat for both dugongs and a wide range of other ecologically sensitive marine biota (Haynes & Michalek-Wagner 2000:429).

Throughout the early stages GBRMPA had refrained from exercising any direct intervention since Oyster Point and the Hinchinbrook Channel are entirely within Queensland State waters. The Oyster Point development, however, is directly adjacent to the Great Barrier Reef World Heritage Area, and since it could present a threat to World Heritage values the Minister for the Environment Senator John Faulkner in October 1994 sought a voluntary moratorium on development by the company until further environmental impact assessment had been undertaken. When the developer objected, the minister, acting under the Commonwealth *World Heritage Properties Conservation Act 1983*, halted all work to seek further assessment. The project remained in limbo until August 1996 when, following change to a conservative federal government in late 1995, the new Minister for the Environment Robert Hill approved continuation of the development, provided that certain safeguards were observed. His approach was a clear demonstration of 'wait-and-see'.

Even so, the matter has never rested. The Senate committee of inquiry, composed of members of all the major political parties, investigated the problem in depth and issued a comprehensive report in 1999 which made it abundantly clear that from the beginning the project was carelessly planned, poorly executed and inadequately supervised by the developer as well as the State and federal governments. Duty of care for the marine environment, it commented, was secondary to the quest for profits and the Queensland government's desire for further development of the Reef coast. Compounding the problem was the change of intention by the developer: no longer was it to be an integrated resort, but restructured as a gigantic real estate housing development that would exacerbate even further environmental stresses on the adjacent World Heritage Area.

In its concluding chapter the committee found that from the outset 'Unsatisfactory environmental assessment procedures had been followed at Port Hinchinbrook'; that stringent 'upfront' environmental impact assessment should be a mandatory planning instrument; and that 'transparent rules for public input' should be employed. It rejected the 'wait-and-see' approach and recommended the precautionary principle (now routinely applied by GBRMPA) that 'lack of scientific certainty should not be an excuse for allowing development that may have serious or irreversible environmental impacts'; and indicated an obvious 'need for better regional planning policies to give clear ground rules to developers about what sort of developments will or will not be acceptable' (Senate *Report* 1999:133–36). Its final observation was that the committee was 'confident all would agree that we do not want to see the Port Hinchinbrook debate repeated up and down the Australian coast'.

Scarcely had the Senate *Report* appeared than it was followed by a comprehensive, well researched and presented position paper on coastal management in Queensland issued by the Queensland Environmental Protection Agency which recognised that the coastal zone had been 'subject to pressure from population growth and human activity over the past 200 years . . . [and that] it requires careful management to ensure that its most valued elements are protected and conserved'. One of its significant demands is that in all future planning the State plan must 'draw on the principles and other strategies and agreements such as the National Strategy for the Conservation of Australia's Biodiversity and the Intergovernmental Agreement on the Environment' along with World Heritage, Ramsar, and Migratory Birds Conventions (Queensland Environment Protection Authority 1999:3). Throughout the document are signs that greater integrated planning between federal, State and local governments should be ensured.

The dominant issue of continued degrading of water quality was taken up the following year when Environment Australia, in association with GBRMPA and both the North Queensland and Central Queensland Local Government Associations, yet again conducted a study of the problem. This time, however, there was a major difference: local government bodies that had been basically land management oriented became involved in a study of the influence their regions were exerting on Reef waters. Numerous activities in addition to siltation were compounding Reef degradation: clearing of melaleuca and mangrove wetlands, dredging for marinas and harbour canals, construction of housing estates and roadworks on adjacent foreshores, not only on the mainland but also on the continental islands that are primarily under the jurisdiction of Queensland.

At last there were definite moves to begin planning for integrated management of the Reef and the coastal hinterland. Even more significantly, efforts were exerted to bring non-government organisations and conservation groups into the planning process, covering five major areas: tourism, industry, fishing, boating, and residential development. That was a remarkable initiative exploring totally new territory: it was the first serious effort to bring together the 21 local government areas bordering the Reef to explore ways of incorporating World Heritage values into land management. Previously the Queensland Department of Environment (QDoE) had grouped those local government areas, all of which were experiencing strong population growth, into nine coast management regions in an effort to begin coordinating the strong demand for urban infrastructure and services, but with no reference to their impact on Reef waters.

The novelty of the situation was not lost at the outset of the new program when it was realised that so far no studies 'in any comprehensive or systematic manner' on the impacts of coastal urban development on the World Heritage values of the Reef had been conducted (Environment Australia 1998:30). Encouraging initiatives were reported in the *Preliminary Study* which indicated that Port Authorities were beginning to come together with local and Queensland government agencies in an effort to solve issues associated with coastal urban development. At the same time it was heartening to note 'the welcome evidence of a trend towards the development of partnerships between government and non-government agencies' and the willing involvement of local conservation groups in raising community awareness (45).

A fundamental obstacle to effective holistic management of the area from Divide to Drop-off was identified: a sequence of 'single issue' decisions on activities that led to mass accumulation of what often become conflicting practices. Single issue decisions had been the standard operating procedure in GBRMPA management, whereby applications for permits had been considered on a case-by-case basis, following land management practices from which they had been adapted. Unfortunately, ocean currents and marine species never respect human-imposed boundaries. While in earlier years this may have provided an effective means of dealing with isolated developments on a small scale, with increased population pressures it can produce a ramifying effect known as the 'tyranny of small decisions' that can vitiate any environmentally sound structure. The proper solution has to be large scale planning to coordinate development and institute integrated, long term management practices in order to ensure sustainable use of the coastal zone. The 'tyranny of small decisions' since 1975 has affected many areas of the Reef lagoon: the Oyster Point development, isolated moorings and other structures, permits for fishing, zoning of numerous specific localities, are ready examples.

In the case of the World Heritage Area, as late as 1998 it was reported by the QDoE that 'despite some excellent initiatives, there is [still] no overall, consistent and coordinated approach to the management of impacts arising from [adjacent] coastal urban developments', that 'local governments do not perceive coastal management as a core business function', and that 'planning scheme reviews do not consider coastal values and issues in as much depth as they probably should'. Moreover, the 'protection of coastal values . . . [is] rarely translated into plan implementation in any structured or systematic way' (Environment Australia 1998:46). The final conclusion was aimed directly at the Commonwealth government, stating that 'the issue of coastal urban development and the value of the Great Barrier Reef World Heritage Area is an area of considerable Commonwealth Government interest and responsibility – this includes the responsibility that the Commonwealth has to monitor the condition of the World Heritage Property as required under the World Heritage Convention' (46).

A further issue that is rarely taken into account in the commercialisation of the Reef is the impact on local communities. The rapid growth of tourism has changed the economy and conventional social structures of many towns and cities along its coastline. Where European settlement once displaced indigenous communities during the last two centuries, now an inflow of outside capital is drastically transforming coastal life. In a number of growth centres like Port Douglas, Cairns and Airlie Beach, there is now a high level of dependency on tourism with an emphasis on financial returns as the chief aim of social life at the expense of traditional values. Anonymity, crowding and loss of communal identity are evident as opportunistic developers and commercial operators become dominant in reconstructing not only the environment but also the social landscape.

ISSUES FOR RESOLUTION: COOPERATIVE MANAGEMENT OF THE REEF

The first 25 years, from the *Great Barrier Reef Marine Park Act* of 1975 to the end of the twentieth century, witnessed a great experiment, never attempted anywhere on earth before: an effort to conserve and manage the world's largest marine park while imposing minimal restrictions on general public usage and exploitation. By 2000, however, it had become clear that the original intentions of the experiment had proved unworkable: usage had multiplied way beyond expectation while adjacent land-based development had also increased rapidly, with a profound and deleterious impact on lagoon waters. Compounding those problems were the inevitable conflicts among some twenty Commonwealth and Queensland management bodies, all with some degree of responsibility for the Reef, guided by innumerable pieces of legislation.

A major underlying issue that had always created antagonism between States and the Commonwealth was the question of intrusion into State sovereignty. In Queensland more than any other Australian state, the issue of presumed trespass was most keenly felt after the passage of the *Great Barrier Reef Marine Park Act* along with four other Commonwealth-declared World Heritage Areas – Wet Tropics of Queensland, Riversleigh, Fraser Island and the Central Eastern Rainforest Reserves – all of which acted as a continuing irritant. No other State has had to suffer the dominating presence in State affairs of the Commonwealth – now with its legal power under the Australian Constitution bolstered by a wide array of international agreements and covenants – enclosing and managing almost all of its eastern coastline. Although the years of intense hostility to Commonwealth governments begun by Bjelke-Petersen in the 'Save the Reef' years, and sustained by his government for over a decade, have been moderated, none the less Queensland has never rested easily with the outcome of the *Great Barrier Reef Marine Park Act* and the formation of the Authority. The Emerald Agreement of 1979 and its ancillary agreements may have assuaged its bitterness: it has never removed it. To this day many Queensland authorities believe their efforts in effective management of the Reef – and other World Heritage Areas – by their numerous Queensland agencies, chiefly Parks and Wildlife, Boating and Fisheries Patrol and the Fisheries Service are not simply undervalued, at times they are even ignored.

Whatever their political ideology, all governments in Australia today are forced to recognise the priority of forward planning. Where difficulties lie is in the essential character of the management and planning processes themselves, compounded by still unresolved tensions between federal and State governments, none more so than in regard to the Reef. Generally these stem from the time when Queensland's position began to weaken after a change of government in Queensland in December 1989. With the departure of Premier Bjelke-Petersen from the Ministerial Council and Co-ordinator General Schubert from the executive Authority (also known as the Board) – Schubert himself becoming chief executive of the Daikyo Corporation which has major tourism interests in the Reef – the heavy politicisation of relations with GBRMPA was replaced by management more sensitive to the environment. The succeeding Labor government in Queensland sent to both the Ministerial Council and the Authority officers of less senior rank, drawn variously from the Queensland departments of Environment, Tourism, National Parks and Wildlife, and Primary Industry. One unintended consequence, however, was the gradual erosion of Queensland's equal status on the Authority where its nominee, being lower in seniority, was unable to present a whole-of-government view and became increasingly influenced by GBRMPA as the agent of Commonwealth power over Reef policy and management. That situation became exacerbated throughout the 1990s.

In order to demarcate areas of responsibility among many Queensland government departments, one assumes a role as 'lead agency' in which it takes primary responsibility in the course of designing and implementing the large number of interacting management plans. A major altercation over respective responsibilities occurred in 1996. Following a 1989 GBRMPA Workshop Conference to discuss the problem of decline in fish stocks, CSIRO and the Queensland Department of Primary Industry conducted a five year experimental comparative study of fish populations in closed and open zones. In response to their report in 1996 a Trawl Management Advisory Committee (TrawlMAC) was established. Before the final report was made public in January 1999, Senator Hill, however, expressed dissatisfaction with the management plan produced by the Queensland Fisheries Management Authority (today Queensland Fisheries Service) – the lead agency – for not meeting ecological sustainability criteria, and threatened to override it with a plan to be devised by the Commonwealth.

A similar situation occurred in the same period in respect to the Vessel Monitoring System introduced by Queensland during 1997–98 whereby it provided over $3 million funding to equip commercial fishing vessels with global positioning system (GPS) transmitting devices so that all vessels can be continually monitored in order to maintain a close watch on fishing operations. The fishing industry itself welcomed the program and some 670 trawlers fitted with VMS equipment were able to be monitored by the Queensland Fisheries and Boating Patrol (QFBP). Senator Hill then decided that GBRMPA should have access to that information. This caused yet another furore since, as VMS devices operate in real time, some fishers believed that sensitive information on fishing grounds being trawled could be made instantly available to their competitors (Sturgess 2000:48, 61–62).

Those events created serious strains and indicated to the Queensland government that the principle of shared cooperative management of the Reef was becoming decidedly one-sided. The Commonwealth was seen as arrogating supreme authority to itself, vesting it in GBRMPA, the lead agency for Reef management as defined in the *Great Barrier Reef Marine Park Act*. Underlying much conflict was not only the presence of the past, exemplified in the bitter conflict between Queensland and the Commonwealth over the Wet Tropics Heritage listing in 1987 and the dismissal of Queensland's appeal by the High Court in 1989 (Queensland v Commonwealth (1989) 63 ALJR 473), but the fact that the Marine Park over which the Commonwealth has jurisdiction does not include the coastal enclaves enclosed by the larger World Heritage Area. None the less, the Commonwealth holds responsibility for protecting the World Heritage Area and is required to act when activities such as dredging and marina construction within Queensland waters could also impact World Heritage values. While in cases such as the Magnetic Island marina project and the Oyster Point resort development it may be argued that the Commonwealth government has failed to intervene on that account, its considerable powers are of concern to Queensland. Projects currently at the proposal stage, such as building the Nathan Dam on the Dawson river which flows into Keppel Bay at Rockhampton, and mining the oil shale deposits a few kilometres north of Gladstone which extend into coastal waters, raise profound issues of both inevitable contamination by effluent and the powers of GBRMPA to exercise control.

It was the cumulative effect of GBRMPA's increasing independence, the assertion of Commonwealth ministerial control, and perceived denigration of input by the Queensland State agencies which together are responsible for virtually all Marine Park day-to-day management and enforcement tasks, that finally reached flashpoint. In early 1999 the Queensland government commissioned an inquiry by consultant Gary Sturgess, a former Head of the Premier's Department in New South Wales, to examine and report on policy development and field management, on 'the efficacy of the current Great Barrier Reef Ministerial Council arrangements' and 'the relevance and legality of current intergovernmental agreements which underpin the Queensland Government's contribution to the Great Barrier Reef Marine Park Authority' (Sturgess 2000:167).

Released in November of that year, the Sturgess report was a thorough, investigative analysis of all aspects of Reef management. Organised around the primary concept of 'cooperation', it asserted that jurisdictional problems had reached a stage where 'tension ... between the two governments, both at a political level and in personal relations between individual public officials' was having a profound effect on management of the Reef, with the potential to affect World Heritage standing (4,23). The Marine Park legislation was found to have been drafted in very general terms and relied far too much on informal agreements. In the first decade, 1979–89, Kelleher and Schubert had paid little heed to strict demarcation issues and made a conscious decision to work together amicably in the interests of effective management. Subsequently, when pressures on the Reef had exploded and management had become more demanding, the Commonwealth began to insist on more literal interpretations of the Act which were not always consistent with State interests. Since it held the two decision making positions – the Minister for the Environment as Chair (or 'Convenor') of the Ministerial Council, and a Commonwealth nominee as Chair of the Authority – policy making was easily dominated by the Commonwealth, and this became manifestly irritating to Queensland in the later 1990s. In addition, since the Commonwealth maintained the Secretariat it was able to preserve a tight hold on power and information. Insensitivity to State concerns, and keeping GBRMPA 'tightly coupled' to the Commonwealth minister, Sturgess reported, provided 'fertile ground for misunderstanding and misrepresentation' (10).

To improve the power sharing relationship and ensure the maintenance of World Heritage status, to the advantage of both the Commonwealth and Queensland, the fundamental reform recommended by Sturgess was to rotate the Chair of the Ministerial Council – currently the Commonwealth Minister for the Environment – between the two governments, and for Queensland to establish its own independent Secretariat which would provide independent policy advice to the Council and to the Executive Authority. If that arrangement were to be effected it would preclude the Chair becoming solely the instrument of the Commonwealth and give Queensland 'a neutral meeting ground where it can freely and safely debate policy issues with the Commonwealth' (11). Such a change would require the Commonwealth to amend the *Great Barrier Reef Marine Park Act*.

The year 1999 was a watershed in Australian environmental policy development and Reef management. Soon after coming to office in 1996 the Coalition government in Canberra had sought to redefine the Commonwealth's role in environmental issues by calling together the Council of Australian Governments in an effort to resolve problems of conflicting legislation and policy coming from the Commonwealth, States and territories on environmental policy and management. After two and a half years of planning the discussions eventually came together in a new omnibus Commonwealth *Environmental Protection and Biodiversity Conservation Act, 1999* which replaced five separate Acts passed by previous Commonwealth governments. Repealed were the *Environment Protection (Impact of Proposals) Act 1974*, *National Parks and Wildlife Act 1975*, *Whale Protection Act 1980*, *World Heritage Properties Conservation Act 1983* and the *Endangered Species Protection Act 1992*. The intention was to provide for a cooperative approach to the environment between governments, the community, landholders and indigenous communities, thereby producing 'a seamless integration of Commonwealth and State laws through a transparent mechanism for Commonwealth accreditation of state processes' (Hansard, Representatives V.227, 1999: 7767).

In both houses of the parliament it drew strong criticism from the Opposition parties, mainly because it reduced the Commonwealth role in environmental policy, and inhibited it from dealing with State or local matters outside its areas of concern. Greens Senator Dee Margetts spoke forcefully that it would be 'potentially a disaster for environmental protection and biodiversity conservation in Australia' since 'the underlying theme . . . is still to hand over the Commonwealth's environmental responsibilities' (Hansard, Senate V.195, 1999: 4350). The major concern of the Opposition parties was the provision that the Commonwealth would no longer unilaterally declare World Heritage Areas except in special cases. It means, in effect, that there may be no more actions like the Commonwealth Government's use of the *World Heritage Properties Conservation Act 1983* to prevent flooding of the Gordon River by the Tasmanian Hydro-Electric Commission in 1983, and the logging of the tropical rainforests of North-East Queensland in 1987. The full effect of the legislation is yet to be experienced. In respect to the Reef, however, although the legislation is firmly in place, the Commonwealth has no intention of leaving its management solely to Queensland, despite the asperity of the Sturgess Report and passage of the *Environmental Protection and Biodiversity Conservation Act*.

Having brought the Authority's *State of the Great Barrier Reef World Heritage Area 1998* to the public, Dr McPhail left the Authority on 12 April 1999 to become executive director of the renamed Queensland Parks and Wildlife Service which has extensive responsibility for Reef management since all of the islands and vegetated cays remain Queensland terrirory. The Queensland Parks and Wildlife Service, in common with the

Queensland Fisheries and Boating Patrol and the Queensland Fisheries Service, effect all the routine tasks throughout Reef waters and discharge their duties with dedication and commendable professionalism. In his short tenure as Chairperson McPhail had become a vital renewing force in Reef management, effecting significant changes with his 'issues based' approach, and 'stringent standards' of performance for GBRMPA staff.

Following his departure the Commonwealth reinforced its control of Reef policy and management when, as critics noted, without customary and prior consultation with the Queensland government, on 1 July 1999 Environment Minister Hill appointed The Hon. Virginia Chadwick to the vacancy. A former school teacher who became Minister for Education in the conservative government of New South Wales in the early 1990s, Ms Chadwick assumed office from 19 July to become chair and chief executive with responsibility for leading Reef management into the twenty-first century.

From the same political party and with a close personal relationship to both Prime Minister Howard and Environment Minister Hill, Virginia Chadwick was able to bring Great Barrier Reef issues to prominent notice within the Commonwealth government. An active politician, she quickly familiarised herself with the conservative electorates along the Reef coastline, and soon introduced major changes in trawling and aquaculture practice. With the largest trawling fleet in the world, the Revised East Coast Trawl Management Plan of January 2001 dealt with the difficult issue of conserving fish stocks by capping catch effort at 1996 levels through structural adjustment, licence surrender, and the closing of an additional 96 000 square kilometres of marine park to prevent expansion of trawl areas. Ms Chadwick was also successful in bringing one of the most serious of all GBRMPA management concerns, that of water quality, to priority notice on the new Commonwealth Sustainable Environment Committee, chaired by the Prime Minister.

With the appointment of a politician as the Chief Executive it was now expressly recognised that statutory authorities cannot act independently of government. Whether directly by strong pressure of the governing party, or indirectly through political appointments to the governing board, funding appropriations, or other restrictions, the party in power can exercise its will. Whereas the original *Great Barrier Reef Marine Park Act* of 1975 envisioned a scientific professional as Chief Executive, and the Ministerial Council as ancillary political overseers, the reality has been that policy and management are, in essence, political processes. During the early decades of the 1980s and 1990s, GBRMPA policy was heavily influenced by Bjelke-Petersen and the Queensland Coordinator-General Schubert, with a counterbalancing effort by Prime Minister Fraser and then Hawke. The only significant change has been the greater assertion of Commonwealth control of the Reef through a close, overt linkage of the Authority with the Prime Minister and the Minister for the Environment, and further limits on policy input by Queensland.

HERITAGE MANAGEMENT IN A WARMING WORLD

Whatever specific environmental and conservation initiatives for Reef management and coral protection are followed in coming decades at a local level, the task will become increasingly difficult and could well be completely undermined if global warming trends continue at their present rapid rate. In this respect, the overall environmental position of the conservative Commonwealth government in the final years of the twentieth century was disturbing. After gaining office at the end of 1995 it engaged in a series of controversial activities: approving the Oyster Point development and allowing uranium mining to proceed within an enclave in the Northern Territory Kakadu World Heritage Area being significant examples. Even more disturbing was the Commonwealth record in response to threats posed by greenhouse issues when the environmental credentials of the new government were displayed to the world in 1997 at the Kyoto Conference on Climate Change. In 1990 the previous Labor government had proposed moving to a 20 per cent reduction in carbon based emissions by the year 2005, leading to agreements finalised at the Rio Intergovernmental Conference on Sustainable Development in 1992. These came into force on 21 March 1994 seeking to limit emissions of the six most dangerous greenhouse gases: carbon dioxide, nitrous oxide, methane, the perfluorocarbons, chlorofluorocarbons (CFCs) and hexafluoride.

There was sustained opposition from business interests, particularly the main industrial producers of greenhouse gases, chiefly the oil and coal industries, represented by their powerful lobby group, the Business Council of Australia. As the world's leading coal exporter, the Australian economy would clearly be subject to some reductions under the current restrictions, and so economic growth policies prevailed. In a series of international meetings throughout the second half of the 1990s the intransigent position of the Australian government, siding with the great global polluters of the United States, Canada, Russia and the cartel of the Organisation of Petroleum Exporting Countries (OPEC), continued to the disappointment of more enlightened nations. At the third Conference of Parties of the Framework Convention on Climate Change held in Geneva in 1996 in preparation for the Kyoto Conference, Australia refused to sign the prepared negotiating text.

Despite the warnings and agreements reached during the Greenhouse decade of the 1980s, and the Rio agreement of 1993, global emissions of the dangerous gases have continued to rise, and in 1990 the Australian Bureau of Agricultural and Resource Economics (ABARE) reported predictions of a 62 per cent increase by 2010 in developing countries from 28 to 46 billion tonnes (Brown et al. 1999:3). Regrettably, no mechanisms were established for enforcement and at Kyoto, despite evidence of mounting climate change, Senator Hill, under instructions from a Cabinet that did not share his concern for the environment, was constrained to fight vigorously, and successfully, not simply against the conference recommendation to hold Australia's emissions to current levels, but actually for increasing them by 7 per cent (with 'offsets') in the interests of the national economy. When the 160 nations involved met at The Hague in November 2000 to implement the Kyoto Protocols, not only were they unable to reach agreement on how to reduce greenhouse emissions, they broke up in acrimonious conflict. It was resolved to attempt a solution by reconvening in Germany in 2001, but that too proved disappointing.

As if to foreshadow the future, 1998 was the warmest global year on record. By 2000 the Yukon glaciers were melting: of the 150 active glaciers in Glacier National Park in 1850, only 50 remained. In the tropics, coral reefs were noticeably affected. Then, in January 2001 when 400 scientists of the world Intergovernmental Panel on Climate Change (IPCC), under the aegis of the World Meteorological Bureau and the United Nations Environment Program convened in Shanghai they issued a unanimous warning that even with the best efforts of all world governments, due to processes that had already raised average temperatures by 0.7°C by the end of the twentieth century, world temperatures will continue to rise between 1.4°C and 5.8°C by the end of the twenty-first century, with the higher figure more likely. What is never acknowledged by governments is the fact that the global average temperature is currently 14.2°C, having already risen from 13.69°C in 1929 (*World Almanac* 2002:164), and a rise of several degrees will make a disastrous impact. All members of the IPCC accepted the conference verdict.

In what became a tragic landmark event for world environmental prospects, a new Republican Party government took office in the United States in January 2001, under the presidency of George W. Bush who immediately rejected any serious checks on United States economic growth. When the Kyoto Conference reconvened in Bonn in July 2001, the world's greatest polluter, the United States, emitting 25 per cent of all noxious gases, refused to sign the protocol as 'not serving its national interests', and Australia, along with other major polluters, chiefly Japan and Canada, chose to follow suit. Despite some face-saving but inadequate amendments the protocol was effectively sidelined for the foreseeable future. In February 2002 President Bush remained adamant that the United States rejected the Kyoto Protocol and would initiate its own greenhouse control program, although no striking evidence of that has emerged since his first reference to it the previous year. In the same month, the Australian government announced it would follow the lead of the United States and also reject the Kyoto Protocol.

It must have been extremely difficult for the Australian Minister for the Environment to oppose the Kyoto Protocols which seek to restrain global warming and sea level rise, when he was also responsible for the Great Barrier Reef World Heritage Area where its coral reefs in the same period had begun, like canaries in nineteenth century coal mines, to issue a warning about rising levels of environmental distress. Coral polyps in normal tropical waters cannot live for long in water above 30°C. In higher temperatures they expel their symbiotic algae, and begin to bleach – losing the rich coloration that gives them their essential attraction – and then, deprived of essential nutrition if the sea temperature does not drop soon enough for the algae to return, they die.

Throughout the 1980s coral bleaching was reported across the tropic reef zone. Evidence had come in 1982–83 not only from the Galapagos, where an exceptionally high temperature phase raised the warm tropic waters more than 5°C to 31°C resulting in 95 to 100 per cent mortality of corals, but also from Costa Rica where a 50 per cent loss was reported, and on the Caribbean side of the isthmus of Panama where it reached 80 per cent loss (Wells & Hanna 1992:54f.). In 1987 extensive coral bleaching was observed in Puerto Rico and along the Florida coast. Observations of that pattern of reef destruction continued throughout the following decade. In 1990 David Kobluk of the University of Toronto found extensive bleaching and coral mortality on the reefs of Bonaire in the Netherlands Antilles off the Caribbean coast of Venezuela where tourist activity is low, and in the same year around 50 per cent mortality was also found in the southern Arabian Gulf. Further incidents of widespread bleaching and mortality continued through the 1990s: in 1991 the phenomenon was observed in Phuket in Thailand, in the Seychelles, in French Polynesia, in the Society Islands, and on the Caribbean shores of Florida (Leggett 2000:37–38, 79).

Throughout the first observed phase in the 1980s, coral bleaching and mortality was a puzzle. Branching scleractinian hard corals appeared the most susceptible and various hypotheses were advanced: chiefly sedimentation, eutrophication and tourist pressure. Global warming was also proposed by some investigators and that hypothesis received strong confirmation in October 1990 when the United States Senate Committee on Commerce, Science and Transportation concluded from a decade of data compiled by the National Oceanic and Atmospheric Administration that global warming was almost certainly the main factor. In the same period Tom Goreau Jr. at the Discovery Bay Marine Laboratory in Jamaica examined those data and confirmed a positive correlation between coral bleaching and higher sea temperatures. Coral cores obtained from 400 year old reefs showed no evidence of previous bleaching, yet when compared with current reefs in the same location, in regions still relatively untouched by humans, bleaching was evident in numerous cases. In 1999 Charles Sheppard reported marked decline in coral cover in one of the most remote archipelagoes in the world, Chagos in the central Indian Ocean, which he had been studying for twenty years. The archipelago consists of several uninhabited atolls and a number of submerged reefs, including the large atoll of Diego Garcia, the centre of which is the Grand Chagos Bank where he found after the warming event of 1998 that 'only 12 per cent of the substrate is living coral, compared with 50–75 per cent before' (Sheppard 1999:472).

At the same time the Greenpeace environmental group also turned its attention to issues of coral warming after the peak year of 1998 and commissioned a highly experienced reef scientist, Ove Hoegh-Guldberg, then director of the Coral Reef Research Institute at the University of Sydney and the One Tree Island Research Station, to investigate and report on coral bleaching. In 1999 he presented much the same evidence as Sheppard, confirming his own earlier findings of 1994 that 'every coral reef examined in Southeast Asia, the Pacific and Caribbean showed the same trend' of severe mortality.

From his data he predicted that 'bleaching events are very likely to occur annually in most tropical oceans by the end of the next 30–50 years' and that the Great Barrier Reef, already affected, would be no exception. He warned that 'the rapidity and extent of these projected changes, if realised, spells catastrophe for tropical marine systems everywhere and suggests that unrestrained warming cannot occur without the complete loss of coral reefs on a global scale' (Hoegh-Guldberg 1999:1).

These are issues of profound importance. If, as predicted, rapidly rising world temperatures and ocean warming unprecedented in human experience do eventuate, and the entire global ecosystem is moved away from relative equilibrium towards instability, the cumulative effects, as chaos theory now shows, are likely to be far greater than our politicians and economists expect. Global warming will not only continue to melt mountain glaciers and polar icecaps, a process already measurable in the thinning of Arctic ice and the collapse of large parts of the Antarctic shelf into the sea, it is also predicted to increase the frequency and violence of storms and other atmospheric disturbances. In the United States tornadoes and cyclones are increasing in number, and their range is now extending inland further north from the Caribbean. Coral reefs are likely to be the first to disappear and with them many of the related organisms that form the reef ecosystem, without hope of recovery if the rate of sea level rise proves faster than remaining corals in waters of correct ambient temperature can grow. Even more destructively, a negative feedback process is then stimulated since the loss of coral reef and ocean biota worldwide would reduce the range of organisms able to extract free carbon from the atmosphere, adding to the greenhouse effect. With the main polluter nations obdurate against reasonable controls, much less a radical change of direction, the scenario for global warming does not hold out any long term future for any coral reef, including the Great Barrier Reef.

CONSERVING BIODIVERSITY: THE INNOVATIVE REPRESENTATIVE AREAS PROGRAM

During the final decade of the twentieth century, in the context of worldwide and local concern over the alarming loss of biodiversity, GBRMPA moved to undertake a radical improvement in protection measures within the World Heritage Area (GBRWHA). With the 1992 Commonwealth plan to create a National Representative System of Marine Protected Areas as a stimulus, in the innovative *Twenty-Five Year Strategic Plan for the World Heritage Area* in 1994 GBRMPA identified as one of its five year objectives the need 'To protect *representative biological communities* throughout the Area to act as *source areas*, *reference areas*, and areas of *biodiversity* and species abundance' (GBRMPA 1994:16, its emphasis). Then in a major proactive move, in 1996 planning commenced for a comprehensive investigation and rezoning of the entire World Heritage Area on these lines in what was designated as a Representative Areas Program (RAP).

A similar complementary State-wide project was also proposed by the Queensland Environmental Protection Agency, and a further stimulus came when Australia in 1998 adopted an Oceans Policy that called for the implementation of a 'representative areas network'. This was followed in 1999 by a Strategic Plan of Action devised by the Australian and New Zealand Environment and Conservation Council to move actively into identification of marine areas needing protection.

Biodiversity protection was now being understood as a far more complex concept than attempting to maintain a wide range of species simply by identifying species at risk and implementing various site or species specific protective measures. A more sophisticated understanding was presented in May 1999 when GBRMPA released its preliminary *Overview of the Great Barrier Reef Marine Park Authority Representative Areas Program* which defined biodiversity as 'the variety of life forms at the level of ecosystems, species, and gene pools' (GBRMPA 1999:18). What marked a major departure was the move towards an ecological perspective in Reef planning, with the recognition that whole communities, representing the full range of species and their habitats within the Reef ecosystem must now be protected – not simply isolated reefs but also the connecting waters and the sea bed with its wide range of supportive habitats.

After twenty-five years of scientific investigation the Reef ecosystem was discovered to hold a mind-boggling inventory of species. Within the World Heritage Area of nearly 350 000 square kilometres containing 2904 coral reefs built out of 359 species of hard corals, researchers had identified 1500 species of fish, 800 of echinoderms, 5000 of molluscs, 2200 of native plants, 80 of sea pens, and six of the world total of seven turtles, all interacting within a marine region containing 2000 square kilometres of mangroves and 3000 of seagrass meadows (GBRMPA 1999:2).

The original zoning approach in the 1980s had been to divide the Park into a small number of administrative Sections and within them to identify some Marine Park or 'no take' Green zones which allow public access and indigenous traditional activities, along with more restrictive 'no go' Pink and Orange zones allowing entry only by permit for scientific research. Based on land planning models, these are inadequate for broadscale planning to protect biodiversity in marine parks where most boundaries are, literally, fluid, crossed regularly by ocean currents and by mobile species such as fish or turtles in the course of their life cycles, and in the case of sedentary adult corals, by their fertilised eggs (planulae) which can drift long distances to find suitable settlement sites. Indeed, rising sea levels in the future may test human ingenuity in providing such locations.

Although those few protected areas in zoning plans have been supplemented by additional habitat protection in the management plans for the high tourism Whitsunday and Cairns areas, and by a series of inshore Dugong Protection Areas designated in 1997, along with Queensland Fisheries Service limits on fishing and associated management procedures, their inadequacy has been acknowledged by GBRMPA. The early 'no take' zones emphasised protection of the better understood – and publicised – coral reefs, covering

20 per cent of these, and only 4 per cent of all other habitats, with less than 5 per cent of the total World Heritage Area highly protected. It is also important to note that these protection plans do not extend to the northern areas of the Reef that lie in Torres Strait outside the Great Barrier Reef Marine Park and the World Heritage Area, although they form a significant part of the entire Reef ecosystem.

The RAP reforms are aimed at overcoming earlier planning deficiencies. Its objectives are to help maintain biological diversity and, as stated by GBRMPA, to provide 'an ecological safety margin against natural and human-induced damage', by seeking to develop 'a solid ecological base from which threatened species or habitats can recover or repair themselves', as well as maintaining 'ecological processes and systems and connections between different habitats' (GBRMPA, *RAP Update*, May 2000).

In implementing the Program, GBRMPA at the outset introduced a smoothly organised public participation and information program, guided by findings from two CRC social research programs, with the explicit intention of minimising conflict with the public, and securing the willing cooperation of commercial, recreational and indigenous users. A primary task was also – for the first time – to identify and map an 'indicative' series of ecological bioregions throughout the World Heritage Area. Defined as an area of land and/or water which in its plant and animal communities, as well as physical features, is significantly different from other areas in the Park, the bioregion is intended to give an indication of biodiversity in the Area. When finally designated, using improved versions of existing computer planning programs, and new purpose-designed software, the bioregions will provide a basis for a network of highly protected 'no take' representative areas, to be identified 'within each habitat of each region' (GBRMPA 1999:5).

After completing a classification of most bioregions by late 1999, GBRMPA reported in May 2000 that maps, based on analysis of some forty levels of data, had been completed for 70 bioregions: 30 reef and 40 non-reef, with some others not classified. Within a year the complex World Heritage Area bioregions map had been updated and public comments invited for the ongoing review process (GBRMPA Map C03501; 19 March 2001).*

Meanwhile, work had commenced on the RAP Identification Phase, with input from a panel of experts and a Scientific Steering Committee, to identify prospective candidate areas and develop operational principles for linking habitat types, based on geomorphological, oceanographic and biogeographical data. This was followed by a Selection Phase, guided by a Scientific Committee, and from January 2000 by a Social, Economic and Cultural Steering Committee, with the task of applying available data in selecting suitable sites to maximise biodiversity protection and minimise problems for users and local communities (GBRMPA, *RAP Update,* May 2000).

In contrast to previous practice of rezoning each Section individually, the last being the Far Northern Section in 1992, the aim was for the entire Great Barrier Reef Marine Park

---

* www.gbrmpa.gov.au/corp_site/key_issues/conservation/rep_areas/updates.html

to be zoned simultaneously in an integrated plan, when finalised, using the bioregions as the basis for identifying suitable Representative Areas. With wisdom gained from earlier management difficulties the rezoning process will incorporate two new fundamental standards: the 'Precautionary Principle' which demands that protection measures should not be inhibited by lack of scientific certainty, and the 'CAR Principle' that requires the network areas to be Comprehensive, Adequate and Representative. The major initiative of implementing the RAP to protect biodiversity is to be introduced throughout the coming years. It has already been welcomed by a World Wide Fund for Nature spokeswoman Imogen Zethoven who applauded the RAP program as ensuring 'huge advances in marine conservation of the area' (*The Australian* 27 December 2001:5).

Another long term research project was initiated to monitor the large number of reefs across the World Heritage Area for abundance of Reef fish, the density of benthic organisms and the still mysterious problem of Crown of Thorns outbreaks that continue unchecked, showing a noticeable southward drift and defying all control approaches. Hampering attempts at starfish control is the absence of any historical evidence. No certain records of past epidemics have been found in undisturbed fossil deposits, indigenous folklore, or in early accounts since European occupation. In 1999 the most intensive analysis of the history of the issue appeared, setting out the conflicting evidence for both the natural cycle and anthropogenic causation theories (Sapp 1999). Evidence had been presented by scientists from James Cook University, using radiocarbon dating of skeletal remains of the calcareous spines in sediments recovered in the palaeographic column retrieved from Green Island, John Brewer Reef and Heron Island Reef that suggested starfish had been part of the reef ecosystem for the past 8000 years (Walbran *et al.* 1989:847). Their evidence, in turn, was questioned by AIMS scientists who observed that sediments are often disturbed in deposition by small burrowing shrimps and that no sequential certainty can be ascertained (Keesing et al. 1992:79–85). The disturbing conclusion now generally favoured is that mass invasion by starfish is a modern phenomenon and that proliferation has been facilitated by an ecosystem profoundly altered by removal of other regulatory species (Jackson et al. 2001:631).

Throughout the year 2000 the AIMS long term monitoring team recorded a significant rise in the number of reefs affected. Starfish numbers, it reported, increased substantially in the Cairns, Innisfail and Townsville sectors, with large outbreaks also observed in the most southerly region of the Swains (AIMS Annual Report 2000–2001:10). By late 2001 starfish had invaded the premier tourist resort region of the Whitsunday Group in epidemic proportions. It was becoming clear that environmental stresses had reached disturbance levels of which the starfish enigma was but one symptom. Indeed Reef biodiversity concerns mounted as 2002 joined 1998 as the two warmest years on record, now with wider bleaching of Reef corals and up to 90 per cent mortality on some reefs (Luntz 2002). It gave urgency to worldwide trials of artificial reefs such as the biorock 'arks' of Thomas Goreau and Wolf Hilbertz where corals are proving more resistant to stress (Henderson 2002).

HERITAGE: A SUSTAINABLE IDEAL?

As we look back over more than two centuries of European contact with the Reef, and nearly three decades of active attempts at conservation, it is clear that major efforts have yet to be made in sustainable management of the Reef, especially with global warming now evidently confirmed by the two record warm years of 1998 and 2002. It is imperative to learn from the past. The *State of the Great Barrier Reef World Heritage Area 1998* report was a significant account that presented a number of issues, the central one being that despite more than twenty years of management and vast sums spent on research, for most environmental categories it was still 'not possible to say with certainty if they are in satisfactory or unsatisfactory condition'. Unfortunately, criteria against which those judgments were made have not been available. At the time of the 'Save the Reef' campaign in the 1960s and the *Great Barrier Reef Marine Park Authority Act* of 1975 the lack of baseline data was explicitly recognised: the Senate Select Committee Report of 1971 and the Royal Commission Report of 1974 made that lack abundantly clear.

Our records of European impact on the Reef since Cook, which this narrative has presented, indicate continuous resource utilisation from the early days of whaling through to the resource raiding of later decades that led to the collapse of the pearling, trochus and bêche-de-mer industries. With every assault on the integrity of any ecosystem – and the Reef is no exception – species population dynamics are altered, some decline and become endangered, others become extinct, others expand in greater concentrations. Every intervention rearranges the structure of the web of life. Endean was pilloried by some of his colleagues throughout the 1970s for his hypothesis on the Crown of Thorns outbreak, yet it could well have been the case that cumulative effects since the early period of excessive resource raiding had been altering Reef ecology to the point that starfish outbreaks were able to begin. We will never know. Meanwhile, pollution and relentless processes of resource extraction – which includes tourism – have continued and even accelerated throughout the twentieth century.

In the confrontational decade of the 1960s during the height of the 'Save the Reef' campaign, as noted earlier, the question was asked: what exactly were the protestors trying to save? No one really knew. Certainly there was a profound, widespread image of the wonders of the Reef and an emotional reaction against visible despoliation from oil slicks fouling beaches and sea birds, but of its complex ecology almost nothing was known. Even the GBRC lacked an integrated data base, and in any case most of its research in the years from 1923 to 1937 had been directed by Richards into geological inquiry, with a focus on basaltic foundations, sedimentary basins and oil-bearing deposits.

Once GBRMPA had been established in 1976 its primary task was to plan and coordinate management of the marine park governed by the concept of multiple use: today better defined as 'rapidly multiplying use'. Its data collection was always reactive to pressures, in contrast to AIMS which in the first decade was preoccupied with fundamental research.

In the closing years of the twentieth century both bodies were directed by government pressure to accelerate their efforts toward greater economic exploitation of the Reef. Even as the world's environmental condition continued to worsen, the same processes were being encouraged by considerable funding, first within the new framework of sustainable development and then, belatedly, of ecological sustainability – although the focus was still not on conserving the Reef ecosystem but on sustaining maximum use and economic extraction.

The one outstanding human characteristic is wisdom after the event. Hindsight is our greatest gift. It also has been least used in planning the future. Looking back on the 1975 *Great Barrier Reef Marine Park Act* and the formation of the Authority, they were, in context, well planned responses to heightened public concern and political agitation at the time, serving as the first warning checks in a period when Australia, in common with the rest of the developed world, was moving into its ugliest phase of exploitative capitalism, as the entire record of the burgeoning environmental protest literature attests. The consequent rise of the concepts of National Estate and Heritage in that period were responses to the rapid degradation of first the built, and then the natural environment. Paradoxically, in one sense declaration of the Great Barrier Reef Marine Park and subsequent World Heritage listing of the Reef served to accelerate the pressures on it. Once labelling occurs any phenomenon gains heightened identity and in the case of the Reef its mystique was raised to international iconic status, attracting a rising flood of visitors and providing more alluring advertising copy for promoters.

What is rarely appreciated, however, is that while heritage can be protected, it can never remain permanently unchanged. Despite their listing as World Heritage monuments the Pyramids and Sphinx continue to erode: the Reef, like any other environmental feature is subject to natural processes and we must accept the reality that its heritage status cannot be frozen in time. From the beginning of history changing levels and types of use, accelerated by technology and evolving social and political cultures, have reshaped the natural environment. Our most effective response is to accept that process and develop strategies of management.

We may wish to see the Sphinx with its face intact in the time of Khephren, or the Parthenon in all its golden Periclean glory before the Turks blew up part of it and Lord Elgin took off with its tympanum statuary, or sail with Flinders when he found the coral reefs 'equalling in beauty and excelling in grandeur the most favourite *parterre* of the curious florist', but alas, we remain constrained by the present and can only make imaginative reconstructions. All heritage documents tell us that we must preserve the present for future generations, but no matter how hard we try, no children in the future are likely to see a red cedar such as the one Phillip Parker King recorded in 1819 with a diameter of 10 feet (3 m) and a crown disappearing beyond the canopy above. More likely they will see little more than rows of plantation striplings, as will be the case in Queensland if its forests and woodlands continue to be cleared at current rates which, as a further assault on the planet, will add to global warming by removing a sink for atmospheric carbon.

In managing the Reef then, we cannot realistically aim at keeping it intact in some kind of vast aquatic time capsule. What we can do is to understand the historical processes that have led to its present condition and, using the best research available to us, now enhanced with satellite data and computer modelling, attempt to predict future effects, either to avoid them or to slow the deteriorative processes to a minimum. Certainly from a present human viewpoint the Reef is moving inexorably through processes of accelerated change. We must recognise that change is a natural process, but our task today is to prevent, or at least seek maximum retardation of those processes: in effect, mainly to save the Reef from ourselves.

To that end we must accept that unless we change our approaches to Reef management by exerting more stringent controls on rural, coastal and marine activities, the Reef's World Heritage status will be at risk and future generations will not see even the Reef we experience today. And, more tragically, unless the nations of the world are able to control global warming and sea level rise, by next century there may well not be a living Reef at all to maintain as heritage. It is easy to understand the encouraging vision of the Reef being preserved 'in perpetuity' and the continued optimistic tone expressed throughout GBRMPA's early public statements: what else could we expect from a management authority operating in the full glare of public scrutiny? At the same time, the extreme vulnerability of corals and other Reef communities in the face of threats looming ahead is a disturbing prospect.

Yet, on a much longer time scale, we can only look with awe at the massive limestone strata marking now the majestic survival of reef building corals across the world, through vast periods of time and cataclysmic change since their emergence in the fossil record as a group during the Triassic more than 200 million years ago. After surviving a major extinction event at the end of the Triassic and proliferating to their greatest diversity by the end of the Jurassic about 140 million years ago, corals persisted through periods of traumatic environmental change over the ensuing millennia, from the high temperatures and carbon dioxide levels of the mid Cretaceous, to the falling temperatures and sea levels at the end of that period, when corals emerged again from near extinction to continue through the Ice Ages and reach their distribution of today (Veron 1995:114f.).

It is tragic then, that the latest threat to corals, in the potentially devastating extinction phase now under way, is due entirely to humans, a species that has emerged in only the last minutes of the long day of life on earth. For us, there are reasons for humility and none for complacency in that story. Recovery times for corals after such disasters in previous eras are measured over a very long time scale and the price for the survival of hard corals in high temperatures of the past may have been to move to cooler ocean depths, losing their symbiosis with algae and thus their ability to build the colourful reefs we admire so much today.

In terms of heritage for the future, the challenge is clear: to respect a wider range of heritage values, natural and cultural, and to renew our efforts to maintain our Reef communities and endangered species, at least in their present state, for the coming generations that we hope may be wiser and more provident than ours. On a global scale, given the compelling attraction of the Reef for world tourism, and the international reputation for world's best practice in reef and marine management, now established jointly by GBRMPA and Queensland's regulatory authorities, we can hope that the Great Barrier Reef will provide a model to reinforce international measures for climate control and responsible environment protection.

A heartening sign for change is the growing competence of research achievements and management procedures in preserving Reef ecosystems – both biota and habitats – while allowing responsible use and public enjoyment of that vast area. Equally heartening is the dedicated enthusiasm and maturity of outlook shown by the many cooperating authorities, both Commonwealth and State, and by a growing number of indigenous communities, ecotourism operators, voluntary conservation groups and the wider population in their concern to protect the Reef for the future.

REFERENCES

From the extensive range of studies consulted, only works cited in the text are listed below. In citing sources a modified Harvard system has been employed in which the original work is given by author, year of publication and locus: succeeding references to the same source are given in page numbers only.

Newspaper reports, parliamentary debates, and direct references from journals are not listed since their documentation is complete within the narrative itself.

Aboriginal Coordinating Council (ACC) (1992) *'On Keeping it Great: A 25 Year Strategic Plan for the Great Barrier Reef World Heritage Area 1992–2017': Draft Submission*. Cairns: Aboriginal Coordinating Council

Adams, Brian (1986) *The Flowering of the Pacific: Being an Account of Joseph Banks' Travels in the South Seas and the Story of his Florilegium*. London: Collins/British Museum (Natural History)

Agassiz, Alexander (1898) *A Visit to the Great Barrier Reef of Australia in the Steamer* 'Croydon', *during April and May 1896*. Cambridge, Mass.: Museum of Comparative Zoology

Agassiz, George R. (1913) (ed.) *Letters and Recollections of Alexander Agassiz, with a Sketch of His Life and Work*. Boston & New York: Houghton Mifflin Company

Alexander, Michael (1971) *Mrs Fraser on the Fatal Shore: The True Story Behind the Legend*. London: Michael Joseph

Anderson, Charles (1936) 'Charles Hedley', *Proceedings*, Linnean Society of New South Wales, 61:1936, 209–14

Andrews, Ernest Clayton (1902) 'Physiography of the Queensland coast and its relationship to the Great Barrier Reef', *Proceedings*, Linnean Society of New South Wales, Part 2, 1902, 146–85

Attorney-General's Department, Commonwealth of Australia (1980) *Offshore Constitutional Settlement: A Milestone in Co-operative Federalism*. Canberra: Australian Government Publishing Service

Australian Conservation Foundation (1969) *The Future of the Great Barrier Reef: Papers of an Australian Conservation Foundation Symposium*. Special Publication No. 3. Sydney, 3 May 1969. Melbourne: Australian Conservation Foundation

Australian Institute of Marine Science (*c*.1997) *Strategic Directions*. Cape Ferguson: Australian Institute of Marine Science

Bach, John (1987) *The Bligh Notebook: Rough Account – Lieutenant Wm Bligh's Voyage in the Bounty's Launch from the Ship to Tofua & from thence to Timor, 28 April to 14 June 1789*. Sydney: Allen & Unwin, with National Library of Australia

Badger, Geoffrey Malcolm (1978) *Science and Technology in Australia 1977–78. A Report to the Prime Minister by the Australian Science and Technology Council (ASTEC)*. 2 vols. Canberra: Australian Government Publishing Service

Badger, Geoffrey Malcolm (1979) *Marine Sciences and Technologies in Australia: Immediate Issues. A Report to the Prime Minister by the Australian Science and Technology Council (ASTEC)*. Canberra: Australian Government Publishing Service

Banfield, Edmund James (1908a) *The Confessions of a Beachcomber: Scenes and Incidents in the Career of an Unprofessional Beachcomber in Tropical Queensland*. London: T. Fisher Unwin

Banfield, Edmund James (1908b) 'Dunk Island: its General Characteristics', *Proceedings*, Royal Geographical Society of Australasia, Queensland, xxiii (1908), 51–64

Banfield, Edmund James (1911) *My Tropic Isle*. London: T. Fisher Unwin

Banfield, Edmund James (1918) *Tropic Days*. London: T. Fisher Unwin

Banfield, Edmund James (1925) *Last Leaves from Dunk Island*. Ed. Alec Chisholm. Sydney: Angus & Robertson

Banks, Joseph (1770, 1771) *The* Endeavour *Journal of Joseph Banks*. 2 vols. Ed. John Cawte Beaglehole (1962) Sydney: Trustees of the Public Library of New South Wales, with Angus & Robertson

Barlow, Nora (1933) *Charles Darwin's Diary of the Voyage of H.M.S. Beagle*. (Ed. from the original manuscript.) Cambridge: Cambridge University Press

Barlow, Nora (1958) *The Autobiography of Charles Darwin 1809–1882*. With original omissions restored. Ed. with Appendix and Notes by his grand-daughter. London: Collins

Barrett, Glynn (1988) *The Russians and Australia*. Vol. 1 of *Russia and the South Pacific 1696–1840*. Vancouver: University of British Columbia Press

Bartlett, Norman (1954) *The Pearl Seekers*. London: Andrew Melrose

Beaglehole, John Cawte (1968) *The Voyage of the* Endeavour *1768–1771* (The Journals of Captain James Cook on his voyages of discovery). Repr. 1st edn 1955. Cambridge: Cambridge University Press for the Hakluyt Society

Beckett, Jeremy (1987) *Torres Strait Islanders: Custom and Colonisation*. Cambridge: Cambridge University Press

Bell, Peter & Veron, Charlie (1998) *AIMS: The First Twenty-Five Years*. Townsville: AIMS

Bladen, Frank Murcott (1893) Foreword: *Historical Records of Australia*. Series I: vii, viii, ix, x; January 1809–December 1822. Sydney: Library Committee of the Commonwealth Parliament

Bligh, William (1789) *The Bligh Notebook: Rough Account – Lieutenant Wm Bligh's Voyage in the Bounty's Launch from the Ship to Tofua & from thence to Timor, 28 April to 14 June 1789*. Ed. John Bach. Sydney: Allen & Unwin, with National Library of Australia, 1987

Blunt, Wilfrid & Stearn, William T. (1994) *The Art of Botanical Illustration*. Rev. & enlarged edn. Woodbridge, Suffolk, England: Antique Collectors' Club with Royal Botanic Gardens, Kew

Botsford, Louis W., Castilla, Juan Carlos and Peterson, Charles H. (1997) 'The management of

fisheries and marine ecosystems', *Science*, vol. 277, 25 July 1997, 509–15
Bowen, Sir George Ferguson (1889) *Thirty Years of Colonial Government: Selection from the Despatches and Letters of the Right Hon. Sir George Ferguson Bowen*. 2 vols. Ed. Stanley Lane-Poole. London: Longmans, Green & Co.
Bowen, James (1994) 'The Great Barrier Reef: towards conservation and management', ch. 11 in Stephen Dovers (ed.), *Australian Environmental History: Essays and Cases*. Melbourne: Oxford University Press
Bowen, Margarita (1981) *Empiricism and Geographical Thought: from Francis Bacon to Alexander von Humboldt*. Cambridge: Cambridge University Press
Brabourne Papers, The (1886) 'Pamphlet containing a summary of the contents.' Sydney: Charles Potter, Government Printer
Branagan, David (1996) (ed. with introd.) *Science in a Sea of Commerce: Journal of a South Seas Trading Venture (1825–1827) by Samuel Stutchbury*. Privately printed by David Branagan
Brennan, Frank (1993) 'Mabo and its implications for Aborigines and Torres Strait Islanders', in *Mabo: A Judicial Revolution*, ed. Margaret Stephenson & Suri Ratnapala. Brisbane: University of Queensland Press
Broadbent, Beverley (1999) *Inside the Greening: 25 years of the Australian Conservation Foundation*. Elwood, Vic.: Insite Press
Brockway, Lucile (1979) *Science and Colonial Expansion: The Role of the British Royal Botanic Gardens*. New York: Academic Press
Broderick, D., et al. (1994) 'Genetic studies of the Hawksbill turtle *Eretmochelys imbricata*: evidence for multiple stocks in Australian waters', in *Pacific Conservation Biology*, I:123–31
Brooks, Paul (1985) *Rachel Carson at Work: The House of Life*. Boston, Mass.: G. K. Hall & Co.
Brown, Robert (1810) *Prodromus florae Novae Hollandiae, et insulae van Diemen*. Facsimile 1960, intro. and trans. William T. Stearn. Weinheim/Bergstr. H.R. Englemann, & New York: Hafner Publishing
Brown, Stephen, et al. (1999) *Economic Impacts of the Kyoto Protocol*. Australian Bureau of Agricultural and Resource Economics. Canberra: Commonwealth of Australia
Brown, Theo (1972) *Crown of Thorns: The Death of the Great Barrier Reef?* Sydney: Angus & Robertson
Browne, Janet & Neve, Michael (1989) 'Voyage of the *Beagle*', *Journal of Researches*, ed. and abridged, Janet Browne & Michael Neve. London: Penguin, 1989
Brundtland Report (1987) World Commission on Environment and Development: *Our Common Future*. Oxford & New York: Oxford University Press
Burkhardt, Frederick et al. (eds) (1985 *et seqq.*) *The Correspondence of Charles Darwin*, vols 1–9. Cambridge: Cambridge University Press
Burmester, Henry (1984) 'Australia and the law of the sea: the protection and preservation of the marine environment', ch. 18 in K. W. Ryan, *International Law in Australia*, 2nd edn. Sydney: Law Book Company
Cannon, Lester Robert Glen & Silver, Howard (1986) *Sea Cucumbers of Northern Australia*. Brisbane: Queensland Museum
Carr, Archie (1967) *So Excellent a Fishe*. American Museum of Natural History. Reprinted, New York: Doubleday, Anchor Books, 1973
Carr, Archie (1969) 'Sea turtle resources of the Caribbean and Gulf of Mexico', *Bulletin*, International Union for the Conservation of Nature, 2.74–5

Carrington, Hugh (ed.) (1948) *The Discovery of Tahiti: A Journal of the Second Voyage of HMS Dolphin Round the World by George Robertson*. London: Hakluyt Society

Carson, Rachel (1962) *Silent Spring*. New York: Houghton Mifflin

Chamisso, Adelbert von (1821) 'On the Coral Islands', in Otto von Kotzebue, *Voyage of Discovery into the South Sea and Beering's Straits*. Trans. H. E. Lloyd. (Vol. 3, 331–36) London: Longmans, Hurst, Rees, Orme & Brown

Chapman, Ivan (1975) *Iven G. Mackay: Citizen and Soldier*. Melbourne: Melway Publishers

Chisholm, Alexander Hugh (1969) *The Joy of the Earth*. Sydney: William Collins

Christesen, C. B. (c.1936) *Queensland Journey*. Brisbane: P. A. Meehan Publicity Service for Queensland Government Tourist Bureau

Churchward, W. B. (1888) *Blackbirding in the South Pacific*. London: Swan Sonnenschein

Claringbould, Roy, Deakin, Jane, and Foster, Pat (1984) *Data Review of Reef-Related Tourism 1946–1980*. Townsville: GBRMPA

Cole, A. E. (1933) 'Organised travel: the work of the Government Tourist Bureau', *The Queen State: A Handbook of Queensland Compiled under the Authority of the State*. Brisbane: John Mills

Collingridge, George (1895) *Discovery of Australia*. Sydney: Hayes Brothers

Commonwealth of Australia Department of Science and the Environment (1980) *The Great Barrier Reef – the Commonwealth Government Role*. Canberra: Australian Government Publishing Service

Cook, James (1770, 1771) *The Voyage of the* Endeavour *1768–1771* (The Journals of Captain James Cook on his voyages of discovery). Ed. John Cawte Beaglehole. Repr. 1st edn, 1955. Cambridge: Cambridge University Press for the Hakluyt Society, 1968

Cook, James (1774) *The Voyage of the* Resolution *and* Adventure *1772–1775* (The Journals of Captain James Cook on his voyages of discovery). Ed. John Cawte Beaglehole. Cambridge: Cambridge University Press for the Hakluyt Society, 1961

Cotter, Maria, Boyd, Bill and Gardiner, Jane (2001) *Heritage Landscapes: Understanding Place and Communities*. Lismore: Southern Cross University Press

Cresswell, I. D. & Thomas, G. M. (eds) (1997) *Terrestrial and Marine Protected Areas in Australia 1997*. Canberra: Environment Australia

Croucher, Paul (1989) *A History of Buddhism in Australia, 1848–1988*. Sydney: University of New South Wales Press

Cullen, E. A. (1895) *List of Lighthouses, Lighted Beacons and Floating Lights on the Queensland Coast* (corrected to 31 December 1894). Brisbane: Government Printer

Cumbrae-Stewart, Francis William Sutton (1930) *The Boundaries of Queensland, with special reference to the Maritime Boundary and the 'Territorial Waters Jurisdiction Act 1878'*. Brisbane: University of Queensland & Government Printer

Currie, George & Graham, J. (1966) *The Origins of CSIRO: Science and the Commonwealth Government 1901–1926*. Melbourne: CSIRO

Dakin, William (1913) *Pearls*. Cambridge: Cambridge University Press

Dalrymple, Alexander (1786) *A Memoire Concerning the Chagos and Adjacent Islands*. Privately published

Darwin, Charles (1838) 'On certain areas of elevation and subsidence in the Pacific and Indian Oceans, as deduced from the study of coral formations', *Proceedings*, Geological Society of London, 2 (1838):552–54

Darwin, Charles (1839a) *Narrative of the Surveying Voyages of His Majesty's Ships* Adventure *and* Beagle *Between the Years 1826 and 1836, describing their Examination of the Southern Shores of South America, and the* Beagle's *Circumnavigation of the Globe*. Volume III. 'Journal of researches into the geology and natural history of the various countries visited by H.M.S. *Beagle* 1832–1836.' London: Henry Colburn

Darwin, Charles (1839b) *Voyage of the* Beagle. (Charles Darwin's *Journal of Researches.*) Ed. and abridged, Janet Browne & Michael Neve. London: Penguin, 1989

Darwin, Charles (1839c) 'The parallel roads of Glen Roy', *Philosophical Transactions*, Royal Society of London, I. 1839, 39–81

Darwin, Charles (1842) *The Structure and Distribution of Coral Reefs*. London: Smith & Elder

Darwin, Charles (1859) *The Origin of Species by Means of Natural Selection or the Preservation of Favoured Races in the Struggle for Life*. London: John Murray

Darwin, Charles (1874) *The Structure and Distribution of Coral Reefs*. 2nd edn, revised. London: Smith Elder & Co.

Darwin, Charles (1876) *The Autobiography of Charles Darwin 1809–1882*. With original omissions restored. Edited with Appendix and Notes by his grand-daughter, Nora Barlow. London: Collins, 1958

Darwin, Charles (1985–98) *The Correspondence of Charles Darwin*. Vols. 1–9. Ed. Frederick Burkhardt et al. Cambridge: Cambridge University Press

Darwin, Francis (1887) *Life and Letters of Charles Darwin*. (Containing the abridged *Autobiography*.) London: John Murray

Darwin, Francis (1908) *Charles Darwin: His Life Told in an Autobiographical Chapter, and in a Selected Series of Published Letters*. London: John Murray

Davenport, Winifred (1986) *Harbours and Marine: Port and Harbour Development in Queensland 1824–1985*. Brisbane: Department of Harbours and Marine

David, Andrew (1995) *The Voyage of HMS* Herald, *to Australia and the South-west Pacific 1852–1861 under the Command of Captain Henry Mangles Denham*. Carlton: Miegunyah Press, Melbourne University Press

David, Tannatt Edgeworth (1904) Section IV. 'Narrative of the Second and Third Expeditions', *The Atoll of Funafuti: Borings into a Coral Reef and the Results*. Report of the Coral Reef Committee of the Royal Society. London: The Royal Society of London

Day, Max (1971) *Marine Science in Australia: Report of the Interim Council of the Australian Institute of Marine Science*, July 1971. Melbourne: CSIRO

de Quatrefages, A. (1857) *Rambles of a Naturalist on the Coasts of France, Spain, and Sicily*. Trans. E. C. Otté. 2 vols. London: Longman, Brown, Green, Longmans & Roberts

Dexter, Ralph W. (1972) 'Historical aspects of Louis Agassiz's lectures on corals and coral reefs', *Proceedings of the Symposium on Corals and Coral Reefs*. Ed. C. Mundandan & C. S. Gopindha Pillai. Cochin: Marine Biological Association of India, 489–96

di Gregorio, Mario A. (1984) *T. H. Huxley's Place in Natural Science*. New Haven: Yale University Press

Docker, Edward Wybergh (1970) *The Blackbirders: The Recruiting of South Seas Labour for Queensland*. Sydney: Angus & Robertson

Douglass, William A. (1995) *From Italy to Ingham*. Brisbane: University of Queensland Press

Ebes, Hank (1988) *The Florilegium of Captain Cook's First Voyage to Australia 1768–1771*. Melbourne: Ebes Douwma, Editions Alecto, British Museum (Natural History)

Ellis, John (1755) *An Essay towards a Natural History of the Corallines, and Other Marine Productions of the Like Kind*. London: John Ellis

Environment Australia (1998) *Preliminary Study of Potential Impacts on the Great Barrier Reef World Heritage Area from Coastal Urban Development*. Environment Australia, GBRMPA,

Livingstone Shire Council, Townsville City Council (with North Queensland Local Government Association, Central Queensland Local Government Association, Queensland Department of the Environment). Townsville: GBRMPA

Field, Barron (1819) *First Fruits of Australian Poetry*. Reprint. Sydney: Barn on the Hill, 1941

Fitzgerald, Lawrence (1984) *Java la Grande: The Portuguese Discovery of Australia*. Hobart: privately published

Fitzgerald, Ross (1984) *A History of Queensland, from 1915 to the 1990s*. Brisbane: University of Queensland Press

Fletcher, Joseph James (1900) 'On the rise and early progress of our knowledge of the Australian fauna.' Presidential Address, Section D, *Report of the Eighth Meeting of the Australasian Association for the Advancement of Science*, Melbourne 1900, pp. 69–101. Published by the Association

Flinders, Matthew (1814) *A Voyage to Terra Australis, Undertaken for the Purpose of Completing the Discovery of that Vast Country and Prosecuted in the Years 1801, 1802 and 1803*. 2 vols with an Atlas. London: G. & W. Nicol

Foley, John (1982) *Reef Pilots: The History of the Queensland and Torres Strait Pilot Service*. Sydney: Banks Bros. & Street

Foley, John (1990) *The Quetta*. Brisbane: Nairana Pty Ltd

Forster, George R. (1778) *Observations Made During a Voyage Around the World, on Physical Geography, Natural History and Ethnic Philosophy*. Section IV: 'The Theory and Formation of Isles'. London: G. Robinson

Frankel, Edgar (1978) *Bibliography of the Great Barrier Reef Province*. Canberra: Australian Government Publishing Service

Frost, Alan (1993) *Sir Joseph Banks and the Transfer of Plants to and from the South Pacific*. Limited edn. Melbourne: Colony Press

Fry, Howard T. (1979) 'Alexander Dalrymple and Captain Cook: the creative interplay of two careers', in Fisher, Robin & Johnston, Hugh (eds) (1979) *Captain James Cook and His Times*. Canberra: Australian National University Press

Fulton, Thomas W. (1911) *The Sovereignty of the Sea*. Edinburgh & London: William Blackwood & Sons

Fuson, Robert H. (1987) (trans.) *The Log of Christopher Columbus*. Southhampton: Ashford Press

Ganter, Regina Josefa (1994) *The Pearl Shellers of Torres Strait: Resource Use, Development and Decline, 1860s–1960s*. Carlton: Melbourne University Press

Geen, Gerry & Lal, Padma (1991) *Charging Users of the Great Barrier Reef Marine Park*. Canberra: Australian Bureau of Agricultural and Resource Economics

Gill, James Connal Howard (1988) *The Missing Coast: Queensland Takes Shape*. Brisbane: Queensland Museum

Goddard, Ernest James (1933) 'The economic possibilities of the Great Barrier Reef', *The Queen State: A Handbook of Queensland Compiled Under the Authority of the State*. Brisbane: John Mills

Good, Peter (1802) *The Journal of Peter Good: Gardener on Matthew Flinders' Voyage to Terra Australis 1801–1803*. Transcribed & ed. Phyllis Edwards (1981). London: Bulletin of the British Museum (Natural History), Historical Series, vol. 9

Goode, John (1977) *The Rape of the Fly*. Melbourne: Nelson

Goreau, Thomas F. (1959) 'The physiology of skeleton formation in corals. I. A method for measuring the rate of calcium deposition by corals under different conditions.' *Biological Bulletin*, Marine Biological Laboratory, Woods Hole, vol. 116, 59–75

Goreau, Thomas F. (1961) 'Problems of growth and calcium deposition on coral reefs', *Endeavour*, vol. 20, 32–9

Goreau, Thomas F., Goreau, Nora J., Goreau, Thomas J. (1979) 'Corals and coral reefs', *Scientific American*, vol. 241, August 1979, 110–20

Gould, Stephen Jay (1987) *Time's Arrow: Time's Cycle: Myth and Metaphor in the Discovery of Geological Time*. Cambridge, Mass.: Harvard University Press

Great Barrier Reef Marine Park Authority (1981) *Nomination of the Great Barrier Reef by the Commonwealth of Australia for inclusion in the World Heritage List*. Townsville: GBRMPA

Great Barrier Reef Marine Park Authority (1994) *Keeping it Great: The Great Barrier Reef. A 25 Year Strategic Plan for the Great Barrier Reef Heritage Area 1994–2019*. Townsville: GBRMPA

Great Barrier Reef Marine Park Authority (1996) *The Great Barrier Reef: Science, Use and Management*. Proceedings, 2 vols. Townsville: GBRMPA

Great Barrier Reef Marine Park Authority (1998) *State of the Great Barrier Reef World Heritage Area 1998*. Townsville: GBRMPA

Great Barrier Reef Marine Park Authority (1999) *An Overview of the Great Barrier Reef Marine Park Authority Representative Areas Program*. Townsville: GBRMPA

Great Barrier Reef Marine Park Regulations (Amendment), Statutory Rules 1991, No. 296; Reg. 26A, Pt. 4A

Groeben, Christiane (1985) 'Anton Dohrn – the statesman of Darwinism'. *Biological Bulletin*, 168 (Suppl.) (June 1985) 4–25

Halley, Edmond (1696) *Correspondence and Papers of Edmond Halley*. Ed. Eugene Fairfield MacPike (1932) Oxford: Clarendon Press

Hardstaff, Reginald John (1995) *Leadline to Laser: the Hydrographic Service, Royal Australian Navy 1920–1995*. Sydney: R. J. Hardstaff

Harper, P. & Powell, T. E. (1991) (compilers) *Catalogue of the Papers and Correspondence of Sir Charles Maurice Yonge F.R.S (1899–1986)*. Held in General Library, Natural History Museum, London

Harrison, Peter L., Babcock, R., Bull, G., Oliver, J. K., Wallace, C., Willis, B., (1984) 'Mass spawning in tropical coral reefs', *Science* 223:16 March 1984, 1186–89

Haswell, William Aitcheson (1891) 'Recent biological theories', Presidential Address, Section D, *Report of the Meeting of the Australasian Association for the Advancement of Science*, Melbourne 1891. Vol. 3, 173–92. Published by the Association

Hawkesworth, John (1773) *An Account of the Voyages Undertaken by the Order of His Present Majesty for Making Discoveries in the Southern Hemisphere*. 4 vols. 1 May 1773. London: W. Strahan & T. Cadell. Second edition, September 1773

Hayman, Roslyn (1979) *Episodes: A Glimpse of Australia-Japan Relations, 1859–1979*. Canberra: Embassy of Japan in Australia

Haynes, David & Michalek-Wagner, Kirsten (2000) 'Water quality in the Great Barrier Reef World Heritage Area: past perspectives, current issues and new research directions', *Marine Pollution Bulletin*, vol. 41, nos. 7–12, 428–34

Heatwole, Harold (1981) *A Coral Island: the Story of One Tree Reef*. Sydney: William Collins

Heatwole, Harold (1984) 'The Cays of the Capricornia Section, Great Barrier Reef Marine Park, and a history of research on their terrestrial biota', in *The Capricornia Section of the Great Barrier Reef, Past Present and Future*. Brisbane: Royal Society of Queensland & Australian Coral Reef Society

Heatwole, Harold (1987) *Sea Snakes*. Sydney: New South Wales University Press

Hedley, Charles (ed.) (1896a) *Memoir III, 1896–1900: 'The atoll of Funafuti: zoology, botany, ethnology and general structure'*. Australian Museum, Sydney

Hedley, Charles (1896b) *Memoir III, 1896–1900: 'General account of the atoll of Funafuti'*. Australian Museum, Sydney, 1–73

Hedley, Charles (1917) 'The economics of *Trochus niloticus*', *Australian Zoologist*, I.4, 69

Henderson, Caspar (2002) 'Electric Reefs', *New Scientist*, 6 July 2002: 38–41

Hilder, Brett (1980) *The Voyage of Torres: The Discovery of the South*. Brisbane: University of Queensland Press

Hill, Dorothy (1985a) 'The Great Barrier Reef Committee: 1922–1982: The first thirty years', *Historical Records of Australian Science*, vol. 6, no.1

Hill, Dorothy (1985b) 'The Great Barrier Reef Committee: 1922–1982: The last three decades', *Historical Records of Australian Science*, vol. 6, no.2

Historical Records of Australia. Series I: vii,viii,ix,x; January 1809–December 1822. Sydney: Library Committee of the Commonwealth Parliament

Historical Records of New South Wales. (1893–1901) vol.2. *Grose & Paterson, 1793–1795*; vol.4. *Hunter & King, 1800, 1801, 1802*; vol.5. *King, 1803, 1804, 1805*; vol.7 (1901). Frank Murcott Bladen (ed.) Sydney: Government Printer

Hoare, Michael E. (1969) 'All things are queer and opposite', *Isis*, vol.60, 198–209

Hoare, Michael E. (1976) *The Tactless Philosopher: Johann Reinhold Forster 1729–1798*. Melbourne: Hawthorn Press

Hoare, Michael E. (ed.)(1982) *The 'Resolution' Journal of Johann Reinhold Forster (1772–1775)* London: Hakluyt Society

Hoegh–Guldberg, Ove (1999) *Climate Change, Coral Bleaching, and the Future of the World's Coral Reefs*. Greenpeace (Internet) Report

Hoffmeister, J. Edward (1924) *Some Posthumous Papers of A. G. Mayor, Relating to His Work at Tutuila Island and Adjacent Regions*. Vol. XIX. Washington DC: Papers from the Department of Marine Biology, Carnegie Institution of Washington

Hoffmeister, J. Edward (1925) *Some Corals from American Samoa and the Fiji Islands*. Vol. XXII. Washington DC: Papers from the Department of Marine Biology, Carnegie Institution of Washington

Hope, (Justice) Robert (1974) *Report on the National Estate*. Report of the Committee of Inquiry into the National Estate. Canberra: Australian Government Publishing Service

Hughes, Terence P. (1994) 'Catastrophes, phase shifts, and large-scale degradation of a Caribbean coral reef', *Science*, vol. 265, 9 September 1994, 1547–50

Humboldt, Alexander von (1850) *Views of Nature: or, Contemplation on the Sublime Phenomena of Creation with scientific illustrations*. (1st publ. 1808.) Trans. E. C. Otté & Henry G. Bohn. London: Henry G. Bohn

Humboldt, Alexander von (1852) *Personal Narrative of Travels to the Equinoctial Regions of America during the years 1799–1804 by Alexander von Humboldt and Aimé Bonpland*. (1st publ. Paris 1814–25.) Trans. & ed. Thomasina Ross. London: George Routledge

Humphries, C. J. & Fisher, C. T. (1994) 'The loss of Banks's legacy', *Philosophical Transactions*, Royal Society of London, B (1994) 344, 3–9

Hutton, James (1788) 'Theory of the Earth', *Transactions*, Royal Society of Edinburgh, I:209–305

Hutton, James (1795) *Theory of the Earth with Proofs and Illustrations*. Edinburgh: William Creech

Huxley, Julian (ed.) (1935) *T. H. Huxley's Diary of the Voyage of HMS Rattlesnake*. London: Chatto & Windus

Ingleton, Geoffrey C. (1944) *Charting a Continent*. Sydney: Angus & Robertson

Ingleton, Geoffrey C. (1986) *Matthew Flinders: Navigator and Chartmaker*. Guildford, Surrey: Genesis Publications

Iredale, Tom (1969) 'Charles Hedley', *Proceedings*, Royal Zoological Society of New South Wales, 1967–68, 24 April, 26–31

Jack, Logan (1922) *Northmost Australia*. 2 vols. Melbourne: George Robertson & Co.

Jackson, Jeremy B. C. et al. (2001) 'Historical overfishing and the recent collapse of coastal ecosystems', *Science*, vol. 293, 27 July 2001, 629–37

James, Peter (1976) *Requiem for the Reef*. Brisbane: Foundation Press

Jameson, H. Lyster (1912) 'Biological science and the pearling industry', *Knowledge*, November 1912, 421–31

Jensen, H. I. (1923) 'The probable oil formations of North-Eastern Australia', *Proceedings*, Pan-Pacific Science Congress, Melbourne & Sydney, Australia 1923; 1924:1254–75. Ed. Gerald Lightfoot. Melbourne: Government Printer

Jones, Raymond Marshall (1989) *Seagulls, Cruisers and Catapults: Australian Naval Aviation 1913–1944*. Taroona, Tasmania: Pelorus Publications

Jukes, Joseph Beete (1847) *Narrative of the Surveying Voyage of H.M.S. Fly, Commanded by Captain F. P. Blackwood in Torres Strait, New Guinea, and Other Islands of the Eastern Archipelago, During the Years 1842–1846*. 2 vols. London: T. & W. Boone

Jupp, David L. B. (1986) *The Application and Potential of Remote Sensing in the Great Barrier Reef Region*. Townsville: GBRMPA

Keesing, J. K., Bradbury, R. H., DeVantier, L. M., Riddle, M. J., De'ath, G. (1992) 'Geological evidence for recurring outbreaks of the crown-of-thorns starfish: a reassessment from an ecological perspective', *Coral Reefs*, 11 (1992), 79–85

Kelleher, Graeme (1987) Foreword to 'The Crown of Thorns Starfish' in *Australian Science Magazine*, Issue 3

King, Phillip Parker (1827) *Narrative of a Survey of the Intertropical and Western Coasts of Australia Performed Between the Years 1818 & 1822*. 2 vols. London: John Murray

King, Phillip Parker (1839) *Narrative of the Surveying Voyages of His Majesty's Ships* Adventure and Beagle, *Between the Years 1826 and 1836*. London: Henry Colburn

Knibbs, G. H., Commonwealth Statistician (1911) *Census of the Commonwealth of Australia*, taken for the night between the 2nd and 3rd April, 1911. Volume I. Statistician's Report, including appendices. Melbourne: Minister of State for Home and Territories

Lack, Clem L. (1959) 'The taming of the Great Barrier Reef', *Journal*, Royal Australian Historical Society, Queensland, 23 April 1959, 130–54, 138

Ladd, Harry Stephen, Ingerson, E., Townsend, R. C., Russell, M., Stephenson, H. K. (1953) 'Drilling on Eniwetok Atoll', *American Association of Petroleum Geologists*. 37: 2257–80

Ladd, Harry Stephen (1968) 'Preliminary report on conservation and controlled exploitation of the Great Barrier Reef'. Brisbane: Mines Department

Lamarck, Jean-Baptiste Pierre Antoine de Monet de (1809a) *Zoological Philosophy: An Exposition with regard to the Natural History of Animals*. Trans. Hugh Elliot. London: Macmillan 1914

Lamarck, Jean-Baptiste Pierre Antoine de Monet (1809b) *Philosophie Zoologique*. Paris: L'Edition de Schleicher frères, 1907

Lang, John Dunmore (1847) *Cooksland in North-Eastern Australia; The Future Cotton Field of Great Britain; Its Characteristics and Capabilities for European Colonisation*. London: Longman, Brown, Green & Longmans

Lassig, Brian & Kelleher, Graeme (1991) 'Crown-of-thorns starfish on the Great Barrier Reef', in Australian Science and Technology Council, *Environmental Research in Australia: case studies*. Canberra: Australian Government Publishing Service

Leggett, Jeremy (2000) *The Carbon War*. London: Penguin

Lenox-Conyngham, Gerald & Potts, Frank Armitage (1925) 'The Great Barrier Reef', *The Geographical Journal*, April 1925, 314–23

Lightfoot, Gerald (1926) 'Marine Biological Economics and Fishery Investigations'. Mimeographed report to the Commonwealth Institute of Science and Industry, CSIRO Archives, I.82/0/1

Limpus, Colin J. (1992) 'The Hawksbill Turtle *Eretmochelys imbricata*, in Queensland: population structure within a Southern Great Barrier Reef feeding ground', *Wildlife Resources*, 19:1992, 489–506

Linnaeus, Carolus (Carl von Linné) (1735) *Systema naturae*. Lugdunum (Lyons): G.D. Ehret

Linnaeus, Carolus (Carl von Linné) (1758) *Systema naturae: Regnum animale*. Photographic facsimile, London: British Museum (Natural History), 1956

Loney, Jack (1993) *Wrecks on the Queensland Coast 1791–1992*. Yarram. Vic.: Lonestone Press, Oceans Enterprises

Love, Rosaleen & Grasset, Nike (1984) 'Women scientists and the Great Barrier Reef: a historical perspective', *Search*, vol. 15, no. 9–10, October/November 1984, 285–87

Lovejoy, Arthur O. (1936) *The Great Chain of Being: A Study of the History of an Idea*. Cambridge, Mass.: Harvard University Press; repr. Harper Torchbook, 1960

Lucas, Percy (Bing) (1997) *The Outstanding Universal Value of the Great Barrier Reef*. Townsville: GBRMPA & James Cook University

Luntz, Stephen (2002) 'Coral bleaching worsens', *Australasian Science*, July 2002:7

Lyell, Charles (1830, 1832, 1833) *Principles of Geology, Being an Attempt to Explain the Former Changes of the Earth's Surface*. 3 vols. London: John Murray

Lyell, Katherine (1881) *Life and Letters of Sir Charles Lyell, Bart*. 2 vols. London: John Murray

Mabberley, David J. (1985) *Jupiter Botanicus: Robert Brown of the British Museum*. Braunschweig: Verlag von J. Cramer

MacGillivray, John (1852a) *Narrative of the Voyage of HMS* Rattlesnake. 2 vols. London: T. & W. Boone

MacGillivray, John (1852b) *Voyage of H.M.S.* Herald *under the command of Captain H. Mangles Denham*. 2 vols. (Private journal in Admiralty, London; microfilm copy, Oxley Library, Brisbane, Film 0025)

Mackay, John Anderson (1908) Chairman: *Royal Commission Appointed to Inquire into the Working of the Pearl-Shell and Beche-de-Mer Industries Report*. Brisbane: Government Printer

Macknight, Charles Campbell (1976) *The Voyage to Marege: Macassan Trepangers in Northern Australia*. Carlton: Melbourne University Press

Maclaren, Charles (1841) 'The glacial theory of Professor Agassiz of Neuchâtel', *Scotsman*. Edinburgh: The Scotsman Office. Reprinted, *American Journal of Science*, 42:346–65

Maclaren, Charles (1843) 'On coral islands and reefs, as described by Mr Darwin', *Edinburgh New Philosophical Journal*, 34 (1843):33–47

Macmillan, David Stirling (1957) *A Squatter Went to Sea: The Story of Sir William Macleay's New Guinea Expedition (1875) and His Life in Sydney*. Sydney: Currawong Press

Mahony, D. J. (1923) 'Petroleum prospects in Victoria', *Proceedings*, Pan-Pacific Science Congress, Melbourne & Sydney, Australia, 1923:1251–52. Ed. Gerald Lightfoot. Melbourne: Government Printer

Malakoff, David (1997) 'Extinction on the high seas', *Science,* vol. 277, 25 July 1997, 486–88

Marr, David (1980) *Barwick.* Sydney: George Allen & Unwin

Marsh, Helene (1996) *The Status of the Dugong in the Southern Great Barrier Reef.* Townsville: GBRMPA, Research Publication No. 41

Marshall, Sheina M. (1930) 'The recent expedition to the Great Barrier Reef of Australia', *Proceedings*, Royal Philosophical Society of Glasgow (1930), vol. 58, 84–100

Mather, Patricia (1998) 'The new coral reef science', *Proceedings of the Australian Coral Reef Society 75th Anniversary Conference, Heron Island October 1997.* Ed. J. G. Greenwood & N. J. Hall. Brisbane: School of Marine Science, University of Queensland

Mayer (Mayor), Alfred Goldsborough (1918) *Ecology of the Murray Island Coral Reef.* Vol. IX. Washington DC: Papers from the Department of Marine Biology, Carnegie Institution of Washington

Mayor, Alfred Goldsborough (1924) *Some Posthumous Papers of A. G. Mayor, Relating to His Work at Tutuila Island and Adjacent Regions.* Vol. XIX. Washington DC: Papers from the Department of Marine Biology, Carnegie Institution of Washington

McCoy, Frederick (1869) Lecture 1, 28 June 1869: 'The order and plan of creation', *Lectures delivered before the Early Closing Association* [for limiting shop closing hours to 6 p.m.] *1869–1870*. Melbourne: Samuel Mullen, pp. 1–10

McCoy, Frederick (1870) Lecture 2, 4 July 1870: 'The order and plan of creation', *Lectures delivered before the Early Closing Association* [for limiting shop closing hours to 6 p.m.] *1869–1870*. Melbourne: Samuel Mullen, pp. 11–32

McCracken, K. G. & Kingwell, J. (eds) (1988) *Marine and Coastal Remote Sensing in the Australian Tropics.* Canberra: CSIRO Office of Space Science and Applications

McIntyre, Kenneth (1982) *The Secret Discovery of Australia.* Rev. & abr. edn (1st publ. 1977, rev. 1982). Edn cited, Sydney: Pan Books 1987

McKinnon, Ken R. (1989) *Oceans of Wealth? A Report by the Review Committee on Marine Industries, Science and Technology.* Executive Summary. Canberra: Australian Government Publishing Service

McMichael, Donald F. (1969) 'Marine National Parks and Reserves' in L. J. Webb (ed.), *The Last of Lands*, pp. 115–22. Brisbane: Jacaranda Press

McPhail, Ian (1996) 'Partnerships and collaboration: management of the Great Barrier Reef World Heritage Area, past, present and future', *The Great Barrier Reef: Science, Use and Management.* Proceedings. Townsville: GBRMPA, I.5–8

Monkman, Kitty (Getela) (1975) *Over and Under the Great Barrier Reef.* Cairns: *Cairns Post*, for Kitty Monkman

Moore, David R. (1979) *Islanders and Aborigines at Cape York.* (An ethnographic reconstruction based on the 1848–1850 *'Rattlesnake'* Journals of O. W. Brierly and information he obtained from Barbara Thompson.) Canberra: Australian Institute of Aboriginal Studies

Moorhouse, Frank W. (1933) 'Notes on the Green Turtle (Chelonia mydas)', *Reports of the Great Barrier Reef Committee*, vol. 4, part 1, no.1, 1–22

Moran, Peter J. & Johnson, David B. (1990) *Final Report on the Results of COTSAC Ecological Research: December 1985 – June 1989.* Townsville: Australian Institute of Marine Science

Moresby, John (1876) *Discoveries and Surveys in New Guinea and the D'Entrecasteaux Islands.* London: John Murray

Morrison, William L. (1974) *Towards an Australian Science Council.* Canberra: Australian Government Publishing Service

Morton, Brian (1992) 'Charles Maurice Yonge, 9 December 1899 – 17 March 1986', *Biographical Memoirs of the Royal Society*, vol. 38, 1992, 379–412

Morwood, Michael J. (1993) 'Cause and effect: Pleistocene aboriginal occupation in the Quinkan region, southeast Cape York Peninsula', in M. A. Smith, et al., *Sahul in Review: Pleistocene Archaeology in Australia, New Guinea and Island Melanesia: Occasional Papers in Prehistory*, No.24, Canberra: Research School of Pacific Studies, Australian National University

Moseley, Henry Nottidge (1880) *Notes by a Naturalist during the Voyage of HMS* Challenger. 3rd edn. London: John Murray

Moyal, Ann (1986) *A Bright and Savage Land*. Ringwood, Vic.: Penguin

Mullins, Steve (1992a) 'Queensland's quest for Torres Strait', *Journal of Pacific History*, 27.2 (1992), 165–80

Mullins, Steve (1992b) 'The Torres Strait bêche-de-mer fishery: a question of timing', *Journal, Australian Association for Maritime History*, 14.1 (1992), 21–30

Mullins, Steve (1995) *Torres Strait: A History of Colonial Occupation and Culture Contact, 1864–1897*. Rockhampton: Central Queensland University Press

Murray, John (1880) 'On the structure and origin of coral reefs and islands', *Proceedings*, Royal Society of Edinburgh, vol. X, no. 107, 1879–80, 505–18

Murray, Keith (1957) *Report of the Committee on Australian Universities*. Canberra: Australian Government

Muscatine, Leonard & Cernichiari, Elsa (1969) 'Assimilation of photosynthetic products of Zooxanthellae by a reef coral', *Biological Bulletin*, December 1969, 137:306–23

Musgrave, Anthony & Whitely, Gilbert (1926) 'From sea to soup: an account of the turtles of North-West Islet', *Australian Museum Magazine*, vol. 2, 1926, 331–36

Napier, S. Elliott (1928) *On the Barrier Reef: Notes from a No-ologist's Pocket-Book*. Sydney: Angus & Robertson (7th printing cited, 1939)

Noonan, Michael (1983) *A Different Drummer: The Story of E. J. Banfield, the Beachcomber of Dunk Island*. Brisbane: University of Queensland Press

Noonan, Michael (1989) (ed.) *The Gentle Art of Beachcombing: A Collection of Writings by E. J. Banfield*. Brisbane: University of Queensland Press

'Oceanic Grandeur' Report (1970) *Report on the Grounding of the Oil Tanker 'Oceanic Grandeur' in the Torres Strait*. Brisbane: Government Printer

Odum, Harry T. & Odum, Eugene P. (1955) 'Trophic structure and productivity of a windward coral reef community on Enewetok Atoll', *Ecological Monographs*, 25:291–320

Osborne, Elizabeth (1997) *Torres Strait Islander Women and the Pacific War*. Canberra: Aboriginal Studies Press

Ottesen, Peter (ed.) (1994) *Hulls, Hazards and Hard Questions: Shipping in the Great Barrier Reef*. Townsville: GBRMPA

Pan-Pacific Scientific Congress 1920, Honolulu, Hawaii (1920) *Proceedings*. Honolulu: Bishop Museum & Honolulu Star-Bulletin

Pan-Pacific Scientific Congress 1920, Honolulu, Hawaii (1920) *Resolutions*. Honolulu: Bishop Museum & Honolulu Star-Bulletin

Pan-Pacific Science Congress, Melbourne & Sydney, Australia 1923 [1924?] *Proceedings* of the Second Congress. Ed. Gerald Lightfoot. Melbourne: Government Printer

Pan-Pacific Science Congress, Tokyo 1926 (1928) *Proceedings* of the Third Congress. Tokyo: The Congress

Parker, Jeffery & Haswell, William Aitcheson (1897) *Textbook of Zoology*. 2 vols. London: Macmillan

Parkinson, Stanfield (1773) *A Journal of a Voyage to the South Seas in His Majesty's Ship the Endeavour*. London: Stanfield Parkinson

Pennisi, Elizabeth (1997) 'Brighter prospects for the world's coral reefs?', *Science,* vol. 277, 25 July 1997, 491–93

Pike, Glenville (1979) *Queen of the North: A Pictorial History of Cooktown and Cape York Peninsula*. Mareeba: Pinevale Publications

Potts, D. C. (1981) 'Crown-of-thorns starfish – man-induced pest or natural phenomenon?' Ch. 4 in R. L. Kitching & R. E. Jones (eds), *The Ecology of Pests: Some Australian Case Histories*. Melbourne: CSIRO

Prideaux, Peter (1988) *From Spear to Pearl Shell: Somerset, Cape York Peninsula 1864–1877*. Brisbane: Boolarong Publications

Purchas, Samuel (1613–1626) *Hakluytus Posthumus, or Purchas his pilgrimes, contayning a history of the world in sea voyages and land travells by Englishmen and others*. 20 vols. Glasgow: James MacLehose & Sons, 1905–07

Queensland Environmental Protection Agency (1999) *State Coastal Management Plan: Queensland Coastal Policy*. Brisbane: Environmental Protection Agency

Queensland Geographical Journal (1907–1908) *Proceedings*, Royal Geographical Society of Australasia, Queensland, 23rd Session, vol. XXIII, s.v. 'Papers read: Address on the Great Barrier Reef', 87–90

Queensland Parliament (1970) *Report on the Grounding of the Oil Tanker 'Oceanic Grandeur' in the Torres Strait on 3rd March 1970 and the Subsequent Removal of Oil from the Waters*. Brisbane: Government Printer

Quoy, Jean René Constantin & Gaimard, Joseph Paul (1823) 'The formation of coral islands', 'Mémoire sur l'Accroissement des Polypes lithophytes considéré géoligiquement', *Annales des sciences naturelles*, VI. 273–90, Trans. Mather, K. F. & Mason, S. L. (1939), *A Source Book in Geology*, Harvard University Press; repr. 1970

Ralston, Caroline (1977) *Grass Huts and Warehouses: Pacific Beach Communities of the Nineteenth Century*. Canberra: Australian National University Press

Rattray, Alexander (1868) 'Notes on the physical geography, climate and capabilities of Somerset and the Cape York Peninsula, Australia', *Journal*, Royal Geographical Society of London, 1868, 38:370–411

Ray, Carleton (1962) 'Inshore marine conservation', in A. B. Adams (ed.), *First World Conference on National Parks*. Washington DC: US National Parks Service, 77–87

Raymond, Robert (1986) *Starfish Wars*. South Melbourne: Macmillan

Read, S., Ramsay, J., Blair, S. (1994) *Australian Heritage Commission: Current Work on Cultural Landscape Assessment*, ICOMOS Landscapes Working Group Newsletter, no. 7, March 1994

Resource Assessment Commission (1993) *Coastal Zone Inquiry: Final Report*. Canberra: Australian Government Publishing Service

Richards, Henry Caselli (1922) 'Problems of the Great Barrier Reef', *Queensland Geographical Journal*, XXXVII, 1922, 42–54

Richards, Henry Caselli & Hedley, Charles (1923a) 'A geological reconnaissance in North Queensland', *Transactions*, Royal Geographical Society of Australasia, Queensland. Vol. 1 (new series) 1923, 1–26

Richards, Henry Caselli & Hedley, Charles (1923b) 'Report to the Great Barrier Reef Committee', *Queensland Geographical Journal*, vol. 38, no. 24, 1923

Richards, Henry Caselli (1927) 'An account of the scientific investigations of the Great Barrier Reef

Committee, April 1922 – February 1927.' (Typescript in possession of the author)

Richards, Henry Caselli (1928) 'Report on Investigations, 1922–27', *Reports of the Great Barrier Reef Committee*, vol. 2:vii–xvi, 1928

Richards, Henry Caselli (1938) 'Boring operations at Heron Island, Great Barrier Reef', *Reports of the Great Barrier Reef Committee*, IV.135–42, 1938.

Richards, Henry Caselli & Hill, Dorothy (1942) 'Great Barrier Reef bores, 1926 and 1937: descriptions, analyses, and interpretations', *Reports of the Great Barrier Reef Committee*, vol. 5:1–109, 1942

Ristvet, Byron L. (1987) 'Geology and geohydrology of Enewetak Atoll'. *The Natural History of Enewetak Atoll*. Vol. 1, ch. 4: 'The Ecosystem: Environments, Biotas and Processes'. Oak Ridge, Tennessee: Office of Scientific and Technical Information, US Department of Energy

Robinson, F. C. B. (1864) *Proceedings*, Royal Geographical Society, VIII.4, 11 April 1864, 117

Rodgers, K. A. & Cantrell, Carol (1988) 'Charles Hedley and the 1896 Royal Society expedition to Funafuti', *Archives of Natural History*, 15(3), 269–80

Royal Commissions into exploratory and production drilling for petroleum in the area of the Great Barrier Reef (1974) *Report*, 2 vols. Canberra: Australian Government Publishing Service

Royal Geographical Society of Australasia (Queensland) *Transactions: Reports of the Great Barrier Reef Committee*. 7 vols, 1925–1968. Brisbane: GBRC

Russell, Henry Chamberlain (1888) Presidential Address, Section D, *Report at the First Meeting of the Australasian Association for the Advancement of Science*, Melbourne 1888. V.1, 1–11. Published by the Association.

Sapp, Jan (1999) *What is Natural? The Coral Reef Crisis*. New York, Oxford: Oxford University Press

Saville-Kent, William (1890a) *Pearl and Pearl–Shell Fisheries of Northern Queensland*. Report to the Queensland Parliament. Brisbane: Government Printer

Saville-Kent, William (1890b) 'Presidential Address', *Proceedings*, Royal Society of Queensland, VII.2, 1890, 17–42

Saville-Kent, William (1893) *The Great Barrier Reef of Australia*. London: W. H. Allen

Saville-Kent, William (1897) *The Naturalist in Australia*. London: Chapman & Hall

Saville-Kent, William (1905) *Torres Strait Pearlshell Fisheries*. Report to the Queensland Parliament. Brisbane: Government Printer

Schiebinger, Londa (1999) *Has Feminism Changed Science?* Cambridge, Mass.: Harvard University Press

Schmidt, Karen F. (1997) 'No-take zones spark fisheries debate', *Science*, vol. 277, 25 July 1997, 489–91

Semper, Carl Gottfried (1880) *Die natürlichen Existenzbedungungen der Thiere*. Leipzig. English trans., 1880, *The Natural Conditions of Existence as they Affect Animal Life*. London: Kegan Paul, Trench Trübner & Co.

Senate, Australian Parliament (1970) *Interim Report of the Senate Select Committee on Off-Shore Petroleum Resources*. Canberra: Australian Government Printing Office

Senate, Australian Parliament (1972) *Report of the Senate Select Committee on Off-Shore Petroleum Resources*. Parliamentary Paper no. 201, 1971. 3 vols. Canberra: Australian Government Printing Office

Senate, Australian Parliament (1980) *Standing Committee on Science and the Environment*, Progress Report on Australian Marine Science, mimeographed.

Senate, Australian Parliament (1997) *Marine and Coastal Pollution*. Parliamentary Paper no. 198, 1997

Senate, Australian Parliament (1998) *Access to Heritage: User Charges in Museums, Art Galleries and National Parks*. Parliamentary Paper no.142, July 1998

Senate, Australian Parliament (1999) *The Hinchinbrook Channel Inquiry*. Australian Senate Communications, Information Technology and the Arts References Committee. Canberra: Parliament of Australia

Sharp, Nonie (1992) *Footprints along Cape York Sandbeaches*. Canberra: Aboriginal Studies Press

Shaw, George & Smith, James Edward (1793) *Zoology and Botany of New Holland, and the Isles Adjacent*. London: J. Sowerby

Shaw, George (1800–1801) *General Zoology, or Systematic Natural History*. 4 vols. in 2. London: G. Kearsley

Sheppard, Charles (1999) 'Coral decline and weather patterns over 20 years in the Chagos Archipelago, Central Indian Ocean', *Ambio*, 28.6. September 1999, 472–8

Shineberg, Dorothy (1967) *They Came for Sandalwood: A Study of the Sandalwood Trade in the South-west Pacific 1830–1865*. Carlton: Melbourne University Press

Sissons, C. D. S. (1972) 'Immigration and Australian-Japanese relations 1871–1971', in James A. Stockwin (ed.), *Japan and Australia in the Seventies*. Sydney: Angus & Robertson

Sissons, C. D. S. (1979) 'The Japanese in the Australian Pearling Industry', *Queensland Heritage*, 3 (10):9–27

Smith, Bernard (1985) *European Vision and the South Pacific*. Sydney: Harper & Rowe

Smith, James Edward (1793) *A Specimen of the Botany of New Holland*. London: Printed by J.Davis for J. Sowerby

Smyth, William Henry (1867) *The Sailor's Word: An Alphabetical Digest of Nautical Terms*. London: Blackie & Son

Solander, Daniel (1753–82) *Collected Correspondence*. Ed. & trans. Edward Duyker & Per Tingbrand. Carlton: Miegunyah Press, University of Melbourne 1995

Sollas, William J. (1904) 'Narrative of the expedition in 1896', Section 1, *The Atoll of Funafuti: Borings into a Coral Reef and the Results, being the report of the Coral Reef Committee of the Royal Society*. London: Royal Society of London

Spate, Oskar H. K. (1979) 'Luso-Australia: in maps and verse', Lisbon: *Junta de Investigações Científicas do Ultramar*, CXXIV, 1979

Spender, Percy C. (1969) 'The Great Barrier Reef: Legal Aspects', *The Future of the Great Barrier Reef: Papers of an Australian Conservation Foundation Symposium*. Special Publication No. 3, Sydney, 3 May 1969, 25–41. Melbourne: Australian Conservation Foundation

Sprat, Thomas (1667) *The History of the Royal Society of London for the Improving of Natural Knowledge*. Facsimile reprint, Ed. J. I. Cope & H. W. Hones. London: Routledge & Kegan Paul

Stanbury, Peter J. & Holland, Julian (1988) *Mr Macleay's Celebrated Cabinet: The History of the Macleays and Their Museum*. University of Sydney: The Macleay Museum

Stearn, William T. (1969) 'A Royal Society appointment with Venus in 1769: the voyage of Cook and Banks in the *Endeavour* in 1768–1771 and its botanical results', *Notes and Records of the Royal Society*, 24:1, 64–90

Stearn, William T. (1979) 'Linnaean Classification', in David Black (ed.), *Carl Linnaeus' Travels*. London: Paul Elek

Steers, James Alfred (1978) 'Concluding remarks', *The Northern Great Barrier Reef: a Royal Society Discussion Organized by D. R. Stoddart and Sir Maurice Yonge, F. R. S., held on 28 and 29 January 1976*. London: The Royal Society

Stephenson, T. A., Stephenson, Anne, & Tandy, Geoffrey (1931) 'The structure and ecology of Low Isles and other reefs', *Great Barrier Reef Expedition 1928–29 Scientific Reports*, vol. 3, no. 2.

British Museum (Natural History) 1931

Stokes, John Lort (1846) *Discoveries in Australia, With an Account of the Coasts and Rivers Explored in the Years 1837–1843*. 2 vols. London: T. & W. Boone

Strahan, Ronald, et al. (1979) *Rare and Curious Specimens: An Illustrated History of the Australian Museum*. Sydney: Australian Museum

Sturgess, Garry L. (2000) *The Great Barrier Reef Partnership: Cooperation in the Management of a World Heritage Area*. (Presented November 1999.) Brisbane: Queensland Government

Talbot, Frank (2001) 'Will the Great Barrier Reef survive human impact?', in Eric Wolanski (ed.), *Oceanographic Processes of Coral Reefs: Physical and Biological Links in the Great Barrier Reef*. Boca Raton, Florida: CRC Press

Thomis, Malcolm Ian (1985) *A Place of Light and Learning: the University of Queensland's First Seventy-five Years*. Brisbane: University of Queensland Press

Thomson, Charles Wyville (1877) *The Voyage of the 'Challenger': the Atlantic*. 2 vols. London: Macmillan

Thomson, Charles Wyville & Murray, John (1885) *Report of the Scientific Results of the Voyage of HMS* Challenger *during the years 1873–76*. Vol. 1, part 1: *Narrative of the Cruise*. London: HMSO

Thoreau, Henry David (1854) *Walden, or Life in the Woods*. Reprint. New York: Signet Classics, New American Library, 1980

Toyne, Philip (1994) *The Reluctant Nation*. Sydney: ABC Books

UNESCO (United Nations Educational, Scientific and Cultural Organisation) (1980) *The World Heritage Convention*. Paris: UNESCO

United States Congress, House Committee on Public Works, Subcommittee on Flood Control (1969). *Oil Spillage, Santa Barbara, Calif.* Washington: US Government Printing Office

Veron, John Edward Norwood (1986) *Corals of Australia and the Indo-Pacific*. Sydney: Angus & Robertson

Veron, J. E. N. (1995) *Corals in Space and Time: The Biogeography and Evolution of the Scleractinia*. Sydney: University of New South Wales Press

Veron, J. E. N. (2000) *Corals of the World*. 3 vols. Townsville: AIMS

Vitousek, Peter M., Mooney, Harold A., Lubchenko, Jane, Mellilo, Jerry M. (1997) 'Human domination of earth's ecosystems', *Science*, vol. 277, 25 July 1997, 494–99

Wachenfeld, David Raymond, Oliver, J. K., Morrissey, J. I. (eds) (1998) *State of the Great Barrier Reef World Heritage Area 1998*. Townsville: GBRMPA

Walbran, Peter, Henderson, Robert, Jull, Timothy, Head, John (1989) 'Evidence from sediments of long-term *Acanthaster planci* predation on corals of the Great Barrier Reef', *Science*, vol. 245, 25 August 1989, 847–50

Ward, Barbara and Dubos, René (1972) *Only One Earth: The Care and Maintenance of a Small Planet*. An unofficial report commissioned by the Secretary-General of the United Nations Conference on the Human Environment. London: André Deutsch. Edition cited: Penguin 1973

Watson, Mereen (1988) *Historical Analysis of GBRMPA Research*. (Duplicated document.) Townsville: GBRMPA

Wells, Sue & Hanna, Nick (1992) *The Greenpeace Book of Coral Reefs*. New York: Sterling Publishing

White, Gilbert (1789) *The Natural History of Selborne*. Reprinted, ed. June Chatfield, The Gilbert White Museum. New York: St. Martin's Press, 1981

Whitehead, Peter J. P. (1968) *Forty Drawings of Fishes Made by the Artists who accompanied*

Captain James Cook on His Three Voyages to the Pacific 1768–71, 1772–75, 1776–80, Some Being Used by Authors in the Description of New Species. London: Trustees of the British Museum (Natural History)

Whitehead, Peter J. P. (1978) 'A guide to the dispersal of zoological material from Captain Cook's voyages', *Pacific Studies*, II. Fall, 1978, 53–93

Whitehouse, John F. (1992) *Review of the Magnetic Island Marina Development.* Canberra: Australian Government Publishing Service

Whitehouse, John F. (1997) *Independent Review of the Great Barrier Reef Marine Park Authority: Report to Senator Robert Hill, Minister for the Environment*

Whitlam, Edward Gough (1985) *The Whitlam Government 1972–75*. Melbourne: Viking (Penguin)

Whitley, Gilbert (1970) *Early History of Australian Zoology*. Sydney: Royal Australian Zoological Society

Whitton, Evan (1989) *The Hillbilly Dictator: Australia's Police State*. Sydney: ABC Books

Williams, Fred (1982) *Written in Sand: A History of Fraser Island*. Brisbane: Jacaranda Press

Wilkinson, Clive R. (1993) 'Coral reefs of the world are facing widespread devastation: can we prevent this through sustainable management practices?', *Proceedings*, 7th International Coral Reef Symposium, Guam, vol. 1, 11–21

Wilkinson, Rick (1988) *A Thirst for Burning: The Story of Australia's Oil Industry*. Sydney: David Ell Press

Wilson, James Thomas (1897) Presidential Address, *Proceedings*, Linnean Society of New South Wales, Series 2, vol. 22.4, no. 88, 812–47

Wolanski, Eric (1994) *Physical Oceanographic Processes of the Great Barrier Reef*. Boca Raton: CRC Press

Wolanski, Eric (ed.) (2001) *Oceanographic Processes of Coral Reefs: Physical and Biological Links in the Great Barrier Reef*. Boca Raton: CRC Press

Wood, George Arnold (1922) *The Discovery of Australia*. London: Macmillan

Wood, George Arnold (1922) *The Discovery of Australia*. Rev. edn 1969, J. C. Beaglehole. Melbourne: Macmillan

Wood-Jones, F. (1910) *Coral and Atolls: Their History, Description, Theories of Their Origin Both Before and Since that of Darwin, the Influence of Winds, Tides and Ocean Currents on Their Formation and Transformations, Their Present Condition, Products, Fauna and Flora*. London: Lovell, Reeve & Co.

World Almanac 2002. New York: World Almanac Books

Wright, Judith Arundell (1977) *The Coral Battleground*. West Melbourne: Thomas Nelson

Yonge, Charles Maurice (1928–29) Low Isles Expedition. The six Reports to the GBRC by Yonge in the author's archives are catalogued as follows: Yonge:LIE, followed by the sequence in Roman numerals from I to VI, followed by month and year, and finally page number

Yonge, Charles Maurice (1929) 'Final report on the Great Barrier Reef Expedition', *Nature*, 2 November 1929, 1–12

Yonge, Charles Maurice (1930a) 'Origin, organization and scope of the Expedition', *Great Barrier Reef Expedition 1928–29*, vol. 1, no. 1. London: British Museum (Natural History)

Yonge, Charles Maurice (1930b) *A Year on the Great Barrier Reef: The Story of Corals & of the Greatest of Their Creations*. London & New York: Putnam

Yonge, Charles Maurice (1931) 'The Great Barrier Reef Expedition, 1928–1929', *Reports of the Great Barrier Reef Committee*, vol. 3, 13 January 1931, 1–25

INDEX

Aborigines:
   antiquity of Australian
      settlement 3;
   Cook's admiration 34;
   degradation noted – *Challenger*
      report 130;
   Low Isles expedition 258f.;
   massacres 113–14;
   see also indigenous concerns
Acts of Parliament:
   Coast Islands Act 1879 151;
   *Coastal Waters (State Title) Act
      1980* 356;
   *Continental Shelf (Living
      Natural Resources) Act
      1968* 330;
   *Environmental Protection and
      Biodiversity Conservation
      Act 1999* 416;
   *Immigration Restriction Act
      1901* 148;
   *National Parks and Wildlife
      Conservation Act 1974*
      346, 353–4;
   *Offshore Agreement (1967) on
      petroleum exploration* 340;
   *Pacific Islanders Protection Act
      1872* 150;
   *Pearl Shell and Bêche-de-Mer
      (Extra-Territorial)
      Fisheries Act 1888* 153;
   *Pearl Shell and Bêche-de-Mer
      Fishery Act 1881* 151, 154;
   *Petroleum (Submerged Lands)
      Act 1967* 320;
   Polynesian Labourers Act 1868
      148, 150;
   *Seas and Submerged Lands Act
      1973* 343–4;
   Statute of Westminster 1931
      153;
   *Territorial Waters Jurisdiction
      Act 1878* 153;
   *World Heritage Properties
      Conservation Act 1983*
      410, 416;

Advisory Council on Science and
   Industry (ACSI) 165
Agassiz, Alexander:
   Darwin's reef formation theory
      opposed 199;
   Great Barrier Reef 1896 203–5
Agassiz, Louis,
   promotes Ice Age theory 197
algae 2;
   see also symbiosis,
      zooxanthellae
Andrews, Ernest C.,
   Reef survey 1901 207–8
Aristotle:
   'ladder of nature' 43, 175;
   see also Divine Design
atolls:
   early mention by Laval 12;
   formation theory: Forster 176,
      Chamisso 177, Quoy and
      Gaimard 178, Lyell 182,
      Darwin 186–91
atomic tests,
   Bikini, Enewetak 309–11
Australasian Association for the
   Advancement of Science
   137–8;
   renamed ANZAAS 137fn.
Australia:
   named by Flinders 75;
      confirmed by Macquarie 81
Australian Academy of Science,
   Crown of Thorns report 341
Australian Conservation Foundation
   331–2, 338;
Australian Coral Reef Society,
   248
Australian Fishing Zone 356, 362
Australian Institute of Marine
   Science:
   early planning 359–60;
   research achievements 386–8
Australian Museum, 125;
   see also Hedley; Krefft
Australian Science and Technology
   Council (ASTEC) 361

Bacon, Francis,
   on scientific method 42
Bamford, Frederick,
   Pearling Royal Commission
      1913–16 162
Banfield, Edmund James:
   Dunk Island 'Beachcomber'
      220–4;
   Dunk as wildlife reserve 226–7;
   ethic of concern for nature 225;
   idea of Reef as national park
      230;
   Reef conservation urged 226–9;
   Reef journalist 214
Banks, Joseph:
   accepts 'chain of creation' 48;
   as 'Kew collector' 46, 57;
   Endeavour botanist 27, 44–5;
   Florilegium 51–3, 54;
   Reef collections 49–51;
   zoological specimens lost 54–5
Barrett, Charles,
   Low Isles reporter 265–7
Barwick, Sir Garfield:
   ACF president 331;
   Reef conservation 337
Bauer, Ferdinand:
   artist on *Investigator* 62;
   illustrates botany of New
      Holland 72
beachcomber concept, *see*
   Banfield
Beagle voyages:
   Darwin and the Admiralty
      survey 1831–36 182–3;
   Reef survey by Stokes 1841 86
Beaglehole, John Cawte 16–17;
   editor of Cook's records 35
bêche-de-mer industry 114, 142,
   329
Bikini Atoll,
   atomic tests 309
biodiversity:
   defined by GBRMPA 422;
   International Convention on
      biological diversity 406, 421;

Representative Areas Program
421–4;
biological research on Reef:
by Australian Museum 125;
by MacGillivray and Jukes 92;
Low Isles, 268, 271–3;
Reef ecosystem survey 1999
422
biology:
and ecology 224–5
coined by Lamarck 43
bioregions 423
Bjelke-Petersen, Johannes,
Queensland premier:
oil speculator, 322, 326;
Reef Marine Park opposed 352;
subsequent cooperation 355
see also Emerald Agreement
'black bans', on Reef drilling for oil
335
Bligh, William,
navigation of Reef passage
1789 59
botany:
Banks' primary interest 48;
loss of Brown's specimens 71;
origin of term 45;
puzzle of New Holland
taxonomy 67
Botany Bay:
on Dieppois maps? 15;
Endeavour naturalists study
47
Bowen 108;
township declared 1861 112
Bowen, Emanuel,
cartographer: Pacific map 28
Bowen, Sir George Ferguson:
governor of Queensland 1859
110;
supports settlement and Italian
migration 115
Bowen, Nigel, Attorney-General,
Commonwealth sovereignty
and Reef protection
329–30
Brierly, Oswald,
Rattlesnake artist 96f.
Bright Point, *see* Magnetic Island
British Association for the
Advancement of Science:
Funafuti Expedition 200;
Low Isles Expedition 1928–29
258;
marine research stations 129

Brown, Robert, master botanist:
Banks' curator 50;
botanical collections,
Investigator voyage 68;
New Holland plants taxonomy
problem 72–3;
Reef specimens loss on
Porpoise 71;
reticulant evolution theory 387
Brundtland Report,
sustainable development
379–80
Buchan, Alexander,
Endeavour topographic artist
45
Buddhism,
influence on Banfield 221
Burdon-Jones, Cyril,
marine research station plans
358–9
Büsst, John,
Reef conservationist 327,
331–3

cabinet collecting:
Macleay era, 125f., 134;
vs natural selection 127
Caley, George,
botanist 62
Camm, Ronald, Queensland
Minister for Mines 328;
supports Reef drilling 321, 333
Cape York:
fortified outpost 112f.;
named by Cook 33;
Somerset abandoned 114
Carson, Rachel,
author of *Silent Spring*: 323
Chadwick, Virginia,
Chair of GBRMPA 417
Challenger, HMS:
in Australian waters 130;
oceanographic survey 129
Chamisso, Adelbert von,
atoll formation theory 177
charting of Reef, *see* hydrography
Chevert voyage,
Reef and New Guinea 132f.
Christesen, Clement B.,
tourism writer 288
chronometers, *see* longitude
climate change 380–1, 418–21;
coral reef monitoring 387,
420–1;
see also Kyoto Conference

Coast Islands Act 1879 151
*Coastal Waters (State Title) Act
1980* 356
Commerson, Philibert,
Boudeuse botanist 22
Commonwealth Institute of Science
and Industry 235
Commonwealth-Queensland
relations:
Commonwealth sovereignty
over Reef 332;
Emerald Agreement 355;
Great Barrier Reef Marine Park
Act introduction 352,
New Federalism 1975–79 355;
Offshore Agreement on
petroleum exploration 320,
340;
Offshore Constitutional
Settlement 356;
Royal Commission into oil
drilling 1970–74 336,
348–51;
Sturgess Report 415
see also Gorton, John
Commonwealth Scientific and
Industrial Research
Organisation (CSIRO) 165
conservation of Reef:
early Commonwealth
legislation 330;
foreign resource raiding 329;
Great Barrier Reef Committee
291;
Great Barrier Reef Marine
Park:
national estate 344;
oil drilling controversy
328;
protection of Reef fauna,
forests, 295;
protests on coral limestone
mining 327;
see also Banfield; Crown
of Thorns; GBRMPA;
GBRWHA
continental shelf legislation:
United Nations 320;
Australian 339
*Continental Shelf (Living Natural
Resources) Act 1968* 330
Cook, James, *Endeavour* voyage:
Admiralty instructions 28, 45;
Australian east coast charted
30–1;

Cook, James (*cont.*)
    first report of Reef 11;
    library and scientists aboard 28, 44;
    reef grounding 32;
    use of Dauphin map? 37–40;
    zoological specimens lost 54;
    see also Hawkesworth
Cooktown settled, 117
Cooperative Research Centres:
    introduced 1990 384;
    CRC funding for 'Ecologically sustainable development' of Reef 384–5
corals:
    early studies 173;
    mass spawning 386;
    observed by Banks 48, Flinders 74, Jukes 91;
    polyps confirmed as animals 174;
    reef building hermatypic 2, 174;
    serious bleaching 419–21, 424
coral reefs:
    early account by Laval 12;
    elevated reefs puzzle 84, 87, 234
    formation theories Forster to Darwin 74, 176–82, 187–92;
    Global Coral Reef Monitoring Network 387;
    reef drilling as Darwin proposed 199: Funafuti 1896–99 200, Samoa 1913–20 211, Oyster Cay 1926 243–6, 297–8, Heron Island 1937 298;
    test drilling for oil on Reef 319, 328;
    see also climate change, Cook, symbiosis
Council for Science and Industrial Research (CSIR) 165
Cripps, Captain 80
Crown of Thorns:
    Australian Academy of Science report 326, 341;
    first Reef sighting 324f.;
    rising concern 375, 424f.;
    systematic research funding 377
Cunningham, Alan:
    aboard *Mermaid* 82;
    additions to Reef collections 83;

King's Botanist 78

Daly, Reginald,
    on Mer Expedition 212
Dalrymple, Alexander:
    criticises Hawkesworth 37;
    declines to join *Endeavour* voyage 27;
    possession of Dauphin map 37
Dana, James, American marine scientist:
    advised Agassiz, 205;
    Wilkes Expedition, 217
Darwin, Charles Robert:
    Cambridge studies 184;
    contradicts Lyell's atoll theory 182;
    diary of Beagle voyage 185;
    questions Divine Design theory 126;
    requests Reef observations by Stokes 87;
    species theory challenged by Veron 387;
    supported by Haswell 138, by Yonge 266
Dauphin map, *see* Dieppe cartography
David, Edgeworth,
    geologist: Funafuti Expedition 200f.
De Jussieu family, taxonomy:
    botanical research in Paris 57;
    search for a 'natural system' 51;
    system used by Cunningham 82
Descartes, René,
    rational method in science 42
De Quiros expedition in Pacific 18
Dieppe cartography:
    early maps of Reef? 15, 20;
    Dauphin map used by Cook? 37–8
Dittmer, Felix,
    urges marine station 357
divers:
    dangers to 157;
    Low Isles 269–70;
    pearling industry, 146;
    Royal Commission 163;
    see also Kezalias
Divine Design:
    and natural science 43, 126;
    Banks accepts 'chain of creation' 48;
    Macleay family defence 125;

'missing links' in nature's chain 65;
    Reef biota a separate creation? 50;
    see also Aristotle
dugong 93;
    extinction risk 228, 290, 407–8;
    Protection Areas 1997 408, 422
Dunk Island,
    named by Cook 220;
    see also Banfield

ecological research and planning:
    ecosystem management 404;
    Low Isles research 281;
    Mayor's ecological study on Mer 209;
    Representative Areas Program 421–4;
    ecologically sustainable development concept 384, 395, 398;
    see also Brundtland Report; climate change
ecology:
    and biology 224–5;
    in thought of Gilbert White 216;
    term coined by Haeckel 224;
    wider concept of Humboldt 224–5
economic resources of Reef:
    GBRC program for investigation 239;
    exploitation planned 256, 381f.;
    Yonge's report 1929 278;
    ecological sustainability sought, 384
Ellison Reef,
    mining conflict 327
Embury, E. M.,
    early tour operator 285
Emerald Agreement,
    Marine Park joint management 355
Empire Marketing Board,
    funds Low Isles Expedition 256
Endeavour, see Cook
Endean, Robert,
    Crown of Thorns research 325
environment:
    and ecosystem management 404;
    Brundtland Report 1987 380;
    management planning 338f.;
    toxicity 322, 393, 407;

United Nations conference 1971 342
Environmental Protection and Biodiversity Conservation Act 1999 416
erratic rocks,
  and glaciation 197
eutrophication,
  of lagoon 391
Exclusive Economic Zone (EEZ) 362

fisheries:
  Australian Fishing Zone (AFZ) 362;
  conserving stocks 417
  worldwide depletion 385
Fitton, William,
  geological discoveries 83–4, 178
Fitzgerald, Lawrence,
  Portuguese charting of the Reef, 18
Fitzroy, Robert,
  Beagle commander 183
Flinders, Matthew 59, 61, 76:
  Investigator voyage 62–5;
  names Australia 75;
  names Great Barrier Reef 74;
  Porpoise wreck 70;
  publishes *A Voyage to Terra Australis* 75;
  survey of Moreton Bay 1799 60
Florilegium: see Banks
Fly:
  Reef survey 1842–45 89;
  second Reef survey 91
Forbes, Edward,
  describes molluscs of *Rattlesnake*, 99
Ford, Alexander,
  founds Pan-Pacific Union 231
Forster, Johann Reinhold,
  on atoll formation 176
France:
  annexes New Caledonia 102;
  botanical collections 57;
  competition with British 38;
  early exploration in Pacific 22
Fraser, Eliza,
  capture and rescue 85
Funafuti,
  coral coring expeditions 200–203

Gaimard, Joseph Paul,
  theory of atoll formation 178
Gardiner, John Stanley:
  and the Low Isles Expedition 253;
  opposes Reef drilling 244
geology:
  catastrophism theory 180;
  controversy on reef formation 83–8;
  Pacific sea bed mapping proposed 233;
  Reef surveys 207–8, 238, 240;
  see also coral reefs
Georges, George:
  prevents drilling 334–5;
  Reef conservationist 333
Glen Roy,
  Darwin's 'greatest blunder' 196–8
Global Coral Reef Monitoring Network 387;
  see also AIMS; climate change
global warming, *see* climate change
Goddard, Ernest James 250;
  founds Heron Island station 312
gold:
  discovery and population influx 102, 116
Good, Peter,
  Investigator diary 64, 69
Goreau, Tom:
  confirms polyp symbiosis 273;
  Low Isles experiments examined 282;
  warnings on reef stress 383
Gorton, John Grey:
  Prime Minister 1968 329;
  Royal Commission on Reef 336;
  support for Reef research 333;
  Territorial Sea and Continental Shelf Bill 339
Gray, John Edward:
  British Museum scientist 127;
  Fly zoology collections described 92
Great Barrier Reef Committee:
  conflict with Geographical Society 247;
  conservation measures urged 294
  founded 1922 235–6
  Hedley as scientific director 243;
  Reef biological research 252f.;
  see also Mather; Richards

Great Barrier Reef Marine Park Authority (GBRMPA):
  charging users 390;
  early planning 346–8;
  management principles 353;
  Strategic Plan, for 25-year Reef management 395, 398f., 421
  zoning practices 364f
Great Barrier Reef Marine Park Act:
  early drafting by Mather 339;
  introduced by Whitlam Government 351;
  resisted by Queensland Government 351;
  „*see also* Commonwealth-Queensland relations
Great Barrier Reef World Heritage Area (GBRWHA):
  Reef listing 1981 371;
  status preservation concerns 402f.;
  UNESCO criteria 7, 394;
  see also Lucas Report
Great Chain of Being:
  and *scala naturae* 43;
  impact of Endeavour discoveries 50;
  search for 'missing links' 13, 65–8;
  see also Aristotle
Great South Land 13;
  search for 22
Green, Charles,
  Endeavour astronomer 31
'greenhouse' effect, *see* climate change
guyots,
  in Reef formation 196

Haeckel, Ernst,
  coins 'ecology' 224–5
Halley, Edmond,
  theory of longitude determination 25
Harleian Map, *see* Dauphin Map
Haswell, William:
  at Queensland Museum 135;
  in Sydney 136;
  promotes Darwinism 138
Hawkesworth, John,
  editor of Cook's journal 11, 35, 38

Heath, George Poynter:
    Quetta controversy 121;
    regulation of Reef shipping 120
Hedley, Charles 165, 249;
    and Australian Museum 242;
    Oyster Cay drilling 245–6;
    Reef resource extraction 166;
    Reef surveys 201, 207–8, 238;
    sponge fishing venture 168
Herald,
    Reef survey 1853–60 102–6
heritage:
    access to heritage: charging users 390;
    concern for sustainability 394f., 402, 425f.;
    expanded concept 5;
    see also GBRWHA
hermatypic (reef-building) corals 2, 174, 276
Heron Island:
    geological drilling 1937 298;
    research station founded 1952 312
Hinchinbrook Channel,
    pollution controversy 409;
    see also Oyster Point; dugong
Humboldt, Alexander von,
    observations on Caribbean reefs 88, 179
Hutton, James,
    uniformitarian theory of the earth 179
Huxley, Thomas Henry:
    Rattlesnake activities 98;
    studies of medusae 101
hybridisation,
    in corals 387
Hydrographers Passage,
    charted 1981–84 368
hydrography:
    aerial photography 303;
    Australian Hydrographic Service 392;
    coral sea in 1860s 102;
    early phase completed by Denham 106;
    mechanical and acoustic methods 302,
    safe passages through the Reef needed 78, 85;
    satellite imagery 369;
    survey of *Beagle* 1841 86, *Fly* 1841–45 89, *Rattlesnake* 1847–50 94–9;

twentieth century developments 237;

ice ages:
    Louis Agassiz 197;
    sea level changes 3;
    see also Maclaren
immigration,
    coloured labour issue 116
Immigration Restriction Act 1901 148
indigenous concerns:
    cultural heritage 371, 395;
    land and sea rights 396–7;
    Pacific War racism 307;
    participation in management 396;
    protection of coloured labourers 151;
    see also Mabo judgment
International Union for Conservation of Living Natural Resources (IUCN),
    joined by Australia 1973 344
Investigator,
    voyage to Reef 62, 68–70;
    see also Flinders
isostasy,
    and sea level changes 181, 189

James Cook University 359;
    Orpheus Island research station 360
Japanese:
    fishing practices opposed 1968 329;
    increasing aggression 1934–41 305;
    Japanese divers in Torres Strait 152f.;
    Japex (Japan Petroleum Exploration) lease 322;
    pearling practices condemned 162–5;
    resource exploitation in 1920s 299f.
Jeffreys, Charles:
    conflict with Macquarie 79, 80;
    Reef coast charting completed 79f.
Jukes, Thomas Beete,
    geologist on *Fly* 1842–45 89
Jura mountains limestone,

ancient coral reefs 179;
    see also Humboldt

kangaroo:
    recorded by Cook 32;
    taxonomic puzzle 68
Kelleher, Graeme,
    first Chair of GBRMPA 365
Kennedy, Edmund,
    Cape York 1848 expedition failure 95, 110
Kew Gardens:
    centre for genetic piracy 46;
    propagation for colonial plantations 56–7;
    see also Banks
Kezalias, Nicholas,
    Greek sponge diver 168
'Kidnapping Act' 1872 150
King, Phillip Parker,
    surveys Australian coastline 80
Kotzebue, Otto von,
    Pacific voyage 177
Krefft, Gerard:
    Australian Museum curator 127;
    dismissed by Macleays 128
Kyoto Conference 418–9;
    see also climate change

'Labyrinth',
    Cook's term for Reef 33, 36
Ladd, Harry:
    Enewetak report 309;
    geological survey of Reef 321
Lamarck, Jean Baptiste de:
    challenges Great Chain of Being 51;
    coins 'biology' 43;
    taxonomy as impositions on nature 72
Lang, John Dunmore,
    promotes 'Cooksland' 109
Laval, François Pyrard de,
    account of Maldive atolls 12
Law of the Sea,
    United Nations legislation 320
legislation:
    Australian Fishing Zone 356, 362
    Exclusive Economic Zone 362;
    protection of coloured labourers 151;
    see also Acts of Parliament

lighthouses,
    promoted by Heath 120
Limpus, Colin,
    turtle conservation scientist 400
Linnaeus:
    corals as animal productions 176;
    Endeavour specimens classified 50;
    taxonomic 'system' 43–4
Linnean Society founded 1788 66
Littoral Society of Queensland,
    opposes Reef limestone mining 327
Lizard Island:
    explored by Cook 33;
    Mary Watson tragedy 143;
    research station 315, 317
longitude:
    Halley's theory 25;
    marine chronometer 41
Low Isles Expedition 1928–29 261, 265, 270;
    aerial survey 304;
    planning, 258f.;
    proposed 253;
    significance 281;
    women members 265, 270;
    Yonge as leader 257
Lucas Report on Reef status 402
Lyell, Charles:
    atoll formation theory 181;
    uniformitarian theory of the earth 180–82

Mabo judgment,
    recognition of indigenous title 396
Macfarlane, Charles,
    conflict with William Macleay 133
MacGillivray, John:
    Fly collections, 89, 92;
    Herald naturalist 105;
    Rattlesnake voyage *Narrative* 98
Maclaren, Charles,
    sea level change and reef formation 198
Macleay, Alexander,
    and Australian Museum 101, 125
Macleay, William John:
    Challenger visit 131;
    conflict with Macfarlane 133;
    voyage of the *Chevert* 132f.
Macleay, William Sharp,
    'Quinary System' of taxonomy 100, 126
Macquarie, Lachlan:
    conflict with Jeffreys 80;
    governor of New South Wales 77;
    recommends name 'Australia' 81
Magnetic Island,
    development controversy 373
marine park,
    concept emerges 347;
    see also Great Barrier Reef Marine Park
marine science:
    Commonwealth pressure for economic benefits 381–3;
    national research policy 361;
    need recognised 310f.
    rapid development 369;
    students encouraged 313
Marsh, George Perkins,
    United States conservationist 225
Mather, Patricia:
    and GBRC on Reef mining 328;
    drafts Great Barrier Reef protection Bill 338–9, 346
marine stations, *see* research stations
Matra, James,
    recommends New South Wales for settlement 58
Mayor, Alfred Goldsborough:
    coral surveys in Torres Strait, Samoa 209f.;
    with Agassiz on Reef 203
McCoy, Frederick,
    and Divine Design 126
McIntyre, Kenneth,
    Portuguese charting of Reef 17
McKinnon Report,
    marine industries 381f.
McMahon, William,
    on Territorial Sea and Continental Shelf Bill 342
McMichael, Donald,
    plans Great Barrier Reef Marine Park 347
McPhail, Ian,
    GRMPA management reform 400f.

Mermaid,
    King's vessel for reef survey 81
Michaelmas Reef,
    boring expedition 245f.
Mid-Pacific Marine Laboratory 310
Miklouho-Maclay, Nikolai,
    Sydney marine station 310
missionaries,
    in Torres Strait 133, 149
Monkman, Noel,
    underwater photography 285, 292
Moorhouse, Frank:
    and Low Isles Expedition 265–7, 279;
    trochus project 286
Moresby, John:
    possession of Hayter Island 132;
    warnings on northern defence 131
Moreton Bay:
    navigated and named by Flinders 60;
    opened for settlement 1840 108;
    separated from New South Wales 1859 109–10
Morrison, William,
    Green Paper on science development 361
Moseley, Henry Nottidge,
    Challenger scientist 129
Murray, John:
    opposes Darwin's theory of reef formation 195;
    Challenger scientist 129
Muscatine, Leonard,
    on symbiosis 273
museums:
    Australian, founded 125;
    British (Natural History) 125;
    Godeffroy, Hamburg 124;
    Queensland 135

Napier, Elliott,
    tourism journalist 284
Nares, George,
    Challenger commander 129
Nathan, Matthew:
    Chairman of GBRC 236;
    governor of Queensland 233;
    supports Low Isles Expedition 252

National Estate:
  Hope Committee of Inquiry 345f.;
  rejected by Queensland Government 351
national parks:
  early declarations 290;
  National Parks Association founded 1930 294
  National Parks and Wildlife Conservation Act 1974 346, 353–4
natural selection:
  established in 1890s 138f.;
  theory of Charles Darwin 126–8
Nelly Bay, *see* Magnetic Island
New Federalism,
  of Fraser Government 1975–79 355
New Holland:
  botanical puzzles 67–72;
  conflicting name for Australia 81
New South Wales:
  expansion of colony 77, 107;
  named by Cook 33;
  settlement 1788 58

Oceanic Grandeur grounding 335
Odum brothers,
  at Enewetak 310
Offshore Agreement on petroleum exploration 320, 340
Offshore Constitutional Settlement 356
oil exploration:
  Reef zoned for exploration 321;
  Senate report on Offshore Petroleum Resources 340, 349, 360;
  test drilling on reef 1959–68 319
oil spills:
  early concerns 291;
  Oceanic Grandeur grounding 335;
  Santa Barbara blowout 330;
  Torrey Canyon disaster 327
Onslow, Arthur,
  opponent of Krefft 127
Orpheus Island, *see* research stations
Ortelius, Abraham,
  early maps of East Indies 15

Oxley, John:
  names Brisbane River 108;
  Surveyor-General of New South Wales 78
Oyster Cay,
  Reef drilling site 1927 245–6, 250, 297
Oyster Point:
  development controversy 1990s 408;
  Ramsar wetlands 410;
  see also Hinchinbrook

Pacific Islanders Protection Act 1872 150
Pacific War 1941–45:
  aerial Reef survey 1928 304–8;
  defence of Reef planned 302
Palau (Pelew, now Belau):
  Palau Pearlers Association 164;
  Semper's investigations 193f.
Pan-Pacific Science Congresses:
  Bandung 1929 286;
  Hawaii 1920 232;
  Melbourne 1923 240;
  Tokyo 1926 254f.
Pan-Pacific Union,
  founded 231
Parkinson, Sydney,
  Endeavour artist 45, 51
Pearl Shell and Bêche-de-Mer (Extra-Territorial) Fisheries Act 1888 153
Pearl Shell and Bêche-de-Mer Fishery Act 1881 151, 154
pearling:
  collapse of industry 161–3;
  depletion and regulation 155–7;
  diving dangers 146, 157;
  early exploitation 145–7
petroleum, *see* oil exploration
Petroleum (Submerged Lands) Act 1967 320;
  Senate report on constitutional validity 339
Peyssonnel, André de,
  animal nature of corals 174
Philp, Robert:
  Townsville merchant 218;
  supports Banfield 219
photography of Reef, *see* hydrography; Monkman
pilotage through Reef:
  compulsory areas 1987 392;
  early regulations 122;
  first recommended 1884 122

platypus:
  accepted and classified 68;
  considered a hoax in London 67
pollution:
  atmospheric: 322, 380;
  greenhouse gases: increasing emissions 418;
  Reef waters 409, 411
Polynesian Labourers Act 1868 148, 150
polyps:
  carnivore hypothesis 273;
  early speculation on 173–4, 177–8;
  hermatypic or reef building, 2, 174, 276;
  recognised as animals 174
Port Curtis,
  proposed for settlement 109
Port Essington,
  evacuated 1849 96
Portugal:
  early maritime activity 13;
  possible charting of Reef 17;
  see also Spain
Potts, Frank Armitage:
  fails to lead Low Isles Expedition 255–8;
  Reef research proposal 252–5
Precautionary Principle 410
Providential Channel,
  Cook's entry 48
Purchas, Samuel,
  publication on reefs 12

Queensland:
  boundaries, in 1860 111,
    extended 1872 150;
  Coast Islands Act 1879 151;
  exploration and settlement 110–114;
  Low Isles Expedition 265f.;
  management conflict with GBRMPA 415;
  Museum 136;
  proclaimed colony 1959 110;
  railways 117
  see also Commonwealth-Queensland relations
Quiros, Luis Váez de 18
Quoy, Jean René,
  theory of reef formation 178

railways,
  early construction 117
Raine Island:
  beacon erected on *Fly* voyage 92;
  Challenger visit 130;
  dredging by Murray 195
Ramsar Convention on wetlands 410
Ramsay, Edward,
  recruits Haswell to Sydney 136
Ratcliffe, Francis,
  Reef activist 331, 338
Rattlesnake:
  Reef survey 1847–50 94–9;
  rescue of Barbara Thompson 96
Reef Pelew Islands, *see* Palau
Representative Areas Program,
  conserving diversity 421–4;
  see also GBRMPA
research stations:
  early founding efforts 239, 242, 310;
  Enewetak Marine Biological Laboratory 309;
  Heron Island 312;
  Lizard Island 315;
  One Tree Island 315;
  Orpheus Island 360;
  urged by: Saville-Kent 158, Hedley 232, Senator Dittmer 357, Burdon-Jones 358
resorts:
  first appear 1920s 285;
  Heron Island 312;
  Lindeman Island 1930 287
resource depletion,
  and fishing stress 385
resource raiding 141f.
Richards, Henry:
  at University of Queensland 233;
  geological drilling of Heron Island 296–8;
  opposes marine stations 239, 251;
  origin of term 'Great Barrier Reef' 289;
  supports national parks 294
Roughley, Theo,
  Reef romance writer 288
Royal Australian Air Force:
  aerial photography of Low Isles 264, 303;
  Flight 101 at St Bees Island 264

Royal Australian Navy,
  Survey Service 392
Royal Commissions:
  Bamford and Mackay, on pearling 1913–16 163f.;
  Inquiry into oil drilling on the Reef 1970–74 336, 348–51
Royal Geographical Society of Australasia (Queensland),
  conflict with GBRC 247
Royal Geological Society of London,
  Reef investigation 236
Royal Society of London:
  adopts Baconian empiricism 43;
  and Transit of Venus 25;
  criticised by Babbage 137;
  outrage over Funafuti publications 201;
  supports *Challenger* voyage 129;
  supports Low Isles Expedition 256
Royal Society of New South Wales 125
Russell, Frederick,
  on Low Isles Expedition 265
Rusk, George,
  described species collected on *Rattlesnake* 98
Russia:
  alarm in Australia 102;
  expansion in the Pacific 101, 131

Samoa,
  Mayor drilling expedition 1913 211–12
sandalwood exploitation 141, 145
satellite reef imagery 369
Saville-Kent, William:
  Queensland Commissioner of Fisheries 156–8;
  recommends marine station 158;
  proposes resource exploitation 159
science and scientific method:
  empiricism, rationalism 42
  'presentism' preface
  see also Bacon
scleractinia 2, 386;
  see also coral reefs

sea level changes,
  Maclaren's argument 197
Seas and Submerged Lands Act 1973 343–4, 354;
  amended 355–56
Semper, Carl Gottfried,
  opposes Darwin's theory of reef formation 193
shipping:
  Commonwealth regulations 392;
  development of shipping lines 118;
  mounting losses 78, 102, 112, 119;
  see also oil spills; pilotage
slave trading:
  moderated 1890s 152;
  Pacific Islanders 148;
  Pacific Islanders Protection Act 1872 150
Solander, Daniel Carl:
  edits coral studies of Ellis 52, 176;
  Endeavour voyage 27, 44
Sollas, William,
  at Funafuti 200–202
Somerset, *see* Cape York
Spain:
  disputes with Portugal 13–14;
  expansion into Pacific 18
Spate, Oskar,
  Portuguese charting of Reef 17
species concept:
  Lamarck 72;
  Veron 387
Spender, Sir Percy,
  Commonwealth sovereignty over the Reef 332
sponge diving,
  early efforts 168;
  see also Kezalias
Spöring, Herman Dietrich,
  Endeavour scientist 45
Stanley, Owen,
  Rattlesnake commander 94
Statute of Westminster 1931 153
Steers, James,
  Low Isles geographical survey 276
Stockholm Conference 1971,
  concern for marine environment 342
Stokes, John L.:
  Beagle survey 86;

Stokes, John L. (*cont.*)
  reef elevation evidence for
    Darwin 87
Strickland, Sir Walter,
  Banfield supporter 221
Sturgess report,
  Commonwealth/Queensland
    reef management 415
subsidence:
  and sea level change 196;
  central to Darwin's reef
    formation theory 187–91;
  confirmed by Hedley at
    Funafuti 201;
  opposed by Agassiz 199, 206;
  studied by: Andrews 208,
    Richards 234
sugarcane 115f., 162
symbiosis, polyp-algae 2, 273–4;
  see also coral reefs
Systema naturae of Linnaeus 44,
  176

Taiwanese,
  Reef resource raiding 329
Talbot, Frank:
  One Tree and Lizard Islands
    research stations 314–5,
    317;
  reef systems ecological
    ignorance 334
taxonomy of nature:
  difficulty in applying 44;
  early attempts 43;
  Linnaean system, additions for
    Australian biota 67;
  Macleay's 'Quinary System' 100;
  'natural' approach of de Jussieu
    82, 126
Territorial Sea and Continental
  Shelf Bill 339, 342
*Territorial Waters Jurisdiction Act
  1875* 153
Theophrastus,
  first describes coral 173
Thomson, Charles Wyville,
  oceanographer on *Challenger*
    129
Thoreau, David Henry,
  environmental philosopher
    216–8;
Thursday Island,
  Torres Strait administrative
    centre 114
Torres, Luis Vaez,

passage through Torres Strait
  19
Torres Strait:
  ice age land bridge 3;
  Mayor's study on Mer 208f.;
  need for defence 131, 307;
  outside Marine Park 423;
  survey of waters 1863 113;
  see also Bligh; Flinders; Mabo
tourism:
  early attractions in 1920s 283,
    287;
  excessive pressure on
    environment 401f.;
  rapid growth 324, 389f., 405;
  structures on Reef 373, 388,
    391, 404;
  'user pays' proposed 389
Towns, Robert,
  early Reef merchant 116
Treaty of Tordesillas 14
trepang, *see* bêche-de-mer
trochus:
  early harvesting 163;
  Hedley cultivation project 167;
  Japanese depredations 291
turtles:
  canneries 283;
  cruel tourism use and capture
    283;
  decline warnings 284, 286;
  protection devices 400

United East India Company, *see*
  VOC

Vaughan, Thomas Wayland,
  Reef research planning 254
Vaugondy Map, used by Cook? 38
Venus, transit of,
  expedition 1769 23–26
Veron, John:
  coral taxonomy 386–7;
  Darwin's species concept
    questioned 387;
  symbiosis research 'unknowns'
    282
VOC (United East India Company):
  challenges Portuguese
    domination 20;
  Dutch monopoly in East Indies
    21

Wallis, Samuel:
  on Tahitian customs 28;
  search for Great South Land 22

Watson, Mary,
  Lizard Island tragedy 143
Whewell, William,
  coins 'scientist' 127
White, Gilbert,
  natural science ecological view
    215–6
Whitlam, Edward Gough:
  against Reef drilling or mining
    333;
  government defeated 354;
  legislative program 343–45
Whitley, Gilbert:
  impact of *Endeavour*
    collections 50;
  on Low Isles 258
Wickham, John Clements:
  Beagle commander of survey of
    1837–40 86;
  Fortitude episode 109
Wildlife Preservation Society of
  Queensland,
  opposes Reef mining 327
Wolanski, Eric,
  Reef water circulation 387
Wood, George,
  first Reef charts not Portuguese
    16
Wood-Jones, Frederick,
  polyp-zooxanthellae symbiosis
    studied 272
Wright, Judith,
  Reef protection activist, 328

Yonge, Charles Maurice:
  coral experiments 273;
  Expedition *Reports* 278;
  Low Isles Expedition leader
    257;
  publications 280

zoning of Reef:
  early planning 366–68;
  user manipulation 389;
  see also Representative Areas
    Program
zoology, *see* biology
zoophyte,
  early term for coral polyps 174
zooxanthellate corals 174
zooxanthellae:
  studied by Yonge 282;
  termed 'infesting' by Mayor
    211